T0344878

Facing Hydrometeorological Extreme Events

Other *Hydrometeorological Extreme Events* Titles

Hydrometeorological Hazards: Interfacing Science and Policy
Edited by Philippe Quevauviller

Coastal Storms: Processes and Impacts
Edited by Paolo Ciavola and Giovanni Coco

Drought: Science and Policy
Edited by Ana Iglesias, Dionysis Assimacopoulos, and Henny A.J. Van Lanen

Forthcoming Titles:

Flash Floods Early Warning Systems: Policy and Practice
Edited by Daniel Sempere-Torres

Facing Hydrometeorological Extreme Events

A Governance Issue

Edited by

Isabelle La Jeunesse
University of Tours
Department of Geography
Laboratory CNRS 7324 Citeres
Maison des Sciences de l'Homme
33, allée Ferdinand de Lesseps
B. P. 60449
37204 Tours cedex 3
France

Corinne Larrue
Université Paris-Est Créteil
Ecole d'Urbanisme de Paris Lab'urba
Bâtiment Bienvenüe – A322
14-20 boulevard Newton
Cité Descartes – Champs-sur-Marne
77454 Marne-la-Vallée Cedex 2
France

This edition first published 2020

Registered Office(s)
John Wiley & Sons, Inc., 111 River Street, Hoboken, NJ 07030, USA
John Wiley & Sons Ltd, The Atrium, Southern Gate, Chichester, West Sussex, PO19 8SQ, UK

Editorial Office
The Atrium, Southern Gate, Chichester, West Sussex, PO19 8SQ, UK

For details of our global editorial offices, customer services, and more information about Wiley products visit us at www.wiley.com.

Wiley also publishes its books in a variety of electronic formats and by print-on-demand. Some content that appears in standard print versions of this book may not be available in other formats.

Library of Congress Cataloging-in-Publication Data

Names: La Jeunesse, Isabelle, editor. | Larrue, Corinne, editor.
Title: Facing hydrometeorological extreme events : a governance issue / [edited by] Isabelle La Jeunesse, Corinne Larrue.
Description: First edition. | Hoboken, NJ : John Wiley & Sons Ltd, 2019. | Series: Hydrometeorological hazards : interfacing science and policy | Includes bibliographical references and index.
Identifiers: LCCN 2019017936 (print) | LCCN 2019980282 (ebook) | ISBN 9781119383543 (hardcover) | ISBN 9781119383468 (pdf) | ISBN 9781119383550 (epub)
Subjects: LCSH: Hydrometeorology–Government policy. | Hydrological forecasting. | Flood control. | Drought management.
Classification: LCC GB2803.2 .F34 2019 (print) | LCC GB2803.2 (ebook) | DDC 363.34/92–dc23
LC record available at https://lccn.loc.gov/2019017936
LC ebook record available at https://lccn.loc.gov/2019980282

Cover Design: Wiley
Cover Image: © Quintanilla/Shutterstock

Set in 10/12pt Times by SPi Global, Pondicherry, India

Printed and bound in Singapore by Markono Print Media Pte Ltd

10 9 8 7 6 5 4 3 2 1

Contents

List of Contributors

Meghan Alexander
School of Earth & Ocean Sciences, Cardiff University, Wales, United Kingdom

Brice Anselme
Institut de Géographie, Laboratory CNRS 8586 PRODIG, Université Paris 1 Panthéon-Sorbonne, Paris, France

Alba Ballester
Autonomous University of Barcelona-Institute of Government and Public Policies, Barcelona, Spain

Suzanne Boyes
Institute of Estuarine and Coastal Studies, University of Hull, United Kingdom

Hans Bressers
Department of Governance and Technology for Sustainability, University of Twente, Enschede, The Netherlands

Nanny Bressers
Former Project Leader of the European DROP Project at the Water Authority of Vechtstromen, now a Consultant at Vindsubsidies. Uthrecht, the Netherlands

Alison Browne
Department of Geography, Sustainable Consumption Institute, University of Manchester, Manchester, United Kingdom

Elie Chevillot-Miot
Laboratory CNRS 6554 LETG-Nantes, University of Nantes, Nantes, France

Claudia Cirelli
Laboratory CNRS 7324 Citeres, University of Tours, Tours, France

Ann Crabbé
University of Antwerp, Faculty of Social Sciences, Centre for Research on Environmental and Social Change, Antwerp, Belgium

Axel Creach
Laboratory CNRS 8185 ENeC, Paris Sorbonne University, Paris, France

Paul Durand
Laboratory of physical geography CNRS 8591 LGP, Université Paris 1 Panthéon-Sorbonne, Paris, France

Virginie K.E. Duvat
Laboratory CNRS 7266 LIENSs, La Rochelle University, La Rochelle, France

Michael Elliott
Institute of Estuarine and Coastal Studies, University of Hull, United Kingdom

Marie Fournier
Laboratoire Géomatique et Foncier (GeF) – Équipe ERADIF (Aménagement, Droit immobilier, Foncier), École Supérieure des Géomètres et Topographes (CNAM), Le Mans, France

Mauro Galluccio
External Speaker to the European Commission, DG COMM, Brussels, Belgium
EANAM – European Association for Negotiation and Mediation, Brussels, Belgium

Lydie Goeldner-Gianella
Laboratory of physical geography CNRS 8591 LGP, Université Paris 1 Panthéon-Sorbonne, Paris, France

Brian Golding
Met Office, Exeter, United Kingdom

Mathilde Gralepois
Laboratory CNRS 7324 Citeres, University of Tours, Tours, France

Yves Henocque
Maritime Policy and Governance, French Research Institute for the Development of the Sea, IFREMER, Paris, France

Giorgos Kallis
Institut de Ciència i Tecnologia Ambientals (ICTA), Universitat Autonoma de Barcelona and Institució Catalana de Recerca i Estudis Avançats (ICREA), Barcelona, Spain

Stefan Kuks
Department of governance and technology for sustainability, CSTM, University of Twente, Enschede, The Netherlands and Vechstromen Water Authority, The Netherlands

Abel La Calle
Autonomous University of Barcelona, Institute of Government and Public Policy, Barcelona, Spain

Isabelle La Jeunesse
Laboratory CNRS 7324 Citeres, University of Tours, Tours, France

Ruta Landgrebe
Ecologic Institute, Berlin, Germany

Corinne Larrue
Lab'Urba, Ecole d'Urbanisme de Paris, Université Paris Est Créteil, Marne-La-Vallée, France

Esmeralda Longépée
Laboratory 228 Espace-DEV, University of Mayotte, Mayotte, France

Alexandre K. Magnan
Institute for Sustainable Development and International Relations, Sciences Po, Paris, France

Hannelore Mees
Department of sociology, University of Antwerp, Antwerp, Belgium

Denis Mercier
Laboratory CNRS 8185 ENeC, Paris Sorbonne University, Paris, France

Virginia Murray
Public Health England, London, United Kingdom

Alexandre Nicolae-Lerma
BRGM Nouvelle-Aquitaine, French Geological Survey, Pessac, France

Lila Oriard Colin
Lab'Urba, Institut d'Urbanisme de Paris, Université Paris Est Créteil, Marne-La-Vallée, France

Gül Özerol
Department of Governance and Technology for Sustainability, University of Twente, Enschede, The Netherlands

Sophie Pardo
Economics and management laboratory LEMNA, University of Nantes, Nantes, France

Sally Priest
Flood Hazard Research Centre, Middlesex University, London, United Kingdom

Thomas Schellenberger
European Center for Research on the Law of Collective Accidents and Disasters (CERDACC), IUT de Colmar, Université de Haut-Alsace, Colmar, France

Ulf Stein
Ecologic Institute, Berlin, Germany

Jenny Tröltzsch
Ecologic Institute, Berlin, Germany

Ethemcan Turhan
Laboratory of Environmental Humanities, KTH Royal Institute of Technology, Stockholm, Sweden

Rodrigo Vidaurre
Ecologic Institute, Berlin, Germany

Thomas Waite
Public Health England, London, United Kingdom

Mark Wiering
Radboud University, Nijmegen, The Netherlands

Christos Zografos
Environmental science and technology Institute ICTA, Autonomous University of Barcelona, Barcelona, Spain

Editors

Isabelle La Jeunesse is Lecturer Habilitated to direct research in Environmental Geography at the University of Tours and in the laboratory CNRS Citeres. She defended a thesis carried out at Ifremer on the anthropization of the phosphorus geochemical cycle and its implications for the eutrophication management of the Thau coastal lagoon. Since then, she has focused her research on the impact of human activities on water and the needs of integrated management of this resource at the catchment scale. Her participation in European research programmes has fuelled the necessarily interdisciplinary vision of water management from environmental sciences to social sciences in the context of existing pressures, including climate change and hydrometeorological extreme events.

Corinne Larrue (1957) is Professor of urban and environmental planning at the University of Paris Est Créteil since 2013, and previously was Professor at University of Tours for 20 years. Within the University she has been (2014–2018) co-director of the Paris School of Planning, one of the most important institutes for urban planning in France and is currently member of the Lab'urba research center.

Her major field of research, teaching, and expertise is policy analysis with emphasis on environmental and regional policies. She has coordinated and participated in several comparative research projects within the European Union Framework Programmes, dealing with implementation of environmental policy issues. She has published numerous academic papers and books devoted to environmental policy analysis and on public policy management.

The Series Editor

Philippe Quevauviller began his research activities in 1983 at the University of Bordeaux I, France, studying lake geochemistry. Between 1984 and 1987 he was Associate Researcher at the Portuguese Environment State Secretary where he performed a multidisciplinary study (sedimentology, geomorphology, and geochemistry) of the coastal environment of the Galé coastline and of the Sado Estuary, which was the topic of his PhD degree in Oceanography gained in 1987 (at the University of Bordeaux I).

In 1988, he became Associate Researcher in the framework of a contract between the University of Bordeaux I and the Dutch Ministry for Public Works (Rijskwaterstaat), in which he investigated organotin contamination levels of Dutch coastal environments and waterways. From this research work, he gained another PhD in chemistry at the University of Bordeaux I in 1990. From 1989 to 2002, he worked at the European Commission (DG Research) in Brussels where he managed various Research and Technological Development (RTD) projects in the field of quality assurance, analytical method development and pre-normative research for environmental analyses in the framework of the Standards, Measurements and Testing Programme. In 1999, he obtained an HDR (Diplôme d'Habilitation à Diriger des Recherches) in chemistry at the University of Pau, France, from a study of the quality assurance of chemical species' determination in the environment.

In 2002, he left the research world to move to the policy sector at the EC Environment Directorate-General where he developed a new EU Directive on groundwater protection against pollution and chaired European science-policy expert groups on groundwater and chemical monitoring in support of the implementation of the EU Water Framework Directive. He moved back to the ECDGResearch in 2008, where he acted as research Programme Officer and managed research projects on climate change impacts on the aquatic environment and on hydrometeorological hazards, whilst ensuring strong links with policy networks. In April 2013 he moved to another area of work, namely Security Research, at the EC DG Enterprise and Industry where he is research Programming and Policy Officer in the fields of Crisis Management and CBRN.

Besides his EC career, Philippe Quevauviller has remained active in academic and scientific developments. He is Associate Professor at the Free University of Brussels and promoter of Master theses in an international Master on Water Engineering (IUPWARE programme), which is under this function that he is acting as Series Editor of the *Hydrometeorological Extreme Events* Series for Wiley. He also teaches integrated water management issues and their links to EU water science and policies to Master students of the EurAquae programme at the Polytech'Nice (France).

Philippe Quevauviller has published (as author and coauthor) more than 220 scientific and policy publications in the international literature, 54 book chapters, 80 reports and 6 books and has acted as an editor and co-editor for 26 special issues of scientific journals and 15 books. He also coordinated a book series for Wiley on *Water Quality Measurements* which resulted in 10 books published between 2000 and 2011.

Series Preface

The rising frequency and severity of hydrometeorological extreme events have been reported in many studies and surveys, including the IPCC Fifth Assessment Report. This report and other sources highlight the increasing probability that these events are partly driven by climate change, whilst other causes are linked to the increased exposure and vulnerability of societies in exposed areas (which are not only due to climate change but also to mismanagement of risks and 'lost memories' about them). Efforts are ongoing to enhance today's forecasting, prediction, and early warning capabilities in order to improve the assessment of vulnerability and risks and to develop adequate prevention, mitigation, and preparedness measures.

This book series, titled 'Hydrometeorological Extreme Events', has the ambition to gather available knowledge in this area, taking stock of research and policy developments at an international level. Whilst individual publications exist on specific hazards, the proposed series is the first of its kind to propose an enlarged coverage of various extreme events that are generally studied by different (not necessarily interconnected) research teams.

The series comprises several volumes dealing with the various aspects of hydrometeorological extreme events, primarily discussing science–policy interfacing issues, and developing specific discussions about floods, coastal storms (including storm surges), droughts, resilience and adaptation, and governance. Whilst the books examine the crisis management cycle as a whole, the focus of the discussions is generally oriented towards the knowledge base of the different events, prevention and preparedness, and improved early warning and prediction systems.

The involvement of internationally renowned scientists (from different horizons and disciplines) behind the knowledge base of hydrometeorological events makes this series unique in this respect. The overall series will provide a multidisciplinary description of various scientific and policy features concerning hydrometeorological extreme events, as written by authors from different countries, making it a truly international book series.

Following a first volume introducing the series and a second volume on coastal storms, the 'drought' volume was the third book of this series. This book, focusing on governance and climate and health aspects, was written by renowned experts in the field, covering various horizons and (policy and scientific) views. The forthcoming volume of the series will focus on floods.

Philippe Quevauviller
Series Editor

Series Preface

Part I

Introduction

1

Governance Challenges Facing Hydrometeorological Extreme Events

Isabelle La Jeunesse[1] and Corinne Larrue[2]

[1] *Laboratory CNRS 7324 Citeres, University of Tours, Tours, France*

[2] *Lab'Urba, Ecole d'Urbanisme de Paris, Université Paris Est Créteil, Marne-La-Vallée, France*

After the preamble on the complementarity of the six volumes of the series on hydrometeorological extreme events edited by Philippe Quevauviller, this chapter aims to provide the readers with two main reading grids. First, it proposes some definitions of the hydrometeorological extreme events considered in this book. Second, it refers to the concept of adaptive governance to introduce the framework proposed for the governance analysis of the three hydrometeorological extreme events considered in this book, namely floods, droughts, and coastal storms.

1.1 Introduction

Introducing this book on the governance of hydrometeorological extreme events necessitates, first of all, defining what is meant by hydrometeorological extreme events for societies and what they are forecast to be in the context of climate change. Then we will introduce the governance issue related to hydrometeorological events in order to frame the specific situations set out throughout this book, which sketch the specific geographical and political contexts within which these governance issues take place.

1.2 Facing hydrometeorological extreme events

Hydrometeorological variability is inherent to terrestrial climate. Hydrometeorological extreme events are part of this variability. Societies are thus naturally exposed to hydrometeorological extreme events and have, through history, developed different strategies to manage their vulnerabilities. Each year, millions of people are affected by hydrometeorological extreme events all over the world, including Europe, with an

Facing Hydrometeorological Extreme Events: A Governance Issue, First Edition.
Edited by Isabelle La Jeunesse and Corinne Larrue.

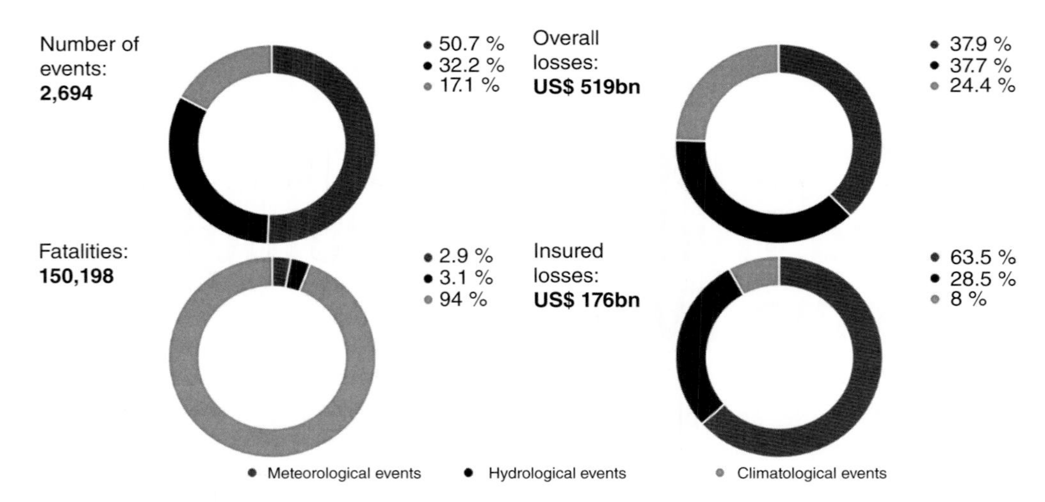

Figure 1.1 Percentage distribution for relevant weather-related losses in Europe over the 1980–2017 period © Munich Re reinsurance database.

observed and reported increase in severity and frequency (Harding et al. 2015). The total amount of floods and economic losses associated with these events have increased over the past decades (Bates et al. 2008; Kundzewicz et al. 2014), as confirmed by the NatCat SERVICE[1] – a database on natural disasters managed by the Munich Re reinsurance agency (Figure 1.1). Overall losses have been assessed to represent more than USD 500 billion.

In 2012, the Intergovernmental Panel on Climate Change (IPCC) produced a Special Report on Managing the Risks of Extreme Events and Disasters to Advance Climate Change Adaptation. This report is commonly referred to as the SREX report. Its main objective is to prevent hydrometeorological extreme events by exploring the physical and social dimensions of weather- and climate-related disasters. Thus, hydrometeorological extreme events must be considered in the context of global warming and its impacts.

According to the IPCC report on extreme events, climate extremes (extreme weather or climate event) means the occurrence of a value of a weather or climate variable above (or below) a threshold value near the upper (or lower) ends of the range of values observed for the variable (IPCC 2012). To simplify, both extreme weather events and extreme climate events are referred to collectively as 'climate extremes' (IPCC 2012). These include any rare, intense, and severe extreme events (Beniston et al. 2007).

All these extreme events – floods, droughts, and coastal storms – originate from climate-system extremes such as persistent anticyclonic conditions or strong gradients in atmospheric pressure and temperature. Thus, if the climate changes, as a consequence, extreme events can also change. Scenarios proposed by the IPCC report on extreme events predict an increase in these events, both in frequency and intensity.

These extreme events are expected to have major impacts throughout Europe, including on water management. However, as regards the impacts on societies, it is always necessary to set these events in a historical context. As shown through the work of climate historians, history testifies to series of extreme events having taken place in

Europe between the sixteenth and twentieth centuries (Garnier 2015). The impacts of past extreme events on societies (Le Roy Ladurie, Rousseau, and Vasak 2011) give important information on what current governance could consider in its risk-assessment process. The Sendai framework for Disaster Risk Reduction, which was established by the UN General Assembly in March 2015 (as a follow up to the Hyogo framework), aimed at increasing the preparedness for climate change impacts through a framework supporting the 'substantial reduction of disaster risk and losses in lives, livelihoods and health and in the economic, physical, social, cultural and environmental assets of persons, businesses, communities and countries' (Sendai Framework 2015). In order to achieve this overall outcome by 2030, the Sendai working groups focused on the following goals: 'Prevent new and reduce existing disaster risks through the implementation of integrated and inclusive economic, structural, legal, social, health, cultural, educational, environmental, technological, political and institutional measures that prevent and reduce hazard exposure and vulnerability to disaster, increase preparedness for response and recovery, and thus strengthen resilience'. These efforts are expected to: (i) generate the information base for the development of Sendai Framework implementation strategies, (ii) facilitate the development of risk-informed policies and decision-making processes, and (iii) guide the allocation of appropriate resources.[2] This must be based on efficient governance for integrated risk management.

This being said, as an introduction to this book on the governance of hydrometeorological extreme events, it seems necessary to provide an overview of the main characteristics of the hydrometeorological extreme events presented and the main impacts and adaptation strategies proposed at European level. Since the main hydrometeorological extreme events occurring in Europe are floods, droughts, and coastal storms, this book focuses on these three types of extreme event.

1.3 Floods

In Europe, flooding is probably the leading natural hazard. Flood lists are mentioned in several webpages and commonly engraved on the piers of river bridges. The European Environment Agency provides a datasheet containing information on past floods in Europe since 1980,[3] based on the reporting of European Union (EU) Member States for the EU Floods Directive (2007/60/EC), along with information provided by relevant national authorities and global databases on natural hazards. The summer of 2016 was the scene of major floods throughout Europe, where Germany and France, along with Austria, Belgium, Romania, Moldova, the Netherlands, and the United Kingdom were highly impacted.

1.3.1 Definition and characteristics

In English, the word flood describes both the natural variability of a river with a maximum flow during the wet period, and the inundation phenomenon due to the river coming out of its bed and filling the floodplains. Thus, a flooding event, as an extreme hydrometeorological event, is a situation in which water temporarily covers land not frequently flooded.

Flood can be described by hydrological characteristics such as:

1. *Flood intensity*, which is characterized by inundation depth and volume;
2. *Flood frequency*, which represents the number of times an area is inundated during a particular time interval; and
3. *Flood duration*, which is the length of time during which a particular area is inundated.

These hydrological characteristics of floods are used to study the series of historical data and to forecast flooding events. Also, in flood management, hydrological characteristics are linked with their consequences both on natural systems and human societies so as to prevent flood damage and design alert systems. Following this, floods can be described according to flow speed, the physical characteristics of the catchment area, and the main causes of flooding. We can distinguish:

1. *Riverine flooding* – the slow increase of the height of the water in the riverbed due to an unusual rainfall event, in spring with snowmelt or in summer if glaciers melt. It occurs in river floodplains in areas with low topographic elevation when the upstream basin experiences heavy rainfalls. With large rivers, the flow speed and rising speed processes are relatively slow. This type of flood can be due to an obstruction in the flow path. These types of floods last the longest amongst flooding events.
2. *Mudflow* – any type of flooding that engages high quantities of sediments. Flash floods due to heavy rainfalls in upper parts of catchments usually mobilize a large quantity of mud. However, mudflow usually concerns steeper watersheds with poor soils highly sensitive to erosion. The impacts are comparable to snow avalanches.
3. *Coastal flooding* – flooding due to particular concomitant physical conditions where the river is in flood and the pressure on the sea does not enable the water to flow into the sea. This usually occurs close to the mouth of the river and the delta area. However, this type of flooding can be induced by storm surges, which are described later.
4. *Urban flooding* or *urban drainage flooding* – is specific as the cause is usually aggravated by a lack of drainage and infiltration in partially impervious urban areas and all these areas are highly vulnerable. This occurs when the drainage infrastructure becomes blocked or overwhelmed due to high-intensity rainfall. Thus, this type of flooding impacts areas close to drainage channels and house basements.
5. *Flash flooding* – a localized flood where water flows in at great speed. The flow speed of the water is mainly determined by the slope of the terrain rained on. The steeper the slope, the faster the water flows. This usually concerns mountain and hilly areas, as well as highly impervious catchments after heavy and localized rainfall. The warning time is very short, less than one hour for some hydro systems. Flash floods can also be induced by structure failures, such as for instance, when a dike or a dam breaks and a large amount of water is released suddenly. The water speed at the breach is similar to the speed of a flash flood, but sometimes involves much larger volumes.
6. *Pluvial flooding or water logging* – due to an accumulation of water exceeding saturation conditions and forming a layer – on agricultural fields and streets alike – provoking significant and unusual runoff. These events are due to excessive rainfall. This issue concerns both urban and agricultural land.

With all this, the height of surface waters, which leads to flooding when a defined threshold is exceeded, can be exacerbated by different natural and anthropogenic factors. Most of the time, floods are caused by an unusual duration of rainfall. However, in some conditions, torrential rains or storms cause flooding. In the summer season, in hydro systems linked to mountain areas, a high average temperature can also result in increased snowmelt with, hence, a high discharge downstream.

Land use and land cover in watersheds are also highly implicated in the variability of the hydrological responses of the catchment to rainfall. The proportion of forest areas in the upper parts of the catchment decreases the runoff generated by heavy rainfalls in these upper parts. Downstream, wetlands buffer the volumes of water and thus have a significant impact on flow peaks. Moreover, depending on how natural meanders have been managed throughout the river's history, the linearity of the river is highly efficient in decreasing flow speed in the riverbed. Last but not least, the proportion of impermeable areas versus soil infiltration capacity has an impact on the height of water runoff and flow peak volumes.

1.3.2 Impacts and adaptation

Both during and after floods, any kind of economic activity may be impacted due to disruptions or possible damage to infrastructure and transportation networks. In many cases, residents of flooded houses are invited to move. It is also well-known that floods have a significant psychological impact, since when someone's house or work place is flooded they can lose everything. However, the impact is considered only if it is negative for the environment and/or human activities. And it all depends on where and when the flooding took place. For instance, flooding has higher impact in cities than in wetlands and is more or less significant in agricultural areas depending on the time of the year and stage of crop growth.

To reduce the impact of flooding, the management options vary from environmental actions to flood defence infrastructures. These may include, for example building a storm basin, restoring river banks, dredging rivers, rehabilitating drained or inefficient wetland areas, and increasing the surface of forests in the upper parts of the catchment. The hydrological modelling facilities are used by national/regional/local communities to assess the effects of either option (e.g. river flow, volume stored in storm basins, and inundated areas).

To prevent flood damage, and in particular to avoid loss of life, authorities need to anticipate and communicate to citizens and economic agents at different decisional levels on the state of the flooding event. In fact, whilst the water level slowly rises in large rivers, officials can first inform citizens as well as decide to evacuate people before the river overflows. However, the area that is flooded can be huge and require local communities, included isolated ones, to be highly organized. Many prevention systems exist, depending on national incentives and local community investment. At the European level, the European Flood Awareness System (EFAS)[4] – developed in the context of several European research projects, is now a single operational European monitoring and forecasting system for floods across Europe. It provides complementary information to the National and Regional Hydrological Services and to the European Response and Coordination Centre operating within the European Commission's Humanitarian Aid and Civil Protection department.

1.4 Drought

1.4.1 Definition and characteristics

Drought refers to a period of abnormally dry weather long enough to cause serious hydrological imbalance (IPCC 2012). Considering this, any region of the globe can suffer from drought. However, drought is a relative term and therefore needs to be clarified. In this book, 'drought', as a hydrometeorological extreme event, refers to the impact of an exceptional lack of rainfall. In contrast to aridity, which is a permanent feature of the climate and is restricted to low rainfall areas, drought is a temporary water shortage condition compared to the average situation. It is usually the consequence of a natural reduction in the amount of rainfall received over an extended period of time, which can be caused or aggravated by other climatic factors, such as high temperatures, high winds, or low relative humidity. Based on this, and depending on the main causes or impacts, some definitions of droughts have been proposed. These are usually grouped into five types (Wilhite and Glantz 1985):

1. *Meteorological drought*, which is mainly due to a long period of no or very low rainfall;
2. *Hydrological drought*, which is characterized by below average river flows;
3. *Agricultural drought*, which refers to a soil moisture deficit affecting crops;
4. *Mega drought*, which is a persistent and extended drought that lasts for a much longer period than normal; and
5. *Socioeconomic drought* has also been considered in order to name droughts induced by human factors, causing, for instance, excessive demands on a supply-demand system. These occur when the demand for water exceeds the supply (Wilhite and Glantz 1985).

The last type of drought makes it possible to distinguish between drought (and drought impacts) and water scarcity. Thus, water scarcity and drought (WS&D) are two interrelated but distinct concepts. Water scarcity may result from a range of phenomena, which may either be produced by natural causes – such as drought, or can also be induced by human activities alone, or, as is usually the case, may result from the interaction of both (Pereira, Cordery, and Lacovides 2002). This explains why the relevant policy at European level is called 'water scarcity and drought policies'.

1.4.2 Impacts and adaptation

Over the past four centuries, Europe has been affected by severe droughts (Garnier 2015). No later than 1921, a severe drought occurred with rainfalls 40% lower than usual, from England to Italy. It affected a large part of Europe and even provoked a famine in Eastern Europe. Later on, the 1976 drought was especially severe in the Northern half of France and affected other parts of North-western Europe (Le Roy Ladurie et al. 2011).

Since then, both policy literature and the popular press have pointed out in the 1990s the potential water wars that reduced water availability, resulting from climate-induced changes, could generate in parallel with demographic growth. Indeed, water scarcity is often cited as a cause of water conflicts, which in turn can threaten water security, enabling us to draw a link between droughts and water security in Europe (Liberatore 2013).

We can already note that the number of people and areas in Europe affected by drought and water scarcity has increased by 20% between 1976 and 2006 (European Commission [EC] 2007). The total cost of these 30 years of drought amounts to € 100 billion (EC 2007). This makes it crucial to deal with drought and water scarcity now, and increase drought resilience before the problem grows even larger. In its 2007 Communication Report, the EC clearly stated that devising effective drought risk management strategies must be regarded as an EU priority.

Out of all hydrometeorological extreme events, drought is undoubtedly the most complex as both its causes and impacts are not very well known or understood, nor are they described and modelled. This partly explains why overall losses are usually more significant than insured losses for climatological events whereas for hydrological events (floods) these losses are very similar, as shown by the extractions of Munich Re's reinsurance database. This can be demonstrated by the two graphs presenting natural disaster losses for Europe in 2016 and 2017. In 2016 there were major floods, mainly in France, Germany, and the Netherlands (Figure 1.2). In 2017, Europe was more affected by freezing events and droughts, where mainly Spain, Italy, and Serbia were affected by dry conditions and heatwaves with low water levels in rivers and reservoirs, which affected crops, fruits, vineyards, and pasture land (Figure 1.3).

In fact, although a lack of rainfall is a frequent phenomenon from a climatic point of view, drought and its socio-economic impacts depend not only on the severity and the spatial extent of the rainfall deficit, but also on several factors such as the state of the environment, as well as social and economic vulnerabilities. Some authors have pointed out that inadequate land use practices, unsustainable management of water resources, and inadequate risk management are key factors in explaining the drought impacts (Vogt and Somma 2000).

The impacts on socio-economic activities are mainly due to losses in agricultural production, which mostly concerns wheat in Europe as it is a non-irrigated crop

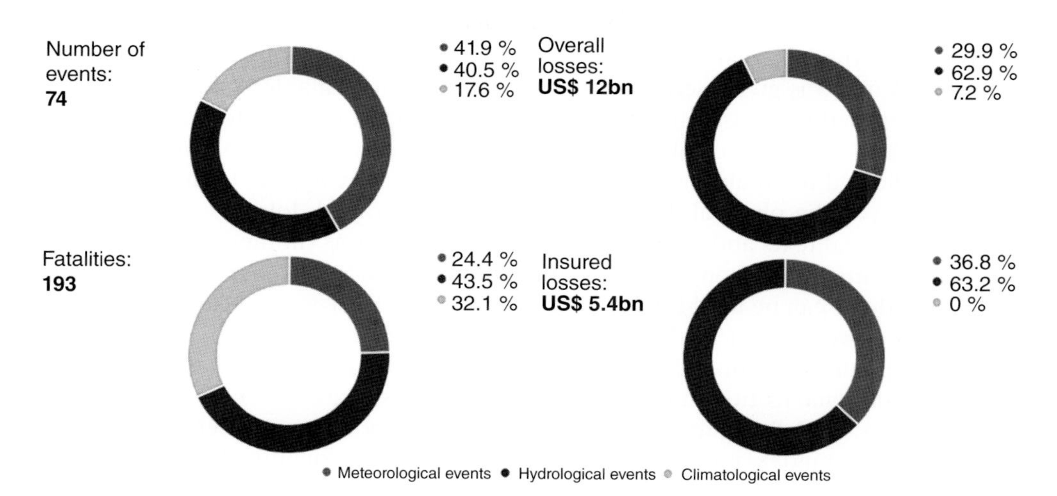

Figure 1.2 Percentage distribution for relevant weather-related loss events in Europe in 2016.

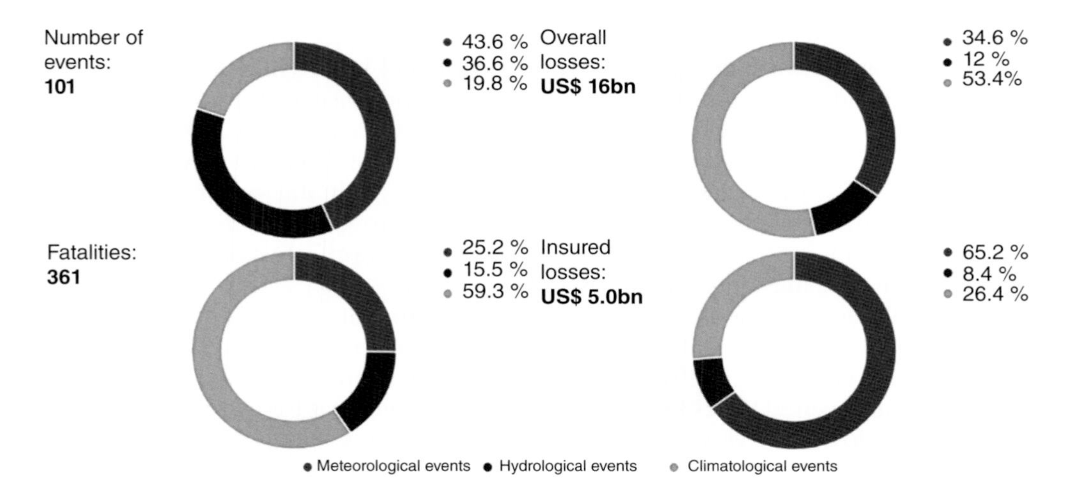

Figure 1.3 Percentage distribution for relevant weather-related loss events in Europe in 2017.

covering a large amount of agricultural areas. Affected non-irrigated crops are considered as really representative of the production impacted by rainfall deficits. However, irrigated crops are also impacted, yet it is more difficult to establish a cause–consequence relation. Drought impacts are also related to reductions in hydropower energy production and environmental degradation. Last but not least, droughts generally provoke public water supply cuts, both because of a degradation of water quality and the quantity of water available. As regards security, another major consequence of droughts is dryness and possible fire outbreaks, as well as the management of heatwave effects, which usually induce an increased water consumption.

From an historical perspective, in the first volume of this series, E. Garnier proposed an interesting drought severity index adapted to available European information from the sixteenth to the nineteenth centuries (Garnier 2015). This index is proposed to be compared to the physical drought classification mentioned above. An index scale between –1 and 5 is given according to the information available in archives. The –1 index is proposed to be used to characterize an event kept in the chronological reconstruction that has insufficient qualitative and quantitative information. Index 1 is mainly dedicated to an absence of rainfall, and thus mainly applies to meteorological droughts as per the physical definition provided earlier, with lots of evidence of the issue in historical texts and materialized by several rogations (in the Christian Church this is a solemn supplication consisting of the litany of the Saints chanted for requiring rainfall). Index 2 can be used for the observation of local low-water in rivers, with notified effects on vegetation. As far as this index concerns the vegetation, it can be compared to agricultural droughts. Index 4 expresses severe low-water marks associated with impossible navigation in rivers, wheat mill lay-offs, search for new springs, forest fire outbreaks, as well as cattle deaths. This level corresponds to a severe hydrological drought. The last index, no. 5, is proposed for exceptional droughts, also referred to as mega droughts, without any possible supply, shortages, sanitary problems, and very high wheat prices, along with several forest fires.

Above all, droughts are expected to increase in the future as a result of climate change. Amongst the impacts is an increase in the frequency and severity of drought periods and water scarcity (EC 2012). In 2007, 11% of the European population and 17% of the European territory were affected by drought (EC 2007).

In this century, despite an increased awareness of drought hazards at European level through the work on the common implementation strategy groups of the water framework directive (WFD) dedicated to drought for instance, the tools, instruments, and management strategies for capitalizing data, risk assessment, forecasting, monitoring, and adapting to potential droughts are not clearly defined. After one of the most widespread droughts affecting over 100 million people in 2003 – one-third of the EU territory, the cost of which was assessed at € 8.7 billion at EU level, the EU Council of Ministers asked the EC to address the challenges of WS&Ds in the EU. This led to the communication of several measures to integrate WS&Ds in river basin management plans in 2007. These were summarized in seven policy options[5]:

1. Putting the right price tag on water;
2. Allocating water and water-related funding more efficiently;
3. Improving drought risk management;
4. Considering additional water supply infrastructures;
5. Fostering water efficient technologies and practises;
6. Fostering the emergence of a water-saving culture in Europe; and
7. Improving knowledge and data collection.

A final report on the Review of the European WS&D Policy was completed in November 2012. This report responds to the 2007 Council request to review by 2012 whether the policy on WS&Ds has achieved its objectives of reducing water scarcity and vulnerability to droughts. It also looks into whether actions taken in the implementation of the WFD[6] helped address WS&D. This report is part of the 'Blue Print for Safeguarding European Waters' adopted by the European Commission on 14 November 2012.

1.5 Coastal storms

Coastal storms can be the most destructive natural hazards. Several coastal storms have hit the European coast over the past 50 years, such as Xynthia in 2010 or Klaus in 2009. The coastal storms, Lothar and Martin, which affected Europe in December 1999, generated such damage inland due to windstorms that the collective memory has denied the very existence of a submersion (Garnier and Surville 2010). In fact, if coastal storms originate from marine waters, the characterization of these events usually depends heavily on what generated the most impacts and what was the most destructive. Furthermore, coastal storms are a combination of multiple specific hydrometeorological conditions.

In addition, coastal storms concern the coastal zone, therefore it seems vital to first characterize the latter and delimit the areas concerned.

1.5.1 Coastal zone delimitation

Coastal zones can be delimited based on several types of criteria: physical, ecological, economical, and societal. Following a general definition proposed by the European Environment Agency on its website, coastal zones can be considered as 'the part of the land affected by its proximity to the sea, and that part of the sea affected by its proximity to the land as the extent to which man's land-based activities have a measurable influence on water chemistry and marine ecology'. However, in order to be operational as regards land use management, the coastal zone is generally defined more specifically in territorial planning documents and can even differ from regions to countries. For instance, the 1991 Planning Act in Denmark defined the landward boundary of the coastal zone as extending 3 km inland from the coast and the seaward boundary as the shoreline; whilst at the same period of time, the Spanish Shores Act (1988) considered the landward boundary extended up to 200 m from the inland limit of the shore (Lavalle et al. 2011).

Regarding the diversity of socio-economic activities generally concentrated in coastal zones, the EU started to define rules for Integrated Coastal Zone Management (ICZM) (Cicin-Sain and Knecht 1998), first through the WFD, which covers the transitional and coastal waters up to one nautical mile from the continental baseline. The ICZM was completed for marine waters by a marine policy initiated in 2008, the environmental part of which was represented by the Marine Strategy Framework Directive (MSFD) (Borja et al. 2010, 2013; Elliott 2013). On this occasion, a definition of the coastal zone was proposed by the European Commission Demonstration Programme on ICZM[7] (EC 1999). A common definition of the coastal zone used by the European Commission is 'a strip of land and sea of varying width depending on the nature of the environment and management needs. It seldom corresponds to existing administrative or planning units. The natural coastal systems and the areas in which human activities involve the use of coastal resources may therefore extend well beyond the limit of territorial waters, and many kilometers inland'. The following criteria – provided in order to consider both the influences of the sea inland and the impacts of human activities on coastal ecosystems – were adopted to define the geographical delimitation of coastal zones:

1. A 10 km buffer from the coastline (obtained from administrative boundaries such as the Geographic Information System of the Commission [GISCO] for the EU);
2. A 2 km buffer from the aggregation of five Corine Land Cover classes: salt marshes, salines, and intertidal flats for coastal waters; and coastal lagoons and estuaries for marine waters.

The first ICZM criterion aims to represent an influence area of the shore, in order to capture its specific ecosystems on one hand, and any urban areas that might generate pressure on the coast, on the other. The aim of the second criterion is to include all inland areas under a direct influence from the maritime environments, these being characterized by transitional waters, where valuable biophysical features (fauna, flora, and geomorphologies) are potentially present.

1.5.2 Definition and characteristics

A general definition of coastal storms could consist in an anomalous set of meteorological conditions comprising extreme winds, large waves with storm surges, and usually heavy rainfalls (Ciavola et al. 2015). Storm surges appear when the sea rises as a result of atmospheric pressure changes and winds associated with a storm. In Europe, storms forming in the ocean and entering into the continent across the North Atlantic usually create extreme sea levels.

Until now, most of the damaging windstorms in Europe have been extra-tropical cyclones. These are low-pressure systems at a synoptic scale, that is to say characterized by an extent of several hundreds to thousands of kilometres and lasting for several days. They include depressions, anticyclones, and barometric troughs. These systems will grow if there is a strong north–south temperature gradient plus pressures that create a vorticity in the unstable frontal waves. In the North Atlantic Ocean, these conditions are met during the months of October to March.[8] They enable extra-tropical cyclones to be formed and have sufficient energy to travel eastwards towards Europe. The path followed by these storms usually curves northwards, which explains why Iceland and northern European countries such as the regions of Ireland, the UK, and Scandinavia are frequently hit. When the storms finally travel further southwards, they can affect countries such as France, Portugal, and Spain.

Moreover, the Mediterranean basin can also generate convective storms and cyclones. These are called 'medicanes', or Tropical-like Mediterranean Storm (TMS), to distinguish them from the hurricanes generated in tropical oceans. These storms are warm-core, low-pressure systems, generating strong winds in the Mediterranean. They are tight, highly convective windstorms without, however, acquiring either size or power, due to the temperature gradient. In the context of climate change, these coastal storms are expected to become more frequent.

Due to climate change, regions in southern Europe are assessed to become the theatre of a gradual and strong increase in overall climate hazards (Forzieri et al. 2016).

1.5.3 Impacts and adaptation

1.5.3.1 Impacts The impacts of climate change will not be distributed evenly or equitably across the territory. With regard to natural hazards, climate change will mainly result in widespread pressure on lower coasts, through coastal erosion or submersion, due to the expected increase in sea level. The specificities of coastal storms, combining violent winds with important movements of water both from the oceans and atmosphere, explain why coastal-storm impacts are most often associated with flooding issues.

Also, so far, fast-rising coastal storms usually give little time to raise the alert and frequently engage massive volumes of water. In general, storms occur in North and North-Western Europe all year round, but mainly between October and March in central Europe, as already mentioned. Storm surges can provoke temporary increases in sea level, rising above tide level, and often cause coastal flooding. Storm events can thus have large impacts on vulnerable systems such as transport, forestry, and energy infrastructures, as well as on human safety. Recordings of storm surges causing floods in large coastal areas date back to 1953, which was the most damaging surge. In the Netherlands and in Eastern England, respectively, 1800 and 307 people died and around 200 000 ha and 160 000 ha of land was flooded (EEA 2017).

If the impacts are considered through the analysis of coastal vulnerability rather than the physical characteristics of the extreme event, these can refer to a spatial concept that considers people and places in the coastal zone likely to be impacted by coastal hazards such as coastal storms (Bevacqua, Danlin, and Yaojun 2018). Socio-economic and demographic data are usually used to assess the social vulnerability of coastal areas. In this sense, Europe's coastal vulnerability is a reality. In Europe, the length of the coastline for the 22 EU Member States with a sea border is estimated at 136 106 km. The European coastal regions account for 43% of the total EU-22 area and population (Eurostat 2009). The population living in coastal municipalities in the EU was approximately assessed at 70 million people (16% of the EU-25 population, EEA 2006).

The impacts of climate change will not be distributed evenly or equitably across the coastal zones in Europe. With regard to natural hazards, climate change will mainly result in widespread pressure on lower coasts, through coastal erosion or submersion, due to the expected increase in sea level. From a geographical point of view, some coastal regions could therefore be significantly affected by the projected changes.

However, long-term extreme sea level increase trends along the European coasts will mostly result from mean sea level changes rather than wave and storm surge changes, which mostly contribute to inter-annual and decadal variability (Weisse et al. 2014). Forzieri et al. (2016) have demonstrated that keys hotspots are located along coastlines and in floodplains because of their significant vulnerability due to high population densities and their economic pivotal role. In these areas, floods and windstorms could be critical in combination with other climate hazards. Moreover, exposure is projected to increase for very extreme events due to their increase in frequency.

1.5.3.2 Adaptation The main adaptation principles proposed for coastal zones are to:

1. Not increase the vulnerability of people, goods and economic activities;
2. Assess the resilience, that is, the ability to adapt to situations, the objective being to develop resilience in the face of climate change hazards;
3. Choose between protection, displacement, or management of temporary disturbances. Beyond the sole socio-economic analysis, these choices to which citizens will be associated will have to take into account the acceptance of the change, its environmental cost, and the available financial resources. Choosing between protection and strategic retreat requires in-depth local studies on a complex dynamic (local trend of rising sea levels, erosion, sediment inputs and availability of materials, as well as, over a slightly longer time frame, the appreciation of the causes of extreme surges);
4. Analyse the vulnerability of territories and populations. The aim being to identify all the issues likely to be damaged in areas subject to potential hazards, including areas protected by structures and territories due to be urbanized in the future. It is then necessary to assess their physical vulnerability and the associated direct and indirect costs, as well as their functional vulnerability in order to determine the

measures to be taken, according to the various hypotheses adopted, social representations, and behaviours;

5. Deploy modelling tools to represent both sea-level rises in temporary or permanent submersions, as well as account for coastal erosion causing territory losses, sediment movements, and the behaviour of existing protection elements (artificial or natural);
6. Develop methods for assessing the economic, social and environmental impact, and the effectiveness of preventive measures; and
7. Adopt a national coastline management strategy and create a national network of coastal changes observations.

National strategies for the sea and coasts and planning documents intend to provide a coastal component meeting each of these objectives, with the final objective of decreasing disaster risks along the coastlines. The main frameworks are the Sendai Framework and the European integrated coastal zone management strategies included in the WFD (2000) and the MSFD (2008).

1.6 Governance issues related to hydrometeorological extreme events

Facing hydrometeorological extreme events therefore raises interesting governance questions. This governance issue is multidimensional and depends on a combination of conditions: public policies, funding opportunities, existing authority and access to knowledge, as well as capacity building at various decision levels, from the citizens up to national and international levels. It can differ from extreme events, insofar as extreme events seem not to be perceived equally both in time and space and depend on the type of actors concerned. Thus, the governance required could be different. For instance, droughts are by definition not sudden extreme events as their severity is linked to a lack of rainfall. On the contrary, floods and coastal storms can be sudden with the necessity to anticipate in order to be ready to face such events. The impact assessment can also be radically different. It seems easier to assess flood and coastal storm impacts than droughts. Thus, the ability to delineate risks and potential impacts is still a major issue in developing governance.

1.6.1 Addressing governance issues

Faced with the rising risks associated with meteorological extreme events, the organization of public authorities and more widely, society as a whole, is a key issue in terms of their ability to cope with these events. As a matter of fact, the importance of extreme event impacts in socio-economic terms depends on how societies anticipate and prepare for these events, as well as on how they manage periods of crisis and post-crisis situations. The risk governance issue is, therefore, at the heart of responses to climate-related risks. In recent years, numerous publications have been drafted in connection with these issues, which can be grouped under the heading of environmental or risk governance.

Speaking of environmental governance first requires clarifying the notion of governance. The term 'governance' refers to the organization of actors and stakeholders involved in a problem considered as public. 'Governance can be defined as a process of coordination between actors of social groups and institutions with a view to achieving objectives defined and discussed collectively' (Le Galès 2006, p. 245). This book will, therefore, focus on the various actors involved in the preparation and management of climate events and their consequences on one hand, and on the interactions between these actors and the regulations they produce on the other hand.

In the environmental field, the concept of governance has also given rise to numerous definitions. We shall retain the one summarized in Chaffin (Chaffin, Gosnell, and Cosens 2014, p. 1) in their review paper on adaptive governance: 'In short, environmental governance is the system of institutions, including rules, laws, regulations, policies, and social norms, and organizations involved in governing environmental resource use and/or protection'. The paper groups the various approaches into three types of governance system: (i) top-down, (ii) centralized, and (iii) bottom-up. The first system is state-based, the second functions via top-down directives, and the third usually emerges through groups of local actors or social networks. The authors explain the interests and difficulties of each approach. A 'new' approach that emerged over the past ten years is called adaptive governance, in the sense of robust governance (Dietz, Ostrom, and Stern 2003). It is a theoretical approach searching for ways to manage the uncertainty and complexity of environmental issues, in other words: social-ecological systems (Berkes and Folke 1998; Costanza et al. 1998).

Applied to hydrometeorological extreme event issues, the question of governance refers first of all to the actors' ability to cope with natural events (or those considered as such). Together with Kooiman, we will consider in this book the 'interactive arrangements in which public as well as private actors participate aimed at solving societal problems' (Koimann 2003), or more precisely, to 'a range of interactions between actors, networks, organizations, and institutions emerging in pursuit of a desired state for social-ecological systems' (Chaffin et al. 2014). This book will show and illustrate the diversity of arrangements between actors in terms of their coping abilities faced with natural events through time, as well as according to the political systems specific to the countries and regions studied, the place given to citizens, and the distribution of responsibilities and resources between these different actors. This book is based on recent research outcomes stemming from European research projects implemented in the field of natural hazards studied from a geographical and/or political point of view and focusing specifically on European contexts.[9]

The question tackled in this book is thus that of the actors of this governance, their perceptions and representations of the problems to be dealt with, the instruments mobilized, as well as the coordination set up between decision and time scales. As a matter of fact, governance related to natural and especially extreme risks is specific: on the one hand it applies to problems that go beyond institutional boundaries, which traditionally organize relations between subnational and supra-national actors. On the other hand, the uncertainty that characterizes the occurrence of these events makes organizing and coordination between actors a complex process. Finally, the importance of potential damage, as well as the relative scarcity of their occurrence, reduces the capacity for anticipation and preparation.

These two last characteristics argue for the use of the notion of a social-ecological system (Berkes and Jolly 2001) to address governance issues related to hydrometeorological extreme events. The concept of the social-ecological system is based on the recognition of the relationships between social and ecological systems: Folke showed that 'social and ecological systems are deeply interconnected and co-evolving across spatial and temporal scales' (Folke 2007, pp. 14–15). Thus, Folke's analysis of the management of the social-ecological system is particularly useful for positioning the stakes of a governance able to cope with hydrometeorological extreme events. Indeed, from this approach emerged the notion of adapting institution: Boyd and Folke define 'adapting institutions' as 'those that allow a variety of actors to deal with complexity, uncertainty and the interplay between gradual and rapid change' (Boyd and Folke 2011, p. 3).

From this first approach emerged the notion of adaptive governance: 'Adaptive governance is an outgrowth of the theoretical search for modes of managing uncertainty and complexity in *Social-Ecological Systems'* (Dietz et al. 2003; Walker et al. 2004 *in* Chaffin et al. 2014; Folke et al. 2005; Folke 2006).

1.6.2 Analytical framework based on adaptive governance

This approach is based on the observation that the traditional top-down 'state-based orientated' approach is not able to address the uncertainties associated with global environmental changes and does not have the flexibility to 'provide effective solutions for highly contextualized situations'. Huitema et al. (2009) summarize the key characteristics of adaptive governance and management. These include notions of polycentricity, evidence of public participation, experimental approaches to resource management, and management at a bioregional scale to match the scale of the problem (as opposed to relying on administrative boundaries) (Fournier et al. 2016).

The analytical framework provided by the adaptive governance approach makes it possible to identify several characteristics required for this type of governance. The following will be taken into account:

1. Multilevel governance, with a balance between top-down and bottom-up decision-making processes;
2. Multi-actor (formal and informal) networks, including active participation from citizens;
3. Flexibility in governance arrangements enabling mitigation measures tailored to local conditions;
4. Governance arrangements supporting management at the appropriate scale of the problem; and
5. Opportunities for experimentation, along with social and institutional learning.

1.6.3 Analytical framework based on risk governance

Adaptation to climate change is sometimes mixed up with disaster risk reduction as they both address issues of prevention and reduction of vulnerability and impacts. Thus, the governance of hydrometeorological extreme events can also be approached through

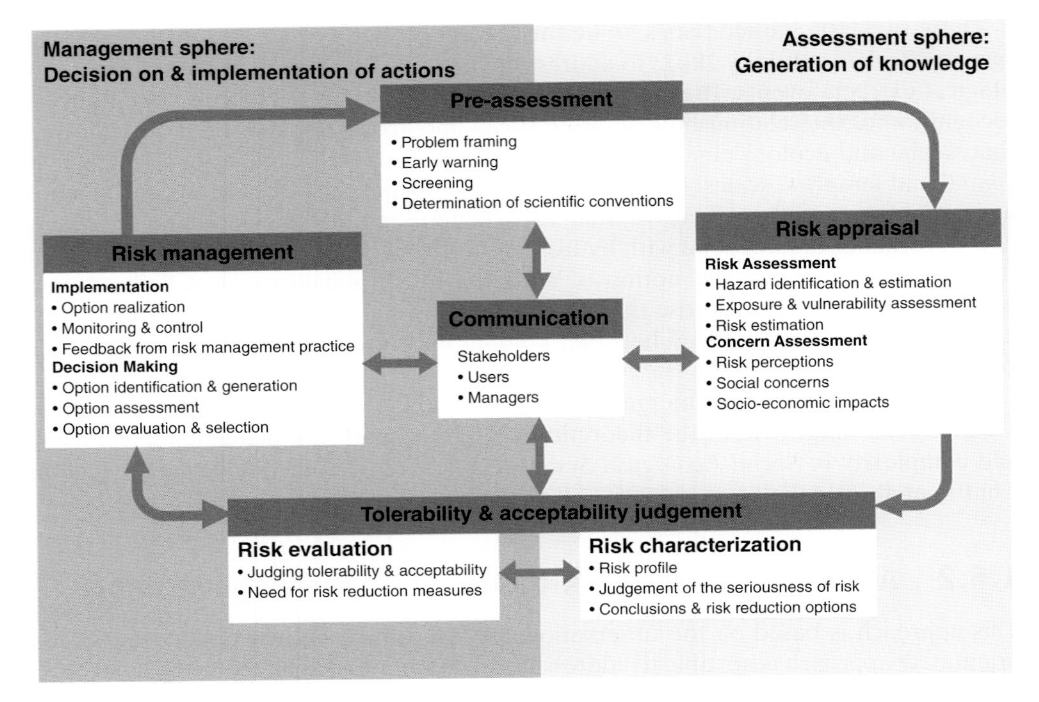

Figure 1.4 Framework for risk governance adapted from the International Risk Governance Council (IRGC, 2006).

the analysis of risk governance, for which the International Risk Governance Council (IRGC) has developed a framework of analysis. This framework (Figure 1.4) comprises five stages (IRGC 2006):

1. *Risk pre-assessment.* In this phase, the early warning and framing of the risk is conducted in order to provide a structured definition of the problem.
2. *Risk appraisal.* This stage combines a scientific risk assessment (of the hazard and its probability) with a systematic concern assessment (of public concerns and perceptions). It provides a concern assessment at different levels of society.
3. *Characterization and evaluation.* The representation of risk and its appropriation make it possible to assess its tolerability which supports the evaluation needs to reduce risks. This necessitates defining solutions to decrease risks and render them acceptable, tolerable (requiring mitigation), or intolerable (unacceptable).
4. *Risk management.* All the actions and remedies initiated to avoid, reduce, transfer, or retain risks are part of risk management.
5. *Risk communication.* In this phase, the authors consider all the solutions used by stakeholders and civil society to understand the risks and participate in the risk governance process.

Frameworks of both environmental and risk governance focus on actors' roles in the multilevel governance structure and their involvement in creating knowledge and implementing actions, from local to larger spatial scales. This explains why analysing the management of hydrometeorological extreme events addresses the following issues:

1. What is the available information on hydrometeorological extreme events?
2. How do actors represent hydrometeorological extreme events by considering all stakeholder categories?
3. What are the strategies and instruments implemented? What measures are taken before and after the events? Which instruments are favoured? Why?
4. What does this governance change in the effective management of past, present, and future uncertain hydrometeorological extreme events?

Based on the adaptive governance framework and risk governance framework, by combining theoretical approaches with national and local case studies, this book aims to point out success stories of multilevel governance decision-making processes for managing hydrometeorological extreme events, as well as present lessons learnt from difficult implementation processes. In doing so, this book presents a variety of points of view: legal, socio-political or more practical, without however attempting to model this wide range of diversified situations and points of view in a global model. Finally, it will issue some recommendations stemming from the cross analysis of the cases presented.

Notes

1. https://www.munichre.com/en/reinsurance/business/non-life/natcatservice/index.html. According to this service, hydrometeorological extreme events are part of meteorological events (tropical storms, extratropical storms, local windstorms) for coastal and windstorms; hydrological events (Flooding: river floods, flash floods and storm surges, wet mass movements) for floods and storm surges; climatological events (extreme temperature: both heatwaves, freezes and extreme winter conditions, droughts, wildfires) for droughts. This book does not concern geophysical events (earthquakes, tsunamis, volcanic activity, dry mass movements).
2. http://www.preventionweb.net/drr-framework/sendai-framework-monitor/indicators
3. https://www.eea.europa.eu/data-and-maps/data/european-past-floods
4. https://www.efas.eu
5. http://eur-lex.europa.eu/legal-content/en/ALL/?uri=CELEX%3A52012DC0672
6. Directive 2000/60/EC, OJ L 327, 22.12.2000, p.1.
7. http://ec.europa.eu/environment/iczm/demopgm.htm
8. http://www.europeanwindstorms.org/expl
9. The papers presented in the book stem from the following European research projects (amongst others):
   Climb: www.climb-fp7.eu
   Starflood: http://www.starflood.eu
   Capflo: http://capflo.net
   DROP: http://www.nweurope.eu/about-the-programme/our-impact/challenge-5/the-drop-project

References

Bates, B.C., Kundzewicz, Z.W., Wu, S. et al. (eds.) (2008). *Climate Change and Water*, IPCC report. Geneva: IPCC.

Beniston, M., Stephenson, D.B., Christensen, O.B. et al. (2007). *Climatic Change* 81: 71–95.

Berkes, F. and Folke, C. (1998). *Linking Social and Ecological Systems: Management Practices and Social Mechanisms for Building Resilience*, 459 p. New York: Cambridge University Press.

Berkes, F. and Jolly, D. (2001). Adapting to climate change: Social-ecological resilience in a Canadian western Arctic community. *Conservation Ecology* 5 (2): 18.

Bevacqua, A., Danlin, Y., and Yaojun, Z. (2018). Coastal vulnerability: evolving concepts in understanding vulnerable people and places. *Environmental, Science, Policy* 82: 19–29.

Borja, A., Elliott, M., Andersen, J.H. et al. (2013). Good environmental status of marine ecosystems: What is it and how do we know when we have attained it? *Marine Pollution Bulletin* 76 (1–2): 16–27.

Borja, A., Elliott, M., Carstensen, J. et al. (2010). Marine management – towards an integrated implementation of the European marine strategy framework and the water framework directives. *Marine Pollution Bulletin* 60: 2175–2186.

Boyd, E. and Folke, C. (2011). *Adapting Institutions: Governance, Complexity and Social-Ecological Resilience*, 312 p. Cambridge: Cambridge University Press.

Ciavola, P., Ferreira, O., Van Dongeren, A. et al. (2015). Prediction of storm impacts on beach and dune systems. In: *Hydrometeorological Hazards, Interfacing Science and Policy* (ed. P. Quevauviller), 227–246. Chichester: Wiley Blackwell.

Chaffin, B.C., Gosnell, H., and Cosens, B.A. (2014). A decade of adaptive governance scholarship: synthesis and future directions. *Ecology and Society* 19 (3): 56. https://doi.org/10.5751/ES-06824-190356.

Cicin-Sain, B. and Knecht, R.W. (1998). *Integrated Coastal and Ocean Management. Concepts and Practices*, 517. Washington, DC: Island Press.

Costanza, R., d'Arge, R., de Groot, R. et al. (1998). The value of the world's ecosystem services and natural capital. *Ecological Economics* 25 (1): 3–15.

Dietz, T., Ostrom, E., and Stern, P.C. (2003). The struggle to govern the commons. *Science* 302: 1907–1912.

European Commission, EC (2007). Addressing the challenge of water scarcity and droughts in the European Union, Communication de la Commission au Parlement européen et au Conseil, COM(200) 414. Brussels: European Commission.

European Commission, EC (1999). Demonstration Programme on ICZM, vol.1 and vol.2, 35 p. and 92 p. Brussels: European Commission.

European Commission, EC (2012). Report on the Review of the European Water Scarcity and Droughts Policy, SWD (2012) 380 Final, COM(2012) 672 Final, 10p., Communication from the commission to the European parliament, the council, the European economic and social committee and the committee of the regions. Brussels: European Commission.

European Environment Agency (EEA) (2006). Annual report 2006, 59p. Copenhagen: European Environment Agency.

European Environment Agency, EEA (2017). European Environment Agency, Climate change adaptation and disaster risk reduction in Europe. Enhancing coherence of the knowledge base, policies and practices? EEA Report No 15/2017, ISSN 1725–9177, 171 p. Copenhagen: European Environment Agency.

Elliott, M. (2013). The 10-tenets for integrated, successful and sustainable marine management. *Marine Pollution Bulletin Editor* 74: 1–15.

Eurostat – Agriculture and Fisheries (2009). Statistics in focus, Key figures for coastal regions and sea areas, 47/2009 12p. Luxembourg: Eurostat.

Folke, C. (2006). Resilience: the emergence of a perspective for social-ecological systems analyses. *Global Environmental Change* 16: 253–267.

Folke, C. (2007). Social-ecological systems and adaptive governance of the commons. *Ecological Research* 22 (1): 14–15.

Folke, C., Hahn, T., Olsson, P. et al. (2005). Adaptive governance of social-ecological systems. *Annual Review of Environment and Resources* 30: 441–473. https://doi.org/10.1146/annurev.energy.30.050504.144511.

Forzieri, G., Feyen, L., Russo, S. et al. (2016). Multi-hazard assessment in Europe under climate change. *Climatic Change* 137: 105–119.

Fournier, M., Larrue, C., Alexander, M. et al. (2016). Flood risk mitigation in Europe: How far away are we from the aspired forms of adaptive governance? *Ecology and Society* 21 (4): 49.

Garnier, E. (2015). Strengthened resilience from historic experience. European societies confronted with hydrometeors in the sixteenth to twentieth centuries. In: *Hydrometeorological Hazards – Interfacing Science and Policy* (ed. P. Quevauviller), 4–25. Chichester: Wiley Blackwell.

Garnier, E. and Surville, F. (2010). La tempête Xynthia face à l'histoire. Submersions et tsunamis sur les littoraux français du Moyen Âge à nos jours. Saintes: Le Croît vif, 174 p.

Harding, R., Reynard, N., and Kay, A. (2015). Current understanding of climate change impacts on extreme events. In: *Hydrometeorological Hazards – Interfacing Science and Policy* (ed. P. Quevauviller), 27–48. Chichester: Wiley Blackwell.

Huitema, D., Mostert, E., Egas, W. et al. (2009). Adaptive water governance: assessing the institutional prescriptions of adaptive (co-)management from a governance perspective and defining a research agenda. *Ecology and Society* 14 (1): 26.

Intergovernmental Panel on Climate Change, IPCC (2012). Managing the risks of extreme events and disasters to advance climate change adaptation. A special report of Working Groups I and II of the Intergovernmental Panel on Climate Change (SREX Full Report). Cambridge: Cambridge University Press, 594 p.

International Risk Governance Council, IRGC (2006). White paper on risk governance towards an integrative approach, White paper number 1. Geneva: IRGC, 157 p.

Koimann, J. (2003). *Governing as Governance*, 256 p. London: Sage.

Kundzewicz, Z.W., Kanae, S., Seneviratne, S.I. et al. (2014). Flood risk and climate change: Global and regional perspectives. *Hydrological Sciences Journal* 59 (1): 1–28. https://doi.org/10.1080/02626667.2013.857411.

Le Galès, P. (2006). Gouvernance. In: *Dictionnaire des politiques publiques* (ed. L. Boussaguet, S. Jacquot and P. Ravinet), 244–252. Paris: Presses de Sciences Po.

Lavalle, C., Rocha Gomes, C., Baranzelli, C.et al. (2011). Coastal zones: policy alternatives impacts on European Coastal Zones 2000–2050, JRC Technical notes, 107 p. Luxembourg: Publications Office of the European Union.

Le Roy Ladurie, E., Rousseau, D., and Vasak, A. (2011). *Les fluctuations du climat de l'an mil à aujourd'hui*, 321 p. Paris: Fayard.

Liberatore, A. (2013). Climate change, security and peace: The role of the European Union. *Review of European Studies* 5 (3): 83–94.

Marine Strategy Framework Directive (2008). The Marine Directive adopted on 17 June 2008. http://eur-lex.europa.eu/legal-content/EN/TXT/?uri=CELEX:32008L0056 (accessed 31 January 2019).

Pereira, L.S., Cordery, I., and Lacovides, I. (2002). *Coping with Water Scarcity*, IHP-VI, Technical Documents in Hydrology, no. 58. Paris: UNESCO, 272 p.

Sendai Framework for Disaster Risk Reduction 2015–2030 (2015). Geneva: United Nations Office for Disaster Risk Reduction, 37 p.

Vogt, J. and Somma, F. (2000). *Drought and Drought Mitigation in Europe*, 325 p. Dordrecht: Kluwer Academic Publishers.

Walker, B., Holling, C.S., Carpenter, S.R. et al. (2004). Resilience, adaptability and transformability in social-ecological systems. *Ecology and Society* 9 (2): 5.

Water Framework Directive (2000). The Water Framework Directive adopted on 23 October 2000. http://eur-lex.europa.eu/legal-content/EN/TXT/?uri=CELEX:32000L0060 (accessed 31 January 2019).

Weisse, R., Bellafiore, D., Menéndez, M. et al. (2014). Changing extreme sea levels along European coasts. *Coastal Engineering* 87: 4–14.

Wilhite, D.A. and Glantz, M.H. (1985). Understanding the drought phenomenon: The role of definitions. *Water International* 10 (3): 111–120.

2

Overview of the Content of the Book

Isabelle La Jeunesse[1] and Corinne Larrue[2]
[1] *Laboratory CNRS 7324 Citeres, University of Tours, Tours, France*
[2] *Lab'Urba, Ecole d'Urbanisme de Paris, Université Paris Est Créteil, Marne-La-Vallée, France*

As mentioned by the editor of this book series dedicated to hydrometeorological extreme events, Philippe Quevauviller, the series offers two additional items of information on hydrometeorological extreme events. First, the series addresses a broader coverage of various extreme events whilst a significant amount of individual publications exist on specific hazards that are generally studied by different research teams in different research projects. Second, it provides an interdisciplinary analysis at a European scientific level.

As the fourth book of the series, this book attempts to determine the state-of-the-art of adaptive strategies to address various hydrometeorological extreme events such as floods, droughts, and coastal storms, including storm surges.

The first prerequisite to develop strategies to face hydrometeorological extreme events is to define what stakeholders consider the risks and vulnerabilities to be in a pertinent area. For hydrometeorological extreme events, and in the context of climate change, this means how stakeholders are addressing complexity and uncertainty, which requires opportunities to shift the status quo situations that can control environmental governance.

In order to answer these questions, this book proposes to explore how societies are organizing themselves to face hydrometeorological extreme events whether before, during, or after the hydrometeorological extreme events and at local, regional, national, and international levels. The different chapters deal with the strategies of integration for hydrometeorological extreme event risks known in governance, and more specifically in public policies. In each part dedicated to one type of extreme event, the authors present (i) the actors' roles, (ii) the strategies used and developed, and (iii) the empirical knowledge stemming from in-depth investigations into several case studies.

The book is divided into three parts, each one devoted to a type of hydrometeorological extreme event: floods, droughts, and coastal storms.

Facing Hydrometeorological Extreme Events: A Governance Issue, First Edition.
Edited by Isabelle La Jeunesse and Corinne Larrue.

In each part of the book, the chapters first overview the actors involved in the management of the specific hydrometeorological extreme event: floods, droughts, or coastal and wind-storms. Then, the authors will present strategies, instruments, and resources used to face this specific hydrometeorological extreme event. Lastly, national and/or regional and local case studies will be presented.

2.1 Floods

The part dedicated to flooding issues first presents the actors playing an important role at two levels: first at European level, for which T. Schellenberger (Chapter 3) points out the different ways in which European bodies act; second at a more local level, for which L. Oriard (Chapter 4) presents how local bodies in several countries are capable of implementing flood policies and involve inhabitants.

As far as the strategies and instruments are concerned, the first chapter in the part offers an overview of the change in the perception of the problems and the strategies formulated on a European scale (Crabbé [Chapter 5]), whereas the second chapter presents an analysis of the various instruments implemented in five European countries (Gralepois [Chapter 6]).

Finally, these general analyses are completed by a special focus on four specific countries: England (Alexander and Priest [Chapter 7]), The Netherlands (Wiering [Chapter 8]), France (Fournier [Chapter 9]) and Belgium (Mees [Chapter 10]).

2.2 Droughts

The same structure is followed for drought issues. The first chapters of the part are dedicated to the actors at European level (Stein and Landgrebe [Chapter 11]) and local level (Ozeröl [Chapter 12]), pointing out the involvement variability on each scale.

Then, strategies and goals ambitions (La Jeunesse [Chapter 13]) and instruments (Bressers et al. [Chapter 14]) are presented and analysed for six countries and case studies, with an inter-comparison between north and south Europe (La Jeunesse et al. [Chapter 13]) aiming at highlighting some general strategy and instrument patterns in these countries and cases.

The final chapters in the part discuss the specific situation in some countries/cases: Flanders in Belgium (Troeltzsch [Chapter 15]), Eifel Rur in Germany (Vidaurre [Chapter 16]), in Italy (Cirelli and La Jeunesse [Chapter 17]), in Turkey (Turhan et al. [Chapter 18]) in Spain (Ballester and La Calle [Chapter 19]) and an intercomparison between France and the UK (La Jeunesse et al. [Chapter 20]). All these chapters focus on the specificities of the cases studied.

2.3 Coastal storms

The part dedicated to coastal storms starts with a presentation of the actors involved in coastal storm management and a description of what sustainable communities and multilevel governance are (Henocque [Chapter 21]). The author then uses the definition of the socio-ecosystem in the specific context of integrated coastal zone management to explain the multilevel governance strategies regarding the multilevel actors involved.

Then, the chapters tackle the strategies, instruments, and resources used to address coastal risk prevention through two angles. Boyes and Elliott (Chapter 22) present an overview of the European strategy and framework to cope with coastal and wind storms, going from the DPSIR to the DAPSIR chain of causes-consequences used for describing the interactions between society and the environment.

Goeldner-Gianella and Longépée (Chapter 23) place what perceptions of extreme coastal events can be in a broader context by presenting the results of an investigation into the French Atlantic and Mediterranean coasts.

Finally, lessons learnt from four situations are described. Mercier et al. (Chapter 24) present the state-of-the-art of coastal storm prevention on the French Atlantic coast eight years after Xynthia, which occurred in February 2010 and is qualified by the authors as the most severe coastal surge in the recent history of the French Atlantic coast in terms of hazard, disasters, fatalities, and consequences for the coastal management policy. Anselme et al. (Chapter 25) provide recommendations to improve the operational response for the management of coastal areas by exploring the risk-management analysis along the French Mediterranean coast. Through an analysis of the Bejisa cyclone in Reunion Island in 2014, Chapter 26, by Duvat and Magnan, denounces the existing gap between cyclone impact analyses and stakeholders' short-term responses, along with the implications of these responses on medium- to long-term changes in the vulnerability to sea hazards. Golding et al. (Chapter 27) describe the changes in coastal and windstorm impacts since these were first observed by analysing both exposure and vulnerability, as well as the governance and management strategies. They conclude that the avoidable impacts of the disasters of 1953 and 1987, for instance, arose primarily from weaknesses in warning procedures and communication.

Part II
Floods

II.1

Actors Involved in Flood Risk Management

3
European Actors Facing Floods Risks

Thomas Schellenberger
European Center for Research on the Law of Collective Accidents and Disasters (CERDACC), IUT de Colmar, Université de Haut-Alsace, Colmar, France

Flood management is a cross-cutting issue affecting several European Union (EU) policies. Flood-management measures are undertaken in the fields of environmental protection, public health, agricultural policy and cohesion, as well as civil protection. It also affects very specific areas. For example, following the Fukushima disaster, the European Commission initiated an inspection process of nuclear power plants in EU Member States to assess their resistance to floods.[1] As a cross-cutting issue, the management of floods thus gives rise to the coexistence of different actors with varying powers at the EU level.

At the same time, there is growing evidence that flood-management actors within the EU are developing, which means that the EU is increasingly active in this area of action. These European actors are involved in a wide variety of areas related to flood management. This article draws on the legal basis on which European actors exercise their flood management powers in order to shed light on their political roles and powers. From the legal basis on which an actor undertakes measures, one can understand the scope of its missions and its powers of action. For example, when the EU takes a decision to protect nuclear facilities from flooding, it does not have the same room for manoeuvre as it has to deal with the adverse consequences of flooding on the agricultural sector. An analysis by the institutional competences of the actors also sheds light on the general landscape of the actors of flood management in the EU. Indeed, some EU actors carry out their tasks on the basis of competences under which the EU is limited to providing support for coordination between the Member States. In this case, the EU's room for manoeuvre is a priori limited. In other cases, some actors intervene in areas of shared competence between the Member States and the EU. In this case, the power of action of the European players can be much more important. In the area of flood management, there are actors in charge of civil protection (EU supporting competence: 'the Union can carry out actions to support, coordinate or supplement Member

Facing Hydrometeorological Extreme Events: A Governance Issue, First Edition.
Edited by Isabelle La Jeunesse and Corinne Larrue.

States' actions', Treaty on the functioning of the European Union [TFEU], art. 2), actors responsible for environmental protection (shared competence between the EU and the Member States: 'Member States cannot exercise competence in areas where the Union has done so', TFEU, art.2), and other actors responsible for the EU's agricultural policy (shared competence but with a high level of political integration).

Shared competences are in principle more in favour of integrative policies at the EU level than actions undertaken under supporting competences whose role is to ensure only cooperation between the Member States. However, the EU flood-management actors work more generally on cooperation between the EU Member States than on their actual integration within a common policy.

First, it will be shown that the management of civil security is developing strongly within the EU but this is based on an EU coordination competence. Consequently, Union actors are limited to missions to assist cooperation between Member States. Second, when it comes to environment, the EU acts on the basis of shared competence with the Member States, but paradoxically the tasks of the EU bodies tend to remain constrained in coordination functions, which are therefore not ambitious in terms of political integration. Third, although the EU's agricultural policy is based on shared competence, it leads to greater political integration than does the environment, which in turn strengthens the powers of EU actors.

3.1 European actors in the field of civil security: A competence which develops within a strict framework of cooperation between the Member States

Under Article 6 of the TFEU, civil protection is a supporting competence. The EU can, therefore, only act to support, coordinate, or complement the actions of the EU Member States.

It should be stressed that civil protection actors at the EU level are expanding but their powers are limited by the strict framework of cooperation between Member States (Pauvert 2017). This means that the actors at Union level will coordinate the actions of the national actors rather than implement EU policy themselves. Indeed, European civil security actors are developing, but their missions are systematically restricted by the legal limits of intergovernmental cooperation. Consequently, the European civil protection organizations are mainly responsible for coordinating the action of the national services of the Member States. In this function of coordinators, the Union bodies play an increasingly active role in flood management, but with a very relative margin of decision.

Historically, civil security was absent from the EU's remit. The EU constitutive treaties, traditionally oriented towards the economy, did not contain any provision relating to the protection of the populations, so that an autonomous policy of civil security could not come into existence. The Union has nonetheless undertaken actions in the field of civil protection in the name of achieving the common market and strengthening political union, but indirectly through the principle of solidarity between European citizens.

In 1987, the European Commission was given the task of implementing a first series of Community cooperation measures in the field of civil protection in order to 'enhance the effectiveness of individual crisis management measures'. To this end, it was decided by the Council to put the national actors of civil security in a network and thus organize a sharing of information on crisis management. From that date onwards, the Union has regularly promoted the Commission's initiatives in the field of civil security, but it does not naturally go beyond their initial objectives of supporting the Member States, coordinating national actions, and ensuring mutual assistance between Member States.

In 1997 the Community action programme for civil protection was set up to coordinate national civil protection actors within a network, through a committee composed of representatives of the Member States and chaired by the Commission. But it was clearly stated that 'This programme excludes any measures aimed at the harmonisation of the laws and regulations of the Member States or of the organising of the national preparedness of the Member States'.[2] The absence of legislative harmonization is a clear sign that the Member States retain their full political decision-making power. This shows that the European Commission can only act in the field of civil security in the strict framework of intergovernmental cooperation and not with an integrative approach at EU level. The implementation of this programme led in 2001 to the establishment of the Monitoring and Information Centre (MIC). The role of this body is to organize the exchange of requests and offers of assistance between Member States and to ensure operational monitoring. From 2002 onwards, the MIC was financed by the Union Solidarity Fund. More generally, the mission of the MIC is to coordinate the implementation of the European Civil Protection Mechanism set up in the same year, which aims to facilitate the cooperation of the Member States in the field of civil security and to put in place strategies for response to disasters.

Since the Treaty of Lisbon, which entered into force on 1 December 2009, the EU's civil protection action is enshrined in the primary law of the Union, which makes it possible to set up a common policy based on Article 196 of the Treaty on the TFEU on civil protection (Pauvert 2017). However, since civil protection is only a coordination competence of the EU under Article 6 of the TFEU, the prerogatives of the EU will not be substantially different than earlier. Since these prerogatives will have to respect the restrictive framework of intergovernmental cooperation, civil protection actors at the EU level will not be able to have a greater decision-making power. Article 196 of the TFEU specifies that 'the Union shall encourage cooperation between Member States in order to improve the effectiveness of systems for preventing and protecting against natural or man-made disasters'. It is also stated that 'Union action shall aim to: (a) support and complement Member States' action at national, regional and local level in risk prevention; (b) promote swift, effective operational cooperation within the Union between national civil-protection services'. Article 196 thus clearly lays down a classic framework for cooperation between the Member States, specifying in addition that any legislative and regulatory harmonization between States is excluded.

It is within this legal framework that the Operational Centre for the Emergency Response Coordination Centre (ERCC), which replaced the MIC from 2008, operates today. The ERCC originally belonged to the European Commission's Directorate-General (DG) for Environment, but now it relies on DG ECHO (European Civil protection and Humanitarian aid Operations), specifically dedicated to civil protection and humanitarian aid operations. Within DG ECHO, the ERCC oversees the operations of the European Civil Protection Mechanism. The ERCC coordinates the interventions of the participating countries in order to avoid duplication and inconsistencies that an uncoordinated action of the member states of the Union would imply. Whilst the missions and financial resources of the ERCC go well beyond those of the former

MIC, it is nevertheless clearly recalled in the texts that the Member States retain their full sovereignty to ensure the protection of their populations: 'The Union Mechanism shall promote solidarity between the Member States through practical cooperation and coordination, without prejudice to the Member States' primary responsibility to protect people, the environment, and property'.[3]

Within the European Civil Protection Mechanism, a European Emergency Response Capacity (EERC) has also been established on a voluntary basis.[4] It consists of a voluntary pool of pre-committed response capacities of the Member States and include modules, other response capacities, and experts.[5] The EERC (or 'voluntary pool') bring together a range of relief teams, experts, and equipment, made available by several European countries. In the EERC, for example, flood rescue and containment modules have been set up.[6] This mechanism also functions through the activities of the Working Party on Humanitarian Aid and Food Aid (COHAFA), which is the main European forum for strategic and policy debate on humanitarian aid between EU Member States and the European Commission.

Since 2014, European initiatives in the field of civil security have developed. For example, a European Voluntary Humanitarian Aid Corps (EU Aid Volunteers Initiative) was formed in 2014,[7] followed by a European Medical Corps in 2016. Finally, in the same year an exceptional emergency assistance mechanism was set up, which can be activated by the Council of the EU on a proposal from the Commission.[8]

Even if they develop, European civil protection actors are generally restricted in their actions by the strict framework of intergovernmental cooperation, enabling the national services of each Member State to decide on their policy in this field. However, a proposal for a more autonomous EU civil protection policy has emerged in the past, but it has not been implemented. Indeed, in 2006, M. Barnier proposed the establishment of a more inclusive EU-wide framework of intervention with the setting up of exclusively European means managed by the EU (Barnier 2006). However, this initiative has been abandoned, in particular because of the cost that the integrated civil protection system would entail for the Member States (Pauvert 2017). The reluctance of member states to engage in a common civil security policy could also be explained by the fact that the provision of relief to the population is a symbol of state sovereignty.

3.2 European actors in the field of the environment: Powers that are paradoxically limited

Within shared competences between the EU and the Member States, there is great variability in the powers of EU actors depending on the policy in question. Indeed, it may be considered that in the field of environmental protection competences are exercised in such a way that they significantly promote cooperation between Member States to the detriment of genuine integration at Union level.

3.2.1 The competence of EU actors

The environment is a shared competence that could in principle give rise to relatively integrated policies and thus to rather autonomous actors at EU level to implement these policies with their own powers. However, European environmental policy tends to rely heavily on national governments that have significant room for action. This is due to the principles of subsidiarity and proportionality, which are applied in such a way as to leave maximum freedom of action to the Member States.

Since the Single European Act of 1987, the Union's policy on the environment has been defined by a specific title in the TFEU (Articles 191–193). Under Article 4 TFEU, the environment is a shared competence, which means that the EU and the Member States are jointly empowered to legislate and adopt binding acts in this area and of public policies. This implies that Member States shall exercise their competence to the extent that the Union has not exercised its competence (TFEU art. 2).

In principle, an area of shared competence gives rise to integrated policies at Union level, in contrast to areas in which the Union has only a supporting competence. Nevertheless, the degree of integration of EU policy, and therefore the degree of autonomy of the actors involved at Union level, depends on how the principles of subsidiarity and proportionality are applied. These two principles constitute the main criteria for the distribution of a shared competence between the EU and the Member States. The respective margins of manoeuvre of the actors of the Union and those of the Member States stem from the implementation of these two principles.

The principle of subsidiarity means that competence must be exercised at the territorial level most suited to the issue. This means that the bodies of the Union act to the extent that the objectives pursued by a policy can be better achieved at Union level than at the level of the Member States. Two cumulative conditions must be met in order to ensure that the Union's action is legally justified in the light of the principle of subsidiarity. On the one hand, 'the Union shall act only if and in so far as the objectives of the proposed action cannot be sufficiently achieved by the Member States, either at central level or at regional and local level' (art. 5 TUE). On the other hand, the EU can act if it is established that these objectives, 'by reason of the scale or effects of the proposed action', can be better achieved at Union level (art. 5 TUE). There must be, on the one hand, a lack of national action, making EU intervention necessary and, on the other hand, a better capacity for action on the part of the EU. The principle of proportionality requires that 'the content and form of Union action shall not exceed what is necessary to achieve the objectives of the Treaties' (art. 5 TUE). In other words, the constraints imposed on Member States under EU legislative acts must not be excessive in relation to the objectives pursued.

Because of their flexibility, these two principles are subject to differentiated implementation according to the areas of shared competence. Indeed, it has been observed that the environment field gives rise to application of the principles of subsidiarity and proportionality favouring the national scale to the detriment of the European scale (De Sadeleer 2012). The result is an increase in government room for manoeuvre to the detriment of joint action at EU level. At first sight, this idea may seem paradoxical, since environmental issues, which often go beyond the borders of each Member State, should in principle justify ambitious actions at Union level under the principle of subsidiarity. However, it can be seen that environmental policies tend to be highly decentralized, that is to say, left to a very large extent at the discretion of national administrations.

An illustration of this phenomenon can be found in the fields of water and flood management. At first glance, it is now generally accepted that problems related to water management must be managed at the level of the river basin which often has a transnational and therefore a European dimension. Under the principle of subsidiarity, neither water nor flooding should be a priori managed at local or even national levels, as the action of the EU seems justified in the light of these transnational issues. Despite the adoption of the Water Framework Directive (WFD)[9] and the Floods Directive (FD)[10] on the basis of Article 191 of the TFEU, it can be seen that water management (De Sadeleer 2012) and flood management (Priest et al. 2016) are based on a highly decentralized approach, transferring the main room for manoeuvre to national actors and conferring only a coordinating role on actors at Union level. This is reflected in minimal legislative harmonization, and therefore limited European integration, as European legislators tend to prefer to use directives, which are flexible texts, rather than strict regulations. Similarly, in the case of directives, Framework Directives are preferred to Directives containing detailed and binding measures in the name of the principle of proportionality. The WFD is a text that sets general and flexible objectives with substantial exceptions for their implementation by the Member States (De Sadeleer 2012). The same applies to the FD, which contains almost no binding measures (Priest et al. 2016).

Of course, the European Commission retains, at least formally, its full power to monitor the proper application of Union law by the Member States. This power authorizes the Commission to take binding preventive measures and to bring proceedings against a Member State which does not transpose or incorrectly transpose one of these Directives. From a formal and institutional point of view, this remains the primary role of the European Commission. Nevertheless, since the abovementioned texts, which are not very restrictive, in fact leave considerable margins of appreciation for the Member States responsible for their transposition, the role of the European Commission in its function of monitoring the correct execution of the law automatically becomes marginal (De Sadeleer 2012).

3.2.2 The missions of EU actors

It will therefore be shown here that, at EU level, the actors of flood management in the field of the environment are well developed, but these tend to be limited to a coordinating role for national actors (dissemination of information, support, networking, etc.). Particularly since 2001, the organization of actors for the transposal of the WFD has given rise to a common implementation strategy (CIS)[11] covering several WFD-related texts. This CIS therefore encompasses the transposition of the 2007 Flood Directive.

It may be pointed out first of all that the very existence of the CIS seems to be justified (i) by the fact that the texts to be transposed are not very restrictive and (ii) by the decentralization which results from this legal flexibility. When a text lays down precise obligations in the form of a precise and clear European regulation, it is not normally necessary to organize the harmonization of the transposition of the text by the Member States. In such a case, the text is self-sufficient and gives the established actors specific obligations and powers. The constraint and precision of a text is inherently capable of guaranteeing a certain harmonization in its transposition. This general observation is reflected in the missions entrusted to EU actors within the CIS.

The CIS is organized by the European Commission DG for Environment and consists essentially of an informal coordination platform. The European Commission has set up a joint group of Water Directors, which is responsible for steering the CIS. However, it should be stressed that the organization of the CIS does not increase or modify the Commission's usual powers: 'the organisational structure is not intended as a new formal forum for decision-making'. As requested by the European Commission at the launch of the CIS, 'competence for the implementation resides exclusively with the individual Member State. The joint group of EU and Commission Water Directors should function as an important initiator and driver of the process but without attaining the character of a board with formal competence'. Similarly, the work of the Water Directors within the framework of the CIS should not lead to any additional constraint for the Member States in transposing the texts: 'Results from the process, e.g. in terms of guidance documents would have an informal and non-legally binding nature, creating a common working basis for the implementation'.

Under the responsibility of the Water directors, a Strategic Coordination Group is set up, whose role is to coordinate the action of the multiple working groups created according to key issues of the directives' transposal. Today, there is a working group specializing in flood management, which is responsible for harmonizing the transposition of the FD. In particular, this working group has produced a guide on the technical aspects of the FD's transposal.[12]

The three objectives of the CIS clearly show the exclusively coordinating and non-binding function of this system. The aim is to share information amongst Member States, to develop technical guides (and not standards), and to carry out technical experiments in the field. This desire to coordinate the many national actors at different scales (national or local administrations, companies, public) also gave rise to the CIRCABC (Communication and Information Resource Centre for Administrations, Businesses and Citizens, under the Directorate General for Informatics at the European Commission [DG DIGIT]). At the same time, the European Commission (DG Environment) and the European Environment Agency (EEA) have created the Water Information System for Europe (WISE). This body is responsible for networking information and expertise related to water management and the transposition of the WFD.

Although the Commission formally has the power to monitor the application of the WFD and the FD, and to impose sanctions against Member States that fail to transpose these texts correctly, the EU is generally limited to the dissemination and sharing of information related to flood management. The aforementioned actors essentially fulfil information-sharing and coordination tasks between the actors of the Member States and consequently have no real political autonomy because of the considerable room for manoeuvre granted to the Member States which in fact refuse to be under the control of the Union in these areas.

3.3 European actors in the field of agriculture: Could there be specific powers to deal with floods?

Within the Common Agricultural Policy (CAP), which is also based on shared competence, it can be assumed that European actors possess more binding flood management tools that can be imposed directly within Member States.

According to Article 4 of the TFEU, agriculture is a shared competence between the Union and the Member States. In principle, therefore, the principles of subsidiarity and proportionality should be applied as well as in the environmental field. However, it can be assumed that the CAP, which contains certain instruments to deal with floods, gives rise to the application of the principle of subsidiarity which favours integration at EU level and reduces the powers of national authorities. Thus, the CAP could illustrate the variability of the distribution of powers within a shared competence, in comparison with the environmental policy.

The CAP is a special shared competence. In practise, the CAP could be a policy area more integrative than the environmental one. It is indeed important to note that the CAP has long been regarded as a competence held exclusively by the EU. This was supported by the literature (Bianchi 2014; Michel 2013; Petit 2002), by the European Commission,[13] and by the Court of Justice of the EU.[14] It was not until 2009, following the Lisbon Treaty, that the Union's agricultural policy was included in the sphere of shared competence between the EU and the Member States.

Although it is now formally a shared competence, the CAP can be regarded as a de facto exclusive competence (Michel 2013). The practise of the principle of subsidiarity in this area would, therefore, remove the CAP from the spheres of shared competence to bring it closer to an exclusive competence. To understand this possibility, it is useful to clarify the implications of the principle of subsidiarity:

> *When the Treaties confer on the Union a competence shared with the Member States in a specific area, the Union and the Member States may legislate and adopt legally binding acts in that area. The Member States shall exercise their competence to the extent that the Union has not exercised its competence. The Member States shall again exercise their competence to the extent that the Union has decided to cease exercising its competence. (TFUE art. 2)*

It follows from this text that, to implement a policy, Member States may adopt binding acts only if and to the extent that the Union has not exercised its power. This means that the more the EU adopts comprehensive measures, and in particular strict binding measures with a broad field of application, the more the Member States' room for manoeuvre is restricted. However, in the area of the CAP, and contrary to the environmental policy discussed above, the Union has favoured regulations rather than directives, that is to say, legally binding acts which are directly binding on litigants in the Member States (De Sadeleer 2012; Bianchi 2014). In these conditions, the power of national authorities to implement the CAP could be consequently more limited than in the field of environment. This could have several consequences.

With regard to flooding in particular, several binding measures may fall under the CAP. Above all, it is possible to mention the conditionality of financial aid granted under the CAP. The conditionality, introduced since 2005, makes the payment of certain aids subject to compliance with basic requirements, in particular about the environment, good agricultural and environmental conditions (Habran 2015). The main environmental conditionality rules are adopted by European regulations and are imposed directly in the Member States. In this respect, the payment of aid to farmers may be conditional on the promotion of soil protection and the maintenance of permanent pastures and thereby improving the capacity of soils for water retention.[15] Similarly, within the framework of the CAP rural development policy, the European Agricultural

Fund for Rural Development (EAFRD) could finance adaptation measures to climate change and aid for the restoration of agricultural potential damaged by of natural disasters.[16]

If the European legislature implements the CAP with regulations directly applicable in the Member States, the European Commission plays in principle a more important role in monitoring the proper application of the rules for the implementation of the CAP in the Member States. The DG for Agriculture within the European Commission, with the help of the various agricultural committees, therefore has more powers to carry out and control the CAP than in areas governed solely by framework directives. The position of the European Commission must nevertheless be put into perspective, because of the power of the Member States in the agricultural committees and the obvious leeway left by the CAP regulations in the Member States. However, there could be a clear and interesting difference between the field of agriculture and the environment as regards the powers and scope for action of national actors in relation to EU actors.

3.4 Conclusion

There is a variability in the missions and margins of action of the European actors according to their legal competences. It is more precisely the practise of these competences that differs and leads to disparities in the powers of European actors from one policy to another. Overall, flood management actors at EU level are developing. They also have increasingly diverse missions in different policy areas. However, European actors are rather restricted in coordination missions between the Member States. Overall, this means that Member States retain significant room for manoeuvre to implement their own national policies. Inter-State cooperation policies exist and are developing, but Member States are still rarely constrained by precise and binding measures from the EU. As a result, national actors are in a strong position vis-à-vis actors at EU level. There is not yet a comprehensive and integrated approach of flood management by EU bodies, even if opportunities for political integration at the EU exist theoretically in the field of environment and agriculture. This contrasts with the nature and severity of flood-related issues. On the one hand, floods are by nature transnational phenomena which have long required a European approach. On the other hand, with climate change now aggravating floods on a global scale, Member States' reluctance to deepen European integration in flood management seems increasingly difficult to justify.

Notes

1. COMMUNICATION FROM THE COMMISSION TO THE COUNCIL AND THE EUROPEAN PARLIAMENT on the interim report on the comprehensive risk and safety assessments ('stress tests') of nuclear power plants in the European Union /* COM/2011/0784 final */.
2. 98/22/EC: Council Decision of 19 December 1997 establishing a Community action programme in the field of civil protection, Official Journal L 008, 14/01/1998 P. 0020–0023.
3. DECISION No 1313/2013/EU OF THE EUROPEAN PARLIAMENT AND OF THE COUNCIL of 17 December 2013 on a Union Civil Protection Mechanism (JO L 347 du 20.12.2013, p. 924), Official Journal of the European Union, 20.12.2013, L 347/924.

4. DECISION n° 1313/2013/UE, op cit., (11), art. 9, art. 11.
5. DECISION n° 1313/2013/UE, op cit., art. 11.
6. http://ec.europa.eu/echo/files/aid/countries/factsheets/thematic/emergency_response_capacity_fr.pdf
7. REGULATION (EU) No 375/2014 OF THE EUROPEAN PARLIAMENT AND OF THE COUNCIL of 3 April 2014 establishing the European Voluntary Humanitarian Aid Corps ('EU Aid Volunteers initiative').
8. COUNCIL REGULATION (EU) 2016/369 of 15 March 2016 on the provision of emergency support within the Union, Official Journal of the European Union, 16.3.2016, L 70/1.
9. Directive 2000/60/EC of the European Parliament and of the Council of 23 October 2000 establishing a framework for Community action in the field of water policy.
10. Directive 2007/60/EC of the European Parliament and of the Council of 23 October 2007 on the assessment and management of flood risks.
11. COMMON IMPLEMENTATION STRATEGY FOR THE WATER FRAMEWORK DIRECTIVE (2000/60/EC), STRATEGIC DOCUMENT, 2 may 2001.
12. Guidance for Reporting under the Floods Directive (2007/60/EC) Guidance Document No. 29, http://ec.europa.eu/environment/water/water-framework/facts_figures/guidance_docs_en.htm
13. The Principle of Subsidiarity. Communication of the Commission to the Council and European Parliament. SEC (92) 1990 final, 27 October 1992.
14. CJEC, 14 juill. 1994, Rustica Semences, aff. C-438/92, Rec. I. 3519.
15. Communication from the Commission to the Council, the European Parliament, the European Economic and Social committee and the Committee of the Regions - Flood risk management - Flood prevention, protection and mitigation /* COM/2004/0472 final.
16. REGULATION (EU) No 1310/2013 OF THE EUROPEAN PARLIAMENT AND OF THE COUNCIL, of 17 December 2013 laying down certain transitional provisions on support for rural development by the European Agricultural Fund for Rural Development (EAFRD) (...)

References

Barnier, M. (2006). *Pour une force européenne de protection civile: europe aid*, Report. Brussels: European Commission.

Bianchi, D. (2014). *La politique agricole commune, Précis de droit agricole européen*, 2e. Brussels: Bruylant 89p.

Habran, M. (2015). *Politique agricole commune, Analyse transversale de la conditionnalité environnementale*. Brussels: Bruylant.

Michel, V. (2013). Union européenne (Traité de Lisbonne). In: *Répertoire du droit européen*. Paris: Dalloz.

Pauvert, B. (2017). Sécurité civile. In: *Jurisclasseur Administratif*. London: LexisNexis.

Petit, Y. (2002). Agriculture. In: *Répertoire du droit européen*. Paris: Dalloz.

Priest, S., Suykens, C., Van Rijswick, H.F.M.W. et al. (2016). The implementation of the Floods Directive: the role of flood risk management plans in increasing resilience to flood risk. *Ecology and Society* special issue.

de Sadeleer, N. (2012). Particularités de la subsidiarité dans le domaine de l'environnement. *Droit et société* 1 (80): 73–90.

4

Multi-actor, Multilevel Assessment of Social Capacity for Community Engagement in Flood Risk Preparedness: Results of Implementation in Five European Cases

Lila Oriard Colin
Lab'Urba, Institut d'Urbanisme de Paris, Université Paris Est Créteil, Marne-La-Vallée, France

4.1 Introduction

4.1.1 Towards a multi-stakeholder flood risk management approach

The importance of citizens and communities engagement in a multi-stakeholder flood risk management system has been highlighted by previous research by Chouinard, Plant, and Martin 2008; Howgate and Kenyon 2009; Jha, Bloch, and Lamond 2011; Nye, Tapsell, and Twigger-Ross 2011; Thaler and Levin-Keitel 2016; research groups funded by the European Commission focusing on integrated risk management and capacity building such as Steinführer et al. (2009), Kuhlicke and Steinführer (2010, 2015), and Kuhlicke et al. (2011) and supported by international treaties such as the Hyogo Framework for Action 2005–2015 and the European Flood Directive 2007. This social-oriented approach to flood risk management, contrasts with a more traditional institutional-centred one, mainly based on technical

Facing Hydrometeorological Extreme Events: A Governance Issue, First Edition.
Edited by Isabelle La Jeunesse and Corinne Larrue.

structural measures. The multi-stakeholder approach promotes collective mobilization to prepare for a flood event at all levels, including the community (citizens) level.

Some experiences during natural disaster crises, especially in developing countries, have shown that people and communities are able to self-organize in innovative ways and unexpected levels of mobilization of multiple resources and abilities to help deal with the situation (Quarantelli 1981). The display of capacities is thus a critical source of support, to be used also in the prevention and recovery stages; they are, as well, an important source of social resilience.

From a research perspective some work has been done in relation to this topic, even if it remains under-researched. An action-research was conducted by Chouinard et al. (2008) to understand how to integrate communities, and community members, to flood risk management. This work involved researchers interacting directly with community members to work on the co-production of knowledge, both scientific and experienced-based, and promote participation in decision-making processes on flood risk preparedness. The research was carried out in the context of adaptation to climate change in two coastal communities in Atlantic Canada. The study revealed that in one community scientific information on the risk was needed, whilst in the second, the community had knowledge and information but required help to position themselves for the projects to come. The research showed the importance of developing collaborative relations between communities and researchers to develop local capacity. It also showed to what extent local dynamics and needs differ from one community to another. The conclusion from this work is that what communities may need to develop their capacities is also specific and should be addressed accordingly (Chouinard et al. 2008).

Recent research has been conducted to identify important capacities present in communities with higher level of engagement in the United Kingdom. The work done by Thaler and Levin-Keitel (2016), analysed two communities with different levels of community engagement, they found that more-engaged communities have a sense of self-responsibility, resources such as knowledge, time, financial, interest, and social and cultural capital, which all together, accordingly to these authors, translates into 'local capacity to act' (Thaler and Levin-Keitel 2016).

The above mentioned literature, Chouinard et al. (2008) and Thaler and Levin-Keitel (2016) focused mainly in the communities and their capacities but not at their relationship to and within the flood risk management structure. Indeed, the development and use of the capacities of citizens and communities in flood risk preparedness are related to, and cannot be detached from, the capacity of the flood risk management structure to elucidate, enhance, and mobilize them. The use of the capacity-building framework was particularly useful to cover this gap in the literature taking into account both, capacities in the communities, and those of institutions including local authorities. To cover this gap, a research consortium under the name of CAPFLO was formed between five research insitutions in Europe to understand how to assess and build capacities with local communities, for more details see Oriard et al. 2016 and Larrue and Oriard 2016.

4.1.2 Social capacity building towards resilience

Capacity building is an approach developed in the 1990s widely used in practice amongst international organizations working on international cooperation for development. Whilst the concept is still under-defined from a theoretical perspective, it is understood in practice as 'helping people to help themselves' (Eade 1997). In practice, the aim of capacity building is to improve and develop skills, knowledge, and abilities to do a task competently (UNCEPA 2006). Initially it targeted mainly institutional actors, such as local authorities and non-governmental organizations (NGOs) in developing countries to set a strategy to help them develop managerial tasks to solve challenging local problems with the help of international organizations. Nowadays, the concept is used in developed countries, especially to build social and community resilience.

An interesting critique of the capacity building concept is that it implies, and overlooks, aspects such as power relations and inequality. Indeed, as in the case of international cooperation, despite good relationships between donors and receptors, the ones receiving support have to position themselves as those who need help, which is not a simple one (Eade 1997). In the case of communities getting prepared for flood risk, the relation between institutions and communities tends to create a division between the ones who have the resources (skills, financial, technical), i.e. institutions and experts, and those who do not, i.e. people and communities. This division that should be avoided, since for a capacity building process to happen there needs to be the establishment of trustful, equal relationships between different actors.

To avoid unequal relationships between 'those who know' and 'those who don't' it is necessary to recognize the experiences and knowledge of citizens and communities as valuable assets. This is easily said, but in practice it is far more difficult to implement, e.g. in some cases the knowledge acquired by citizens and communities is not acknowledged by the communities themselves, and requires to be unpacked.

Capacity building has recently been associated with resilience, especially related to disaster management and climate change. Resilience refers to the capacity of a system to overcome a shock whilst keeping its basic functions (Holling 1986). In socio-ecological systems, resilience refers to the capacity of a forest, to give an example, to be able to recover after a fire if it has the properties that allow the process of recovery to happen (Berkes, Colding, and Folke 2003). The importance of local communities is highlighted by this approach, they have knowledge of the conditions that are required for the ecological system to live as they have a close relationship with the environment, they have a daily contact with it and past experiences related to it. Indeed, communities have knowledge on, and are part of, those aspects that are required for the socio-ecological system to resist, adapt, and recover from a shock. The resilience approach promotes the integration of the communities' knowledge and practices to the institutional management structures in order to develop a more robust, sensible, and adaptable structure (Berkes et al. 2003).

Research work on resilient communities, especially in developing countries, has shown how civil society and communities mobilize to bring valuable help in dealing with the crisis. This was the case of the 1985 earthquake in Mexico City. In this case, resilience was built as a result of the way the communities and neighbourhoods were formed historically, the way social interactions were established based on solidarity and collaboration despite conflict (Davis 2004; Oriard 2015; Quarantelli 1995). The extent to which this process can be triggered by authorities and institutions outside the communities through a capacity building process is still an unanswered question, especially in societies working mainly on an individual basis, where communities are not confronted to deal with local problems on their own.

Communities in this work refer to groups of people living in a specific area such as a neighbourhood or a risk-prone area; it also refers to groups of people linked by a common interest that is directly related to flood risk directly or indirectly such as groups or associations related to river leisure activities, sustainable practises, river alert systems, neighbourhood committees, amongst others. This is a practical definition that helped us work with people and groups in different cultural contexts.

The resilience approach looks at cycles of learning and adapting in long periods of time. Resilient societies have the capacity to learn how to deal with a changing environment, not only to face a crisis situation, by capitalizing past experiences to act better in the future, even if this means rearrangement and change. To build social resilience is the ultimate objective of the social capacity building process for disaster management approach.

4.2 Social capacity building framework for community engagement

4.2.1 Social capacity definition

In the context of social capacity building for disaster management, social capacity was defined by Kuhlicke and Steinführer (2010, 2015) and Kuhlicke et al. (2011), a consortium of eight research institutions in European countries, as the set of specific capacities and resources embedded in different social actors which can be institutional, i.e. government representatives, water bodies, city authorities, NGOs, and non-institutional, i.e. citizens and communities.

> *By social capacity we mean all the resources available at various levels (e.g. individuals, organizations, communities, institutions) that can be used to anticipate, respond to, cope with, recover from and adapt to external stressors (e.g. a hazardous event). These resources include skills, knowledge, social networks as well as institutions, structures and knowledge of how to elicit and use them. (Kuhlicke and Steinführer 2010, p. 16)*

Following this definition, an operational framework was developed by the consortium led by C. Kuhlicke and A. Steinführer from 2009 to 2012. This framework is the result of an important effort to identify the specific capacities required to grow the social capacity required for a multi-stakeholder disaster risk reduction. This definition was taken as a starting point to define specific capacities for community engagement for flood preparedness, develop an assessment tool, and implement it in the study cases.

The framework highlights five different dimensions of capacities, a type of categories related to knowledge, motivation, networks, finance, and governance. The final category was changed to 'participation' for practical terms. Each of the dimensions includes specific capacities and resources. Whilst the framework developed by Kuhlicke et al. focuses more on the dimensions than the specific capacities, the present research deepened in the identification of the specific capacities applied to flood-risk preparedness with special emphasis on civic capacity. A definition of this concept is given before entering into the details of the capacity-assessment tool.

4.2.2 Using civic capacity in flood-risk preparedness

Civic capacity is a concept coming from a different body of literature. The concept is used in various aspects of society, for example to improve children's' formal education through parents and communities involvement in schools working together with professors and authorities. The research in this area of knowledge has shown how important and efficient it is to involve civil society in providing support to carry out tasks that before were conceived as exclusive from formal institutions. We used this concept as an analytical tool to promote social capacity building, especially through collaboration between communities and authorities.

The concept is defined as: the ability of local actors to act in concert around a matter of community-wide import (Stone 2001). The concept is used for the first time in the context of flood-risk preparedness in the research presented in this chapter. A difference with other local aspects, such as school education, is that flood risk implies a complex system of institutional actors organized throughout different territorial scales: area, city, basin, region, and nation, in general unknown for the local population. Flood risk requires decision-making centralization to be efficient especially during a crisis as quick, informed decisions from the top have to be made to ensure the security of the population at risk. Moreover, the topic requires a minimum of technical knowledge not related to the population's daily life. For these reasons, flood risk is not a topic that citizens and communities appropriate naturally, as could happen for aspects such as children education, especially if they are not confronted to flooding on a regular basis.

Furthermore, communities have their own dynamics and perceptions of problems that might be different from that of the authorities. It is not always easy for institutional actors, including local authorities, to start a collaboration process with the communities. In the cases where there are governance problems around local sensible topics, even if not related to flood risk, civic capacity for flood-risk preparedness might be difficult to implement.

Thus, communities' civic capacity has to be developed and in close relationship with and within the institutional multi-actor flood risk management system, in general disconnected from the local population. From this perspective, one critical capacity is the integration of communities to the institutional flood-risk management networks from a bottom-up linked approach, i.e. local initiatives linked to higher levels of governance (Eizaguirre et al. 2012). This implies a double capacity-building process, on one hand the capacity of communities to elucidate and develop their own capacities, and on the other, the capacity of the management system, mainly institutional, to promote and integrate communities' capacities. Therefore, social and civic capacities were assessed in both groups, institutional actors on one hand and citizens and communities on the other. More details on the specific capacities will be provided in the assessment tool.

The research had two main stages: the design and implementation of the capacity assessment tool and the design and implementation of the participatory capacity building tool based on the first one, where specific actions and participatory mechanisms were implemented. In this chapter I give details of the design of the assessment tool and its implementation in the study cases. It is worth saying that the assessment is not disconnected from the experimental stage in which actions where implemented with the participation of local stakeholders. The assessment process was also important in establishing a first contact with the communities and authorities and for them to start collaborating together for a common aim. Indeed, working together was the best method to learn how to work together.

4.3 The capacity assessment tool

The questions that the research addressed concerning the assessment tool were: Which are the capacities, in both institutions and communities, that are required for a community engagement in flood risk preparedness to happen? and accordingly, which is their level of development?

The first step to design the assessment tool, was to set the specific capacities of each of the dimensions already defined by the social capacity building framework, the second step was to set a methodology that allows the implementation of the assessment in different cultural contexts whilst having a common evaluation criteria for all the cases.

Twelve indicators were defined to evaluate these specific capacities. Each of the indicators included a set of evaluative questions to guide the researchers on which information they need to obtain. Four methods are proposed as examples to get the information but researchers are encouraged to use and adapt the methods that better fit to the local context, they are encouraged to use a mix of quantitative and qualitative methods.

The capacity assessment tool was applied to five cases: Spanish case, four municipalities in the Ribera Alta along the Duero River; Italian case, Rivergaro close to the Po River; German case, two cities, Ulm in Baden-Württemberg and Neu-Ulm in Bavaria along the Iller River; Dutch case, two cities, Itteren and Borgharen in the municipality of Maastricht along the Meuse River; and French case, Vitry-sur-Seine in the Paris metropolitan area along the Seine River. The cases belong to the same action-research project.

The assessment focuses mainly in preparedness in ex-ante stage (preparedness), however the approach recognizes that the capacities built in this stage are useful throughout the different phases of a flood event. For example, a group of citizens that have worked together during a flood event to help each other can keep the momentum and keep work together for preparedness strategies.

To start the assessment, a detailed description of the river stretch, the flood risks related to it and the system of actors involved in flood-risk management was prepared. Then, 12 indicators were used to evaluate the specific capacities in communities and flood-risk management institutions. To do this, a mix of qualitative and quantitative methods was used. The assessment was done with the guidance of a set of evaluative questions addressed to the experts applying the assessment.

All the cases included a set of interviews at different levels of the flood-risk management system, especially at basin and municipalities scales and of other relevant stakeholders such as officials at the local cultural institutions. From the side of the citizens and communities, the cases included questionnaires, in-depth interviews, and/or focus groups.

The answers to the evaluative questions were obtained by applying at least one of the four main methods mentioned earlier. The assessment tool includes an evaluation guide for setting the level of capacities from zero to three, for more details on the assessment tool see the capacity assessment tool report (Oriard et al. 2016).

Table 4.1 shows the capacity assessment tool divided in five dimensions: knowledge, motivation, networks, finance, and participation. Each of the dimensions have specific capacities indicated as K.1, K.2, ... Each of the indicators is explained and divided into abilities and resources. The level of development of the capacities is indicated with a number from zero to three and a tone of red colour; the more the red is strong, the capacity has higher levels of development. The tool was applied twice, first for communities and then to institutions at multiple levels and scales in each of the cases.

4.4 Indicators and case findings

The indicators and their application that we considered for the assessment are described in the following paragraphs. Regarding knowledge, the first indicator (K.1) recognizes the importance of citizens and communities to have knowledge on floods. This indicator also considers the authorities communication capacity to share information to the general public and establish a two-way communication. The second indicator of the knowledge dimension looks at the learning capacity of the actors involved in the assessment (K.2).

In relation to the knowledge dimension, indicators K.1 and K.2, we found that in the cases where floods are not recurrent there is a tendency in citizens and communities to have less knowledge on floods such as the French case. In other cases, such as the Dutch case, even though there is a strong culture of floods, the population tend to have a distorted idea of the risks based on the past experiences. Furthermore, the protection measures recently constructed make people feel over-protected and as a consequence, overlook the risk of flood. In general, all the cases presented an important gap in terms of knowledge transmission to the population, especially vulnerable categories such as those living in poverty, the less educated, and migrants not speaking the languages, which tend to think that an event will not occur and have other priorities.

The capacity to communicate was revealed as a critical aspect to build flood knowledge. Institutional actors related to flood-risk management are used to communicate between themselves but not directly with communities, which requires them to change their position from experts and policy doers to one more accessible to the public. The role of researchers and NGOs working on pedagogic approaches to risk was revealed to be important as shown in the French case. However, the most important challenge of the communication strategy was to make people interested in the topic and participate in the focus groups. Indeed, in many cases flood risk is not an attractive theme for the population, especially when they are not confronted regularly by flooding.

The focus group in the French case showed that citizens who participated, which represents a minority of motivated, engaged, and educated citizens, were able to express clearly their needs, feelings, and the gaps in the management structure. For example, they expressed how important it was for them to have the authorities present during the May–June flooding, they said the ones who attended the urban walk activity organized by the authorities felt more comfortable during the event as they knew what to expect. They asked where they can get reliable information, one that is meaningful to them, for example what does having six metres of water above the normal level mean:

Table 4.1 Use Likert scale below to quantify indicators. When implemented, the intensity of red colour in the boxes of the implemented table reflects the level of development of the capacities within 4 identified levels: Level 0, Level 1, Level 2, Level 3.

CAPFLO Capacity Assessment Tool								
Knowledge		Motivation	Networks		Finance		Participation	
K.1 – Flood risk knowledge		M.1 – Motivation to mitigate flood risk	N.1 – Network performance		F.1 – Flood insurance		P.1 – Community participation in flood decision making	
Resources	Abilities	Abilities	Resources	Abilities	Resources	Abilities	Resources	Abilities
Presence of communities with flood-risk knowledge embedded in local culture	Existence of a two-way communication strategy between authorities and communities	Communities' motivation to prepare for flood events	Presence of networks collaborating on flood risk mitigation or on other issues	Ability of networks to collaborate amongst members and with other actors	Presence of individuals insured against floods	Understanding of individuals about the importance of being insured against floods	Power of communities to influence flood decision making in the form of political/ financial leverage, leadership and negotiation capacity, high level of organization	Ability of communities to participate proactively in FRM decisions

K.2 – Learning capacity	M.2 – Motivation to work collectively	N.2 – Network autonomy	F.2 – Financial resources for community action		P.2 – Participation dynamics
Abilities	Abilities	Abilities	Resources	Abilities	Abilities
Ability of the flood risk management structure to integrate communities experience-based knowledge into the structure (capacity of the structure to learn and adapt)	Communities' motivation to collaborate intra-group (bonding) and inter-groups (bridging)	Self-organization capacity of networks	Availability of funds for community action	Abilities of community organizations and networks to obtain and manage funds	Ability of the flood risk management structure to enable proactive participation of communities in flood decisions by supporting communities inclusion and redistribution of roles and tasks
	M.3 – Motivation dynamics		F3 – Financial resources dynamics		
	Abilities		Resources	Abilities	
	Change in communities' perception and level of motivation to prepare for flood events		Change in funds allocation and availability for community action	Change in capacity of communities to obtain and manage funds	

do they have to evacuate or are they safe to stay? In fact, the level of water is not meaningful for them unless they can relate it to what they have to do. This case was also interesting to see how citizens and municipality tensions affect flood risk management, for example the lack of presence of the municipality in the riverfront whilst the water level was rising, bought back old communication and trust problems with the municipality related to completely different subjects. Thus, the unresolved tensions between the citizens and the municipality, which are not related to floods, might restrain the capacity-building process, as the municipality has a leading role to play.

Regarding learning capacity, we found that every actor has its own learning pace – some actors in the system are more receptive to learning than others. In most cases the valuable experiences of citizens and communities are collected and shared by the municipalities who represent them in those spaces; the participation of citizens is limited. This is due to the fact that the study cases belong to representative democracies. However, this is a problem of learning as citizens and communities are not directly associated to the networks and spaces that feedback the process.

Another relevant finding is that the more the communities get to know the researchers and flood risk actors over time, i.e. after several visits, interviews, and actions, individuals and communities are slowly willing to get collectively more involved as they start considering outsiders as becoming part of the community. Thus, knowledge transmission and learning are capacities that require the construction of a relationship over time and time investment from the different actors.

The motivation dimension has three indicators. The first is related to the citizens' and communities' motivation to undertake actions to prepare for a flood event (M.1), the second is related to the motivation of citizens and communities to work together with different flood risk actors (M.2), and finally, the capacity of actors to change their perception of risk, in order to move from passive actors to active ones able to undertake their own initiatives (M.3).

In relation to motivation to prepare for a flood event (M.1), it was observed that those cases which have frequent flood events, such as the Italian and Spanish cases, communities tend to have more knowledge on floods and more motivation to prepare for an event; people are directly affected and feel concerned. In the cases in which episodes are rare, motivation to prepare is low, and the experimentation showed that in those cases it is not easy to build motivation, especially amongst the population that are not aware of the question.

During the capacities' assessment and later stages of the project, three relevant groups of communities were identified in relation to motivation: first, those who are not interested for flood risk nor motivated to participate in the actions, second, those who are interested in receiving information by e-mail and eventually assist to an information session, and finally, those motivated to actively participate in the actions and to exchange information and ideas. The last group of people were identified and involved successfully in the actions in the experimental part of the project and used as an important resource to share knowledge on the community dynamics and spread information.

The question about how to build motivation in non-interested communities remains under-researched. For example, the researchers in the French case stayed one hour in the Vitry market and only got five interviews, none of the persons interviewed wanted to be contacted afterwards to participate in the actions, even though they were told that this is a major issue in the area. This situation shows the low level of motivation and the difficulties to work on building it in non-interested communities. It is worth mentioning that if motivation is not present in the communities then a participatory capacity-building process cannot take place.

In relation to the indicator M.2, motivation to work collectively, in all the cases it was found that people have capacities to undertake initiatives and work with others during a flood crises; however, this capacity is difficult to transfer to a preparation stage. Indeed, only few cases have communities working on flood-risk preparedness, namely the Dutch and Spanish cases. In the cases in which there are communities involved, such as in the Spanish case, it was found that there were difficulties of collaboration amongst communities as they have contradictory approaches on how to prepare for flood risk. Thus, looking at the possible conflicts is an important aspect to be taken into account when doing the capacity assessment.

The French case is interesting to look at indicator M.3, motivation dynamics related to change perception of risk in order to enable action. The May–June 2016 flood events in the Paris metropolitan area gave a clear message that floods can actually happen and that the population was not prepared for it. The hypothesis was that the period after the event was a good moment to educate citizens and communities about the risk, however, only one year after the events the population had forgotten the events and possible consequences. Therefore, it is important to consider in the assessment the mechanisms established to keep memory, for example books, photographs, events, and articles.

The networks dimension is a crucial one as it allows the flow of resources and capacities amongst the different connected actors. Two indicators were considered. The first, indicator N.1 is about the existence of networks working on flood risk, and the second one, N.2 is about the level of autonomy of the network related to its capacity to self-organize.

The cases have networks working on flood preparedness. In the Dutch case there are networks involving institutional actors and communities, these networks were reinforced by the project implemented in the flood plain. The Spanish case also has community networks but not properly connected to the institutional networks. Others, namely the French and German cases, have only institutional networks working on flood preparedness in which citizens and communities are only represented by the local authorities but not directly involved due to their tradition as a representative political system. In these two cases, both indicators (N.1 and N.2) are non-existent or very low.

The assessment tool involves the identification of potential capacities that could be transferred to flood risk preparedness. In the French case it was interesting to look at local communities, i.e. associations, and their collaborative networks especially with the city cultural centre. Even though these communities and networks are not working currently on flood risk, they are already organized and could be used to propagate information and to join flood risk networks if a member is interested in taking the role of representative, especially associations related to solidarity networks, neighbourhoods, ethnic and cultural activities. This example shows the importance of looking at civic capacities beyond flood risk.

The dimension related to financial resources is important to ensure that the actions related to preparedness can actually take place. This dimension has three indicators. The first one, indicator F.1 is related to personal home insurance. In all the cases, there was a system in place to help individuals to overcome material loses. The second one, indicator F.2 is about available funds to finance community initiatives and incipient organizations. The third one, F.3, is about changing current funding schemes towards more community-supportive ones.

All the cases have compulsory personal home insurance as in the French case or collective funds as in the Italian case for personal losses in the recovery stage (indicator F.1). The assessment tool did not get into the details of how to get these funds or difficulties found by individuals, it only assesses the availability of the resources. In contrast, community funds to undertake initiatives regarding flood risk preparedness (indicator F.2) were lacking in all the cases except from the Dutch case. This is an important gap as these funds are important to motivate communities to undertake actions. Funding innovation (indicator F.3) was also lacking in all the cases or had a low level of development.

The last dimension refers to participation. Two indicators were considered. The P.1 indicator related to the participation of communities in flood-risk management, especially the decision-making process; and, the indicator P.2 related to the evolution of the flood risk management structure towards enabling more participation of citizens and communities. Four of the five cases present low levels of participation of communities in the decision-making processes (indicator P.1) and evolution of flood risk management structure towards enabling communities participation (indicator P.2), only the Dutch case has medium levels.

One of the German cases, situated along one side of the river, have higher levels of participation than the city on the other side of the river. The factor that makes a difference is the political will and role of the city to engage citizens and communities in a participatory approach to flood-risk management. This case shows the importance of the political dimension to support participatory approaches to flood-risk preparedness.

4.5 Conclusions

The integration of citizens and communities in flood-risk management systems is important but not an easy task to achieve, especially in societies in which flood risk has been managed traditionally by institutions and considered by citizens as a government affair. Citizens and communities have many capacities, including a daily contact with the environment, local networks, valuable experiences of floods in the past, and interest to reduce flood impacts, which are not always identified and valued in the institutional networks. On the other hand, institutional actors and networks do not have the mechanisms to integrate this kind of civil society actors in their institutional networks.

The capacity assessment is an important part of the participatory capacity building process, not only to identify and evaluate the social and civic specific capacities to be built in the next steps of the process, but also to start building a collaborative and trustful relationship with local authorities and people including citizens and communities. Indeed, the proposed assessment tool uses a set of indicators applicable to all the cases but users are highly encouraged to adapt the specific methods to obtain information and take into account the local context.

To start a participatory capacity building process it is crucial to establish, first of all, a good relationship with the city authorities; their support and active participation throughout the process is essential. In some cases this relationship is established in a very straightforward way, like in the Italian case, and in others it is established gradually as the project unfolds and trust is built gradually, as in the French case. In this case at some point the municipality kept its distance for political reasons but it was important to keep trying to involve them in every step of the project until a collaborative relationship was established.

From the first stages of the capacity assessment, it is important to have in mind the kind of topic flood risk is in the specific context and how it relates to people. In some cases it is a matter of survival and people feel highly concerned, such as the Spanish and Italian cases. In others, it is the opposite, something that will not occur and thus it makes no sense to people to invest themselves in preparation, as in the Dutch, German, and French cases. As explained earlier, it is not easy to motivate non-interested citizens and communities, in these particular cases, the identification of the different communities and their level of motivation is necessary as well as the development of a communication strategy tailored to those communities. It is recommended to link flood risk actions with local festivals and to involve local artists and cultural centres to design the communication and actions to be more effective in attracting people to get involved in the actions.

Once knowledge and motivation are built, the next step is to link citizens and communities to flood-risk management institutional networks. In some of the cases, namely the Dutch case, this process was facilitated by the implementation of a green area project in the river plain. In other cases this process was difficult as flood-risk management is dominated by institutional actors; citizens and communities experiences are only transmitted by city authorities which represent them in the network. The cases that have this kind of configuration present an important disconnection between institutional actors and citizens, such as the French and German cases. These cases also present an important difference between the level of capacities of institutional actors, in general higher, and those of the citizens and communities which are lower. The inclusion of citizens and communities in flood-risk networks is particularly important in these cases to ensure the transfer of capacities to the latter.

In sum, from the capacity assessment we observe that each case presents specific challenges for a capacity-building process. Communities in places confronted regularly by flooding will easily develop knowledge on flood risk and motivation to prepare and work collectively. On the contrary, those places where risk is not frequent are facing problems to engage citizens and communities in the process and rise motivation, a critical capacity for community engagement. On the other hand, we observed that those cases in which the representative tradition is stronger, citizens and communities are facing difficulties to join flood-risk management networks as they do not consider themselves as competent or do not feel directly concerned. Whilst in other cases communities do form networks and engage easily in flood risk management. Lastly, the cases in which the municipality engage more strongly have more chances to implement a successful participatory capacity-building process. In most of the cases there is a lack of community funds and innovative funding schemes to support community initiatives, a theme that could be further developed.

References

Berkes, F., Colding, J., and Folke, C. (2003). *Navigating Socio-Ecological Systems: Building Resilience for Complexity and Change*. Cambridge, UK: Cambridge University Press.

Chouinard, O., Plante, S., and Martin, G. (2008). The community engagement process: a governance approach in adaptation to coastal erosion and flooding in Atlantic Canada. *Canadian Journal of Regional Science* XXXI (3): 507–520.

Davis, D. (2004). Reverberations: Mexico City's 1985 earthquake and the transformation of the capital. In: *The Resilient City* (ed. L. Vale and T. Campanella). Oxford, UK: Oxford University Press.

Eade, D. (1997). *Capacity Building an Approach to People-Centred Development.* Oxford, UK: Oxfam Development Guidelines.

Eizaguirre, S., Pradel, M., Terrones, A. et al. (2012). Multilevel governance and social cohesion: bringing back conflict in citizenship practices. *Urban Studies* 49 (9): 1999–2016.

Howgate, O.R. and Kenyon, W. (2009). Community cooperation with natural flood management: A case study in the Scottish Borders. *Royal Geographical Society* 41 (3): 329–340.

Holling, C.S. (1986). The resilience of terrestrial ecosystems; local surprise and global change. In: *Sustainable Development of the Biosphere* (ed. W.C. Clark and R.E. Munn), 292–317. Cambridge, UK: Cambridge University Press.

Hyogo Framework for Action (2005–2015). Building the resilience of nations and communities to disasters. Extract from the final report of the World Conference on Disaster Reduction, United Nations, International Strategy for Disaster Reduction. Geneva United Nations. https://www.unisdr.org/files/1037_hyogoframeworkforactionenglish.pdf (accessed 5 February 2019).

Jha, A.K., Bloch, R., and Lamond, J. (2011). *Cities and Flooding: A Guide to Integrated Urban Flood Risk Management for the 21st Century*. Washington D.C: The World Bank.

Kuhlicke, C. and Steinführer, A. (2010). Social capacity building for natural hazards. A conceptual frame. CapHaz-Net WP1 Report. Leipzig and Braunschweig: Helmholtz Centre for Environmental Research – UFZ and Johann Heinrich von Thunen Institute. https://www.preventionweb.net/publications/view/18214 (accessed 3 March 2019).

Kuhlicke, C. and Steinführer, A. (2015). Building social capacities for natural hazards: Emerging field for research and practice in Europe. *Natural Hazards and Earth System Sciences* 15: 2359–2367.

Kuhlicke, C., Steinführer, A., Begg, C. et al. (2011). Perspectives on social capacity building for natural hazards: Outlining an emerging field of research and practice in Europe. *Environmental Science & Policy* 14: 804–814.

Larrue, C. and Oriard, L. (2016). Task C4: Cases Comparative Analysis. Marne la Vallée: CAPFLO Project, Lab'Urba, 30 p.

Nye, M., Tapsell, S., and Twigger-Ross, C. (2011). New social directions in UK flood risk management: Moving towards flood risk citizenship. *Journal of Flood Risk Management* 4 (4): 288–297.

Oriard, L. (2015). Street vending and its ability to produce space: The case of the Tepito market in Mexico City downtown area. PhD thesis. The Bartlett Development Planning Unit University College London.

Oriard, L., Larrue, C., Hubert, G. et al. (2016). Task B Capacity Assessment Tool, Action B. Marne la Vallée: CAPFLO Project, Lab Urba, Ecole d'Urbanisme de Paris.

Quarantelli, E. (1981). *An Agent Specific or an all Disaster Spectrum Approach to Sociobehavioral Aspects of Earthquakes*. Delaware: University of Delaware Disaster Research Centre.

Quarantelli, E.L. (1995). *Disaster Planning, Emergency Management and Civil Protection: The Historic Development and Current Characteristics of Organized Efforts to Prevent and Respond to Disasters*. Newark: Disaster Research Center, University of Delaware.

Steinführer, A., Kuhlicke, C., De Marchi, B. et al. (2009). *Local Communities at Risk from Flooding: Social Vulnerability, Resilience and Recommendations for Flood Risk Management in Europe*. Leipzig: Helmholtz Centre for Environmental Research – UFZ.

Stone, C. (2001). Civic capacity and urban education. *Urban Affairs Review* 36 (5): 595–619.

Thaler, T. and Levin-Keitel, M. (2016). Multi-level stakeholder engagement in flood risk management – a question of roles and power: Lessons from England. *Environmental Science and Policy* 55 (Part 2): 292–301.

United Nations Committee of Experts on Public Administration (2006). Definition of basic concepts and terminologies in governance and public administration. New York: United Nations Economic and Social Council. http://unpan1.un.org/intradoc/groups/public/documents/un/unpan022332.pdf (accessed 5 February 2019).

II.2

Strategies, Instruments, and Resources Used to Face Floods

5

Flood Risks Perceptions and Goals/Ambitions

Ann Crabbé
University of Antwerp, Faculty of Social Sciences, Centre for Research on Environmental and Social Change, Antwerp, Belgium

5.1 Introduction

Future flood risks in Europe are likely to increase due to a combination of climatic and socio-economic drivers (Alfieri et al. 2015; Feyen et al. 2012). Climate change is expected to result in sea-level rise and to induce more extreme weather events (Mitchell 2003). Even without taking climate change into account, the potential consequences of these extreme weather events will be intensified due to population growth, economic growth, and urbanization. Although quantitative projections for flood frequency and intensity are uncertain, the contribution of climate change to the damage costs from natural disasters is expected to increase in the future due to the projected increase in the intensity and frequency of extreme weather events in many regions in Europe (EEA 2017).

Even though awareness on flood risks grows – not in the least stimulated by recent extreme weather events causing floods in many parts of Europe – there is neither a straightforward nor automatic response by the public and policy-makers to start managing the increased flood risks. In this chapter we follow the line of reasoning of John W. Kingdon (Kingdon 1995). Kingdon stated that three 'streams' must be aligned for a matter to be dealt with in the public policy arena: the problem stream (is the condition considered a problem?), the policy stream (are there policy alternatives that can be implemented?) and the political stream (are politicians willing and able to make a policy change?). When these three streams come together, Kingdon assumes a 'window of opportunity' will open and action can be taken on the subject at hand.

In the following section, we start with describing, in Kingdon's terms, the problem stream. Based on a literature review, we present frequently used narratives on what is the nature of the problem, what causes it, and which other problems relate to it. We then focus on the policy stream. What are perceived to be suitable solutions to deal with increased flood risks? For example, we describe the historical evolution in preferred

Facing Hydrometeorological Extreme Events: A Governance Issue, First Edition.
Edited by Isabelle La Jeunesse and Corinne Larrue.

types of solutions. In turn, in the following section, we discuss the political stream: which driving factors have influence on how high flood risks are ranked on the political agenda? Finally, we provide some conclusions on the problems stream, the policy stream, and the political stream in flood-risk management.

5.2 The problem stream: Perceptions on increased flood risks

Whilst until recently floods were often perceived as natural disasters, caused by extreme forces of nature, human action is nowadays blamed more and more as the cause of cause floods (Prokopf 2016). For example, insight has grown that flood hazards are triggered by rapid urbanization, leading to more concrete surfaces that seal the soil and do not allow water to seep in. Flood hazards are also increased by techniques aimed at managing water. For example, dikes help contain the water but they also cut off the river from the floodplain. The floodplain helps to retain some of the floodwater. Cutting the river off from it, means that this natural storage is gone. The consequent water pressure on the dikes leads to new 'man-made' disasters. Flood hazards are also stimulated by other human actions, such as deforestation and fast drainage measures.

Floods have lost their exceptional character. People are more aware that floods are no longer something extraordinary but rather iterative (Prokopf 2016). The number of catastrophic floods have increased significantly in the past decades, endangering both human lives and the environment, and causing severe economic losses (Smith and Petley 2009) This change enables new responsibilities, making floods – and in consequence the human governance of nature – a political problem that has to be governed (Prokopf 2016). The occurrence of yet another flood allows for existing flood-related perceptions on human responsibility to become more and more politically relevant.

There are huge differences between countries in terms of the significance of flooding, and related to this, the presence of recent experience with flood events (Hegger et al. 2013). In Poland, floods constitute the main natural disaster (Ostrowski and Dobrowolski 2000). In recent history, major floods of large economic and social significance occurred in 1980, 1997, and 2010. The July 1997 inundation was an all-time high in Poland as far as economic losses are concerned. On the other extreme, in Sweden there is limited experience with disastrous floods (but there is a potential for flooding which is expected to increase due to climate change) (Swedish Civil Contingencies Agency 2012). The Netherlands, The United Kingdom, Belgium, and France seem to fall between both extremes. The large 1953 flood struck The Netherlands (1800 casualties), The United Kingdom (308 casualties), and Belgium (14–22 casualties). The latter event acted as a shock event for all three countries, although the precise responses to the event vary (Hegger et al. 2013).

The notion of *climate change is more and more linked to the occurrence of floods*. As it remains today, no single extreme event or flood can attributed directly to climate change, but the assumption that climate change will increase the likelihood of extreme events has gained legitimacy, even though the perception of the influence of climate change varies across countries and regions.

Climate change is expected to result in sea-level rise and to induce more extreme weather events. As a result, modifications in frequency, severity, and duration of hydro-meteorological hazards will occur (IPCC 2011). In Europe, it is likely that rising temperatures will intensify the hydrological cycle, leading to more frequent and intense floods in many regions. Although quantitative projections for flood frequency and intensity are uncertain, the contribution of climate change to the damage costs from natural disasters is expected to increase in the future due to the projected increase in the intensity and frequency of extreme weather events in many regions (EEA 2017). Water-level patterns in rivers are expected to change and the flash flood hazard is expected to increase in frequency and severity. Urban areas in particular face increasing flood risks.

Even when mankind mitigates climate change by dramatically curbed emissions of greenhouse gases, still extreme weather events are more likely to occur. The potential consequences of these extreme weather events are intensified due to population growth, economic growth, urbanization, and, in some cases also, soil subsidence (Mitchell 2003). These consequences will be suffered by the whole European Union, due to interdependencies between economic sectors and between regions.

In summary, the contemporary problematization entails recognition of both the limits to structural flood defence, and of the potential for increased flood risk due to climate change and pressures from developments in demography, economy, land use, and so on (Johnson and Priest 2008). These aspects of the problem are, in part, tied to rationalities about sustainable development. How can we strive for prosperity for this generation and the next, without surpassing the limits of the ecological system and at the same time empower people to deal with the bigger risks?

Although there is a broad consensus that the probability and potential impact of flooding are increasing in many areas of the world, the conditions under which flooding occurs are still uncertain in several ways (Doorn 2016). First, lack of observation records, both in time and in space, hinder the development of a solid knowledge base on climate evolutions. Second, climate-change effects will vary significantly across time and space, making it hard to predict what the impacts will be locally. Third, hydrologic and hydraulic models use quite a lot of approximations or assumptions, influencing the solidness of the models' output. Fourth, the dynamic nature of social systems and social vulnerability to flooding changes permanently: developments in and near floodplains change the flood risks, as does land surface alteration (Morss et al. 2005).

Further, there often seems to be a discrepancy between, on the one hand, the 'objective' risk assessment and risk analysis by scientists such as climate change experts and hydrologists and on the other hand the 'subjective' risk perception by people. In particular when people have not been struck by flooding, notwithstanding the fact that they are living in an area that experts consider floodprone, it is hard to convince them about the objective flood risks they face. People also make interest and value judgements in which, for example, buying cheap building plots is preferred over building houses in water-safe but more expensive plots.

In the terminology of policy sciences, flood risk governance is a typical example of a *wicked problem*. That is, a problem that is difficult or impossible to solve because of incomplete, contradictory, and changing requirements that are often difficult to pin down (Brunner et al. 2015). Wicked problems are characterized by ambiguity with regard to problem definition, uncertainty about the causal relations between the problem and potential solutions, and a wide variety of interests and associated values (Rittel and Webber 1973).

5.3 The policy stream: Perceptions on the solutions needed to deal with increased flood risks

The perception on the *type of solutions needed* to deal with flood risks has changed significantly during the past decades. In the nineteenth century, flood management received a scientific component with engineers taking the lead in anticipating and controlling hazards. The most preferred solution for dealing with flood risks was taking *technical, engineering measures*. A corps of engineers, part of state administrations, took responsibility and had the support of political leaders and citizens to implement engineered solutions like building dikes, dams, embankments, and so on. The main reasons to take engineered action against flooding were providing security and sustaining and developing economic prosperity.

From the 1970s onwards, opposition against the hegemonic position of civil engineers dealing with flood risk grew. Water issues became framed in terms of safety versus ecology. Water engineers were criticized for their 'short-sighted' technocratic solutions that were often harmful to the environment and the landscape (Doorn 2016). The support for 'building with nature' instead of 'building against nature' grew. It was believed that the lack of room for the river – due to, for example, embankments and the construction of buildings on floodplains – was a primary cause of riverine flooding. Hard preventive measures were considered less desirable than soft spatial measures, giving more room to the river. This discourse on 'room for the river' was accompanied by an 'ecological turn' in flood policies (Disco 2002). In parallel with the development of the environmental movement in the 1970s, awareness grew that, for example, wetlands had to be protected and that restoring or even building new nature could help protect the land from flooding.

In the late 1980s the ecological turn in water management was further enforced by the introduction of a discourse on *integrated water management*. Integrated water management stimulated water managers to think in terms of interconnectedness between land and water systems. Integrated water management proposes to manage water on the scale of water bodies (river basins, catchments, watersheds, and so on), taking into account the pressures of land use and the complex interrelations between a river's morphology, ecological quality, flood risk, and so on. The basic idea in integrated water management is that water systems have problems when the (big) diversity of water uses is not well-coordinated. As the number and significance of these 'water-related conflicts' increases, it becomes necessary to address the divergent uses together. River-basin management planning is a widely used approach to come to coordination and better governance of multiple stakeholders and numerous water authorities. The Water Framework Directive (WFD) (2000/60/EC) boosted the importance of river-basin management in Europe by putting river-basin management planning to the foreground.

Since the 1990s, another approach to flood governance has gained momentum, broadly marked by a move from 'flood defence' to 'flood risk management'. It is argued that flood risks can no longer be dealt with by focusing solely on flood defences (building dikes, dams, embankments, and so on). More and more actors in flood management wish for and make efforts for a diversification of flood-risk management strategies in which multiple strategies are applied simultaneously and linked together. These strategies include proactive spatial planning (building permits), flood mitigation in various ways (e.g. green urban infrastructures, adaptive buildings), flood preparation, and flood recovery. Literature suggests that such a diversification of flood-risk management strategies may lead to more resilience to flood hazards (Aerts et al. 2008; Van den Brink, Termeer, and Meijerink 2011; Innocenti and Albrito 2011).

The *distribution of responsibilities* (on *who* should do *what*) has undergone a lot of changes during the past decades. Since the 1990s, European flood-risk policies have further undergone a 'governance turn'. Until the late twentieth century, safety against flooding was seen as a purely economic good and the responsibility for managing flood risks was seen as the exclusive task of the state. In the past decades, this government approach is increasingly replaced by a more flexible and adaptive *governance approach* (Doorn 2016). The term governance stems from political science and it is used to refer to the way in which authority is exercised and shared between different actors in order to come to collectively binding decisions. Applied to flood risks, governance refers to the interplay of public and private institutions involved in decision making on flood-risk management (Asselt and Renn 2011). The governance approach in flood-risk management (in short: flood-risk governance) ascribes more responsibility to private actors and decentralized governmental bodies (Meijerink and Dicke 2008). For example, insurance companies become more involved: they do not only do they pay for recovery costs, they have increased interest in flood prevention, availability of flood risk maps, and so on.

Building on the 'governance turn', but stimulated by the European Floods Directive (FD) (2007/60/EC) and budgetary cuts since the financial-economic crises as from 2008 to 2009, a discourse on *multi-layered water safety* has emerged (Kaufmann et al. 2016). The concept of multi-layered water safety not only advocates diversified strategies for flood-risk management (prevention, protection, and preparedness). It also stimulates rethinking and further reconsidering the distribution of responsibilities between private and public actors. One narrative in the multi-layered water safety discourse is that public authorities cannot by all means protect every single property, but that – complementary to the public and collective defence measures – private house owners should invest in their own safety, for example by taking prevention measures like building on stilts on flood-prone residential land plots, moving electric appliances to the first floor in case of acute flood risk, and so on.

This redistribution of responsibilities has prompted some urgent *moral questions* (Doorn 2015; Mostert and Doorn 2012). How should the money available for minimizing the risk of flooding be distributed? How should the responsibilities pertaining to flood-risk management (both between private and public actors and between several governmental bodies or countries sharing a water course) be distributed? How should environmental impact be taken into account?

Flood-risk management has a long tradition of working with predictive approaches, but recently these approaches are complemented with adaptive or resilience-based approaches (Dessai and Van der Sluijs 2007). The predictive approaches focus on scenarios to assess the impact of changing conditions. However, given the deep uncertainty involved in climate change, the predictions for the long term may vary significantly along the scenarios chosen. That makes it hard to build robust policies: policies that keep their relevance and effectiveness on the long term. For that reason, policy-makers have made a shift from using probabilistic to *possibilistic* knowledge (Betz 2016). Probabilistic approaches rely heavily on climate predictions, the adaptive and resilience-based approaches focus on vulnerability and the adaptive capacity of a system, and on the measures required to improve its resilience and robustness (Carter et al. 2007). This adaptive capacity is assessed by looking at the social factors that determine the ability to cope with climatic hazards; the outcomes are partly based on qualitative data (experiences of stakeholders, expert judgements, and so on). Although the probabilistic approach is still dominant, *adaptation and resilience* are getting more and more attention.

Comparing types of solutions is often not easy. Whilst hard infrastructure measures can be assessed quantitatively, for example in terms of economic costs avoided by dikes or storm surge barriers, it is much harder to assess quantitatively what the advantages are of more soft interventions, such as ecological river restoration. In comparing both types of measures, the most one can do is say that the soft measure is more preferable than a hard (infrastructure) measure 'from an ecological point of view' (Doorn 2016). The latter implies that prioritization of measures has become more and more a matter of *prioritization of values*: do we think ecological aims are important to take into account? The same impossibility of quantification holds for other values, such as socio-cultural ones. Do we aim for absolute safety against floods (as in the Netherlands) or do strive for a 'reasonable' level of safety without exceeding economic costs? Do we want to safeguard land for agriculture or do we want to use that land for water retention and protect flood-prone residential areas downstream? This suggests that flood-risk governance, apart from being an object of technocratic decision-making, also is an object of political discussions on values and stakes. The political nature of flood-risk governance urges the call for participatory approaches in which deliberation processes can be organized with several stakeholders involved on values and stakes (Doorn 2016).

5.4 The political stream: Willingness to take action

Inspired by Kingdon's policy window model (Kingdon 1995), we assume we need at least three conditions to be fulfilled in order to initiate change in flood risk management. Firstly, there is a need for awareness and clear definition of the problem. Second, ideas on how to solve the issue should be available in the format of clear and mobilizing policy alternatives. Third, the issue should get on the political agenda: politicians should be willing to take action, based on society's pressure to do so.

In contrast to how it might appear, there is *not always a high sense of urgency* when it comes to flood risks. As long as no extreme weather events cause floods, the societal and political pressure to come to action is relatively low. The more intense a flood and the more victims or economic damage it causes, the greater the call from society and the greater the willingness of politicians to undertake policy action. As already indicated, now that the frequency of floods is rising, the problem of flooding becomes harder to ignore. Flood events offer, what Kingdon would call, windows of opportunity. Policy change is more likely to occur when extreme weather events 'shock' a society and when politicians, supported by the public support, see political opportunity to take action in the aftermath of a flood.

Of course, waiting on a disaster in order to induce change, is a very *reactive approach* to dealing with the increased flood risks. A *preventive approach* seems more sensible and proactive. Swart et al. (2009) researched the numerous factors that contributed to

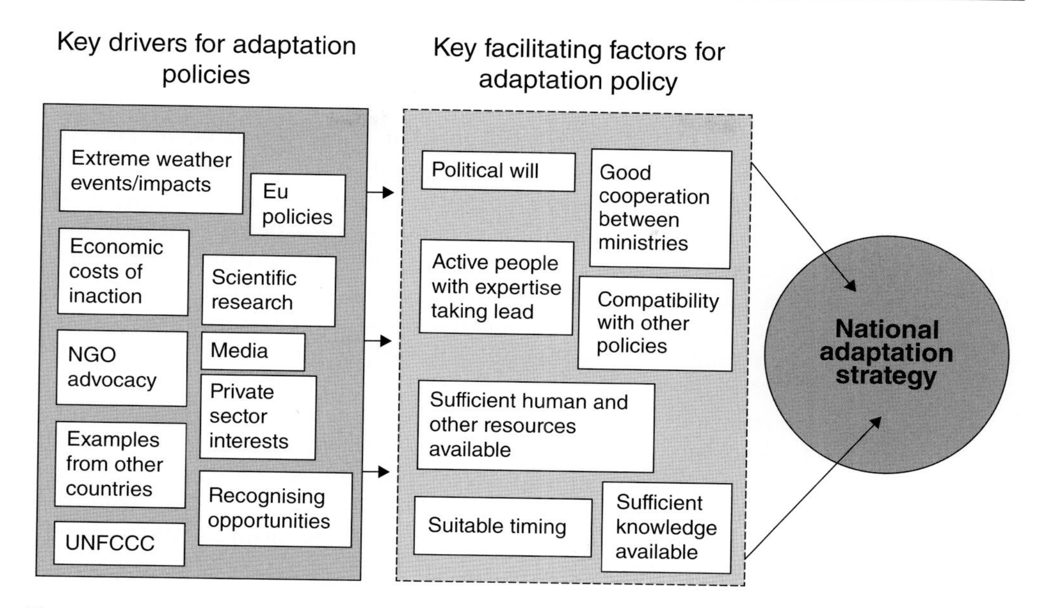

Figure 5.1 Key drivers and key facilitating factors for political commitment for climate change adaptation. Source: Swart et al. (2009).

drafting national adaptation strategies in several European countries. They discerned between 'key drivers' and 'key facilitating factors' contributing to willingness to take political action (see Figure 5.1). The first group of key drivers encompasses various sorts of pressure that have a direct impact on politicians' willingness to commit to action. The second group of key facilitating factors enable the key drivers to have an impact on political commitment. Hereafter we discuss, inspired by Swart et al., key drivers and key facilitating factors for flood-risk policies in Europe; flood-risk policies being an important part of national adaptation policies.[1]

There are numerous policy initiatives taken all over the world dealing with flood-risk management. Hereafter, we provide a short and therefore by no means intended to be exhaustive overview of such initiatives. First, certain United Nations (UN) initiatives are discussed then some European Union (EU) initiatives are described. We believe these initiatives can be considered important driving forces to get flood risk policies on the national agendas.

5.5 International policies

The UN General Assembly adopted the International Strategy for Disaster Reduction (UNISDR) in December 1999 and established the UN office for Disaster Risk Reduction, as the secretariat to ensure its implementation. The UN office for disaster risk reduction, is also the focal point in the UN system for the coordination of disaster risk reduction and the implementation of the international blueprint for disaster risk reduction – the 'Hyogo Framework for Action 2005-2015: Building the Resilience of Nations and Communities to Disasters'. UNISDR currently has several campaigns, amongst which 'Making cities resilient' which aims at engaging all stakeholders in reducing their risk to disasters such as floods (Bakker et al. 2013).

In January 2005, 168 governments adopted the Hyogo Framework for Action (HFA) at the World Conference on Disaster Reduction, held in Kobe, Hyogo, Japan. The HFA was a global blueprint for disaster risk reduction efforts during the decade between 2005 and 2015. It offered guiding principles, priorities for action, and practical means for achieving disaster resilience for vulnerable communities. Responsibilities for monitoring the HFA were assigned mainly to governments, but they were also identified for regional organizations and institutions, international organizations, and partners in the UNISDR. The HFA had obvious links to flood-risk management, since floods are one of the main hazards affecting millions of people all over the globe every year (Bakker et al. 2013).

The *Sendai Framework* is the successor instrument to the HFA. It is the outcome of stakeholder consultations initiated in March 2012 and inter-governmental negotiations held from July 2014 to March 2015, which were supported by the UNISDR upon the request of the UN General Assembly. The Sendai Framework is a 15-year, voluntary, non-binding agreement which recognizes that the state has the primary role to reduce disaster risk but that responsibility should be shared with other stakeholders including local government, the private sector, and other stakeholders.

Further, the stimulating role of the Organization for Economic Cooperation and Development (OECD) may not be underestimated. OECD countries come together to discuss and share their experience in a High Level Risk Forum. OECD researches and provides good-practice support for efforts to build resilience to major shocks and promote adaptation to climate change. Recently, the OECD published a report on the financial management of flood risk, providing an overview of the approaches that economies facing various levels of flood risk and economic development have taken to managing the financial impacts of floods (OECD 2016).

5.6 European directives and policy documents

The EU *Water Framework Directive* established a legal framework to protect and restore clean water across Europe by 2015 and to ensure the long-term sustainable use of water. Whilst a stated goal of the WFD was to reduce the impact of floods, flood protection is not explicitly addressed in the WFD. However, the River Basin Management Plans due under the WFD may take into account the impacts of climate change and the aim is to have these plans fully climate-proofed. Further, the implementation of the WFD is often seen as an opportunity to optimize sustainable flood protection with an ecological orientation (Dworak and Hansen 2003, cited in: Bakker et al. 2013).

The EU Directive on the assessment and management of flood risks (2007/60/EC), often referred to as the Floods Directive (FD), entered into force on 26 November 2007. The main aim of the FD is to reduce and manage the risks posed by floods to human health, the environment, cultural heritage, and economic activity. European Member States are required to follow a certain process to coordinate their flood-risk management practice in shared river basins, and should not undertake measures that would increase the flood risks in neighbouring countries. In addition, the implementation of the FD should be carried out in coordination with the implementation of the WFD, especially through the coordination of the flood-risk management plans and the river-basin management plans, and through coordination of the public participation procedures in the preparation of these plans (Bakker et al. 2013).

The *Blueprint to Safeguard Europe's Water Resources* was issued in November 2012. It outlines actions that concentrate on better implementation of current water legislation, integration of water policy objectives into other policies, and filling the gaps in particular as regards water quantity and efficiency. As part of the Fitness Check, the entire relevant body of legislation – including the WFD and the FD – has been analysed to identify potential improvements (Bakker et al. 2013).

5.7 Experiences with flood risk management in other countries

The example of countries that are (re)developing their flood risk policies may have an influence on the policies of other countries. However, this influence varies. Policies in neighbouring countries or in countries with similar geographic, cultural, or socio-economic conditions are generally of more interest and are paid more attention than developments in countries that are more remote and do not have so much in common with the country in question. Another source of inspiration, is the networks that cities from various countries form, for example in the context of Mayors Adapt, the Covenant of Mayors initiative on adaptation to climate change.

5.8 Research on impacts and adaptation

Several European countries have undertaken scientific impact assessments and research programmes related to the impacts of and adaptation to climate change. Publication of results from national and international research programmes and projects related to climate change impacts and climate change adaptation either directly at the request of policy makers or receiving attention through the media have also pushed countries towards developing national adaptation policies and re-orienting their flood-risk policies. They have generated new knowledge about future impacts of climate change in the countries as well as the countries' vulnerabilities and needs to adapt. However, because of the complex interactions between science and policy (Scott et al. 2005), it is not always apparent how research influences policy-making.

5.9 Economic costs (of inaction)

Insurance companies have published several studies calculating the costs of extreme weather events in several countries (EEA 2007). Floods, storms, and other hydrometeorological events account for around two-thirds of the damage costs of natural disasters, and these costs have increased since 1980, according to recent EEA assessments of climate change impacts in Europe (EEA 2012).

Growing awareness of the economic costs of inaction in relation to climate change can be recognized as one of the background drivers for adaptation policies. One of the most important recent studies that has boosted public discussion on the economic dimensions of climate change and the costs of inaction is the Stern Review on the Economics of Climate Change (Stern 2007). The report stresses the importance of early action, in relation to both mitigation and adaptation, in order to avoid the most severe consequences of climate change. The economic costs of climate change impacts and the costs of inaction can be regarded as an important factor contributing to the development of both mitigation and adaptation policies.

5.10 Facilitating factors

Separate from the main drivers, Swart et al. (2009) also distinguish a set of other factors that are required to convert the drivers into action, characterized here and in Figure 5.1 as facilitating factors.

On the one hand, facilitating factors can be seen as a 'catalyst' through which re-orientation of flood risk policies takes place. For example, enough background knowledge on the potential impacts of climate change in a certain country is needed in order to (re)formulate flood risk policies, sufficient resources and expertise have to be available, and ideally flood-risk policies that are suggested should be consistent with other policy goals, for example in spatial planning.

On the other hand, the 'catalyst' can also become a barrier for the development of flood policies, if crucial facilitating factors are missing or are too weak. This may help to explain why the (re)development process of flood risk policies has started more slowly in some countries than in others. For example, when there have been no significant flood events in recent history, the pressure to change policies is often low.

5.11 Factors contributing to agenda-setting

Figure 5.1 presents a simplified model of drivers and facilitating factors that are thought to have influenced the development of national adaptation strategies in Europe. As flood-risk management is an important part of these adaptation strategies, we see the parallel with factors contributing to the agenda setting of flood-risk management. A crucial factor is objective knowledge on the (increased) flood risks faced by communities, in combination with a subjective perception of increased flood risks, often due to the experience of increased frequency and severity of floods. Further, recognizing opportunities and creating win-win solutions for several policy fields/domains facilitates measures for flood-risk management, as does increasing legislative pressure to deal with flood risks (differently) and growing interests of private actors in flood prevention/protection.

5.12 Conclusions

In this chapter, we aimed to make a condensed overview of the current dominant perceptions on the problem of flood risks in Europe, how to deal with them, and what it takes to get flood risk management on the political agenda.

With regard to the latter – in contrast to how it may appear – there is not always a high sense of urgency when it comes to flood risks. Policy change with regard to flood-risk management often seems only likely to occur when extreme weather events 'shock' a society and when politicians, supported by the public's support, see political opportunity to take action in the aftermath of a flood. Of course, waiting for a disaster in order to induce change, is a very reactive approach to deal with the increased flood risks. A preventive approach seems more sensible and proactive. Inspired by Swart et al. (2009), we discussed several drivers, facilitating factors and their interplay, contributing to a political willingness for flood-risk policies. We highlighted the importance of international policies of the UN and the EU, and also considered 'good practices' from similar countries and similar cities, research on climate change impacts and adaptation, and information on economic costs as relevant drivers for inducing intensified flood risk policies.

Based on our reflections, however, it appears extremely relevant to also refer to the importance of problem framing and framing of potential policy alternatives, to explain the political willingness to deal with increased flood risks.

With regard to the framing of the problem, we see reason for some optimism: current perceptions of flood risks stimulate the putting of flood risks more on the societal, political, and policy agendas. (i) Whilst until recently floods were often perceived as 'natural' disasters, caused by extreme forces of nature, human action is nowadays blamed more and more as a cause for floods. This helps to incite human-driven action to prevent and accommodate floods. (ii) Floods are no longer perceived as an extraordinary but as an iterative phenomenon – this will regularly bring flood risks high on the societal and political agendas. (iii) Further, megatrends like population growth, economic growth, urbanization, and climate change are more and more discursively linked to the increased flood risk. They are considered 'drivers', stimulating the likelihood, severity, and intensity of floods, which urges society, policy makers, and politicians to put flood-risk management higher on the agenda.

With regard to the framing of potential policy alternatives, the ecological turn has stimulated the introduction of new policy approaches in dealing with flood risks. Whilst technical-engineering solutions were absolutely dominant until the 1960s, the rise of ecological concerns from the 1970s onwards, have triggered this ecological turn in dealing with flood risks. Soft, more ecological measures are more often considered valuable, next to the hard, technical solutions. The discourse on integrated water management helped to sustain the ecological approach and it also helped to (re)introduce the now rather dominant perception that river basins are a good scale for thinking on measures and actions. Further (and this is a break with the long-standing tradition of predictive approaches in flood-risk management), adaptive or resilience-based approaches have been introduced, indicating that policy makers acknowledge society's vulnerability to increased flood risks. The ecological turn thus offered a policy alternative that seemed valuable to deal with flood risks appropriately.

As also discussed earlier, we see a lot of 'agenda-setting potential' in the recently introduced narratives on multi-layered water safety. This concept broadens the scope of actors that are assumed responsible for taking action: not only government should do something; also private actors like individual house owners, insurance companies, non-governmental organizations (NGOs), and others need to take up tasks (Mees et al. 2016). This redistribution of responsibilities urges intensified 'co-production' in developing and implementing flood-risk policies in the near future: not only consulting all relevant actors is important, also involving them to – jointly – develop and implement good flood-risk management is crucial to deal with flood risks appropriately in the near future.

Note

1. Part of these key drivers and key facilitating factors might also influence the willingness to take political action at the local level. For example, it is known that *local* floods play a pivotal role in increasing the local flood awareness of politicians and citizens. It is also known that creating win-win solutions (recognizing opportunities), at suitable moments, stimulated by matchmaking figures are – also on the local level – important levers for action.

References

Aerts, J.C.J.H., Botzen, W., Van der Veen, A. et al. (2008). Dealing with uncertainty in flood management through diversification. *Ecology and Society* 13 (1): 41.

Alfieri, L., Feyen, L., Dottori, F. et al. (2015). Ensemble flood risk assessment in Europe under high end climate scenarios. *Global Environmental Change* 35: 199–212. https://doi.org/10.1016/j.gloenvcha.2015.09.004.

Asselt, M. and Renn, O. (2011). Risk governance. *Journal of Risk Research* 14: 573.

Bakker, M.H.N., Green, C., Driessen, P. et al. (2013). Flood risk management in Europe: European flood regulation. In: *STAR-FLOOD Consortium*. The Netherlands: Utrecht. ISBN: 978-94-91933-04-2.

Betz, G. (2016). Accounting for possibilities in decision making. In: *The Argumentative Turn in Policy Analysis. Reasoning about Uncertainty* (ed. S.O. Hansson and G. Hirsch Hadorn), 135–169. Cham: Springer.

Brunner, R.D., Steelman, T.A., Coe-Juell, L. et al. (eds.) (2015). *Adaptive Governance: Integrating Science, Policy, and Decision Making*. New York: Columbia University Press.

Carter, T.R., Jones, R.N., Lu, X. et al. (2007). New assessment methods and the characterisation of future conditions. In: *Climate Change 2007: Impacts, Adaptation and Vulnerability. Contribution of Working Group II to the Fourth Assessment Report of the Intergovernmental Panel on Climate Change* (ed. M.L. Parry, O.F. Canziani, J.P. Palutikof, et al.), 133–171. Cambridge: Cambridge University Press.

Dessai, S. and Van der Sluijs, J.P. (2007). *Uncertainty and Climate Change Adaptation – A Scoping Study*. Utrecht: Copernicus Institute for Sustainable Development and Innovation.

Disco, C. (2002). Remaking 'nature': the ecological turn in Dutch water management. *Science, Technology, & Human Values* 27 (2): 206–235.

Doorn, N. (2015). The blind spot in risk ethics: managing natural hazards. *Risk Analysis* 35 (3): 354–360.

Doorn, N. (2016). Reasoning about uncertainty in flood risk governance. In: *The Argumentative Turn in Policy Analysis. Reasoning about Uncertainty* (ed. S.O. Hansson and G. Hirsch Hadorn), 245–263. Cham: Springer.

Dworak, T. and Hansen, W. (2003). The European flood approach. In: *Towards Natural Flood Reduction Strategies*, Kotowski, W., Oswiecimska-Piasko, Z. and Sobocinski, W (ed.), 104. Conference Programme and Abstracts. Warsaw, 6–13 September 2003, Warsaw.

EEA (2007). Climate change: the cost of inaction and the cost of adaptation (EEA Technical report No. 13/2007). Copenhagen: European Environment Agency.

EEA (2012). Climate change evident across Europe, confirming urgent need for adaptation. Copenhagen: European Environment Agency.

EEA (2017). Indicator assessment. Economic losses from climate-related extremes. Copenhagen: European Environment Agency http://www.eea.europa.eu/data-and-maps/indicators/direct-losses-from-weather-disasters-3/assessment (accessed 5 February 2019).

Feyen, L., Dankers, R., Bódis, K. et al. (2012). Fluvial flood risk in Europe in present and future climates. *Climatic Change* 112 (1): 47–52.

Hegger, D.L.T., Green, C., Driessen, P.P.J. et al. (2013). Flood risk management in Europe: Similarities and differences between the STAR-FLOOD consortium countries. WP1 deliverable report STAR-FLOOD project. https://dspace.library.uu.nl/handle/1874/314882 (accessed 5 February 2019).

Innocenti, D. and Albrito, P. (2011). Reducing the risks posed by natural hazards and climate change: the need for a participatory dialogue between the scientific community and policy makers. *Ecology and Society* 21 (4): 53.

IPCC (2011). *Summary for Policymakers of Intergovernmental Panel on Climate Change Special Report on Managing the Risks of Extreme Events and Disasters to Advance Climate Change Adaptation*. Cambridge: Cambridge University Press.

Johnson, C.L. and Priest, S.J. (2008). Flood risk management in England: a changing landscape of risk responsibility? *International Journal of Water Resources Development* 24 (4): 513–525.

Kaufmann, M., Mees, H., Liefferink, D. et al. (2016). A game of give and take: the introduction of multi-layer (water) safety in the Netherlands and Flanders. *Land Use Policy* 57: 277–286. https://doi.org/10.1016/j.landusepol.2016.05.033.

Kingdon, J.W. (1995). *Agendas, Alternatives, and Public Policies*, 2e. New York: Longman.

Mees, H., Tempels, B., Crabbé, A. et al. (2016). Shifting public-private responsibilities in Flemish flood risk management. Towards a co-evolutionary approach. *Land Use Policy* 57: 23–33.

Meijerink, S. and Dicke, W. (2008). Shifts in the public-private divide in flood management. *International Journal of Water Resources Development* 24 (4): 499–512.

Mitchell, J.K. (2003). European river floods in a changing world. *Risk Analysis* 23 (3): 567–574.

Morss, R.E., Wilhelmi, O.V., Downton, M.W. et al. (2005). Flood risk, uncertainty, and scientific information for decision-making. Lessons from an interdisciplinary project. *American Meteorological Society* https://doi.org/10.1175/BAMS-86-11-1593.

Mostert, E. and Doorn, N. (2012). The European flood risk directive and ethics. *Water Governance* 6: 10–14.

OECD (2016). Financial management of flood risk. Paris: Organization for Economic Cooperation and Development. http://www.oecd.org/daf/fin/insurance/Financial-Management-of-Flood-Risk.pdf. (accessed 23 May 2017).

Ostrowski, J. and Dobrowolski, A. (eds.) (2000). *Monografia katastrofalnych powodzi w Polsce w latach 1946–1998*. Warsaw: IMGW(CD-ROM).

Prokopf, C. (2016). International river governance. Extreme events as a trigger for discursive change in the Rhine river basin. In: *Environmental Politics and Governance in the Anthropocene: Institutions and Legitimacy in a Complex World* (ed. P. Pattberg and F. Zelli), 145–164. Abingdon,UK and New York, USA: Routledge.

Rittel, H.W.J. and Webber, M.M. (1973). Dilemmas in a general theory of planning. *Policy Sciences* 4 (2): 155–169.

Scott, A., Holmes, J., Steyn, G. et al. (2005). *Science meets Policy: Next Steps for an Effective Science-Policy Interface*. Conclusions of the conference held as part of the UK's presidency of the European Union. London: NERC, EA, DEFRA.

Smith, K. and Petley, D.N. (2009). *Environmental Hazards: Assessing Risk and Reducing Disaster*, 383 p. Abingdon, UK: Routledge.

Stern, N. (2007). *The Economics of Climate Change: The Stern Review*. Cambridge: Cambridge University Press.

Swart, R., Biesbroek, R., Binnerup, S. et al. (2009). *Europe adapts to climate change: Comparing national adaptation strategies*. PEER Report No 1. Helsinki: Partnership for European Environmental Research.

Swedish Civil Contingencies Agency (2012). *Översvämningar i Sverige 1901–2010*. Karlstad: Swedish Civil Contingencies Agency.

van den Brink, M., Termeer, C., and Meijerink, S. (2011). Are Dutch water safety institutions prepared for climate change? *Journal of Water and Climate Change* 2 (4): 272–287.

6

Instruments for Strategies to Face Floods through Prevention, Mitigation, and Preparation in Europe: The Age of Alignment

Mathilde Gralepois
Laboratory CNRS 7324 Citeres, University of Tours, Tours, France

6.1 Introduction

6.1.1 Recurring floods in Europe and risk management: A mounting dilemma

Extreme hydrological events have been witnessed over the years in Europe, bringing about substantial flood damages. As a result, issues of vulnerability and risk have come under increased scrutiny amongst policymakers (Fuchs et al. 2013). In an attempt to address this issue, various risk-management strategies have been implemented by governments. Many other European countries, have over time tried several methods, going from land drainage (1930s–1970s), defence and prevention (1980s–1990s) (Johnson, Tunstall, and Penning-Rowsell 2005), to spatial planning (1990s–2000s) (Gralepois 2012), whilst combining today a whole range of flood-risk strategies in a move towards more diversification (Dieperink et al. 2016).

Although the range of available strategies has been significantly broadened since the end of the 1990s, over the past decade European countries have faced recurring flood disasters, especially in dense urban territories as well as in predictable flood-prone areas (November 2010). However, several factors make it particularly difficult for a clear and common flood-risk strategy to emerge within the European Union. Such a common framework would first have to address a whole variety of situations, spanning a range of

Facing Hydrometeorological Extreme Events: A Governance Issue, First Edition.
Edited by Isabelle La Jeunesse and Corinne Larrue.

flood types going from coastal and tidal to pluvial and flash floods, not to mention the specificities of urban flooding. A second layer of complexity comes from the necessity to accommodate the diversity of legal, institutional, political, and social contexts observed from one country to the other as well as within each specific country (Hegger et al. 2013).

Twelve years after the Flood Directive 2007[1] was first implemented, this chapter builds on recent findings from the STAR-FLOOD project[2] to analyse some of the difficulties met by flood-risk management strategies as well as the specific role played by policy instruments.

6.1.2 From instruments to strategies: Definitions

Flood management is composed of several 'strategies' (Hegger et al. 2014), which themselves have to be distinguished from 'instruments'. Bearing this in mind, the chapter will focus on the specific role of instruments in the concrete implementation of flood risk strategies.

A strategy can generally be defined as 'a combination of measures and instruments as well as the necessary resources for actions to implement the basic long-term goals of a business organisation' (Whipp 2001). Applied to flood management, strategy is understood as 'a combination of long-term goals, aims, specific targets, technical measures, policy instruments and processes which are continuously aligned with the societal context' (Floodsite 2005). Therefore, in this process, instruments stand as means of application, involving a certain use of resources and techniques to attain a set of goals within the framework of a specific strategy (Howlett 2010). Mostly described as legal or financial tools, their definition is itself a subject of debate within the academic field.

Nowadays however, policy scholars tend to agree that the way instruments are selected and used is never neutral. Indeed, amongst the number of instruments available in the public policy toolkit, there is room for choice (Hood 1986; Lascoumes and Le Galès 2004). However simple it might be, this observation has strong implications. Indeed, it shows that instruments are key to understanding strategies and policies. Put differently, the latter can partly be explained by looking at their instruments. Halpern (2010) even goes as far as considering strategies as 'a sedimentation of instruments'. This chapter will thus argue that studying flood management instruments can be critical to gain a better understanding of flood management.

6.1.3 Five flood-management strategies

Building on the legal definition given in the 2007 Flood Directive, the following sections focus on five different strategies: prevention, preparation, mitigation, defence, and recovery, whilst paying a specific attention to the instruments that concretely allow their implementation. We refer to these notions according to a set of definitions crafted by Hegger et al. (2014) (Table 6.1).

1. *Prevention* is understood in two different ways: planning and communication. It consists first in adapting future developments to the risk of flooding, therefore allowing the construction of houses, activities, services, and industries in flood-prone areas. It also involves promoting appropriate agricultural and forestry practises as well as relevant land-use plans. This strategy has one motto:

Table 6.1 Flood risk management strategies: definitions from Hegger et al. (2014).

Strategy	Explanation
Flood defence	Flooding can be prevented by infrastructural works, such as dikes, dams, embankments and weirs, upstream retention or giving more space to the river within its current embankments ("keeping water away from people"), mostly referred to as "flood defence" or "structural measures". Main actors: generally governmental water management actors: at national/regional level.
Flood risk prevention	Negative consequences of flooding can be avoided by proactive spatial planning or land use policies ("keeping people away from water"), aimai at building only outside areas that are prone to flooding. Main actors: actors involved in planning processes (governmental actors, private parties). Flood insurance companies may influence planning decisions, for instance by (not) insuring properties in high-risk areas or the use of risk-based premiums.
Flood risk mitigation	Consequences of floods can be mitigated by a smart design of the flood-prone area. Measures include spatial orders, constructing flood compartments, or (regulations for) flood-proof building. Main actors: citizens, project developers, water managers and other public and private actors.
Flood preparation	Consequences of floods can also be mitigated by preparing for a flood event Measures include developing flood warning systems, preparing disaster management and evacuation plans and managing a flood when it occurs. Main actors: governmental organisations like the meteorological office, flood forecasting centres, local and regional governments.
Flood recovery	This strategy facilitates a good and fast recovery after a flood event Measures include reconstruction or rebuilding plans as well as compensation or insurance systems. Main actors: national governments establishing disaster relief funds, insurance companies as well as the affected citizens themselves.

'keeping people away from water'. Prevention consists also in communication-based objectives including consultations with stakeholders and the public, debates, campaigns, forums…

2. *Mitigation* strategy aims at decreasing the likelihood and/or magnitude of flooding via measures to accommodate water. It relies on structural and non-structural measures to reduce flood events and/or their impact in specific flood-prone areas using smart design (Fournier et al. 2016). It is a strategy focused on 'reducing the water of flood' (absorption, canalization…).
3. *Preparation* strategy is mostly concerned with the concrete steps to be taken in the event of a flood. It falls into three parts: emergency management, preparedness, and response to contemporary threats (Gilissen et al. 2016). This strategy aims at minimizing the consequences of flooding via measures that strengthen communities' social capacity to prepare and respond to flood events.
4. By *defence*, we mean a specific strategy that aims to decrease the likelihood and/or the magnitude of flooding by keeping water away from people through more resistant and specifically designed infrastructural works such as dikes, dams, and on so (Gralepois et al. 2016).
5. Finally, *recovery* strategies are focused on facilitating good and fast recoveries after a flood event has occurred. Typical measures include reconstruction or building plans as well as compensation or insurance systems.

6.1.4 The hard problem. Understanding the different combinations of flood-management instruments behind flood strategies in Europe

Although research in flood-risk management has been prolific in recent years, the specific role of instruments received little attention especially from socio-political, governance, or comparative perspectives. For example, Fitova (2014) explored the role played by market-based instruments in flood-risk management using micro-economic modelling to evaluate the respective influence of individual behaviour and market forces. However, her mathematical arguments do not fall within the field of social science. Sterner (2003) also looked at policy instruments in environmental resource management. If his approach is largely comparative, he focuses his analyses on economic principles lying behind environment policy and does not consider governance issues. Therefore, there is clearly a gap to be filled. This positions social science as a new field of study, with the potential to generate new understandings of flood management in general. Put differently, adopting an empirical as well as a comparative perspective, this chapter aims to cast some light on the specific role played by instruments and their influences on the way strategies have evolved.

By doing so, the chapter will look at a range of European countries affected by recurring floods and will answer the following research question: what are the different combinations of instruments behind flood strategies? What does this say of European countries' political capacity to deal with flood management issues? In order to answer these questions, the chapter studies the differences and the similarities between flood management instruments implemented in three European countries: England, the Netherlands, and France. Facing recurring floods, these also happen to evolve within the same European legal contexts. Whereas the three of them have implemented flood-management strategies for quite a long time and are based on the same tradition of bureaucratic authority, their strategies are based on different types of instruments. When combined, these represent most of the flood policies that were implemented from the 1980s to the 2010s. They also enable to identify distinct country patterns, which will be described and analysed.

The empirical data used in this chapter are based on country reports and case studies written for the STAR-FLOOD project, a research consortium that conducted extensive interviews with stakeholders (50 interviews per country on average). It also built a substantial database including observations, policy and legal document analyses as well as minutes from workshops with practitioners. Although the STAR-FLOOD dataset already received scholarly attention, it can still lend itself to fruitful analyses if viewed through the lens of flood-management instruments. In the following, we analyse the specific role these instruments happen to play across five flood-management strategies in England, France, and the Netherlands, taking concrete examples from the three national country reports of the STAR-FLOOD database. A more detailed presentation of flood management in each country is available to the public via the project reports of England (Alexander et al. 2015), the Netherlands (Kaufmann et al. 2015) and France (Larrue et al. 2015).

This chapter is divided into five parts. After this introductory section, the second part establishes the conceptual framework that we will use to understand the political role of policy instruments. The third section describes the differences and similarities observed in the way instruments are implemented for the five strategies (prevention; mitigation; preparation; defence and recovery) in England, the Netherlands, and France. In our fourth section, we further discuss the choices made around these instruments, the various understandings of each specific strategy they entail, and how this helps characterize flood management's recent evolutions. The fifth section will conclude by looking at success or failure factors in flood-risk management and how these relate to policy instruments.

6.2 Conceptual framework

In order to gain a clearer understanding of the role played by instruments in flood management at the European level, the chapter draws a conceptual framework from a socio-political body of literature whilst relying specifically on the work of Lascoumes and Le Galès (2004). Indeed, the latter developed a model aimed at analysing policy instruments through a multidimensional typology highlighting the links between various instruments, legitimacy, and political relations types. The model was improved and translated into English in 2007 (Table 6.2).

According to Lascoumes and Le Galès' main arguments (2004), public-policy instruments appear as both technical and social devices. Their generic purpose is to carry a concrete notion drawn from a specific policy and sustained by a regulation concept. Therefore, there is no way instruments can be either neutral, transparent, or democratic. As stated by Dehousse (2004), the pervading lack of agreement in public policy at large on what its very goals should be has led research to narrow down its focus and concentrate on discussing the objectives of specific public-policy instruments. However, if we are to see policy as a combination of instruments, we cannot but agree that instruments are both meaningful and powerful (Howlett 1991). Indeed, they partly determine the way in which actors are going to behave. They can include or exclude actors or options (Le Bourhis and Lascoumes 2014). They drive society at large in the way it frames or represents its problems (Halpern and Le Galès 2008).

Table 6.2 Typology of policy instruments.

Type of instrument	Type of political relations	Type of legitimacy
Legislative and regulatory	Social guardian state	Imposition of a general interest by mandated elected representatives
Economic and fiscal	Wealth producer state, and redistributive state	Seeks benefit to the community; Social and economic efficiency
Agreement-based and incentive-based	Mobilizing state	Seeks direct involvement
Information-based and communication-based	Audience democracy	Explanation of decisions and accountability of actors
De facto and de jure standards; Best practices	Adjustments within civil society; Competitive mechanisms	Mixed: scientific/technical, democratically negotiated and/or competition, pressure of market mechanisms

Source: Lascoumes and Le Galès (2007).

Public policy instrumentation is 'the set of problems posed by the choice and the use of instruments (techniques, methods of operation, devices...) that allow governmental policies to be made material and operational' (Lascoumes and Le Galès 2007, p. 4). Their properties, justifications and applicability are not secondary issues. Depending on its own structure and logic, every instrument constitutes a condensed form of knowledge and power. Elaborating on Hood's reference work (Hood, 1986), Lascoumes and Le Galès (2004) have identified five different types of instrument as follows:

1. *Legislative and regulatory instruments* are traditional tools of state interventionism based on legal forms such as laws, acts, rules, and so on. Legal instruments are considered as coercive.
2. *Economic instruments* are similar to the previous regulatory instruments: their legitimacy often comes from their legal dimensions. Their peculiar feature is to resort to monetary techniques in a way either to redistribute resources (taxes, fees) or to redirect them (subsidies or expense allocations).

The three other instrument types appear as less interventionist forms of public regulation. They are classified hereunder in decreasing order of power coercion, understood as the capacity to decide or strongly influence someone's decision.

3. The 'Govern by contract' model (Gaudin 2004) stands as a framework of agreements coupled with sets of incentives. It draws its name from what has become a general injunction nowadays, in a context strongly critical of bureaucracy and more and more averse to the rigidity of legislative and regulatory rules.
4. De facto *standards* are made by governmental actors to organize power relations within civil society (economic actors, non-governmental organizations, consumer associations, environmentalist groups and so on). Standards are well known to frame methods or conditions of production for services produced (e.g. safety standards).
5. *Communication and information-based* instruments are part of what is generally referred to as the 'open' or 'grassroot' democracy, a relatively participative public space in a political sphere traditionally based on representative power.

We concentrate our analysis on those five instrument types (legal, economic, incentive, standards, and informational), whilst looking at five strategies (prevention, mitigation, preparation, defence, and recovery) in three European countries (England, the Netherlands, and France). Lascoumes and Le Galès' typology proves particularly useful to understand the political role played by government actors and the type of legitimacy supposed to be achieved whilst using these instruments.

Lascoumes and Le Galès develop three key arguments that this chapter will take on and apply to flood management in Europe. First, the authors assume that policy instruments have political effects of their own linked to the specificities and constraints they entail. They assume that policy instruments can quite thoroughly reveal types of public policies outcomes 'in [their] meaning [and] in the cognitive and normative framework' they set (Lascoumes and Le Galès 2007, p. 16). However, they also contend that policy instrumentation can reflect specific governance choices and the ways in which they evolve.

The next section presents the similarities and differences between flood-management instruments in England, the Netherlands, and France as well as the way they are used from one strategy to another. It will thus provide us a good basis to discuss how instruments relate to issues of political relations and governance.

6.3 Comparison. Similarities and differences in flood instruments' implementation in Europe

Longitudinal analyses of the instruments used to face floods can bring fresh and new understandings of flood management strategies because 'policy instrumentation may relate to the fact that actors find it easier to reach agreement on methods than goals' Lascoumes and Le Galès 2007, p. 16). Whilst instruments are often seen as secondary administrative techniques, they may in fact have a huge potential to provide a better understanding of power relations and governance choices, especially in a comparative perspective. We study successively: prevention, mitigation, preparation, defence, and recovery.

6.3.1 Prevention, a spatial planning oriented strategy

In all the three countries, prevention strategy in flood management mostly revolves around spatial planning. Its main objective is to influence the location of concrete activities as well as future spatial and economic developments. In comparison, little attention is paid to communication-based instruments. In either case, prevention strategy is based mostly on legal instruments, whether in England, the Netherlands, or France.

In England, legal instruments for spatial planning are organized at three levels, with hierarchical articulations between them:

- First, a *National Planning Policy* sets out general guidelines on how flood risks should be incorporated into the overall planning system, with specific provisions regarding climate change since 2016. Despite its prospective and strategic status, Local Plans[3] integrate its recommendations.
- Secondly, *Land Planning Permissions* are drafted at the level of each specific borough or district council. These local entities bear the responsibility of assessing risks at the local level whilst ensuring flooding is taken into account in local planning decisions. In cases where future developments are contemplated in flood-prone areas, local authorities have a statutory requirement to consult the Environment Agency.
- The third level is this of *Flood Risk Assessments* which any land developer is required to conduct according to the rules set out in Local Plans. These assessments are then to be included in Land Planning Applications. A specific feature of the English system is called the *Sequential test*. Put simply, this mechanism aims at reducing developments in floodplains. Evidence that a sequential test has been conducted has to be joined to Flood Risk Assessment files. 'The sequential test compares the site you are proposing to develop with other available sites to find out which has the lowest flood risk' (Department for Environment, Food and Rural Affairs, Environment Agency 2017). In other words, developers have to prove that they cannot build their project somewhere else.

As for communication-based instruments, a *Register of flood risk* assets, structures, and features has been set up, with the aim of identifying flood-risk areas and helping with the drawing of *Flood maps* flagging up specific hazards. Realized by the Environment Agency, it is supposed to supply relevant information by leveraging on local knowledge. Other than that, no incentive-based, economic-based, or standards policy instrument has been implemented in England so far.

In the Netherlands, 'water' is considered as a major twenty-first-century challenge in terms of spatial planning (V&W 2000). Just as in England, legal instruments happen to be predominant. However, rules are less embedded in spatial planning legislation. *Strategic spatial development plans* (*structuurvisies*) and legally binding *Spatial zoning plans* (*bestemmingsplannen*) have to feature flood reduction provisions. These are supposed to include specific bans or restrictions applicable to constructions, as well as relevant expropriations or re-allotments.

On top of that, two agreement-based incentives can be found. Designed as bridging mechanisms, these are supposed to align spatial planning measures with flood prevention policies. The most famous of them is called the *Water test* (*watertoets*). As an advisory construction mechanism, it is used to foster water management and is supposed to ensure that: 'the consequences for safety and water-related problems [are] explicitly addressed as a separate section in the explanatory policy document and form part of the integrated assessment' (Environmental Protection Department 2007). The second agreement-based incentive tool operating at the sub-river basin level is the *Flood risk management plan*. Drafted and piloted by a 'steering group on flooding', it supports information exchange and cooperation between emergency managers, spatial planners, and water managers.

In terms of information-based instruments, *Flood hazard maps* and *Flood risk maps* are publicly available via the www.risicokaart.nl website. Public consultation sessions are supposed to be organized. In reality, no trace of events held on flood prevention can be found. However, the Dutch Government still considers that the criterion of the Flood Directive Article 9 requiring an 'active involvement of interested parties' has been met (Kaufmann et al. 2015). Finally, no standards instruments exist in the Netherlands.

To summarize, the Netherlands make a more diversified use of flood prevention instruments, especially via the water test incentive-based instrument which is a pillar of the flood prevention system.

As for France, it is worth noting that the country has a long tradition in terms of flood prevention. Dating back to the nineteenth century, it originated with the idea of floodplain preservation, the first flood-planning documents being introduced in 1935 (Billet 2000). Currently, the French flood-prevention system remains characterized by the importance of legal instruments in spatial planning and, to a certain extent, by a specific place given to economic instruments.

Within the array of legal instruments, one specific tool became particularly important, playing a key role in spatial planning. Since the 1980s, *Risk Prevention Plans* have been used by France to take into account natural and industrial risks, as part of a broader planning culture dominated by rigorous restrictions on construction in risk areas (Larrue et al. 2015). Introduced in 1995, the Flood Risk Prevention Plan reinforces this principle whilst reasserting the state's responsibility in this domain. At more local levels, provisions of the Flood-Risk Prevention Plan are to be embedded in *Spatial strategic plans* (*schémas de cohérence territoriale*) as well as *Local land-use plans* (*plan local d'urbanisme*).

In terms of economic tools, two important mechanisms can be found. As they are connected to each other, they can to some extent be considered one instrument. In some parts of France, at a watershed level, the *Action Program for Flood Prevention (PAPI)* stands as an integration-based programme, with strong incentives focused on prevention and specific funding streams allocated to it.

Whilst dealing with prevention, PAPI also aims to build a global management framework for flood issues, going beyond infrastructure work programmes such as dike construction plans. In this perspective, PAPI must articulate numerous different measures, playing at the same time on knowledge improvement, monitoring, and forecasting, information, planning, protection works, reduction of buildings' vulnerability, crisis management preparedness, and 'lessons learned' programmes. Its main resources come from the *National Fund for Major Natural Hazards*. Funded by taxes on home insurance contracts to the amount of €150 million per year, the fund is mainly funnelled into PAPI with outstanding revenues used to cover the so-called Flood Risk Prevention Plan (PPRI) provisions, including bans on new buildings as well as regulation through planning.

As for communication and information, one instrument should be noted. An obligation for all municipalities under the Flood-Risk Prevention Plan, the *Municipal Information Document on Major Hazards* is meant to explain to inhabitants what a risk is, which types can be found locally, and how to behave in case of disaster. Apart from that, there is no other incentive-based or best-practices instrument.

Prevention strategy is largely based on legal instruments. England and France display particularly hierarchical legal structures in the field of flood prevention (Table 6.3)

Table 6.3 Prevention strategy instruments.

	Prevention strategy		
	England	Netherlands	France
Legal instruments	National Planning Policy – Land Planning Permission – Flood Risk Assessment – Sequential test	Strategic spatial development plans – Spatial zoning plans	Risk Prevention Plan – Spatial strategic plans – Local land-use plan
Economic instruments	—	—	Action Programme for Flood Prevention / National Fund for Major Natural Hazards
Incentive-based instruments	—	Water test / Flood risk management plans	—
Standards & Best practices	—	—	—
Information & communication	Register of flood risk assets – Flood maps	Flood hazards maps – Flood risk maps	Municipal Information Document on Major Hazards

which is systematically diversified in the Netherlands with the water test incentive-based instrument. Many information-based instruments exist in the three countries to inform the general public about hazards and risks. However, these tend to be produced in a very top-down manner. Issued by governmental authorities – both national and local – they are then disseminated without any feed-back to citizens, business partners, communities, and so on, a feature which appears quite strongly in France for instance.

6.3.2 Mitigation, a trend lacking strong instruments

Within the three countries however, a significant shift can be observed over the period between the 1970s and the 1990s, with mostly defence-based flood-management strategies (1970s–1990s) reorganized following a new diversification principle. At the same time, a new emphasis was put on the development of non-structural measures such as mitigation (Fournier et al. 2016). In concrete terms, mitigation is applied to spatial planning developments through an array of adaptive measures. In urban spaces, these comprise of green roofs, living walls, permeable pavements, floor heights designs above flood level… In natural spaces, these range from peatland restoration and wetland creation to tree planting and restoration of riverside corridors. Flood storage areas and retention basins can also be implemented in both contexts. Interestingly, they often lead to multifunctional land-use models described as being 'adapted to' both floods and urban development.

In England, mitigation is a growing strategy. Although numerous and diversified instruments can be observed, these are not however dedicated directly to mitigation but to other purposes such as sewage and drainage. By reinterpreting the sewage and drainage rules through the objectives of flood adaptation, sewage and drainage rules can actually be recorded as mitigation measures.

In terms of legal instruments, mitigation measures clearly originated from drainage policies. Indeed, the rule of *'Keeping the drains clear'*, coming from common law and ensuring that properties do not evacuate water drainage into their neighbour's property, is considered as a step towards mitigation in English flood management (Alexander et al. 2015). In the same vein, *Sustainable Drainage Systems* are to be treated as an additional consideration falling within the existing planning system (Department for Environment, Food and Rural Affairs (Defra) (2014).

Finally, regarding incentive-based measures, *Non-statutory technical contracts* have been produced by the Department for Environment, Food and Rural Affairs (Defra) (2013) to support developers. On top of that, statutory consultees provide the necessary technical advice on major developments (Department for Communities and Local Government 2016). Government also encourages at-risk households to adopt property-level measures against flooding through the *Resist flooding incentive-based program*.

In the Netherlands, mitigation instruments for floods are concentrated on mechanisms such as rainwater run-offs and groundwater surpluses. There is a rather low degree of institutionalization. As for England, numerous instruments fall into the mitigation basket but these are placed under the umbrella of urban water management policy.

The 'retaining, storing, draining' watchword marks a shift from a technical approach based on sewer management to a more sustainable approach centred on the idea of a *Sustainable Urban Drainage System* (Kaufmann et al. 2015). Legal mitigation instruments are incorporated in the *Water Act* (Articles 3.5 and 3.6) whilst being supplemented with provisions on sewerage in the *Environmental Management Act*. In addition, spatial planning instruments are used to implement flood mitigation at the local level, for example through the adoption of coercive criteria such as paving/building percentages or storm water drains in *Spatial zoning plans*.

As it turns out, economic instruments do not prove really valuable. Although some exist to protect built properties such as houses, barns, or cellars from flooding via *Adjusting thresholds*, there is no legal instrument available for public actors to influence landowners' decisions in this respect.

This being said, no clear agreements exist either. The so-called incentives for mitigation prove to be in fact a mixture of preventive and defensive measures (Kaufmann et al. 2015). Increasing flood-mitigation capacity is mainly done through sewage and spatial planning instruments. Nevertheless, some non-binding standards can be found within the framework of the *National Administrative Agreement on Water Issues*. No communication, information-based, nor standards' instruments can be recorded either.

If we look at France, we see that the mitigation theme is infusing public discourses (Larrue et al. 2015; Fournier et al. 2016). However, it does not represent a strategy in itself, standing rather as a set of measures to support other strategies mainly associated again with the water sector.

Therefore, in order to embrace the whole set of French mitigation instruments, one needs to partly reinterpret the outputs of three legal instruments: the Water Management Master Plan (SAGE) which includes financial funds for River restoration, the PAPI which can implement dynamic retention of floods via bypass, ponds, or retention areas in rural zones, and finally the Flood Risk Prevention Plan (PPRI) which sometimes encourages developers to experiment with private or professional resilient buildings. Just as in England or in the Netherlands, local land-use plans can make recommendations for sustainable urban drainage systems, including green roofs, urban green spaces, and permeable pavements. Although at first sight these seem to be solutions for drainage, they can also be presented as mitigation measures.

In the three countries, mitigation instruments mostly consist of rainwater run-off and groundwater surpluses. There is a rather low degree of institutionalization (Table 6.4). As mitigation instrumentation is in many ways a reinterpretation of instruments from other policies, it appears quite difficult to clearly identify best practices and to communicate on them easily.

6.3.3 Preparation, a highly instrumentalized strategy

As it appears, the three countries under study are well equipped with preparation instruments. These can be divided into two sub-strategies: forecasting and crisis management.

In England, preparation strategy turns out to be highly instrumentalized. In terms of legal instruments, preparation is mainly organized at the national level, with some parts devolved to the local level. A *National Storm Tide Forecasting Service* established and operated by the Met Office (*Meteorological Office*) is coupled with a strategic policy framework also called the *National Flood Emergency Framework for England*. Contingency planning is organized by the 2004 and 2005 *Civil Contingencies Acts*. National-level instruments therefore appear to play a key role.

Table 6.4 Mitigation strategy instruments.

	Mitigation strategy		
	England	Netherlands	France
Legal instruments	Keeping the drains clear - Sustainable Drainage Systems -	Sustainable Urban Drainage System - Water Act - Environmental Management Act - Spatial zoning plans	Action Programme for Flood Prevention - Flood Risk Prevention Plan - Sustainable Urban Drainage Systems -
Economic instruments	—	Adjusting thresholds	Water Management Master Plan -
Incentive-based instruments	Non-statutory technical standards - Resist flooding incentive-based programme	National Administrative Agreement on Water Issues -	—
Standards & Best practices	—	—	—
Information & communication	—	—	—

However, economic preparation instruments such as the *Trading Fund for Commercial Business* are also quite powerful. In addition, the 1989 *Local Government and Housing Act* laid the foundations of the Bellwin Scheme, a strong centralized economic instrument with a budget provided by the government to cover unexpected losses. However, it is worth noting that such funding is capped in order to encourage local authorities and developers to act cautiously and build some degree of resilience as well as financial reserves for emergency expenditures (Penning-Rowsell and Wilson 2006).

In addition, several agreement-based instruments exist to propagate local flood warnings such as *Preparation contractual agreements* with the Environmental Agency and communities. At the local scale, members of the community may act as flood wardens. Their role is to provide locally sourced flood information as well as to assist in response efforts and ensure warnings reach vulnerable groups (Alexander et al. 2015).

Finally, communication also has important instruments, especially forecasting ones via the *Flood Forecasting Centre* for instance, a joint venture established in 2009 with the aim to provide forecasting for all. The Met Office also provides a *Public Weather Service* for England, offering free-of-charge forecasts to the general public. On top of that, a *National Severe Weather Warning Service* gives advance notice of weather likely to affect public safety, should it lead to flooding or bear some other risks. Information-based instruments also exist in crisis management with *Public informal networks* which may also act as propagators of 'unofficial flood warnings' (Parker and Handmer 1998) or *Local Resilience Forums* holding meetings on a regular basis to promote risk awareness amongst organizations and the general public.

In the Netherlands, Dutch policy makers started at some point to realize that flood risks could at best be minimized but never be eliminated. From then on, preparation has become more and more independent from the others flood strategies and is now characterized by two main features: it is mainly composed of legal and communication-based instruments and is also mostly organized by emergency planning.

As for legal instruments, so-called *Crisis Plans* (*crisisplannen*) specify operational measures as well as the way responsibilities are to be divided. These have to be aligned with the *Emergency Plans* (*calamiteitenplannen*) set out by each relevant regional water authority. Regarding communication, *Coordination Forums* are organized, bringing together water managers and emergency managers (V&W, VROM, and LNV 2009).

In France, preparation is mostly understood in terms of civil security. Instruments on flood risk policy are mainly produced by the state. According to the provisions of the 2004 Law reforming the civil security sector, two main instruments are part of the preparation strategy. At the national level, *Civil Security Response Organization Plan* (*Plan ORSEC*) stands as an emergency tool. It encompasses all types of risks and sets out very detailed action plans in case of disaster, including training and exercises. At the local level, *the Municipal Crisis Management Plan* (*Plan Communal de Sauvegarde*) is a document developed by municipalities. Articulated to the ORSEC Plan, its goal is to implement the latter at a local level, but this being an obligation for municipalities set out by the Flood Risk Prevention Plan.

Communication instruments mainly fall into the forecasting sub-strategy. A 24-hour monitoring and intervention team, the *Flood Forecasting Service* (SCHAPI) broadcasts flood alerts and informs people living alongside flooded rivers. It also provides cartographic representations of forecasts flagging out various levels of danger. This is made available to a wide public via the Vigicrues National Alert System website (www.vigicrues.gouv.fr).

In the three countries, the preparation strategy is characterized by two types of instrument: legal instrument and information-based instrument (Table 6.5). Legal instruments are mobilized for crisis management and information-based instruments are used for weather forecasting and public warning alerts. There is little room for innovation in terms of incentive-based or best practices, except in England where the range of existing instruments is more diverse.

6.3.4 Defence: A strategy relying on government-based instruments

Historically, the three countries within our survey have all put a strong emphasis on defence-orientated measures aimed at improving the capacity of watercourses or stopping floods through dredging, walls, dikes, or embankments. Nowadays, the costs of maintaining such flood defences are escalating, triggering a search for new solutions or compromises (Gralepois et al., 2016).

England for instance has been very proactive in its efforts to find alternatives to a strict defence approach. Two legal instruments exist that still maintain a defence strategy, the two main ones being incorporated in the *2010 Flood and Water Management Act* and in the *2009 Flood Risk Regulation Act*. The rest of the defence strategy comes from a mesh of legislation streams going from land drainage, water resources development, or environmental conservation to sustainable development and climate change. In England, there is no statutory right for flood protection. Permissive powers can be found, which means that there is no legally designed standard of protection. Economic

Table 6.5 Preparation strategy instruments.

	Preparation strategy		
	England	Netherlands	France
Legal instruments	National Storm Tide Forecasting Service – Strategic Policy Framework – Civil Contingencies Act – Local Government and Housing Act –	Crisis Plans – Emergency Plans	Civil security response organization – Municipal Crisis Management Plan
Economic instruments	Trading Fund for commercial Business – Bellwin Scheme –	—	—
Incentive-based instruments	Local flood warnings contractual agreements –	—	—
Standards & Best practices	—	—	—
Information & communication	Flood Forecasting Centre – Public Weather Service – National Severe Weather Warning Service – Public informal networks – Local Resilience Forums	Coordination Forum	Flood Forecasting Service

instruments are much more important. Capital for defence infrastructures is distributed via instruments based on cost-benefit analysis, an economic prioritization technique for decision making. There are no real incentive-based nor information-based instruments. Flood defence remains very much a state-controlled strategy involving few partnerships, even if discourses are evolving towards a more open management system.

In the Netherlands, flood defence strategy has always been very important. The dominant discourse claims the necessity to 'fight against water' via technical or engineering solutions (Ten Brinke 2007; Van den Brink 2009). Legal instruments are embedded within the Water Act. According to it, the government has a general public duty to establish Legal Safety Standards for primary and secondary defence structures. Such standards are set up by regional water authorities, which have their own tax system (Havekes et al. 2011) and prove to be relatively stable. At the national level, financial means are provided in the guise of specific funds provisioned by the Water Act. Therefore, defence in The Netherlands also proves to be a highly programmatic and technocratic sector, with little room made for information-based instruments or best practices.

French defence strategy is also centralized at the national level, even if some steps towards devolution can also be observed at the level of instruments. For instance, national-level authorities can delegate the management of some defence infrastructures like minor fluvial dikes. Legal and economic instruments in place depend on which local

Table 6.6 Defence strategy instruments.

	Defence strategy		
	England	Netherlands	France
Legal instruments	Flood and Water Management Act 2010 - Flood Risk Regulations.	Water Act	Legal Safety Standards for defence structures
Economic instruments	Cost-Benefit Analysis	—	—
Incentive-based instruments	—	Legal Safety Standards for defence structures -	—
Standards and Best practices	—	—	—
Information and communication	—	—	—

authorities is in charge of the defence structures, meaning that legal safety standards or investment amounts for example can vary from one entity to the other. This devolution trend from the state to local authorities is also starting to trickle down to individuals themselves. The preamble of the 2004 Act on Civil Security states for instance that 'citizens are responsible for their own safety'. Nevertheless, flood defence remains fully funded and regulated at the government level.

Over its long history in flood management, defence strategy appears rooted in expert coalitions and governmental top-down decision-making processes (Table 6.6). No place is given to best practices or information-based instruments. However, we see new trends developing. Devolution of powers gives a greater role to local authorities (Netherlands and France) or to communities (England). The economic crisis tends to decrease the amount of funding available, triggering the use of cost-benefit analyses (England) and a diversification of strategies (Netherlands and France).

6.3.5 Recovery instruments based on political will

Interestingly, recovery strategies and instruments can either rely mostly on private compensation via insurance companies or on national funds. Indeed, this mere observation is full of political implications.

In England for instance, flood insurance is provided as part of general household insurance covering buildings and their contents. It has always been provided via private insurance companies, operating on a pure market basis. Nevertheless, the Bellwin Scheme, a centralized government-funded approach, also provides funding for catastrophic losses in case of major unexpected flood events to reimburse local authorities. The 1989 Local Government and Housing Act sets out the key principles of the Bellwin Scheme, including covered incidents as well as thresholds and recoverable amounts. Otherwise, there is no incentive-based, best practice or information-based instrument.

As for the Netherlands, flood recovery strategy has received little attention over time and is only partly institutionalized. One can find private insurance against damages

Table 6.7 Recovery strategy instruments.

	Recovery strategy		
	England	Netherlands	France
Legal instruments	Local Government and Housing Act 1989	Calamities Compensation Act	Natural Disasters Fund (CAT-NAT)
Economic instruments	Bellwin Scheme - Private Insurance Market	Private Insurance Market	—
Incentive-based instruments	—	—	—
Standards and Best practices	—	—	—
Information and communication	—	—	—

from pluvial flooding but insurance for damages from fluvial flooding is only available to a limited extent. Under specific circumstances, flood damages can also be compensated for on the basis of the Calamities Compensation Act, a type of national solidarity fund that took effect in 1998. Just as in England, only legal and economic instruments can be found for flood recovery strategy.

Finally, recovery appears to be a characteristic pillar of the French welfare system. It is operated as a State domain through the Natural Disasters Fund (CAT-NAT) realized in 1982. From an operational viewpoint, the CAT-NAT is provided by insurance companies and their premiums. Insured clients are its main contributors. Therefore, this instrument is framed and regulated by a public decision-making process, for example a decree of natural disaster must be declared by an inter-ministerial commission. It is the only important instrument in French recovery strategy.

Recovery instruments illustrate a clear division between private insurance and national funds (Table 6.7). Although the alternative may be simple, the choice is tough in so much as politically speaking it translates into balancing private responsibility with public solidarity in the face of risks. The private/public alternative also appears particularly crucial in the way it contributes to determine who will – between private companies or government authorities – be the main legitimate actor for flood recovery.

Building on this comparative description as well as on the typology suggested by Lascoumes and Le Galès (2004), we will now start drawing a few conclusions by looking at the effects of instruments and strategical trends. We will then discuss their meaning in terms of political powers and governance in flood management.

6.4 Discussion. Political effects, power relations, and governance choices in flood management: What do flood instruments teach?

In the following section, our position will be explicitly controversial with the overall ambition to trigger a debate. As a matter of fact, flood-policy instruments are rarely studied from a socio-political angle. We therefore present our key results as starting points for further investigation. This part will present some of the emerging trends that contribute to shape flood instruments – and hence strategies, before unravelling their main implications in terms of public policies outcomes and finally in terms of flood-management governance choices in the European Union.

6.4.1 Emerging trends in flood instruments and strategies

At first sight, the range of instruments currently available in the sector of flood management appears quite large, spanning a whole diversity of tools except for best practices, which have not permeated European flood management practices yet. Regarding the five types of instruments we have reviewed previously (legal, economic, incentive-based, standards, and information-based), the classification presented in Table 6.8 shows how instruments can be articulated to the various strategies we have listed. Some instruments are chosen by the three countries when the three names appear in the table, whereas other instruments are only developed by one or no country.

Interestingly, legal instruments are ubiquitous in every strategy. In Europe, flood management policies used to be highly regulated by rules, especially in prevention and defence. In Europe, flood-management strategy relies more on procedural obligations than on common and shared challenges, even if preparation strategy is beginning to make a commitment to this end. As the conclusion of Priest et al. (2016) states: 'Strictly speaking, this implies that it is possible to satisfy requirements without making any substantial changes' (Priest et al. 2016, p. 15). With no surprise, legal instruments are thus a major source of outputs for every strategy we are looking at. What is in fact more noticeable is the diversification trend in instruments we see shaping up, especially in England and in the Netherlands. Economic instruments can be found in the three countries, even if England has completely integrated them for post-disaster strategies. Preparation, defence, and recovery are based on financial-based instruments. There is a general move towards more incentive-based instruments, even if France has not adopted this alternative yet. This type of instruments is often articulated to the perspective of a pay-off, i.e. 'if you follow the standards you will save money'. Finally, what we see in terms of communication-based instruments demonstrates quite clearly the serious lack of effort on this front. Only two strategies involve information-based instruments: prevention, with its traditional array of legal documents disseminated to citizens – and preparation, with the range of alert warning systems it relies on.

This being said, what can we take away from these first few observations as far as strategies are concerned? First, we can see that prevention and preparation are the top two strategies in terms of the number of instruments, those being mostly legal and information-based. They appear as central strategies for the three countries. As for the mitigation strategy, the influence of which is currently developing (Fournier et al. 2016), we see however that it generated relatively few instruments of its own. Most of the tools it relies on derive from sewage and drainage rules. Defence is still a strongly instrumentalized strategy, whereas recovery displays the fewest number of instruments.

Table 6.8 Classification of instruments per strategy.

	Prevention	Mitigation	Preparation	Defence	Recovery
Legal instruments	England – Netherlands – France	England – Netherlands – France	England – Netherlands – France	England – Netherlands – France	England – Netherlands – France
Economic instruments	France	Netherlands – France	England	England	England – Netherlands
Incentive-based instruments	Netherlands	England – Netherlands	England	England – Netherlands	
Standards and Best practices					
Information and communication	England – Netherlands – France		England – Netherlands – France		

6.4.2 A lack of alignment in flood instruments: What results in terms of flood public policies outcomes?

We observe two trends in flood instruments, which can be set out as contradictions in regards to European flood-management requirements: a lack of diversity and a lack of alignment amongst flood instruments. What does it teach in terms of flood public policies outcomes?

Even if diversity of strategies is said to be a strength since the 2007 European Flood Directive (Hegger et al. 2014; Priest et al. 2016; Directive 2007/60/EC_2007), the concrete implementation of a diversified range of instruments is not so active until now. In comparison with EU environmental policy instruments (Halpern 2010), flood management instrumentation is less innovative and less prolific in general. The lack of diversity of instruments gives at least two insights on evolution in flood-management strategies in Europe.

First, flood management in Europe still appears as strongly institutionalized. Whereas the Flood Directive asks for more adaptive governance (Priest et al. 2016), European flood management is still mostly organized amongst public actors, often at the head of quite pyramidal hierarchies. But we can discern a trend to devolution of power, in accordance with the European principle of subsidiarity, from national authorities to regional or municipal actors. Second, the sector revolves around a coalition of actors and interests which can be described as an 'expert' coalition: hydraulic expertise, civil engineering expertise, architects, and so on (Guevara and Gralepois 2015; Gralepois et al. 2016).

Those two arguments can explain why technical and legal instruments are so omnipresent. However, according to Halpern (2010), this is also true for environmental policy. The difference might lie in the importance given to participation. Indeed, in comparison with environmental policy instruments, and even with water management, flood management has not integrated the logic of integration and cooperation and appears very weak in this regard, except in England, where the role of citizens in flood management – especially in preparation – is increasing. If, in theory, flood management 'necessitates the involvement of (…) a broad range of public, private, and civil society actors' (Mees et al. 2016, p. 1), the study of instruments highlights that, except in England, there are still few participative instruments such as bottom-up communication-based instruments, guidelines, or best practices. Reciprocally, the power of legal rules and the number of top-down information-based instruments demonstrate that public authorities are still the main actors that drive and orient strategies. However, at the same time, authorities are asking individuals to be responsible for themselves during flood events. This section uncovers what looks like a strong contradiction lying at the core of European flood management, in so much as the objectives officially pursued by strategies cannot *de facto* be implemented by the instruments put in place to reach them.

Drawn on those findings, another conclusion can be made concerning the challenge of aligning flood strategies in Europe. This aim does not seem to have been reached, because instruments are not aligned themselves. Institutional views and academic debates in European flood management are dominated by a discourse positioning alignment of strategies as a necessity. Mixing them and ensuring their consistency appears as a determining factor for the success of twenty-first-century future policies.

This 'age of alignment' is reflected quite clearly in European policy, particularly in the EU Flood Directive (Directive 2007/60/EC). However, in pragmatic terms, it is generating substantial difficulties as coordinating the various strategies adds another layer of complexity to flood-management processes and potentially nurtures conflicts of interests (Hegger et al. 2014). Furthermore, there is actually no link between the instruments of the different strategies, each of which seeming to evolve on its own track separately from the others. Given that an alignment of strategies is achievable only if corresponding instruments are linked and coordinated, the truth is that as of today, strategies are neither aligned nor integrated. Hence, amongst the eight coordination mechanisms identified by Dieperink et al. (2016), we would like to add that the five flood-management strategies must also take into account the coherence of the instruments used, not just the actors' game, rules, or discourses.

6.4.3 Flood governance patterns Analysed Through the lens of instruments

Flood instruments choices lead to different governance patterns. We present our findings for each country pattern, with similarities and differences.

Historically, England developed flood management on the basis of legal instruments, whatever the strategy was. This being said, the country also chose to promote the role of local responsibilities and to give more importance to economic instruments (Table 6.9). These are developed quite proactively especially in a post-flood situation, in an approach clearly inspired by cost-benefit analyses. Market-based instruments are characteristic of the recovery strategy. As for communication-based instruments, these are not only considered as a form of top-down communication but also encompass community forums and local agreements. In a word, we can see that all the community-based initiatives as well as the flexibility and innovation oriented tools that the Flood Directive 2007 asks European Union members to implement are already developed in England. Drawing on the typology proposed by Lascoumes and Le Galès's (2004), England can be associated with a redistributive polity type as well as with a legitimacy model based on efficiency.

The Netherlands presents the most diversified pattern of instruments (Table 6.10), reflecting the Dutch multi-layered strategy in flood management (Kaufmann et al. 2016). The defence strategy is strong, however, ongoing possibilities of infrastructure failure and increased flood risks due to both climate change and economic development do not leave much choice to the Netherlands: other strategies have to be activated. Therefore, the Dutch flood-management pattern looks like a patchwork of instruments in each strategy, which hardly helps to associate it with any given polity or legitimacy types.

In France, flood management is mostly based on legal instruments (Table 6.11). Two important economic instruments are put in place for prevention and mitigation strategies and applied to the pre-flood situation. Two top-down information-based instruments can be found for prevention and preparation strategies. Examined through the lens of its instruments, French flood governance is marked by a strong social guardian state, based on a strong need for legitimacy. No particular effort is made in terms of participation or competitive mechanisms. No other special recent trend is noticed recently, except a trend of devolution towards local authorities for defence and preparation.

Table 6.9 Instrumentation pattern in England.

	Pattern by country: England				
	Prevention	Mitigation	Preparation	Defence	Recovery
Legal instruments	National Planning Policy – Land Planning Permission – Flood Risk Assessment – Sequential test	Keeping the drains clear – Sustainable Drainage Systems –	National Storm Tide Forecasting Service – Strategic Policy Framework – Civil Contingencies Act – Local Government and Housing Act	Flood and Water Management Act 2010 – Flood Risk Regulations 2009	Local Government and Housing Act 1989
Economic instruments	—	—	Trading Fund for commercial Business – Bellwin Scheme –	Cost-Benefit Analysis	Bellwin Scheme – Private Insurance Market
Incentive-based instruments	—	Non-statutory technical standards – Resist flooding incentive-based program	Local flood warnings – contractual agreements	—	—
Standards and Best practices	—	—	—	—	—
Information and communication	Flood risk assets Register – Flood maps	—	Flood Forecasting Centre – Public Weather Service – National Severe Weather Warning Service – Informal Public Networks – Local Resilience Forums	—	—

Table 6.10 Instrumentation pattern in the Netherlands.

	Pattern by country: The Netherlands				
	Prevention	Mitigation	Preparation	Defence	Recovery
Legal instruments	Strategic spatial development plans – Spatial zoning plans	Sustainable Urban Drainage System – Water Act – Environmental Management Act – Spatial zoning plans	Crisis Plans – Emergency Plans	Water Act	Calamities Compensation Act
Economic instruments	—	Adjusting thresholds	—	—	Private Insurance Market
Incentive-based instruments	Water test – Flood risk management plans	National Administrative Agreement on Water Issue	—	Legal Safety Standards for defence structures	—
Standards & Best practices	—	—	—	—	—
Information & communication	Flood hazard maps – Flood risk maps	—	Coordination Forum	—	—

Table 6.11 Instrumentation pattern in France.

	Pattern by country: France				
	Prevention	Mitigation	Preparation	Defence	Recovery
Legal instruments	Risk Prevention Plan – Spatial strategic plans – Local land-use plan	Action Programme for Flood Prevention – Flood Risk Prevention Plan – Sustainable Urban Drainage Systems	Civil security response organization – Municipal Crisis Management Plan	Legal Safety Standards for defence structures	Natural Disasters Fund (CAT-NAT)
Economic instruments	Action Programme for Flood Prevention / National Fund for Major Natural Hazards	Water Management Master Plan –	—	—	—
Incentive-based instruments	—	—	—	—	—
Standards & Best practices	—	—	—	—	—
Information & communication	Municipal Information Document on Major Hazards	—	Flood Forecasting Service	—	—

More generally, crossing the results of the three countries, we see that flood management policies – currently characterized by legislative and regulatory instruments – are mostly associated with social guardian state features as well as with a type of power elected to represent a notion of the general interest (Table 6.2). This legitimacy given to the state is understood by the need to defend national interests, security, and equal treatment of citizens, but the debate must be open: how can flood management be opened up to more participatory, more local instruments without calling into question solidarity and national security?

6.5 Conclusion

We have presented in this chapter the critical role played by policy instruments in the way flood management was dealt with in three European countries. Even if a move can be observed towards more diversified toolkits favouring flexibility and participation just as in England for instance, flood management is still mostly based on legal and economic instruments. From a governance perspective, this reflects the strong and undisputed legitimacy of public actors, leaving little room to accommodate more incentive-based instruments or best practices regulation. This creates a gap between dominant discourses on strategies on the one hand and the instruments that are actually put in place to implement them on the other hand. At the same time, many voices are raised in scientific literature encouraging a diversification of flood-risk strategies that may lead to an increase of the overall resilience to floods. In this regard, alignment is supposed to make European countries willing to be more resilient. This shows that flood instruments – hence strategies – are neither adaptive nor aligned: there is still a long a way to go if flood management wants to achieve pragmatically its resilience objectives.

Notes

1. All EU Member States have to implement the 2007/60/EC Directive on the assessment and management of flood risks – also designated as the Floods Directive. Its requirements range from identifying areas at risk of flooding, drawing flood risk maps outlining the various types and levels of risks, as well as developing action plans describing the key measures to be taken in case of major flood events.
2. STARFLOOD is an acronym standing for 'STrengthening And Redesigning European FLOOD risk practices' associated with the following subtitle: 'Towards appropriate and resilient flood risk governance arrangements'. This project (2012–2016) received funding from the European Union's Seventh Programme for Research, Technological Development and Demonstration under Grant Agreement No 308364. It aimed at designing policies to deal more effectively with flood risks. STAR-FLOOD is carried out by an international consortium of six countries.
3. As defined in the Planning Guidance set by the Department for Communities and Local Government, 'local Plans set out a vision and a framework for the future development of the area, addressing needs and opportunities in relation to housing, the economy, community facilities and infrastructure – as well as a basis for safeguarding the environment, adapting to climate change and securing good design'. Department for Communities and Local Government, Planning Guidance: https://www.gov.uk/guidance/local-plans--2

References

Alexander, M., Priest, S., Micou, A.P. et al. (2015). *Analysing and evaluating flood risk governance in England – Enhancing societal resilience through comprehensive and aligned flood risk governance*. London: STAR-FLOOD Consortium. Flood Hazard Research Centre, Middlesex University.

Billet, P. (2000). La zone inondable: essai de typologie de la délimitation juridique des zones soumises au risque 'inondation'. *Géocarrefour* 75 (3): 245–254.

Dehousse, R. (2004). La méthode ouverte de coordination: Quand l'instrument tient lieu de politique. In: *Gouverner par les instruments*, 331–356. Paris: Presses de Sciences Po.

Department for Communities and Local Government (2016). Property level flood resilience local authority guidance. https://www.bitc.org.uk/sites/default/files/berg_-_property_level_flood_resilience_local_authority_guidance.pdf (accessed 26 February 2019).

Department for Environment, Food & Rural Affairs (DEFRA) and Environment Agency (EA) (2017). Flood risk assessment: the sequential test for applicants, Guidance plan. www.gov.uk/guidance/flood-risk-assessment-the-sequential-test-for-applicants. (accessed 5 March 2019) London, DEFRA and EA.

Department for Environment, Food and Rural Affairs (DEFRA) (2013). Managing flood risk, Third Report of Session 2013–2014, Volume 1, HC330. London: House of Commons, DEFRA.

Department for Environment, Food and Rural Affairs (DEFRA) (2014). Working with natural processes to reduce flood risk. R&D Framework Science Report. Report – SC130004/R2. Bristol: Environment Agency.

Dieperink, C., Hegger, D.L.T., Bakker, M.H.N. et al. (2016). Recurrent governance challenges in the implementation and alignment of flood risk management strategies: a review. *Water Resource Management* 30: 4467–4481.

Directive 2007/60/EC (2007). Directive 2007/60/EC of the European Parliament and of the Council of 23 October 2007 on the assessment and management of flood risks. Brussels: European Parliament.

Environmental Protection Department (2007). Review of the International Water Resources Management Policies and Actions and the latest practice in their environmental evaluation and strategic environmental assessment. Final Report November 2007 The Hague: Netherlands Government.

Fitova, T. (2014). Market-based instruments for flood risk management: a review of theory, practice and perspectives for climate adaptation policy. *Environmental Science & Policy* 37: 227–242.

FloodSite (2005). Language of risk, Project FloodSite Integrated Flood Risk Analysis and Management Methodologies, EU FP7 Floodsite Project, Report: T32–04-01 March 2005. Brussels: FloodSite.

Fournier, M., Larrue, C., Alexander, M. et al. (2016). Flood risk mitigation in Europe: how far away are we from the aspired forms of adaptive governance? *Ecology and Society* 21 (4): 49. https://doi.org/10.5751/ES-08991-210449.

Fuchs, S., Keiler, M., Sokratov, S.A. et al. (2013). Spatiotemporal dynamics: the need for an innovative approach in mountain hazard risk management. *Natural Hazards* 68 (3): 1217–1241.

Gaudin, J.P. (2004). *L'action publique*. Paris: Presses de Sciences Po/Dalloz.

Gilissen, H.K., Alexander, M., Matczak, P. et al. (2016). A framework for evaluating the effectiveness of flood emergency management systems in Europe. *Ecology and Society* 21 (4): 27. https://doi.org/10.5751/ES-08723-210427.

Gralepois, M. (2012). *Face au risque d'inondation*. Paris: Éditions d'ULM.

Gralepois, M., Larrue, C., Wiering, M. et al. (2016). Is flood defense changing in nature? Shifts in the flood defense strategy in six European countries. *Ecology and Society* 21 (4): 37. https://doi.org/10.5751/ES-08907-210437.

Guevara, S. and Gralepois, M. (2015). L'adaptation aux risques d'inondation façonnée par les métiers de la ville. *Développement durable et territoires* 6 (3): http://developpementdurable.revues.org/11014 (accessed 26 February 2019).

Halpern, C. (2010). Governing despite its instruments? Instrumentation in EU environmental policy. *West European Politics* 33 (1): 39–57.

Halpern, C. and Le Galès, P. (2008). Europeanisation of policy instruments, NewGov Working Paper 09/D06, http://www.eu-newgov.org (accessed 20 March 2017).

Havekes, H.J.M., Koster, M., Dekking, W. et al. (2011). *Water Governance. The Dutch Regional Water Authority Model*. The Hague: UvW.

Hegger, D.L.T., Driessen, P.P.J., Dieperink, C. et al. (2014). Assessing stability and dynamics in flood risk governance. An empirically illustrated research approach. *Water Resource Management* 28: 4127–4142.

Hegger, D.L.T., Green, C., Driessen, P.P.J. et al. (2013). Flood risk management in Europe: Similarities and differences between the STAR-FLOOD consortium countries. Utrecht: STAR-FLOOD Consortium.

Hood, C. (1986). *Tools of Government*. New York: Chatham House Publishers.

Howlett, M. (1991). Policy instruments, policy styles, and policy implementations: national approaches to theories of instrument choice. *Policy Studies Journal* 19 (2): 1–21.

Howlett, M. (2010). *Designing Public Policies: Principles and Instruments*. Abingdon, Oxon and New York: Routledge, 2011.

Johnson, C.L., Tunstall, S.M., and Penning-Rowsell, E.C. (2005). Floods as catalysts for policy change: historical lessons from England and Wales. *International Journal of Water Resources Development* 21: 561–575.

Kaufmann, M., Lewandowski, J., Choryński, A. et al. (2016). Shock events and flood risk management: a media analysis of the institutional long-term effects of flood events in the Netherlands and Poland. *Ecology and Society* 21 (4): 51. https://doi.org/10.5751/ES-08764-210451.

Kaufmann, M., Van Doorn-Hoekveld. W.J., Gilissen, H.K. et al. (2015). Drowning in safety. Analysing and evaluating flood risk governance in the Netherlands. Utrecht: STAR-FLOOD Consortium.

Kundzewicz, Z.W., Krysanova, V., Dankers, R. et al. (2016). Differences in flood hazard projections in Europe – their causes and consequences for decision making. *Hydrological Sciences Journal* https://www.tandfonline.com/doi/full/10.1080/02626667.2016.1241398.

Larrue, C., Bruzzone, S., Lévy, L. et al. (2015). *Analysing and evaluating Flood Risk Governance in France: from State Policy to Local Strategies*. Tours, France: STAR-FLOOD Consortium.

Lascoumes, P. and Le Galès, P. (2004). *Gouverner par les instruments*. Paris: Presses de Sciences Po.

Lascoumes, P. and Le Galès, P. (2007). Understanding public policy through its instruments. *Governance* 20 (1): 1–144.

Le Bourhis, J.-P. and Lascoumes, P. (2014). Les resistances aux instruments de gouvernement. Essai d'inventaire et de typologie des pratiques. In: *L'instrumentation de l'action publique* (ed. C. Halpern, P. Lascoumes and P. Le Galès), 493–520. Paris: Presses de Sciences Po.

Mees, H., Crabbé, A., Alexander, M. et al. (2016). Coproducing flood risk management through citizen involvement: insights from cross-country comparison in Europe. *Ecology and Society* 21 (3): 7. https://doi.org/10.5751/ES-08500-210307.

Ministerie van Verkeer en Water (V&W), Ministerie van Volkshuisvesting, Ruimtelijke Ordening en Milieu (VROM), and Ministerie van Landbouw, Natuur en Voedselkwaliteit (LNV), (2009). *Beleidsnota Waterveiligheid 2009–2015*. http://publicaties.minienm.nl/documenten/beleidsnota-waterveiligheid-2009-2015 (accessed 5 March 2019).

November, V. (2010). Recalcitrance of risks: a management failure?, Society for Social Studies of Science (4S) conference, University of Tokyo, 25–29 August 2010.

November, V. (2016). Risk, planning and socio-spatio-temporal differenciation. In: *Ecological Risks and Disasters – New Experiences in China and Europe* (ed. L. Peilin and L. Roulleau-Berger), 43–57. London: Routledge.

Parker, D.J. and Handmer, J.W. (1998). The role of unofficial flood warning systems. *Journal of Contingencies and Crisis Management* 6 (1): 45–60.

Penning-Rowsell, E. and Wilson, T. (2006). Gauging the impact of natural hazards: the pattern and cost of emergency response during flood events. *Transactions of the Institute of British Geographers* 31 (2): 99–115.

Priest, S.J., Suykens, C., Van Rijswick, H.F.M.W. et al. (2016). The European Union approach to flood risk management and improving societal resilience:lessons from the implementation of the Floods Directive in six European countries. *Ecology and Society* 21 (4): 50. https://doi.org/10.5751/ES-08913-210450.

Sterner, T. (2003). *Policy Instruments for Environmental and Natural Resource Management.* Washington, DC: Resources for the Future.

Ten Brinke, W.B.M. (2007). *Land in zee: de watergeschiedens van Nederland.* Diemen: Veen Magazines.

V&W (2000). *3e Kustnota, Traditie, Trends en Toekomst.* Den Haag: V&W.

Van den Brink, M. (2009). *Rijkswaterstaat on the Horns of a Dilemma.* Delft: Eburon.

Whipp, R. (2001). Strategy: organizational. In: *International Encyclopedia of the Social & Behavioral Sciences*, vol. 22 (ed. N.J. Smelser and P.B. Baltes), 15151–15154. Amsterdam: Elsevier.

II.3

Lessons from Cases of Flood Governance

7

A House of Cards: The Challenge of Establishing Societal Resilience to Flooding Through Multi-Layered Governance in England

Meghan Alexander[1] and Sally Priest[2]
[1]*School of Earth & Ocean Sciences, Cardiff University, Wales, United Kingdom*
[2]*Flood Hazard Research Centre, Middlesex University, London, United Kingdom*

7.1 Introduction

The need to enhance societal resilience to flooding in a way that is sustainable and adaptive to changing risks is widely recognized across Europe (Driessen et al. 2016). This requires a holistic and cohesive arrangement of policies and legislation, dedicated resources and coordination across a diverse set of actors (public, private, and civil society), and allied policy sectors, across multiple scales. In short, underlying arrangements of governance through which Flood-Risk Management (FRM) is delivered have the potential to facilitate or constrain efforts to improve societal resilience. Given the multi-layered nature of flood-risk governance, this argument can be conceptualized as a house of cards, whereby the integrity of each layer may be either threatened or strengthened by those below as well as above.

This chapter examines this concept in the context of England (UK) where an estimated six million properties are currently at risk from fluvial, coastal, or surface water flooding (Ramsbottom, Sayers, and Panzeri 2012). This figure is predicted to increase under future climate scenarios; for instance, the 860 000 residential properties currently

Facing Hydrometeorological Extreme Events: A Governance Issue, First Edition.
Edited by Isabelle La Jeunesse and Corinne Larrue.

exposed to a significant likelihood of fluvial or tidal flooding (i.e. an annual probability of 1.3%) could rise by 40% with 2 °C of warming (Sayers et al. 2015). Correspondingly, the UK Climate Change Risk Assessment 2017 specifies that more action is needed to mitigate flood risks to communities, businesses, and infrastructure to reduce long-term vulnerability to climate change (CCC 2016; HM Government 2017). Using English flood-risk governance as an example, this chapter assesses the relationship between current governance (as of September 2017) and societal resilience to fluvial and surface water flooding, with a focus on spatial planning, flood defence and mitigation, emergency management, and flood insurance.

7.2 Deciphering multi-layered governance

7.2.1 Theoretical background

Governance research is fundamentally concerned with the interaction between various actors (public, private, and civil society) in the realization of collective goals (Lange et al. 2013). Proponents of this field either conceive governance as the transition from centralized, state-led decision making towards non-hierarchical forms of governance ('government to governance'; Swyngedouw 2005); or as a broadening of the *modes* of governance (e.g. [de]centralized, public-private, and so on), through which goals are realized (Driessen et al. 2012). Although a common theme in governance research pertains to the multi-layered nature of governance, scholars are divided in their articulation of this. From the European school of thought, multi-layered governance is typically expressed through a limited number of nested jurisdictional levels (e.g. international, national, to sub-national scales) (Marks and Hooghe 2004). In contrast, others assert that task-specific arrangements of governance may emerge that are not limited to jurisdictional levels but defined by the task and scale of the problem. Thus, multiple centres of decision making may co-exist independent of one another, leading some to coin the expression of polycentric governance (Marks and Hooghe 2004; Ostrom and Janssen 2004).

Within the context of climate change and natural hazards, multi-level governance is seen as a key characteristic for adaptation and resilience (Daniell et al. 2011; Pahl-Wostl et al. 2013), and regarded by many as best placed to anticipate and respond to the inherent uncertainty, complexity, and ambiguity associated with risk (Renn, Klinke, and van Asselt 2011). Delineating the concept of risk governance (Renn et al. 2011), Alexander, Priest, and Mees (2016a, p. 39) define *flood-risk governance* and Flood-Risk Governance Arrangements (FRGA) as embodying 'the actor networks, rules, resources, discourses, and multi-level coordination mechanisms through which FRM is pursued'. Adopting this definition, this chapter now turns to the study of flood-risk governance in England.

7.2.2 Overview of flood risk governance in England

English flood-risk governance has evolved in a piecemeal way through punctuated evolution, with long periods of stability accompanied by episodes of rapid incremental change (Johnson, Tunstall, and Penning-Rowsell 2005). Whilst key FRM strategies (i.e. to prevent, prepare, respond, and recover) have been established for around 65 years,

Box 7.1 *English risk management authorities (RMAs) identified in the* Flood and Water Management Act 2010

Environment Agency (EA) – public body with strategic and operational oversight for FRM
Lead Local Flood Authority (LLFA) – Unitary Authorities or County Councils
Internal Drainage Boards (IDBs) – local public authorities established in areas of special drainage need in England, chiefly in low-lying areas.
District Councils
Highways Agency
Water Companies (regulated by the Water Services Regulation Authority, Ofwat)

these have become increasingly coordinated within an overarching holistic risk-based approach (Alexander et al. 2016b; Gilissen et al. 2016a). Simultaneously, FRM has become further aligned to other allied policy domains including spatial planning, environment, climate change, water, civil contingencies, welfare, and private market (re) insurance. As such, the network of actors and landscape of 'rules' outlined in both policy and legislation is notably complex, with responsibilities divided across different tiers of government, public bodies, the quasi-commercial water sector, private market insurers, and civil society. In addition to Risk Management Authorities (RMAs) (Box 7.1), riparian property and land owners must also act in accordance with common law (Environment Agency [EA] 2013).

Multi-layered flood risk governance is displayed through clearly-defined jurisdictional levels and hierarchical distributions of responsibilities. Indeed, responsibilities devolved locally (e.g. in spatial planning), must be consistent with national policy. Likewise, under its membership agreement with the European Union, national FRM policy and domestic legislation must also comply with EU law (under the European Communities Act 1972).[1] Correspondingly, the EC Floods Directive 2007 has been transposed into domestic law through the Flood Risk Regulations 2009 and Flood and Water Management Act 2010. Although public–private partnerships do exist and are increasingly being promoted in England, these are not autonomous of these jurisdictional tiers of governance. Recognizing this multi-layered structure, this chapter examines the ways in which this appears to influence societal resilience to flooding.

7.3 Methodology

Efforts to evaluate societal resilience are challenged by the multitude of theoretical perspectives (e.g. engineering, human ecology, political economy, and social-ecological systems) and contested standpoints on what constitutes resilience (Gunderson and Holling 2002; Walker et al. 2004; Folke 2006; Gallopín 2006). Whilst these are not reviewed in depth here, broadly speaking resilience is either conceptualized as a measure of the capacity to (i) resist, (ii) return to normality, (iii) absorb, or (iv) adapt, (or a combination of these), in response to an external shock or ongoing stress.

Table 7.1 Benchmarks for determining the extent to which flood risk governance supports societal resilience.

Evaluation criteria	Benchmarks of success
Capacity to resist	• Measures/projects/or governance arrangements shown to have enhanced the capacity of the social-environmental system to reduce the likelihood and/or magnitude of flooding
Capacity to absorb and recover	• Diversity of measures/projects/or FRM strategies to address risk in a holistic way (i.e. from the likelihood of occurrence [resistance] to the potential range of consequences) • Bridging mechanisms to support integration/coordination between different governance levels • Use of measures/projects/FRM strategies is multi-layered to address risk at different spatial and temporal scales • Measures/projects/or governance arrangements have enhanced social-environmental resilience by reducing consequences, enabling the system to absorb and/or quickly recover
Capacity to adapt	• Opportunities for learning are evident and there is evidence that 'lessons learned' have been implemented • Opportunities are created for innovation and experimentation • The legal framework or legal instruments / plans and programmes are subject to periodic review in order to incorporate new information about climate change and floods • Balance between legal certainty and adequate legal flexibility to enable adjustments • Evidence that future risks/uncertainty (e.g. climate change) are factored into decision-making

Source: Modified from Alexander et al. (2016a).

In the absence of an agreed definition of resilience, Alexander et al. (2016a) argue that in the context of flooding, societal resilience can be isolated into three inter-related facets, relating to societal capacity to *resist, absorb and recover*, and *adapt*. In turn, these can be translated into benchmarks for evaluating flood risk governance (Table 7.1). Although the resistance aspect is often theoretically contested, the authors defend this in the context of public policy, whereby resistance (i.e. capacities to defend and mitigate the likelihood and magnitude of flood hazards) are often framed within resilience discourses (e.g. Defra/EA 2011).

Adopting this framework, this research draws from extensive document analysis of historic and current policy and legislation between 1930 and September 2017. These findings were complemented by 61 semi-structured interviews conducted between December 2013 and April 2015 with past and present flood-risk professionals operating at different scales. Interviewees were purposively sampled to elaborate on distinct aspects of English flood-risk governance, as well as eliciting broader opinions on existing strengths and weaknesses, and specific views of societal resilience. Interviews were recorded, transcribed, and subject to qualitative thematic analysis.

7.4 Flood-risk governance and implications for societal resilience

The following discussion reflects on the extent to which multilevel flood-risk governance influences the construction and distribution of societal resilience in England. In this regard, the research findings focus specifically on key aspects of the English FRGA relating to spatial planning, flood defence and mitigation, emergency management, and flood insurance.

7.4.1 Spatial planning

Flood risk is one of a number of material considerations in spatial planning, enacted under the National Planning Policy Framework (DCLG 2012) as well as formal legislation primarily surrounding Town and Country Planning (e.g. Town and Country Planning Act 1990; The Town and Country Planning (Development Management Procedure) (England) Order 2015). Mechanisms in spatial planning, (e.g. Sequential and Exception Tests) essentially act to minimize the exposure of people and property to flooding. Whereas the Sequential Test aims to steer new development towards areas with the lowest probability of flooding, the Exception Test outlines conditions for regulating development when this is not possible (established under Planning Policy Statement 25; DCLG 2006). Underlying these, Strategic Flood Risk Assessments, produced by Local Planning Authorities (LPAs), outline flood-risk zones against which Local Plans are informed (see paragraph 100 in DCLG 2012). Planning conditions (such as minimum floor heights above the estimated flood level and use of flood resilient materials) are outlined in national policy and planning practise guidance (DCLG 2012; Defra 2017), EA Standing Advice, local byelaws, or specific advice provided by the LPA or EA for individual planning applications.

However, the effectiveness of spatial planning to prevent inappropriate development is threatened by increasing demands placed upon LPAs and the EA at a time of stretched resources, which mean that not all planning applications (particularly smaller developments or adjustments to existing properties) are subject to risk assessments (Grant and Chisholm 2015). Indeed, the EA only has a duty to comment on major developments and in areas with a greater than 1% annual probability of fluvial flooding. Whilst very few planning applications are approved contrary to EA advice, development in areas with a greater than 1% annual likelihood of flooding appears to be increasing and could exacerbate future flood risk (e.g. increased dependency on flood defences and increased exposure to the insurance industry: CCC 2017). Simultaneously, a recurring theme from interviews with FRM practitioners was the issue of enforcement, or lack thereof, and concerns that this could result in the incremental encroachment of piecemeal development. Enforcement of planning conditions is currently part of the discretionary powers of the LPA, but few planning applications are refused in practice, particularly given the general presumption in favour of development (Bell, McGillivray, and Pederson 2013).

In light of Government housing targets, there is a need to better embed the principles of adaptation within spatial planning to reduce the build-up of future risk. Various suggestions were proposed by interviewees, such as incentivizing property owners to invest in property-level resistant and resilience measures (Defra 2016), through grant schemes or insurance mechanisms. Indeed, there is considerable emphasis in England on fostering household and community ownership of risk responsibility (Nye, Tapsell, and Twigger-Ross 2011). However, we contend that risks should not be simply passed onto homeowners, but that those undertaking and profiting from new developments should retain some responsibility and liability.

On this front, some advance has been made in the context of Sustainable Urban Drainage Systems (SUDS). In an effort to mitigate the build-up of surface water flood risk, SUDS are treated as an additional planning consideration and guided by non-statutory technical standards (Defra 2015). Under this revised policy, it is now the duty of a developer to establish a regime for maintaining SUDS, with a clear division of responsibilities (e.g. assigned to the developer, property owner, water companies, local authority, outsourced management company, or combination thereof). For major developments, Lead Local Flood Authorities have acquired new responsibilities as statutory consultees (under the Development Management Procedure Order 2015; also see DCLG 2015). In some regards this has been an important step forwards in terms of embedding surface water management into the planning process.

However, there are also significant concerns. Under Schedule 3 of the Flood and Water Management Act 2010 it was envisioned that SUDS would be a statutory requirement for future development to be overseen and maintained by SUDS Approving Bodies (SABs). However, this was never translated into practice due to difficulties of establishing national standards for SUDS and operational clarity on their functioning. Moreover, there were concerns that this additional governance layer could delay development and conflict with housing needs and economic recovery (House of Commons 2015). Now that SUDS are no longer a legal requirement, enforcement is an expressed concern. Furthermore, the success of SUDS is often dependent upon the effective implementation of a 'chain' of approaches and a series of complementary measures which, when considered as a whole, are able to deliver the reduction of surface water. This conflicts with the case-by-case nature of the application approval process and could lead to a piecemeal approach to drainage. The requirement for SUDS also applies to schemes of 10 properties or more, thus the impact of small developments could have a cumulative effect on surface water run-off (House of Commons 2015). Moreover, research has revealed weaknesses in the transposition of national non-statutory guidance into Local Plans (CCC 2017). Whilst it is still too soon to fully evaluate the impact of this policy change, these are key aspects of spatial planning that could undermine capacities to resist surface water flooding and thus warrant long-term monitoring.

7.4.2 Flood defence and mitigation

Fundamental to societal capacity to resist flooding is the assemblage of measures and schemes to alleviate the likelihood and magnitude of flood hazards, either by acting to resist (i.e. defence) or accommodate water (i.e. mitigation). In England, this is enacted through predominantly centralized modes of governance, with some elements of decentralization and interactive governance (as defined by Driessen et al. 2012; Alexander et al. 2016b). The latter has emerged with the introduction of Partnership Funding in 2012, which stipulates that centrally-administrated Grant-in-Aid (GiA) finances are accompanied by funding sourced at the local level, via Local Authorities, the private sector, and/or civil society (Defra 2011a).

In theory, this funding approach enables more schemes to be developed than in the past, however the completion of these projects is dependent upon securing local funding, as well as satisfying an 8:1 cost–benefit ratio (i.e. demonstrating £8 benefit for every £1 of government spend). A good example of this is in the Lower Thames (south east England), where Partnership Funding has played a key role in the approval of the River Thames Scheme (RTS) after almost two decades of failing to meet a favourable cost–benefit ratio. However, £50m must still be secured by local authorities and other partners in order for the scheme to be implemented (Environment Agency 2014a). Moreover, there are significant challenges with raising this money, especially as distribution of benefits is varied across four Lead Local Flood Authorities. This tension is summarized in the quote below:

> *Before Partnership Funding the RTS would not probably get off the block with the national funding system, [due to the high cost of it]. Partnership Funding opens up the opportunity to reassess the Scheme ... but it has been one of the reasons why little progress has been made since the approval of the RTS in 2011, as efforts have been put mostly in working towards a funding strategy. (Senior representative, Environment Agency)*

This example highlights multiple ways through which societal resilience is not only supported, but constrained through flood-risk governance, and linked to *inequalities* across spatio-temporal scales. Temporally, the evolution of FRM policy and strategy for allocating public money to fund defence works, has at different times prevented, constrained, and enabled the RTS to emerge. From a spatial perspective, as a multi-scale initiative involving the construction of a flood alleviation channel as well as property-level products, the Scheme also delivers variable standards of protection to those living in the Lower Thames area. The latter is an inevitable outcome given the absence of statutory safety standards.

Although standards of protection may vary across England, the existing line of defence has proven successful at times of floods (e.g. Environment Agency 2014b). Furthermore, a six year spending programme has committed £2.3bn towards 1400 defence schemes by 2021 (Defra 2014a), and was further supplemented in response to significant flood events in 2015/16 to £2.5bn (CCC 2017). In an effort to deliver a socially-just approach to funding, a positive deprivation bias is also embedded into the funding calculator (Defra 2011a) to ensure that communities 'least likely to be able to contribute towards the cost of a flood defence scheme and less able to recover after a flood without additional support from the state' (former National-level policymaker), are not disadvantaged.

To further enhance capacities to resist, Government recently launched a National Flood Resilience Review, from which £12.5m was allocated to increase the stock of temporary flood defences (e.g. demountable barriers, high volume pumps) to enhance national readiness to flooding (HM Government 2017). As part of this, particular attention has been paid to the resilience of critical infrastructure (electricity, water, and telecoms), with sectoral commitments to assess and put in place temporary defences for at-risk assets. At the property scale, Defra has also outlined its vision for normalizing the uptake of property resilience measures (Bonfield 2016). This has included £11m (combined central GiA and external sources) investment in property-level measures. Whilst this represents a significant advance compared to other EU countries (Mees et al. 2016), it continues to be a fraction of what is required for effectively managing residual flood risk and appears to be hampered by a lack of awareness and lack of financial incentives from the insurance sector (see Section 4.4; CCC 2017).

Beyond capacities to resist, adaptive approaches are also advocated through the use of 'managed adaptive' approaches to enable the adjustment of defence measures to reflect changing conditions such as climate change (Environment Agency 2016; CCC 2017). Future concerns are also integrated within Catchment Flood Management Plans to support strategic decision-making over a 50–100 year timescale. Another crucial benchmark for adaptive capacity refers to innovation opportunities, of which there are numerous examples. For instance, the River Aire Flood Alleviation Scheme in Leeds (north-east England) will be the first time movable weirs are used in the UK for a flood management function (Leeds City Council 2017).

However, this research has also identified some weaknesses within this aspect of English flood risk governance. First, there was a strong consensus amongst interviewees that funding for defence maintenance is critical. Nationally, a short-fall in the budget for defence maintenance has been widely reported (NAO 2014), meaning that existing defences are not necessarily being maintained to their optimum standards (Grant and Chisholm 2015). In turn, this could undermine England's ability to resist flooding. At the time this research was conducted, the revenue budget for defence maintenance (previously reviewed on an annual basis) had been protected until 2021 (HM Treasury 2015), however there is some uncertainty as to what will happen following the current spending programme. In order to sustain this facet of resilience it is critical that this revenue continues to be protected and given equal consideration to capital funding for new defences.

Another shortcoming in this aspect of governance relates to Partnership Funding and efforts to reduce dependency on public monies. Defra's plan to attract £600 m from private funders over the six year spending period has been called into question, especially as only £40 m of the £148 m in the initial period came from sources beyond local government (Efra Committee 2015). Whilst the Income Tax, Trading, and Other Income Act 2005 (as amended) outlines provisions for tax relief for businesses contributing to Partnership Funding, the limited involvement of the private sector indicates the need for additional incentive mechanisms and research to inform a strategy to promote private sector contributions.

7.4.3 Flood emergency management

Flood emergency management is enveloped in the broader construct of 'emergency' within civil contingencies legislation, namely the Civil Contingencies Act 2004 and the Civil Contingencies Act (Contingency Planning) Regulations 2005 (as amended). This statutory framework outlines the duties of Category 1 Responders (e.g. emergency services, EA, Local Authority, and key health authorities), who are each subject to the full set of civil protection duties to assess, plan, advise, and respond to emerging risks. Acting as 'cooperating bodies', Category 2 Responders (e.g. utility companies, transport operators, Health and Safety Executive, and NHS Clinical Commissioning Groups), have a duty to cooperate and share information and support Category 1 Responders. For flooding, these responsibilities are also outlined in the National Flood Emergency Management Framework (Defra 2013).

There are a number of governance features that are evaluated positively in terms of supporting capacities to absorb and recover from floods. First, a range of activities, such as the dissemination of flood warnings, evacuation, provision of rest centres and temporary housing, not only minimize the risk posed to life, but also encourage behaviours that minimize the damages caused by flooding and in turn prompt faster recovery. Second, there are numerous provisions to ensure the effectiveness of emergency response arrangements. Under the CCC Regulations 2005, Category 1 Responders are required to form Local Resilience Forums (LRF)[2] and attend regular meetings to facilitate multi-agency, joined-up working (Cabinet Office 2011). The development of multi-agency emergency plans and Community Risk Registers, which are essential for proportional risk planning, are facilitated through the LRF. For floods specifically, responders are required to produce a Multi-Agency Flood Plan (Defra 2011b). Recovery planning is also acknowledged as equally important and guided by additional non-statutory guidance (Cabinet Office 2013). These planning documents are regularly tested through exercises which are periodically up-scaled to simulate large-scale emergency events and test the effectiveness of higher-tiers of strategic decision-making as responsibilities shift according to principles of subsidiarity (Cabinet Office 2013). In relation to this, it should be noted that national (and local) exercises are also a tool for enhancing community awareness of risk, such as Exercise Watermark in March 2011 (Exercise Watermark Review Team 2011).

Another facet of resilience concerns the capacity to adapt, which is evidenced through the practice of debriefing and identifying lessons from emergency incidents. This process of critical reflection and constructive learning aims to enhance future response capabilities. Also supporting adaptive capacity at the local scale, certain Category 1 Responders (namely local authorities and the EA) engage with the community to encourage the development of local flood action plans, helping to cultivate self-reliance and adaptive behaviours (Environment Agency 2012). Indeed, such efforts to actively involve civil society are seen as a key strength of the English system (Mees et al. 2016).

Few weaknesses with this aspect of governance were revealed. In fact, Gilissen et al. (2016b) compare the effectiveness of Flood Emergency Management Systems in several EU Member States and present England as an optimal example of emergency planning, arrangements for institutional learning, and community preparedness. However, a potential weakness of the English system relates to the inclusion of the voluntary sector, which is highly variable. Whilst there is no formal obligation to involve voluntary organizations, such networks can help to 'push' warning messages and 'pull' potentially vulnerable individuals towards the authorities in advance of an emergency or major incident (Cabinet Office 2008). Moreover, better working relationships with the voluntary sector (such as their participation within the LRF), could help support response and recovery efforts at times of long-duration flood events or successive flooding where resources become stretched.

7.4.4 Flood insurance

The capacity to absorb and financially recover from flooding is supported through the provision of flood insurance via a private market system. Flood insurance is typically provided as a standard part of composite policies, moreover, buildings insurance is compulsory for those with mortgages, which has created a situation of high market penetration. To ensure access to affordable insurance, a not-for-profit reinsurance fund,

Flood Re, was introduced in April 2016. This new approach enables the formal cross-subsidization of high risk properties and places a cap on insurance premiums (Defra 2014b; Flood Re 2016). This is regarded as a transitionary arrangement to assist the move towards risk-reflective pricing over the next 25 years. In turn, it is anticipated that this will incentivize homeowners to invest in property-level measures to minimize the likelihood of flooding.

However, there was widespread concern amongst interviewees that more could be achieved within the insurance sector to promote adaptation at the household scale. To some extent, the insurance industry provides an incentive mechanism to minimize development on the floodplain or at least encourage adaptive designs by excluding properties built after 1 January 2009 (CCC 2017). However, there continues to be a 'return to normality' culture, which is reinforced through standard reinstatement techniques post-flooding, as opposed to resilient reinstatement. Furthermore, formal incentive mechanisms to promote the uptake of property-level measures are absent from Flood Re. Whilst it is anticipated that the transition to risk-reflective pricing will incentivize autonomous adaptation on behalf of the property owner, in the meantime, the premium cap somewhat undermines this (Surminski and Eldridge 2014). Despite the recent publication of the Property Flood Resilience Action Plan (Bonfield 2016) Flood Re arguably removes the financial incentive to invest in such measures (CCC 2017).

The Association of British Insurers (ABI) and government have produced a Flood Risk Report template for homeowners to declare resistance and resilience measures to their insurance provider, although it is not clear how such measures will be rewarded – in fact, there is a stated caveat as follows; 'Please note that using flood protection measures does not guarantee reduced insurance premiums and excesses' (see EA 2017). The first Transition Plan (to be reviewed every five years) specifies an intention to investigate the role that Flood Re might play in incentivizing the uptake of property-level measures (Flood Re 2016), but it remains unclear what progress has been made to date (CCC 2017).

From the perspective of multilevel governance, Flood Re demonstrates how national-led modes of public–private governance have the potential to both facilitate and constrain societal resilience at the property scale. However, not everyone in high risk areas is eligible for the scheme; this includes small businesses and properties owned by buy-to-let landlords. Thus the conditions placed on Flood Re also have implications for the distributional consequences of flooding and raise issues of social justice (also see Penning-Rowsell and Priest 2015).

7.5 Reflections on the 'house of cards' of flood risk governance

This analysis has highlighted the various ways through which different facets of societal resilience (resistance, absorption, and adaptation) are shaped by English flood-risk governance. In turn, this can inadvertently lead to variations in the construction of resilience across spatio-temporal scales and different groups in society. Overall, this analysis reveals that there are considerable strengths to the English system. In particular, risk is addressed holistically through a diversified suite of FRM strategies that act to minimize the frequency and magnitude of flooding, as well as exposure and consequences should floods occur. These are enacted at multiple scales, whilst being coordinated through national policy to facilitate a strategic and nationally-consistent approach to FRM. However there are also weaknesses that threaten societal resilience, such as enforcement in spatial planning, encouraging private sector investment in defence infrastructure, and incentivizing property-level measures through the use of insurance mechanisms.

Like a house of cards, this research has drawn attention to the role played by multi-layered governance in both constructing and undermining aspects of flood resilience. To strengthen the stability of the overall structure, and solidify the foundations for resilience, governance interventions are clearly required. In light of the increasing threat posed by climate change, amongst other risk-enhancing factors (e.g. population growth, urbanization), and growing consensus that extreme flood events will become more frequent, insights from this type of research can play an important role in efforts to establish flood resilient societies.

Notes

1. Following the European Union membership referendum on 23 June 2016, the UK decided to withdraw its membership from the EU. At the time of writing, the European Union (Withdrawal) Bill 2017–2019 passed its second reading in the House of Commons on 11 September 2017 and awaits further debate from the Committee of the Whole House. The Bill outlines provisions for saving EU-derived domestic legislation and states: 'EU-derived domestic legislation, as it has effect in domestic law immediately before exit day, continues to have effect in domestic law on and after exit day'.
2. A total of 42 LRFs exist in the UK and are defined by the boundaries of Police Areas (with the exception of London, where sub-Regional Resilience Forums are based on groupings of London Boroughs).

References

Alexander, M., Priest, S., and Mees, H. (2016a). A framework for evaluating flood risk governance. *Environmental Science and Policy* 64: 38–47.

Alexander, M., Priest, S., Micou, A.P. et al. (2016b). Analysing and evaluating flood risk governance in England – Enhancing societal resilience through comprehensive and aligned flood risk governance. London: STAR-FLOOD Consortium. Flood Hazard Research Centre, Middlesex University.

Bell, S., McGillivray, D., and Pederson, O.W. (2013). *Environmental Law*. Oxford: Oxford University Press.

Bonfield, P. (2016). Property flood resilience action plan. https://www.gov.uk/government/uploads/system/uploads/attachment_data/file/551615/flood-resilience-bonfield-action-plan-2016.pdf (accessed 9February 2019).

Cabinet Office (2008). *Identifying People Who are Vulnerable in a Crisis – Guidance for Emergency Planners and Responders*. London: Cabinet Office.

Cabinet Office (2011). *The Role of Local Resilience Forums: A Reference Document*. London: Civil Contingencies Secretariat. Cabinet Office.

Cabinet Office (2013). *Emergency Response and Recovery: Non-Statutory Guidance Accompanying the Civil Contingencies Act 2004*. London: Cabinet Office.

Committee on Climate Change (CCC) (2016). *UK Climate Change Risk Assessment 2017 Synthesis Report: Priorities for the Next Five Years*. London: Committee on Climate Change https://www.theccc.org.uk/wp-content/uploads/2016/07/UK-CCRA-2017-Synthesis-Report-Committee-on-Climate-Change.pdf (accessed 28 September 2017).

Committee on Climate Change (CCC) (2017). Progress in preparing for climate change. 2017 Report to Parliament. June 2017. London: Committee on Climate Change https://www.theccc.org.uk/wp-content/uploads/2017/06/2017-Report-to-Parliament-Progress-in-preparing-for-climate-change.pdf (accessed 28 September 2017).

Daniell, K.A., Máñez Costa, M.A., Ferrand, N. et al. (2011). Aiding multi-level decision-making processes for climate change mitigation and adaptation. *Regional Environmental Change* 11: 243–258.

Department for Communities and Local Government (DCLG) (2006). *Planning Policy Statement 25: Development and Flood Risk*. London: DCLG.

Department for Communities and Local Government (DCLG) (2012). *National Planning Policy Framework*. London: DCLG.

Department for Communities and Local Government (DCLG) (2015). *Further Changes to Statutory Consultee Arrangements for the Planning Application Process. Government Responses to Consultation*. London: DCLG.

Department for Environment, Food and Rural Affairs (Defra) (2011a). *Flood and Coastal Resilience Partnership Funding: An Introductory Guide*. London: Defra.

Department for Environment, Food and Rural Affairs (Defra) (2011b). *Detailed Guidance on Developing a Multi-Agency Flood Plan*. London: Defra.

Department for Environment, Food and Rural Affairs (Defra) (2013). *National Flood Emergency Management Framework*. London: Defra.

Department for Environment, Food and Rural Affairs (Defra) (2014a). *Reducing the Risks of Flooding and Coastal Erosion: An Investment Plan*. London: Defra.

Department for Environment, Food and Rural Affairs (Defra) (2014b). *Government Response to the Public Consultation on the Flood Reinsurance Scheme Regulations*. London: Defra.

Department for Environment, Food and Rural Affairs (Defra) (2015). Sustainable Drainage Systems: Non-statutory technical standards for sustainable drainage systems. London: Defra. https://www.gov.uk/government/uploads/system/uploads/attachment_data/file/415773/sustainable-drainage-technical-standards.pdf (accessed 28 September 2017).

Department for Environment, Food and Rural Affairs (Defra) (2016). *The property flood resilience action plan*. https://www.gov.uk/government/uploads/system/uploads/attachment_data/file/551615/flood-resilience-bonfield-action-plan-2016.pdf (accessed 9 February 2019).

Department for Environment, Food and Rural Affairs (Defra) (2017). *Flood risk assessment: Local planning authorities*. https://www.gov.uk/flood-risk-assessment-local-planning-authorities (accessed 9 February 2019).

Department for Environment, Food and Rural Affairs (Defra) and Environment Agency (2011). *The National Flood and Coastal Erosion Risk Management Strategy for England*. London: The Stationary Office.

Driessen, P., Dieperink, C., van Laerhoven, F. et al. (2012). Towards a conceptual framework for the study of shifts in modes of environmental governance – experiences from The Netherlands. *Environmental Policy and Governance* 22: 143–160.

Driessen, P.J., Hegger, D.L.T., Bakker, M.H.N. et al. (2016). Toward more resilient flood risk governance. *Ecology and Society* 21 (4): 53.

Environment Agency (2012). *Flood Plan Guidance for Communities and Groups*. Bristol, UK: Environment Agency.

Environment Agency (2013). *Living on the Edge. A Guide to Your Rights and Responsibilities of Riverside Ownership*. Bristol: Environment Agency.

Environment Agency (2014a). Flood and coastal erosion risk management programmes of work for the financial year 2015 to 2016. Programme of flood and coastal risk management projects.

Bristol: Environment Agency. https://www.gov.uk/government/publications/programme-of-flood-and-coastal-erosion-risk-management-schemes (accessed 9 February 2019).

Environment Agency (2014b). Written evidence, Winter floods inquiry. March 2014. Submitted to the Environment, Food and Rural Affairs Committee (Efra committee). http://www.publications.parliament.uk/pa/cm201415/cmselect/cmenvfru/240/240.pdf (accessed 9 February 2019).

Environment Agency (2016). *Adapting to Climate Change: Advice to Flood & Coastal Risk Management Authorities*. Bristol: Environment Agency.

Environment Agency (2017). *Property flood protection: Flood risk report*. Bristol: Environment Agency. https://www.gov.uk/government/publications/property-flood-protection-flood-risk-report (accessed 9 February 2019).

Environment, Food and Rural Affairs Committee (Efra) (2015). Defra performance in 2013–14. Eighth report of Session 2014–15. London: Efra http://www.publications.parliament.uk/pa/cm201415/cmselect/cmenvfru/802/802.pdf (accessed 9 February 2019).

Exercise Watermark Review Team (2011). Exercise watermark: Final report September 2011. https://www.gov.uk/government/publications/exercise-watermark-final-report (accessed 9 February 2019).

Flood Re (2016). Transitioning to an affordable market for household flood insurance: The first Flood Re transition plan. www.floodre.co.uk/industry/how-it-works/transition-plan (accessed 9 February 2019).

Folke, C. (2006). Resilience: the emergence of a perspective for social-ecological systems analyses. *Global Environmental Change* 16 (3): 253–267.

Gallopín, G.C. (2006). Linkages between vulnerability, resilience, and adaptive capacity. *Global Environmental Change* 16: 293–303.

Gilissen, H.K., Alexander, M., Beyers, J.-C. et al. (2016a). Bridges over troubled water – towards an interdisciplinary framework for evaluating the interconnectedness within fragmented domestic flood risk management systems. *Journal of Water Law* 25 (1): 12–26.

Gilissen, H.K., Alexander, M., Matczak, P. et al. (2016b). A framework for evaluating the effectiveness of flood emergency management systems in Europe. *Journal of Ecology and Society* 21 (4): 27.

Grant, L. and Chisholm, A. (2015). *Breaking the Bank? Funding for Flood and Coastal Erosion Risk Management in England*. London: The Chartered Institution of Water and Environmental Management (CIWEM).

Gunderson, L. and Holling, C.S. (2002). *Panarchy: Understanding Transformations in Human and Natural Systems*. Washington DC: Island Press.

HM Government (2017). *UK Climate Change Risk Assessment 2017*. London: The Stationary Office.

HM Treasury (2015). *Spending Review and Autumn Statement 2015*. London: The Stationary Office.

House of Commons (2015). Living with Water. Report from the Commission of Inquiry into flood resilience of the future. London: HMSO. http://cic.org.uk/admin/resources/cic9605-appg-report.pdf (accessed 28 September 2017).

Johnson, C.L., Tunstall, S.M., and Penning-Rowsell, E.C. (2005). Floods as catalysts for policy change: historical lessons from England and Wales. *International Journal of Water Resources Development* 21: 561–575.

Lange, P., Driessen, P.P.J., Sauer, A. et al. (2013). Governing towards sustainability – conceptualising modes of governance. *Journal of Environmental Policy and Planning* 15 (3): 403–425. https://doi.org/10.1080/1523908x.2013.769414.

Leeds City Council (2017). Leeds flood alleviation scheme Phase 1. Leeds: Leeds City Council.www.leeds.gov.uk/residents/Pages/Leeds-Flood-Alleviation-Scheme-Phase-1.aspx (accessed 9 February 2019).

Marks, G. and Hooghe, L. (2004). Contrasting visions of multi-level governance. In: *Multi-Level Governance* (ed. I. Bache and M. Flinders), 15–30. Oxford, UK: Oxford University Press.

Mees, H., Crabbe, A., Alexander, M. et al. (2016). Coproducing flood risk management through citizen involvement – insights from cross-country comparisons in Europe. *Ecology and Society* 21 (3): 7.

National Audit Office (NAO) (2014). Strategic flood risk management. HC 780, Session 2014–15, 5 November 2014. London: NAO. www.nao.org.uk/wp-content/uploads/2014/11/Strategic-flood-risk-management.pdf (accessed 8 February 2019).

Nye, M., Tapsell, S., and Twigger-Ross, C. (2011). New social directions in UK flood risk management: moving towards flood risk citizenship. *Journal of Flood Risk Management* 4: 288–297.

Ostrom, E. and Janssen, M.J. (2004). Multi-level governance and resilience of socio-ecological systems. In: *Globalisation, Poverty and Conflict* (ed. M. Spoor), 239–259. Dordrecht: Kluwer Academic Publishers.

Pahl-Wostl, C., Becker, G., Knieper, C. et al. (2013). How multilevel societal learning processes facilitate transformative change: a comparative case study analysis on flood management. *Ecology and Society* 18 (4): 58.

Penning-Rowsell, E.C. and Priest, S. (2015). Sharing the burden of increasing flood risk: who pays for flood insurance and flood risk management in the United Kingdom. *Mitigation and Adaptation Strategies for Global Change* 20 (6): 991–1009.

Ramsbottom, D., Sayers, P., and Panzeri, M. (2012). Climate change risk assessment for floods and coastal erosion sector, Defra project code GA0204. http://randd.defra.gov.uk/Default.aspx?Module=More&Location=None&ProjectID=15747 (accessed 8 February 2019).

Renn, O., Klinke, A., and van Asselt, M. (2011). Coping with complexity, uncertainty and ambiguity in risk governance: a synthesis. *Ambio* 40: 231–246.

Sayers, P.B, Horritt, M., Penning-Rowsell, E. et al. (2015). Climate Change Risk Assessment 2017: Projections of future flood risk in the UK. Research undertaken by Sayers and Partners on behalf of the Committee on Climate Change. London: Committee on Climate Change.

Surminski, S. and Eldridge, J. (2014). Flood insurance in England – an assessment of the current and newly proposed insurance scheme in the context of rising flood risk. Centre for Climate Change Economics and Policy, Working Paper No.161. Grantham Research Institute on Climate Change and the Environment, Working Paper No.144 London: LSE.

Swyngedouw, E. (2005). Governance innovation and the citizen: the Janus face of governance-beyond-the-state. *Urban Studies* 42 (11).

Walker, B., Holling, C.S., Carpenter, S.R. et al. (2004). Resilience, adaptability and transformability in social- ecological systems. *Ecology and Society* 9(2)5

8

Understanding Dutch Flood-Risk Management: Principles and Pitfalls

Mark Wiering
Radboud University, Nijmegen, The Netherlands

8.1 Introduction

This chapter discusses the governance of flood risks in a specific case, the Netherlands, by way of explaining the core normative principles that rule the Dutch. To be able to understand the Dutch approach it is important to sketch briefly the historical role of floods and flood management in the country, elaborating some more on how sequential floods left their scars in society and – at the same time – created an ever stronger public domain. Flood-risk management evolved historically into a core task of government with an important role of public responsibilities, leaving only small tasks and duties to businesses and civil society. But this is gradually changing with the increasing pressure on the collective public system, due to climate change as well as increasing economic or human risks but also because of the evolving expectations of contemporary governance. Global principles and supranational regulation urge for diversification of strategies and for more responsiveness of governance agencies.

When we look solely at the physical environment of The Netherlands, it is probably the most vulnerable country in Europe in terms of flood risks. About 26% of the territory lies below sea level, and at some low lying polders houses are still built at 6m beneath the sea level. An additional 29% of the country is flood prone to floods from rivers or the sea. In these vulnerable, low lying areas, nine million people are living and working – and two-thirds of GDP is earned there (OECD 2014, p. 54). However, when we include the system of Dutch flood-risk management, it probably is one of the safest countries in the world. There are billions spent on infrastructural works and other policies and the Dutch are making reservations to prepare for and adapt to climate change in the form of a so-called Delta Programme.

In short, Dutch flood-risk governance is strong, but society is still vulnerable. We will first discuss different principles and characteristics of Dutch flood-risk governance.

Facing Hydrometeorological Extreme Events: A Governance Issue, First Edition.
Edited by Isabelle La Jeunesse and Corinne Larrue.

We will start with the concept of *public interest* and water management: the strong idea of flood safety as a unitary, collective public interest leads to a call for national solidarity. This *solidarity* principle is very strong, but also has some limitations, both for those that fall out of the system and for international solidarity and cross-border governance. As a consequence, (internal) solidarity strengthens the national public infrastructure for flood risks, reinforcing certain national flood-risk strategies, which influence the third principle: the principle of *resilience*.

We could state that new principles are putting older principles to the test. The concept of resilience, for example, is widely discussed in the literature (Folke et al. 2010; Wiering et al. 2015; Hegger et al. 2016) and most scholars emphasize the need for diversification of flood risk strategies and adaptability to reach a state of resilience. But this is also problematic, because some countries have chosen a limited selection of strategies; in The Netherlands the strategy of flood defence is key. We will further discuss some of the tensions between principles.

The series of public choices that the Dutch made, and the normative governance principles that are the product of this institutional history, are increasingly challenged by new driving forces, of which climate change is the most important. The Dutch discuss adaptation to climate change intensively, but have difficulties in taking another course because of these strong principles that rule and guide them. The OECD (2014) has pointed to the so-called Dutch (public) awareness gap: although there is a strong governmental system in place, the people are no longer aware of their greatest societal risks: flooding. We will discuss these principles and challenges – and give possible answers in the sections that follow. But first we have to give a sketch of the battle against floods.

8.2 Historical background

The Netherlands is famous for its water management and those who care to have a closer look at the history of the country will be not surprised at all. Although there were people living in the Southern and higher part of the Lowlands (now Southern Limburg) from 5000 BCE onwards, it was only around 500 BCE that the first residents created artificial dwelling mounds (OECD 2014, p. 53) in the lower lying northern parts, concurred from the North Sea. At that time it was difficult to detect whether territories were belonging to land or sea (Rooijendijk 2009). The first Romans that came to extend their empire looked upon those poor Dutch as 'ship wreckers on low water'. From then on, the Dutch, slow but steadily, and bit by bit, created the country as it is now.

The Middle ages were full of disastrous sea storm floods, a few each century. But around the twelfth century the responses to these floods, taking shape in dike rings and polders, became more organized and professional. Major steps were the creation of the very first water boards (that still exist and are now formally referred to as regional water authorities). Then again, there were new drawbacks, such as the famous Saint Elisabeth Flood of 1404 that hit parts of Flanders and Holland and left its mark especially by destroying a newly created polder near Dordrecht, and swallowed tens of villages never to recover again. This flood is still visible in the form of the *Biesbosch* nature conservation area.

These shock events, relating to overflowing rivers, North Sea surges and the floods creating an inner sea (*Zuiderzee*) lead eventually to 'fighting back' by way of a system of polders, dike rings, dams, and sluices to new land areas re-captured from the sea.

8.3 The concept of public interest

The first principle we would like to put forward has to do with the conception of the public interest, more specific pointing at a unitary conception of public interest (Alexander 2002). Because of the crucial importance of flood-risk management as a societal condition (in fact even more important than economic prosperity, as without flood-risk management there would hardly be any land or economic prosperity) it is a very special public interest. Alexander (2002) nicely categorized different methods of constituting and processing the public interest, referred to as conceptions of the public interest. The conception of a public interest is both defined by the nature of the interest itself and by the political system of which it is part. As an example, some general societal interests will be important for all countries in most cases, e.g. protection from enemies, and consequently need some sort of public organization (in the form of military defence). But some countries will find it crucial to do this by themselves and others see no evil in cooperation with other allied countries. In other cases, e.g. environmental care or nature conservation, although not principally different in nature, the priority for and processing of the interests depends strongly on the political system that deals with these issues. Alexander (2002) mentions four core conceptions of the public interest: the utilitarian, unitary, deontic, and dialogical concept, and relates these to the core political philosophy of a country.

For example, England has a different stance towards flood-risk management as a public interest, still finding it important and giving it political priority sometimes, but the state has essentially no legal obligation to protect its citizens against floods, and crucial decisions are made – at least much more than in the Netherlands – on the basis of cost–benefit analysis of the specific flood risks and areas involved (thus, a utilitarian concept). This explains that Somerset in England (during Christmas 2013- January 2014) could be flooded for a while, as no dike protection was calculated to be necessary, in other words, protection was considered too costly. In the Netherlands there *is* a legal obligation for the state institutions (on different levels) to protect its inhabitants and all people are protected in principle. Although also the Dutch reason from a nation-wide cost–benefit analysis on a very basic level, (e.g. there is stricter level of protection in the densely populated and economically crucial West than in the East), this is not specified in differentiated cost–benefit calculations; generally speaking the principle is that everybody living within the dike-ring system should be equally protected considering the flood risks involved (Wiering and Winnubst 2017). In popular terms, 'all Dutch inhabitants are equal' in terms of flood protection, but within the bounds of the hydro-technological system. The collective system for flood risks represents a claim above all other claims. This basis leads to a strong role of the principle of solidarity as a national governance principle.

8.4 Solidarity and subsidiarity

When it comes to flood risks, all Dutch citizens are asked to act in solidarity with each other. This is reflected by the vast budgets for flood infrastructure (dikes, dams, embankments, coastal defence, sluices, pumps, and so on) and for the flood-risk management of national and regional layers of government. Dutch flood-risk governance, briefly sketched, is functionally organized in two core levels of responsibility for – mainly – fluvial floods: the national level and the regional water authorities (or water boards) next to responsibilities for pluvial management and sewerage of municipalities and regional coordination by provinces.

The general concept of solidarity was already in use in Roman times and refers to 'the unlimited liability of each individual member in a family or other community to pay common debts' (Keessen et al. 2016, referring to Bayertz 1999). In a more contemporary use of the term, there is no need to see it as an obligation or as a balanced reciprocity: it can be aimed at helping those in need or the less fortunate in a country, and it can be voluntary. In other words, solidarity can be one-sided (e.g. I am helping the poor, without obligation, expecting nothing in return) or two-sided (I am helping the poor, but I expect that all people do this, and therefore all contribute to the whole system – with the aim that in the end we are all better off – therefore creating an obligation and some reciprocity).

Water solidarity in The Netherlands is organized via the tax system. The specialized and functional water authorities are financed by general national taxes for national institutions such as the national water agency (*Rijkswaterstaat*), and the taxes of the regional water authorities. Obviously, this is only possible if there is some agreement on solidarity between citizens. This solidarity and the accompanying taxes were originally only a local and regional basis with early water boards making dikes and embankments and creating the first polders – keeping feet dry. This lead to an expansion of smaller water boards over the years (around 2600, even crossing national borders). At the end of eighteenth century and beginning of the nineteenth century the French ruled The Netherlands and centralized (in part) the flood risk system and created *Rijkswaterstaat*. This eventually generated a national system of taxes and compulsory solidarity on a national basis (Keessen et al. 2016). The regional system however still exists and is an institutional regime of its own:

> *Water management costs [in total] are around 6.5 billion Euro per year. Landowners (farmers), nature reserve organisations, businesses and residents pay a water management tax to their water board. Democratic legitimacy is guaranteed through the representation of various categories of stakeholders in the governing bodies of water boards (Van Rijswick and Havekes 2012). In line with the [adagio] 'no taxation without representation', each group can elect the water authority board members and is eligible to take a seat in the board. The water boards' tasks of water quantity control and flood protection are thus carried out on the basis of 'stakeholder participation' and the 'benefit principle'. The financial mechanism exemplifies the importance of the solidarity principle: those who benefit from the activities of the water authority pay taxes for its services and have a (proportionate) say in the assembly in return (the stake-pay-say' triplet). (Wiering et al. 2015)*

A few comments are in place here: first, the regional water system is increasingly loaded with tasks as (similar to France, but less consequential perhaps) there is devolution of responsibilities: Dutch central and regional authorities clearly share the financial burdens ever more. Second, the regional 'stake, pay, say' system is also increasingly under discussion, as it does not mean that 'the more you pay, the more you can say'; households pay higher taxes in greater amounts than agriculture and industry. This is even seen as disproportionate, testing the solidarity between the urban population and rural agriculture.

Triggered by the financial crisis and the rise of neoliberal thoughts on the role of national governance (rolling back the national level), it was agreed upon, in 2012, that the nation state would reduce its contribution to 50% to pay for measures to improve primary flood defences (Keessen et al. 2016). The regional water authorities had to fill this financial gap. In response to this devolution, they decided that the financial system for water management should still be based on collective solidarity: 40% is paid by *all* regional water boards, 10% is left to an individual responsibility of the specific region (Keessen et al. 2016). Actually this is remarkable, because some regions are much more flood-prone than others. We concluded elsewhere that there is both two-sided solidarity (reciprocal and obligatory; to the benefit of all) and one-sided solidarity when regions hardly benefit from expensive flood risk management in vulnerable areas – or perhaps better said: indirect solidarity between the regional water authorities. Despite these differences, this mechanism of regional solidarity was formalized through law. It is clear that the water boards cling to the idea of connecting regions through solidarity.

Interestingly, the call for and importance of solidarity was also reproduced in the debate on Dutch climate adaptation. In 2005, the Dutch Senate asked for a committee on adaptation in light of societal vulnerability to changes in the climate. This so-called new Delta-Committee (*Delta Commissie* 2008), was named after the first Delta committee, which was installed after the storm surge of 1953 (the last disastrous sea flood that hit the South West of the country and made more than 1800 flood victims). The Committee sketched different scenarios of floods and other risks expected to be caused by climate change effects – with floods obviously being the country's most important risk. The Committee made it clear that to *avoid* a disaster, the Dutch should be proactive and strengthen their flood risk infrastructure (amongst other issues as preserving fresh water collectively) in order to be 'climate proof'. Climate change, and especially climate adaptation, in Dutch eyes, is therefore almost synonym with flood risk and flood disasters. In fact, some authors emphasize that climate change and the exogenous nature of climate change (as a force coming from outside) was consciously framed as a threat to the nation's prosperity as a whole. The framing of climate change strengthened to idea of a national threat and the need for a national scope of institutions (Boezeman, Vink, and Leroy 2013; Vink et al. 2013).

> *In 2010, the Dutch government responded to the committee's recommendation by establishing the Delta Programme, the Delta Commissioner, and the Delta Fund, which are all legally rooted in the new Delta Law [incorporated in the Water Act of 2009] (…) and operate under the Minister of Infrastructure and Environment. Interestingly, the Delta Programme is explicitly based on solidarity in the context of flood safety because solidarity contributions from the Delta Fund are justified by fear of societal disruption in case of a flood event (…). Disruption on a national scale, in case of an event, appears to be the main justification for generous solidarity contributions from the state to a region. (Delta Commission 2008) (Keessen et al. 2016)*

Consequently, continuous solidarity was asked from the Dutch people, now not only fighting the water, but even 'fighting climate change' – in the sense of preparing for it.

In sum, the principle of solidarity is very strong for the Dutch – in this specific case of water management – it glues it all together. It is reproduced at the national level, by the climate adaptation response of a 'Delta' programme and related 'Delta institutions'. It is also reproduced at a regional level, by connecting (and not dividing) regional water authorities. The tax system and the work of the water agencies, both regional and national, are generally considered to be very effective and efficient (OECD 2014).

But this collective approach has its downsides, as we shall see later in the 'Challenges and pitfalls' section.

8.5 Resilience

The next principle that is important for contemporary flood-risk management is the concept of resilience. The idea is that our societies are considered to be social-ecological systems that should be reducing their vulnerability (to climate change, most often). When they do, they can be called resilient. We can conclude, as an attempt to summarize the literature, that resilience as a term is ambivalent, as it refers to 1) both stability (keeping a system in its place) and change (being adaptable) – and 2) both flexibility and robustness. An extreme interpretation of resilience referring to the latter debate, is postulated by Liao (2012) who sees resilience as the capacity to bend and return (like the grain bending in the wind) mainly emphasizing flexibility and redundancy of the system that should be preserved, others state that resilience is also about more static, robust features of the system and emphasize resistance and robustness (Klijn, van Buuren, and van Rooji 2004; Folke 2006; Davoudi 2012). The ambivalence here is that resilience cannot be both flexible and stiff.

A pragmatic answer to these important but conceptual debates is that resilience, as a common and popular term, indeed suggests flexibility: bend and return. However, related to a social-ecologic system as a whole – coping with a changing natural environment – it predominantly refers to *whatever is needed* to avoid increasing vulnerability of the system: a resilient society is one that deals with its vulnerability the best it can (Levina and Tirpak 2006) and the way it can – with the best means available. In other words, resilience is a contingent concept and different for every society. In answer to these core ambivalences of the concept of resilience, we can state that both stability and change are part of resilience. Change and adaptability is needed on lower scales of the socio-ecological (societal) system (e.g. small adjustments in flood-risk governance, redundancies on lower scales) but these are done only in the light of stability of the wider socio-ecological system as a whole (see Folke et al. 2010; Wiering et al. 2015). For the second debate (flexibility versus robustness/ stiffness) it would mean that there should be a good balance between a variety of strategies: a robust hydro-technical infrastructural system should be combined with mature alternatives for this approach.

8.6 Challenges and pitfalls

Returning to the empirical case of the resilience of The Netherlands an important question is raised: are the Dutch safe? (Van den Brink, Termeer, and Meijerink 2011).This question on resilience and appropriateness of flood-risk management and related institutions of European countries was also answered in the EU FP7 programme of STAR-FLOOD (www.starflood.eu). One hypothesis within the programme was taken from the literature on resilience: to decrease vulnerability and increase resilience of a country a *diversity of flood risk strategies* is needed (Hegger et al. 2016). The full range of core strategies, as categorized in STAR-FLOOD, are: flood prevention (as proactive planning), flood defence, flood mitigation, flood preparation, and flood recovery).

As said previously, in order to increase resilience of the full socio-ecological system (Dutch society and its natural environment, you might say), the Dutch should therefore not 'bet on one horse' but have different systems and strategies in place and be more flexible. Also, the European Floods Directive asks for a mapping of risks and explicit communication on a variety of strategies to be able to increase the public awareness of risks, whilst the Dutch flood-risk approach has long been avoiding too much public communication on risks. After all, the Dutch have it under control.

However, this is exactly the problem: the Dutch have traditionally avoided to communicate too much about the consequences of floods, so these other strategies are not 'popular' and easy to communicate; also, they have become so dependent and so focused on their historically evolved, sector-based flood infrastructure and the rich and robust institutions taking up this task so efficiently, the Dutch cannot easily move in another direction to strategically and seriously diversify. Although there is a certain variety of flood-risk measures and other alternative strategies *are* seriously discussed, at the end of the day most of the reserved budget for climate adaptation is spent on the infrastructural system of flood protection and not, or hardly, on flood mitigation, preparation, or recovery systems. Those strategies are somewhat strengthened – as we can trace back in the term of 'multi-layered safety' – but this does *not* lead to a very different risk approach whereby people and communities become more resilient. Ideally, in a more diversified, multi-layered flood-risk approach, we would at least expect greater emphasis on flood mitigation (e.g. strong investments in spatial adjustments in urban areas), flood preparation (restructuring polders to create safe zones and evacuation routes or investments in vertical evacuation) and perhaps changes in the recovery systems. But, as the Dutch consider themselves safe behind the dike system, these measures are mostly considered superfluous. In other words, and to put it a bit stronger, the Dutch are heavily path dependent (North 1990) or, with a small twist, strongly dike-dependent (for an elaboration on path dependency mechanism see Wiering, Liefferink, and Crabbé 2018 and the related Special Issue). This could even be seen as an institutional lock-in of the social-ecological system.

8.7 Conclusion and recommendations

Several normative principles have been described and discussed and in this closing section the opportunity is taken to reflect on a few elements of the Dutch governance of flood risks. The first is related to flood risk as an overarching, unitary public interest and the governance system that was produced. The OECD (2014), in its evaluation and assessment of Dutch flood-risk governance highlighted the lack of awareness of flood risks amongst the general public. People feel flood safe because of the strengths of flood defence infrastructure and trust in water agencies and other institutions. They are less aware of the fact that not all problems will and can be solved by the floods infrastructural system and that there are limits to this system. Because of a strong sector-based governance of floods, other (governmental) actors do not feel responsible for flood-related effects of their activities anymore. To be more specific, strong sector-based functional democracies – despite of functioning well, create a risk of division between generic spatial planning and expansions policies of municipalities (or provinces) on the one hand and increasing risk for water management – separately paid for – of the sector-based water agencies. The OECD (2014, p. 155) calls this a 'snowball effect':

> *municipalities and provinces reap the benefits of spatial developments, while the regional water authorities and central government bear the costs. This situation of split incentives drives the costs of water and flood management.*

This is indeed a danger of the sector-based system, which could actually be a victim of its own success. Therefore, it would be wise if municipalities and provinces would be made well aware of, and if possible, co-responsible for important investments in flood-risk safety when they can be expected to benefit strongly from the results.

A second element of reflection is related to future solidarity. The OECD report on Dutch water governance points at the unequal economic development of different regions and puts these regional disparities on the agenda. What about solidarity between regions, through the regional water authorities' financial system, in case of some regions growing and prospering, whilst others shrink and become less economically viable? This is indeed a vulnerability of the collective system, although we could also see this kind of solidarity as a strength – to hold regions together. The recent discussions on adaptation in light of climate change and flood risks lead to a greater transparency of the solidarity system (how does it work?), but also to greater emphasis on its flaws and its exceptions (where does it end?) (Keessen et al. 2016).

Finally, we return to resilience as a principle. The literature on resilience often stresses the need for a variety of measures and scenarios that can be translated to diversification and creating redundancy of risk strategies. This was presented as a dilemma for The Netherlands. There is fear that the more other strategies next to flood defence are highlighted, it would be at the expense of public trust in (and budgets of) the traditional flood defence measures and its implementing agencies. My recommendations would go in this direction: the Dutch can be proud of their infrastructural system and can trust governmental institutions to do their utmost best to protect people from floods and other water nuisance. But the Dutch should also be well aware of the general path dependency mechanisms (Wiering et al. 2018) that are created. The general public, politicians, and the flood-risk managers should acknowledge and overcome this dilemma. No one can expect the Netherlands to leave its core strategy of flood defence, nevertheless, to really avoid disastrous events to happen, the Dutch should take diversification much more seriously: it should not be only talk. Further strengthening nature based solutions (proactive planning for safety), communication on the reality of flood risks, taking spatial measures to mitigate risks and communicate about flood preparation (spatial measures in polders to create safe zones, communication on evacuation strategies, vertical evacuation) are all of utmost importance for community resilience. This creates redundancies but this is not that strange, as resilience asks for certain levels of redundancy in strategies. In short, The Netherlands should increase its adaptability as much as it can to preserve the social-ecological system as a whole to be stable.

In closing this tour along the normative principles of Dutch flood-risk management, it is good to emphasize once more that there are very different ways of resilience. In contrast to other countries, e.g. England or the USA, where resilience is predominantly seen as a feature of citizens or community, Dutch resilience is not so much citizens-resilience (or resilience of local communities) wherein citizens have a way of bouncing back, but it is rather 'system resilience': the *robustness of the collective system* to endure and withstand flood risks is key. The strength of the Dutch, that have created a unique system of flood risk infrastructure over many centuries, is also its weakness: in light of even more threats because of climate change it has become dependent on its system's resilience and is not able to really increase its flexibility and adaptability. If a great disaster would happen, we do not know if the Dutch can bounce back. In this sense it is perhaps comparable to San Francisco, happily building and living in an area susceptible to seismic activity. It is a fact that gradually it is becoming more acceptable to talk about future flood disasters. Recently Dutch (and Flemish) television were broadcasting a series with the title 'When the dikes break' (*Als de dijken breken;* 2016 http://vimeo.com/193859405) about a disastrous sea flood in the West of the country. This illustrates again: though the water wolf might be sleeping, it has not moved elsewhere.

References

Alexander, E.R. (2002). The public interest in planning: from legitimation to substantive plan evaluation. *Planning Theory* 1 (3): 226–249.

Bayertz, K. (1999). Four uses of 'solidarity'. In: *Solidarity* (ed. K. Bayertz), 3–28. The Netherlands: Kluwer, Dordrecht http://dx.doi.org/10.1007/978-94-015-9245-1_1.

Boezeman, D., Vink, M., and Leroy, P. (2013). The Dutch delta committee as a boundary organisation. *Environmental Science and Policy* 27: 162–171.

Davoudi, S. (2012). Resilience: a bridging concept or a dead end? *Planning Theory & Practice* 13 (2): 299–333.

Delta Commissie (2008). Working together with water. A living land builds for its future. [in Dutch] Findings of the Delta Commission. The Hague, The Netherlands, Delta Commissie. http://www.deltacommissie.com/index (accessed 12 February 2019).

Folke, C. (2006). Resilience: the emergence of a perspective for social-ecological systems analyses. *Global Environmental Change* 16: 253.

Folke, C., Carpenter, S.R., Walker, B. et al. (2010). Resilience thinking: integrating resilience, adaptability and transformability. *Ecology and Society* 15 (4): 20.

Hegger, D.L.T., Driessen, P.P.J., Wiering, M. et al. (2016). Toward more flood resilience: is a diversification of flood risk management strategies the way forward? *Ecology and Society* 21 (4): 52. https://doi.org/10.5751/ES-08854-210452.

Keessen, A., Vink, M., Wiering, M.A. et al. (2016). Solidarity in water management. *Ecology and society* 21 (4): 35.

Klijn, F., van Buuren, M., and van Rooij, S.A. (2004). Flood-risk management strategies for an uncertain future: living with Rhine River floods in the Netherlands? *AMBIO: A Journal of the Human Environment* 33 (3): 141–147.

Levina, E. and Tirpak, D. (2006). *Adaptation to Climate Change: Key Terms*. Paris, France: Organisation for Economic Co-operation and Development.

Liao, K.-H. (2012). A theory on urban resilience to floods – a basis for alternative planning practices. *Ecology and Society* 17 (4): 48.

North, D.C. (1990). *Institutions, Institutional Change and Economic Performance*. Cambridge, UK: Cambridge University Press.

OECD (2014). *Water Governance in the Netherlands – Fit for the Future?* Paris: OECD Studies on Water, OECD Publishing, Organization for Economic Cooperation and Development.

Rooijendijk, C. (2009). *Waterwolven* [Water wolves]. Amsterdam/Antwerpen: Uitgeverij Atlas. [in Dutch].

Van den Brink, M., Termeer, C., and Meijerink, S. (2011). Are Dutch water safety institutions prepared for climate change? *Journal of Water and Climate Change* 2 (4): 272–287.

Van Rijswick, H.F.M.W. and Havekes, H.J.M. (2012). *European and Dutch Water Law*. Amsterdam, The Netherlands: Europa Law Publishing.

Vink, M.J., Boezeman, D., Dewulf, A. et al. (2013). Changing climate, changing frames: Dutch water policy frame developments in the context of a rise and fall of attention to climate change. *Environmental Science and Policy* 30: 90–101. https://doi.org/10.1016/j.envsci.2012.10.010.

Wiering, M., Green, C., van Rijswick, M. et al. (2015). The rationales of resilience in English and Dutch flood risk policies. *Journal of Water and Climate Change* 6 (1): 38–54.

Wiering, M., Liefferink, D., and Crabbé, A. (2018). Stability and change in flood risk governance: on path dependencies and change agents. *Journal of Flood Risk Management.* 11 (3): 230–238.

Wiering, M. and Winnubst, M. (2017). The conception of public interest in Dutch flood risk management: untouchable or transforming? *Environmental Science and Policy* 73: 12–19.

9

Flood Governance in France: From Hegemony to Diversity in the French Flood-Risk Management Actors' Network

Marie Fournier

Laboratoire Géomatique et Foncier (GeF) – Équipe ERADIF (Aménagement, Droit immobilier, Foncier), École Supérieure des Géomètres et Topographes (CNAM), Le Mans, France

When we look back to the last 30 years, flood governance in France has changed drastically. Even though it is possible to identify some elements of stability, it is also crucial to point out various incremental changes. Hence, throughout this chapter, we will identify the main changes that took place during the last decades, along with the key characteristics of flood governance in France. We will describe these processes and decipher the network of stakeholders involved in this field.

First of all, French governance is quite specific on various aspects in comparison with the situation observed in many other Western European countries. Indeed, although there is an increasing involvement of private stakeholders in many neighbouring countries (Mees et al. 2016), French public authorities – more precisely central government and local authorities/municipalities – remain the cornerstone of flood-risk management (FRM) governance in France, whilst private stakeholders remain at the margins of this policy. Legal responsibilities are still mainly divided between these two stakeholders. However, it is also worth highlighting the gradual shifts that have been taking place in assigning competences within the last decade. We will point out the main reasons for these changes. Then, the second part of this chapter will focus on the gradual involvement of inter-municipal cooperation within this binary system. Since the beginning of the 2000s, inter-municipal cooperation gradually started to invest in FRM and claim a position within flood governance. Whilst more and more FRM responsibilities have been devolved to French municipalities, the sharing of resources and capacities quickly became a key issue. Today, inter-municipal cooperation benefit from various

Facing Hydrometeorological Extreme Events: A Governance Issue, First Edition.
Edited by Isabelle La Jeunesse and Corinne Larrue.

institutional or spatial assets to appear locally as the most relevant level for management and they have introduced themselves in various FRM strategies. Finally, within French flood governance, it is mainly private stakeholders who are missing, whether companies or individuals. They still have very few ways to get involved within French FRM governance. In the third part of this chapter, we address more specifically the issue of citizens' involvement. We point out some elements of explanation for this situation and draw potential perspectives for future changes.[1]

9.1 Flood-risk management governance: A stakeholders' network still dominated by central government and municipalities

In France, central government remains powerful within FRM policies partly because of a long tradition of centralization. Historically, this tendency can be explained by the fact that risk management is viewed as a public security mission. Hence, opportunities for stakeholders other than the central government (local authorities as well as private actors) to engage in flood management have long remained extremely limited.

Even though the 1807 law pointed out individuals' responsibility for their own safety, the French central government gradually took responsibility for this matter during the nineteenth century, by constructing protection works on major rivers as well as creating the first forecasting systems during the 1860s (Fournier 2010). These developments must be linked to the gradual construction of the French Nation-State (*État-Nation*), after the French Revolution. It was only at the end of the twentieth century, with the renewal of the flood issue (leading to the diversification of strategies against the flood risks), that several changes gradually took place. In particular, various competences and responsibilities were devolved to French municipalities.

9.1.1 The French central government, a major stakeholder for flood-risk management

If we analyse flood-risk management through a set of strategies (prevention, defence, mitigation, preparation, and recovery, as defined by Hegger, Driessen, and Bakker 2016), we can say that French central government remains a dominant actor in several of these strategies.

Defence is the first traditional strategy against floods. Since the seventeenth century, central government has contributed to defence against floods, mainly by funding works on major rivers, as well as through the expertise and knowledge of the civil engineers working for the central government (*Ingénieurs des Ponts et Chaussées*) who contributed to and carried out most of the major flood-protection systems for the past centuries in France. As an example, on the Loire river, most dikes were built under the orders of successive royal authorities and later by the central government in the nineteenth century. In the beginning of the twentieth century, large programmes for the construction of dams were also conducted on the Seine and Rhône rivers by central government services (Bravard 1996). More recently, whilst the safety issue concerning protection works has become more and more debated, central government reassessed its dominant position via the 2007 Dikes Decree (*Décret 'Digues'*) with the institutionalization of danger studies (*Etudes de danger*) on all French dams and dikes and the definition of safety criteria (*niveau de sûreté des digues et barrages*). These studies, which must be carried out by the manager of the hydraulic works, mainly focus on the technical characteristics of the dikes (materials, former accidents, and so on), the protection level of the works, the issues protected, and property rights. In the end, those studies lead to a prioritization by the central government of the hydraulic works to be maintained on the national scale.

If we consider the other traditional strategy – preparation – it is important to recall that the first forecasting services were also created by the French central government's civil engineers, the *Ponts et Chaussées*, at the end of the nineteenth century. Today, most rivers are still monitored by the *Services de Prévision des Crues* and the *Service central hydrométéorologique d'appui à la prévention des inondations* (SCHAPI, Floods Forecasting Services) which is an administrative body coordinated with *Météo France* (central weather forecast services). Finally, preparation also means civil security. As such, it is an intrinsically centralized and multi-risk mission. Operationally, central government services decentralized at departmental level[2] and coordinated by the *Préfet* manage civil security procedures, via the ORSEC Plan (*Organisation de la Réponse de SEcurité Civile*).

The French central government also dominates the prevention strategy. Since 1995, central government services have played a major role in the definition of non-buildable areas at local level via the adoption of *Plans de Prévention du Risque d'inondation* (PPRi). These planning documents map flood-prone areas and potentialities for new urban development in some of them. They also set adaptation rules for existing houses and buildings. This document governs local planning documents. In 1995, the Barnier Act gave back competences for risk mapping to the central government, whereas the former planning documents for risk (*Plans d'Exposition aux Risques*) created in 1982 were to be drafted by the municipalities. Within the field of prevention, central government also controls the National Fund for Major Natural Hazards (NFMNH, also known as the 'Barnier Fund'). More recently, central government also took responsibility for implementing the Floods Directive (2007). As such, central government services defined a National Strategy for Flood Risk Management (SNGRI, *Stratégie Nationale de Gestion des risques d'inondation*) and led the definition of PGRI (*Plans de Gestion des risques d'inondation*) as well as the identification of key river basins and associated coastal areas at risk of flooding (*Territoires à Risque Important*, TRI). Today, they are a key stakeholder in the definition of the local flood-risk management plans, in collaboration with local authorities and inter-municipalities (*Stratégies locales de gestion des risques d'inondation*).

Finally, recovery – which is a characteristic pillar of the French welfare system – still remains a competence of the central government. In France, recovery is based on the constitutional principle of national solidarity. The CAT-NAT system embodies the solidarity principle. CAT-NAT operates via a super-fund – the Natural Disaster Compensation Scheme – collected on residential insurance premiums, involving the national community as a whole. Almost all French homes are protected. The amount collected is used for paying out compensation for flood and natural risk damage. Within this scheme, central government decides on the design and operational mechanisms for the system and intervenes to determine the legal insurance obligations, insurance rates and acknowledges the 'natural disaster' status, which opens the way to insurance compensation.

9.1.2 Municipalities are gaining competences for flood-risk management

If central government remains the cornerstone of FRM, more and more responsibilities have gradually been devolved to municipalities since the first decentralization laws adopted at the beginning of the 1980s. Traditionally, municipalities have been responsible for public safety. As such, they must control the integrity of all protection works (such as dikes and dams) and compel all landowners to maintain the works they are responsible for. In practice, as many dikes belong to the French central government, municipalities may impose its decentralized services to better restore or maintain their belongings. In the field of preparation, municipalities also have large responsibilities. They must monitor water levels in small tributaries and, at least, organize safety measures if a flood occurs. These responsibilities correspond merely to the policing powers of mayors.

Although municipalities traditionally hold these competences, in the fields of preparation, crisis management, and recovery, however, they have gained new responsibilities over the last decade. Following the 2004 law on the reform of civil security, better coordination was organized between the two intervention levels. Municipalities are responsible for floods that can best be handled at a local level, whilst central government services manage floods of greater magnitude (Cans 2014). This law has reinforced the role of Mayors by creating the local Municipal Crisis Management Plan (*Plan Communal de Sauvegarde PCS*). Within this Plan, municipalities are expected to organize a local safety plan. In practice, this document identifies all issues at stake, at the municipal level (inhabitants, companies, public services, infrastructures, and the rest) and sets out in great detail the procedures to be followed in case of emergency (practical information [phone numbers, location of safety devices], identification of responsibilities between the various services [municipality, fire services, local health services, and so on]). Within this plan, the municipality must provide all the answers to potential questions in case of a major event. As such, whilst civil security used to be organized by the central government services, the 2004 law has led to the identification of the mayor and the municipality as the first level of intervention in case of a major event.

Since the beginning of the 2000s, other responsibilities have also been transferred by law to municipalities. More specifically, in the field of prevention, municipalities were identified as responsible for public information, for instance. Since 2004, all the municipalities covered by a PPRi must provide and display a *Document d'Information Communale sur les Risques Majeurs* (DICRIM) (Local Information on Major Risks). The DICRIM is attached to the PCS. This document should be short, descriptive, and easily accessible. It must provide the key information about risk and crisis management to local inhabitants at the municipal level. Some municipalities have decided to distribute it to every household whilst others uploaded it onto the internet. Municipalities also have an obligation to inform inhabitants (via public meetings for instance) of the issues at stake. In a nutshell, municipalities' key role within risk management (including floods) has been gradually acknowledged by law during the first decade of the twenty-first century. Although municipalities already had some responsibilities, mainly within civil safety, the municipal level has been clearly identified as the first step for a better risk management system, adapted to local specificities (Larrue et al. 2016).

It is also worth describing how the involvement of municipalities in mitigation strategies has progressively contributed to the growing acknowledgement of the role of local authorities in flood management. Mitigation measures began to be tested and implemented at the end of the 1990s, along with a growing understanding that protection systems might fail. Within the PPRi, central government already had the opportunity to impose measures to reduce the vulnerability of houses and other constructions (property-level measures). However, these measures have very rarely been imposed, let alone implemented. It is interesting to see that the first initiatives as regards property-level measures were mainly launched by municipalities and groups of municipalities during the first decade of the twenty-first century. As regulations were not so strict, the development of local mitigation strategies became an opportunity for municipalities to challenge the rules imposed by the central government, which were often considered as being too restrictive. For local elected officials, it was crucial to imagine innovative housing policies in areas at high risk of flooding. As a consequence, municipalities started to launch hazard-oriented mitigation measures (retention areas, bypasses, and spillways) in addition to issue-oriented measures. Hazard-oriented measures have tended to be more common in rural areas as a way of accommodating water in areas with a limited number of issues at stake. They are still at an experimental phase in urban areas, however municipalities – and even more inter-municipalities – have become central in this field. To conclude on this aspect, it is interesting to see that because flood mitigation is not as strictly regulated or governed as other strategies can be, it has provided local authorities with some room for innovation and an opportunity to challenge the binding regulations set by central government (Fournier et al. 2016). In order to illustrate this trend, the case of Saint-Pierre-des-Corps is a rather good example. Saint-Pierre-des-Corps is a medium-size town located to the east of Tours. It stretches between the Loire (to the north) and the Cher (to the south). This municipality is very vulnerable to fluvial risks in case of a breach or overflow of the dikes that protect it. The whole territory is flood-prone and has been classified as a medium to high-risk zone in the regulatory mapping of risks (PPRi). Yet, floods are rare (the last one occurred in 1866) and the risk of flooding is not a reality that has been experienced by inhabitants. However, municipal authorities have been highly involved in local information since the 1970s. During the 1990s, as the whole municipality is located in the floodplain, the town started to face a total prohibition of any new form of urbanization. After a very conflicting time with central government, the municipality gradually shifted to a more constructive approach. Elected representatives became greatly involved in flood-risk management at local and national levels. They increasingly began to make a number of innovative proposals, in particular for vulnerability reduction, in order to overcome the constraints imposed by central government. Since that time, the municipality of Saint-Pierre-des-Corps has kept innovating in this field. In 2014, it was identified as one of the key cities in France to experiment on new types of housing in high-risk zones ('*Atelier National sur les territoires en mutation face aux risques*'). All recent housing projects (buildings or houses) are adapted to the gradual increase of the water level.

Finally, in 2014, the definition of a *Gestion des Milieux Aquatiques et Prévention contre les Inondations* (GEMAPI) competence for municipalities by law definitely settled the major role of municipalities within FRM. GEMAPI stands for River, wetlands, and flood management (GEMA and PI). This new competence was created and defined within the MAPTAM Act (27/01/2014) in order to facilitate the integrated management of water and flood issues at local level. Within the MAPTAM Act, municipalities are clearly identified as the key stakeholders within flood risk management. They hold an exclusive and mandatory competence in this field. Even though municipalities were already responsible for water production and delivery, the MAPTAM Act defines several other competences:

- River basin management
- Maintenance and works on rivers, canals, and lakes
- Defence against floods and sea surges
- Protection and restoration of rivers and wetlands.

For instance, responsibility for infrastructures has been devolved to the local level and municipalities are now highly involved in defence systems along rivers (especially secondary rivers). This is a major issue as about one-third of French dikes have no identified owner. As a consequence, municipalities have become responsible for all hydraulic works with no management scheme. Similarly, the MAPTAM Act and the GEMAPI competence now impose on municipalities the obligation to address in a coordinated way flooding issues, river, and land management. Finally, the MAPTAM Act has clarified responsibilities and reaffirmed the role of municipalities for some of these (for which they were already responsible). As a matter of fact, it is interesting to see innovative local projects attempting to address all these issues. For instance, this is the case for the Ile Saint-Aubin, located upstream of the city of Angers (Maine-et-Loire). The Ile Saint-Aubin is a large green area located at the outskirts of Angers City. This area is about 600 ha wide. It is located at the confluence of two rivers and is partly flooded every year. For the city of Angers, it plays the role of a 'natural sponge', a natural retention area upstream of the city centre. However, it is mainly owned by private owners and extensive grazing remains the main activity. Competing uses (anglers, hunters, local environmental NGOs) have also created a conflicting situation in the area, whereas the city of Angers intends to both develop leisure activities and keep it as a flood-prone area. However, it is interesting to see how the municipality of Angers (as well as the inter-municipality of Angers Loire Métropole) has gradually succeeded in designing a development project coordinating leisure activities, environmental issues (floods, biodiversity, landscape, and so on) and economic valorization (Bonnefond et al. 2017).

Hence, since the 1980s, the ongoing process of decentralization has involved a shift of competences and responsibilities amongst the two 'historical' stakeholders of FRM. Central government still dominates traditional strategies as French governments have invested a lot over the centuries to guarantee protection. They took responsibility for various aspects (at least other stakeholders consider the central government as responsible) and central government still benefits from a strong political legitimacy amongst FRM governance stakeholders. Today, central government is still often considered as the main actor responsible for flood management. However, since the 1980s, there has been a vast redistribution of power and competences towards local authorities, namely municipalities. They became responsible for the safety of their citizens and informing them, as well as for establishing the local planning documents which must include the risks and drawing up emergency planning documents (*Plans Communaux de Sauvegarde*). Even in the defence strategy, responsibilities seem to be more and more divided between local authorities and central government.

9.2 Inter-municipalities as new players within the French FRM governance

If flood-risk governance is still very much dominated by central government and municipalities, we can also acknowledge a diversification in the set of actors during the past decade. New stakeholders have become involved and responsibilities distributed amongst them. This redistribution of power and resources to the local level is partly explained by the national decentralization process we have already mentioned, as well as by local actors claiming the lead and proposing innovative solutions in FRM.

9.2.1 Inter-municipalities, a long tradition in France

Because of the traditional divisions of the French municipal landscape (about 36 000 municipalities, of which at least 32 000 have fewer than 1000 inhabitants), French municipalities got into the habit of setting up different forms of cooperation amongst themselves. This began at the end of the nineteenth century with associations to manage technical competences. The law of 22 March 1890 created 'intercommunal' associations with a single purpose (*Syndicat intercommunal à vocation unique*, SIVU) devoted to a specific field of community action for which intercommunal cooperation provides added value. In 1959, these associations were able to involve several competences (*Syndicats à vocations multiples*, SIVOM). In October 1935, a decree also established mixed associations, stipulating in its first article the possibility for *Départements*, municipalities, and other public institutions to group together as associations to manage public services representing an interest for each of the legal entities in question by means of concessions. From the 1960s – and even more so since the 1990s – the creation of many different forms of inter-municipalities was proposed to the individual municipalities. In this case, we can identify two types of inter-municipalities:

- 'integrated' associations of municipalities, with an imposed framework (mandatory and optional competences), the financing of which is ensured by local taxes levied directly by the intercommunal body instead of by the municipalities themselves: 'Communautés urbaines' (1966), 'Communautés d'agglomération' (1992), 'Communautés de communes' (1992), 'Métropoles' (2014). In these cases, there are real transfers of competence from the municipality to the inter-municipal structure.
- Flexible forms of inter-municipality: associations of municipalities (and potentially other public institutions within mixed associations), based on freely determined competences, funded by the contributions of the different members

The setting up of inter-municipal structures has allowed local authorities to deal with environmental issues that were previously difficult to address (Larrue and Fournier 2014). Since the end of the nineteenth century, inter-municipalities have become involved mainly in the field of environmental public services (water services, waste management, energy production, and so on). What is more, environmental questions are often difficult to handle at the municipal level and incompatible with administrative borders. As such, French scholars have pointed out how the inter-municipalities may favour a greater harmonization between problem areas and solution areas (Berdoulay and Soubeyran 2002). Environmental issues, and amongst them flood risks, may be a good example.

If we consider more specifically the flood issue and FRM governance, two main trends can be identified. First, 'integrated' inter-municipalities are now identified as key actors for FRM. Second, specific types of inter-municipality have been created to address water and flood management issues and assist local authorities.

9.2.2 Integrated inter-municipalities: New leading actors within FRM

Following the example of certain pioneer municipalities (such as Saint-Pierre-des-Corps), it is important to point out the ability of 'integrated' inter-municipalities to get involved in a specific aspect of FRM: vulnerability reduction. Since the beginning of the 2000s, vulnerability-reduction measures have gradually cropped up more or less on an autonomous, ad hoc basis, depending on the will of the local authorities. Many examples of 'integrated' inter-municipalities have space for action in this field (both in vulnerability reduction and mitigation).

For instance, since 2005, the inter-municipal body of Nevers (*Nevers Agglomeration*) has strongly invested in FRM. Its commitment is visible in the Nevers Flood Risk Assessment Study, which was launched in 2007 (EGRIAN, *Etude Globale du Risque Inondation sur l'Agglomération de Nevers*). This study addresses the flood issue and aims at building an integrated vision of local FRM, by addressing a wide range of strategies (protection, mitigation, prevention, and crisis management). If both the decentralization process and the depletion of the central government resources may explain the involvement of Nevers agglomeration in FRM, it is also important to explain that Nevers agglomeration uses FRM as a way to increase its political legitimacy locally. Around 10 years ago, it was crucial for such a new institution to identify certain fields of action within which it could prove its competence as well as create synergies/solidarities between municipalities. As such, the inter-municipal institution can compensate for the lack of technical and operational resources of the small municipalities' part of the agglomeration.

Integrated inter-municipalities have also invested in some other fields of FRM, such as vulnerability reduction. In Orléans for instance, the agglomeration launched a large project to diagnose housing vulnerability in 2005. Information was sought by 605 people and 555 properties benefited from a building flood risk diagnosis.

The involvement of integrated inter-municipalities within FRM may also rely on certain strong local initiatives. This is the case for instance in Le Havre. During the 1960s, Le Havre local public authorities had already established an early risk-management policy. Because of the expansion of the harbour and the implementation of industrial activities, the municipality decided to reinforce its preparation plans by investing in communication and crisis exercises. From the 1980s, the flood issue has appeared in the local public agenda. FRM was mainly dealing with flash-floods coming from the Lézarde basin, i.e. agricultural lands from the upper part of the chalk Plateau. As a consequence, skills and good practices capitalized for industrial risk were transferred to flood policy. The first actor of this policy transfer was a new association called Seine

Estuary Major Risk Office (*Office des Risques Majeurs de l'Estuaire de la Seine* ORMES), which was the direct emanation of local authorities. The ORMES facilitated inter-municipal cooperation on flood management. From the 2000s, several municipalities have chosen to gather into the inter-municipal agglomeration of Le Havre (CODAH). The latter is a cooperative institution in charge of several competences such as local economic development, spatial planning, social housing, water facilities, and other technical urban networks (Gralepois 2008). For instance, the inter-municipal agglomeration of Le Havre (*Communauté de l'Agglomération Havraise* CODAH) has long taken responsibilities in the field of flood prevention and especially flood forecasting. Similarly, CODAH has been really proactive in negotiating with the central government services about the local PPRi.

Therefore, via innovative projects, integrated inter-municipalities have gradually gained legitimacy to participate and act within FRM governance. Today, these recent changes tend to be reinforced and legitimized by law. Indeed, even though the MAPTAM Act stated that the GEMAPI competence should be settled at the municipal level, it also imposed its transfer to inter-municipalities (L. no 2014-58, 27 Jan 2014, art. 56, I). Finally, municipalities are thus not considered as the most relevant level of action against flood risks whilst the inter-municipal level facilitates the sharing of resources and fosters better solidarity between municipalities to deal with flood risks. This strategy is very relevant with other legal developments currently taking place in France. Indeed, integrated inter-municipalities are increasingly reaffirmed as being the most relevant level of action for all land- and urban-planning issues (Fournier, Larrue, and Schellenberger 2017).

9.2.3 Defining flexible forms of inter-municipalities to support local FRM strategies

Within FRM, other forms of inter-municipalities have also gradually reinforced their position. This is especially the case for the *Etablissements Publics Territoriaux de Bassin* (EPTB, Public Bodies for River Basin Management). Since the 1960s, certain EPTBs were created by the central government, first to facilitate the implementation of development programmes on rivers (EPTB Vilaine was created in 1961 and EPTB Oisne/Aisne in 1968). Gradually, local authorities (municipalities, *Départements*, *Régions*) also became involved in these institutions, as their competences and responsibilities increased with the decentralization processes. Since the 1980s and 1990s, local authorities have taken the lead within the EPTBs; this is the case for instance on the Loire and on Garonne rivers. EPTBs were institutionalized by law no 2003-699 (30 July 2003) dealing with technical and natural-risk management, and recently, the MAPTAM Act resettled all provisions concerning these public bodies.[3] Today, there are 38 EPTBs in France, covering about three-quarters of metropolitan France.

EPTBs cover a river basin or several small river basins. They support local authorities and their groupings in the field of water management (quantity) and flood management. Until the end of 2017, EPTBs were mixed associations which could bring together municipalities, inter-municipalities, along with departmental and regional authorities willing to get involved in water and FRM issues. Since the beginning of 2018, EPTB can only be mixed associations constituted of inter-municipalities (and potentially departmental or regional authorities). Their missions are defined by paragraph L.213-12 of the Environmental Code. EPTBs facilitate flood prevention, defence against the sea, sustainable management of water, protection, management, and restoration of river ecosystems and wetlands at the basin level or on several smaller river basins. They also potentially contribute to the establishment and implementation of the local Water Development and Management Plan. EPTBs also coordinate activities and operations carried out by the *Etablissement Public d'Aménagement et de Gestion de l'eau* (EPAGE). Above all, EPTBs must facilitate spatial solidarity at the river basin level. In the current context, this is vital as more and more mitigation projects are implemented in an upstream/downstream solidarity principle.

Amongst the French EPTBs, the EPLoire is the territorial public establishment of the Loire basin, recognized as an inter-municipal water actor in the 2006 Water Law. It groups together local authorities in the whole river basin in order to gain a better understanding of the issues related to the planning and management of the river. Many of the structure's tasks concern issues related to flood management. For instance, the EPLoire is strongly committed to the field of vulnerability reduction, mobilizing European and national funding intended for an industrial process in order to reduce the vulnerability of enterprises.

Finally, it is important to recall that the MAPTAM Act has created a new flexible form of inter-municipality specifically dedicated to river and flood management. EPAGEs group local authorities (both municipalities and inter-municipalities) at the level of a coastal river or small river basin in order to facilitate the implementation of flood management and river management activities and operations. EPAGEs are still very new institutions and it is still difficult to describe the future interactions between integrated inter-municipalities, EPAGEs, and EPTBs. However, the MAPTAM Act explains that EPAGEs may focus on the operational aspects of water and FRM, whilst EPTBs must be set up on a larger scale and coordinate local actions.

Therefore, inter-municipal bodies play an increasingly significant role in many Flood Management Strategies in France. This is particularly true in the field of civil security as well as in mitigation strategies. The most recent legal developments (mainly Floods Directive and MAPTAM Act) have reinforced their position in FRM governance. However, the on-going reorganization in this field of action is very much debated today. In particular, competition between inter-municipal bodies may arise, for instance between EPTBs and integrated inter-municipalities.

9.3 Where are citizens in FRM?

In the first parts of this chapter, we pointed out how various institutional stakeholders gradually became involved in French FRM governance. This section aims to address the issue of private stakeholders. As such, although the devolution of responsibilities to communities at risk of flooding and the involvement of citizens is more and more observed in many countries in Europe (Mees et al. 2016), this trend is still not so obvious within the French context. In this part, we first describe how French citizens are involved in flood governance and management and point out certain factors explaining this lack of involvement. More specifically, we show that citizens have often first entered flood-risk governance in a position of conflict, as riparian inhabitants and opponents to

FRM projects launched by public authorities. However, we also see the gradual emergence of innovative approaches promoted by private stakeholders which contribute to the whole FRM policy.

9.3.1 Citizens' involvement in FRM via prevention (vulnerability reduction), information, and crisis management policy instruments

Since the beginning of the nineteenth century and the 1807 law, individuals have been responsible for their own safety. However, we explained previously that public authorities – both central government and local authorities – gradually took the lead in FRM and that populations expect them to act in case of emergency. It is mainly since the mid-1990s that citizens' involvement started to become an issue and was considered as a necessity to face flood risks more effectively.

First, within the prevention strategy, different public information instruments have been created or reinforced since the beginning of the 2000s, in order to make citizens aware of the existing risks. Since 1982, central government's services must elaborate and regularly update a *Dossier Départemental sur les Risques Majeurs* (DDRM) at the departmental level. This document identifies all existing risks (natural and technological) at the departmental level. However, individuals rarely know the existence of these documents and they are not updated as often as requested. The 2004 Act on the Modernization of civil security provided new guidelines for municipalities and departmental actors on how to inform the population about the existing risks. Apart from the DDRM, citizens' information has been greatly improved at municipal level. As already mentioned in the first part of this chapter, all municipalities under the regulation of a PPRi must now establish a DICRIM. This document, which has no binding capacity, is drawn up in order to inform local inhabitants on the existing risks, the first emergency measures, and key information should an event occur. Most often, French municipalities establish short documents and sometimes send these to households. Municipalities must also organize public information meetings regularly. Finally, each individual purchasing a new household must be informed of the potential risks for this property. All these measures share the same objective: the better-informed people are, the less damage there will be. It also expected that better information will make people more responsible for their own safety. Cost-efficiency considerations form a major driver for the increased attention paid to citizens' involvement in FRM delivery. More recently, local authorities have started to make use of artistic initiatives and events to better raise public awareness of flooding issues.

As citizens are better informed, it is also expected that they will contribute to FRM. At the end of the 1990s, various initiatives were launched by different public institutions (water agencies, inter-municipalities, and so on) to convince individuals to reduce their home's vulnerability. *Mesures de réduction de la vulnérabilité* (property-level protections) were promoted, with the growing understanding that large floodplains had been urbanized in the last decades, increasing direct and indirect vulnerability of both cities and homes. Within the frame of the PPRi, central government services also started to impose property-level protection measures. However, most of the time, these measures proved to be rarely acknowledged by landowners – who did not even know about the existence of this planning document (Hubert, Capblancq, and Barroca 2003). The occurrence of a flood event proves to have a great impact on landowners to adapt their homes. Property-level protection measures promoted by local authorities or water agencies also proved to be difficult to implement. At the individual level, landowners were reluctant to pay for the works that were needed and professionals reluctant to modify their building techniques. Broadly speaking, property-level protection measures on existing buildings and houses did not meet a great success in the beginning of the twenty-first century. At last, in coherence with the fact that 'citizens are responsible for their own safety', the 2004 Act on the Modernization of Civil Security enabled local authorities to organize a Municipal Civil Protection Reserve (*Réserve Communale de Sécurité Civile* [RCSC]), constituted of citizens from the municipality. This initiative has proved to be quite successful. The members of the RCSC must help the mayor and security services in the case of a major event, as well as in the preparation phase, including through informing the local populations. Such initiatives intend to improve local solidarities between inhabitants and safety measures when a major event occurs.

Hence, within the French FRM policy, various instruments are implemented to inform citizens and incite them to get involved in FRM. However, it appears that French citizens remain very rarely involved in floods governance. In comparison with other European countries, how can we explain this situation?

Two main reasons can be identified. First, even though we lack global data specifically on this topic, it appears from various case studies that citizens do not feel responsible for their safety. Flood management is a safety issue and a public matter, i.e. a matter for public authorities. Legally and culturally, it is difficult in the French context, and for the central government to encourage citizens to embrace responsibility. The solidarity principle and the constitutional duty of the central government to ensure its citizens' safety remain prominent and conflict with the recent and successive efforts to reshape the relationship between central authorities and citizens. However, some slight changes seem to be on the way and are more and more discussed at the national level. This is particularly true if we consider the CAT-NAT system, within which some modulations of the premium granted were established in case of frequent flood events in areas not covered by a PPRi. Second, as FRM is mainly considered as a safety issue, there are still few opportunities to integrate the flood issue into an overall strategy for floodplain areas. As stated by Pottier et al. (2003), 'flood management policy is not integrated into more general spatial planning and development policy'. However, the on-going developments, as described previously, may lead to major changes to these aspects in the years to come.

9.3.2 Urban planning in flood-prone areas: From conflict to cooperation

Flood management projects, like most planning projects (Subra 2007), are bound to be conflictual especially with the new definition of flood risk and the solutions to tackle it (give space back to the rivers). In France, since the beginning of the 2000s, participatory democracy instruments have multiplied upstream of the decision-making process (via public debates or negotiating tables, and so on). Public consultation through public enquiries is widely established but often criticized (Blatrix 1999). Public enquiries are held too late in the decision-making process to have a real impact. Participation processes are often reduced to communication purposes. However, it is interesting to see how, via the implementation of recent urban schemes planned in flood-prone areas, citizens can contribute to the definition of adaptive projects in high-risk zones.

At the beginning of the 2000s, two medium-size cities on the Loire river basin faced quite a similar issue. What was the possible future for flood-prone areas located in the very heart of their city centres?

In the perspective of the future PPRi to be set up by central government services, the city of Blois wished to tear down a large neighbourhood (more than 100 housing units and about 15 companies) to re-establish the flood expansion field. The project was carried out by the municipality of Blois, and later by the inter-municipality (Agglopolys), under the strict control of the central government. Via a specific urbanistic procedure (*Zone d'Aménagement Différé*), Agglopolys has gradually acquired most properties between 2004 and 2017.

In parallel, in Le Mans, the municipality had the ambition to develop a new housing estate on a brown field site located on the island called 'Ile aux Planches'. However, after the 1995 flood, it became crucial to integrate the flood issue into the project.

In both cases, local opposition from inhabitants was organized (local residents or riparian inhabitants) to defend another vision for these areas than the one promoted by local authorities. In Blois, the 'Association de défense des citoyens de la Bouillie' launched various initiatives. Citizens contested the legitimacy of the project (why destroy this neighbourhood and not the ones nearby? Why not build more dams on the Loire river to better control the water level? Why not make more effort to maintain the river bed, which is more and more covered with vegetation and trees?) as well as the legal procedure implemented, considered as very detrimental for landowners. In Le Mans, the 'Association de défense de l'Ile aux Planches' mainly contested the housing project proposed by the municipality. Instead, citizens were asking for a new green and leisure area which could benefit from the proximity of the Sarthe river.

Faced with this opposition, the local authorities reacted in very different ways. In Blois, the inter-municipality took the decision to visit every household and organized 'kitchen table conversations' with all inhabitants of the area. Its ambition was to convince each person of the validity of the project. In order to minimize the impact of the association, local authorities in Blois decided to convince individuals one by one. The inter-municipality also added financial incentives. They were gradually able to buy land and start destroying houses (Morrisseau 2012).

In Le Mans, the local authorities decided to involve representatives of the association in the definition of the project. Thanks to one of the City council Aldermen in charge of flood risk and water issues, the association could sit at meetings dedicated to the project, starting in 2003. Gradually, a project combining both functions (flood management and green infrastructure) was designed. The island was transformed into a leisure park, and also plays the role of a spillway. A smaller housing estate was also built (Gatien-Tournat et al. 2016).

When we compare these two cases, major differences appear: no houses were destroyed in Le Mans and the central government incentives were not as strong as they were in Blois. The safety issue was much more crucial in the second case. However, it is interesting to see that even though the inter-municipality in Blois succeeded in its project, the local authorities have so far failed to overcome the safety issue and propose a new vision/

urban project for the area. Today, 'La Bouillie' remains a patch of land where the former conflict still lingers in people's memories. In this landscape of empty meadows, we can still see some walls left from the houses destroyed and banners are still present on other houses. Conversely, in Le Mans, the association is still involved in the management and definition of activities on the Ile aux Planches. This green area has gradually become one of the hotspots in Le Mans for festivals and cultural activities. As regards the opponents, they gradually took the position of promoters of an innovative flood management project combining the risk issue with social and environmental purposes.

This last example illustrates how individuals also become key stakeholders in FRM projects. On this matter, various recent projects on the Loire river show this is particularly true when managers do not only focus on the flood issue but try to design a multi-functional project that overcomes the safety problem in flood-prone areas to enhance the variety of potential uses and activities in wetlands and flood retention areas (Bonnefond et al. 2017).

9.4 Conclusion

In France, flood governance has long been dominated by central government and services. Since the beginning of the 1990s, events that caused a national shock have stimulated legislative reforms (flash floods in the beginning of the 1990s in the Southern part of France, AZF accident in 2001, Xynthia storm in 2010, and so on), and local authorities' competences and responsibilities were reinforced. In order to understand these changes, it is important to point out several trends. First, at the national level, decentralization processes and budget cuts may explain the gradual shift in the role of the central government. Second, the diversification of FRM strategies has gradually led to the involvement of new stakeholders (inter-municipalities) and citizens. As such, central government's disengagement is also associated with the strengthening of inter-territorial cooperation.

However, very recently, new questions have come to the forefront, especially with the creation of the GEMAPI competences – amongst them, the identification of the best geographical level to address flooding. Inter-municipalities have become a key stakeholder but are they a relevant level to address the flood issue? Today, some think that the whole French flood governance is quite destabilized, and we might need some more years to define a new stakeholders' network.

Notes

1. This chapter is based on the results of several research projects dealing with flood governance issues conducted over the past decade (*Freude am Fluss* (Interreg 3B), STAR-FLOOD (FP7), TRANS-ADAPT (JPI Climate)). Based on these research projects, we drew elements of analysis and illustrated them with case studies taken from various local situations in France.
2. The French administrative system is organized into three levels: regional (Région), departemental (Département) and municipal (Commune). Central government services are represented at those three levels by sectoral decentralized services, the *Préfets* and the Mayors (at the municipal level). In parallel, via decentralization processes, local authorities are also represented at those three levels and share various competences (Région, Département, Municipalities).
3. ELNET (2018).

References

Berdoulay, V. and Soubeyran, O. (2002). *L'écologie urbaine et l'urbanisme: aux fondements des enjeux actuels*. Paris: La Découverte.

Blatrix, C. (1999). Le maire, le commissaire enquêteur et leur 'public'. La pratique politique de l'enquête publique. In: *CRAPPS-CURAPP, La démocratie locale, représentation, participation et espace public*, 161–176. Paris: Presses universitares de France.

Bonnefond, M., Fournier, M., Servain, S. et al. (2017). La transaction foncière comme mode de régulation en matière de protection contre les inondations. Analyse à partir de deux zones d'expansion de crue: l'Île Saint Aubin (Angers) et le déversoir de la Bouillie (Blois). *Ecology and Environment: Urban Risks* 17 (1): https://doi.org/10.21494/ISTE.OP.2018.0218.

Bravard, J.-P. (1996). La gestion des excès dans les bassins des grands fleuves français. *L'Information géographique* 60 (2): 72–80.

Cans, C. (2014). *Traité de droit des risques naturels*. Paris: Le Moniteur.

ELNET (2018). Dictionnaire Permanent Environnement et nuisances – Établissements publics territoriaux de bassin et agences de l'eau. Updated May. Paris: ELNET.

Fournier. M. (2010). Le riverain introuvable. La gestion du risque d'inondation au défi d'une mise en perspective diachronique. Une analyse menée à partir de l'exemple de la Loire. PhD Thesis in Land Planning, Université de Tours, 431p.

Fournier, M., Larrue, C., Alexander, M. et al. (2016). Flood risk mitigation in Europe: how far away are we from the aspired forms of adaptive governance? *Ecology and Society* 21 (4): 49. https://doi.org/10.5751/ES-08991-210449.

Fournier, M., Larrue, C., and Schellenberger, T. (2017). Changes in flood risk governance in France: a David and Goliath story? *Journal of Flood Risk Management* https://doi.org/10.1111/jfr3.12314.

Gatien-Tournat, A., Fournier, M., Gralepois, M. et al. (2016). *Societal Transformation and Adaptation Necessary to Manage Dynamics in Flood Hazard and Risk Mitigation (TRANS-ADAPT): France Country Report*. Brussels: TRANS-ADAPT JPI Climate project.

Gralepois, M. (2008). Les risques collectifs dans les agglomérations françaises à travers le parcours des agents administratifs locaux. PhD Thesis in Land Planning, Université Paris-Est/ENPC, Paris.

Hegger, D.L.T., Driessen, P.P.J., and Bakker, M.H.N. (eds.) (2016). *A View on More Resilient Flood Risk Governance: Key Conclusions of the STAR-FLOOD Project*. Utrecht: STAR-FLOOD consortium.

Hubert, G., Capblancq, J., and Barroca, B. (2003). L'influence des inondations et des documents réglementaires sur le marché foncier en zone inondable. *Annales des Ponts et Chaussées* 105: 32–39.

Larrue, C. and Fournier, M. (2014). The role of local actors in water and flood management in France: between policy formulation and policy implementation. In: *Studying Public Policy. An International Approach* (ed. M. Hill), 167–178. Clifton, UK: Policy Press.

Larrue, C., Bruzzone, S., Lévy, L. et al. (2016). *Analysing and Evaluating Flood Risk Governance in France: From State Policy to Local Strategies*. Tours, France: STAR-FLOOD Consortium.

Mees, H., Crabbé, A., Alexander, M. et al. (2016). Coproducing flood risk management through citizen involvement: insights from cross-country comparison in Europe. *Ecology and Society* 21 (3): 7. https://doi.org/10.5751/ES-08500-210307.

Morisseau, G. (2012). Le quartier périurbain de la Bouillie (Blois): les nouveaux paysages du risque. Projets de paysage https://www.projetsdepaysage.fr/le_quartier_periurbain_de_la_bouillie_blois (accessed 10 February 2019).

Pottier, N., Reliant, C., Hubert, G. et al. (2003). Les plans de prévention des risques naturels à l'épreuve du temps: prouesses et déboires d'une procédure réglementaire. *Annales des Ponts et Chaussées* 105: 40–48.

Subra, P. (2007). *Géopolitique de l'aménagement du territoire*, 326. Paris: Armand Colin.

10

Flood-Risk Governance in Belgium: Towards a Resilient, Efficient, and Legitimate Arrangement?

Hannelore Mees
Department of sociology, University of Antwerp, Antwerp, Belgium

10.1 Introduction

During the twentieth and twenty-first centuries, Belgium was hit by a number of severe flood events, of which the most disastrous ones took place in 1953 and 1976. Whereas the most harmful floods of the twentieth century had a tidal cause, more recent events were caused by fluvial and pluvial flooding. Belgium has a very high degree of land sealing, which decreases its rainwater infiltration capacity, thereby making it more vulnerable to these types of floods. In the future, the amount of built-up area is expected to increase even further. In addition, climate change is predicted to increase the likelihood of floods, both in winter and summer (VMM 2015).

Confronted with these challenges, policy makers in Belgium have launched a series of initiatives to increase society's resilience to flooding. Actions were taken to decrease the negative impact of spatial planning on the water system, to improve coordination between relevant actors, and to increase citizen engagement in flood-risk governance (FRG).

In this chapter, we describe and evaluate these main developments in Belgian FRG. For this, we use the evaluation framework of Alexander, Priest, and Mees (2016), which evaluates FRG in terms of resilience, efficiency, and legitimacy. What makes Belgium an interesting research subject in terms of FRG is not merely its flood risks, but also its specific government system. Over the past decades, the Belgian federal state has transferred competences related to water management and spatial planning to the regions (Flemish, Walloon, and Brussels-Capital Region, see Figure 10.1) through several state reforms.

Facing Hydrometeorological Extreme Events: A Governance Issue, First Edition.
Edited by Isabelle La Jeunesse and Corinne Larrue.

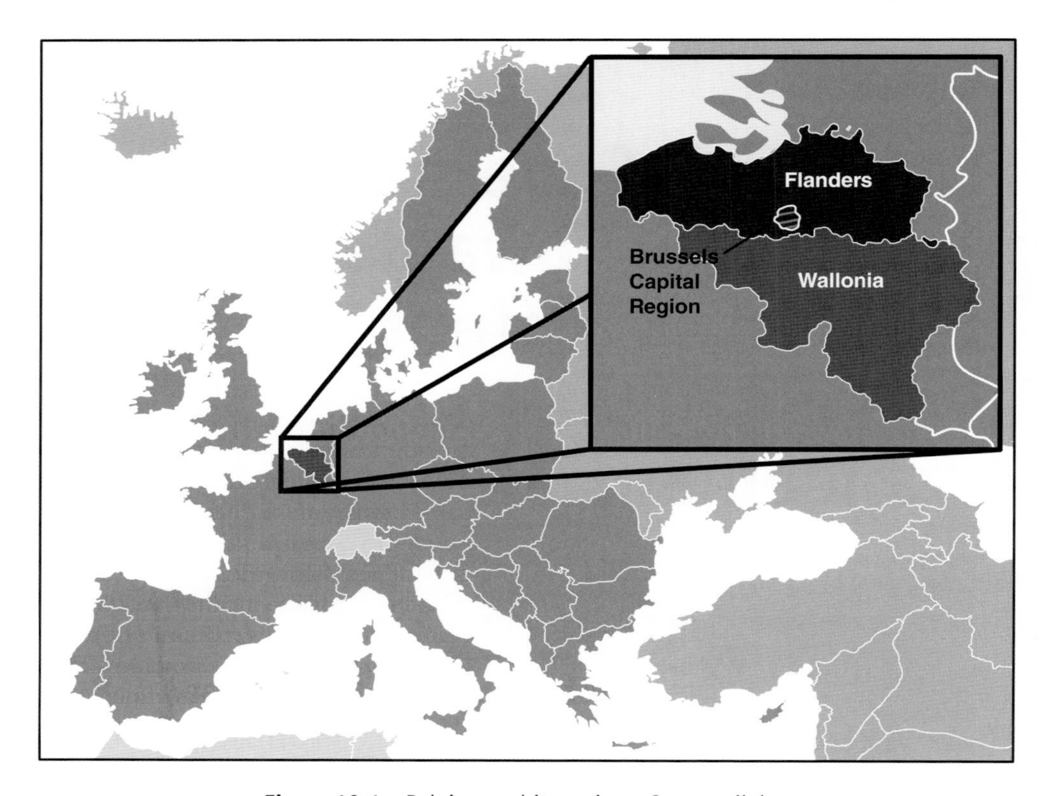

Figure 10.1 Belgium and its regions. *Source*: vib.be.

In each of these regions, FRG is thus subject to different legal frameworks and policies. In contrast, responsibilities on crisis management and insurance issues have largely remained at the federal level. The resulting complex governance arrangement makes Belgium a valuable case for intra-country comparison.

This chapter compares and evaluates developments in Flemish and Walloon FRG between 1995 and 2015. The research was conducted in the framework of the EU STAR-FLOOD project (more information can be found on www.starflood.eu).

10.2 Evaluation framework

This research employs the policy arrangement approach (PAA) to enable a comprehensive evaluation of FRG. The PAA assesses the governance of a particular issue by studying four dimensions: actors, discourses, rules, and resources (Liefferink 2006). Each of these dimensions has its own developments, which mutually influence each other. Together, the actors, rules, discourses, and resources form a flood-risk governance arrangement (FRGA), which is defined as 'the temporary stabilization of the content and organization of a policy domain' (Van Tatenhove, Arts, and Leroy 2000).

Furthermore, the research presented in this chapter adopts the evaluation framework of Alexander et al. (2016). These authors suggest evaluating complex arrangements by a series of building blocks, which gradually add to the understanding of the overall arrangement. These building blocks consist of three main evaluation criteria – resilience, efficiency, and legitimacy – which entail a number of sub-criteria. Effectiveness has not been included as a criterion as such but is considered a precondition for achieving resilience and efficiency.

Resilience is defined as the ability of a system, community, or society exposed to hazards to resist, absorb, accommodate to, and recover from the effects of a hazard in a timely and efficient manner, including through the preservation and restoration of its essential basic structures and functions (UNISDR 2009). Alexander et al. (2016) disaggregate this concept into the capacity to resist, capacity to absorb and recover, and the capacity to adapt. In order to be considered resilient, an arrangement needs to possess all three sub-criteria.

- The *capacity to resist* refers to the strategies societies develop to reduce the probability of flooding, e.g. building dikes, and flood retention zones. Although some authors argue that this should not make part of resilience (e.g. de Bruijn 2004), the prevention of flooding remains a highly effective strategy to 'preserve the essential basic structures and functions of a system' as is prescribed by the UNISDR definition of resilience (see also Bruneau et al. 2003; Walker et al. 2004; Davoudi et al. 2012).
- By building *capacity to absorb and recover*, a system acknowledges that floods cannot always be prevented and additional measures should be in place to respond in a flexible way to flood events when they occur. These include the development of crisis management and financial, material, and psychological recovery systems, and small-scale protection measures. Absorption and recovery measures bring diversity and a certain degree of redundancy into FRG. In its decision-making, these should be well-balanced against concerns of cost-efficiency.
- The *capacity to adapt* is fostered by effective learning processes. In order to be able to cope with a risk environment characterized by uncertainty and unpredictable change, learning procedures must be established that allow a rapid adaptation or even transformation with the emergence of new internal and external stressors. Learning can be stimulated through adaptive governance, which has as main features polycentricity, stakeholder co-management and room for experimentation (Nelson, Adger, and Brown 2007; Huitema et al. 2009). Governance structures should neither exclusively be located at local nor at higher level but exist as a multilevel cooperative structure with a sufficient degree of decentralization (Nelson et al. 2007; Cosens 2010). The involvement of governmental and non-governmental stakeholders should allow for a positive synergy of different types of knowledge.

Efficiency is conceptualized by the OECD as 'the contribution of governance to maximize the benefits of sustainable water management and welfare at the least cost to the society' (OECD 2015). A more general definition refers back to the use of the concept in physical science, i.e. 'the ratio of the output of work to the input of energy' (Larrue, Hegger, and Trémorin 2013). This input/output ratio does not merely relate to financial resources, but also to expertise, human engagement, technologies, and so forth (Alexander et al. 2016).

Legitimacy refers to 'the validity of an organization's authority to govern, whether conferred by democratic statute or acquired through social acceptance' (Lockwood et al. 2010). A FRGA can be built both on input and output legitimacy (Scharpf 1997). Input legitimacy focuses on the process by which policy is made; the governing institutions and process are accepted as representative, accountable, transparent, and inclusive. Output legitimacy refers to the outcome of the decision-making process; the policy output is accepted as a sufficiently effective solution.

A number of factors can be derived that enable legitimacy, namely accountability, transparency, procedural fairness, and a perceived effective and fair policy output (Lockwood et al. 2010; Schmidt 2013; Alexander et al. 2016; Pettersson et al. 2017). In this study, we focus primarily on how citizens are involved into FRG decision-making and implementation.

10.3 Methods

The research for this chapter was carried out in the framework of the EU-funded FP7 project STAR-FLOOD (www.starflood.eu), which compares FRG in six EU Member States. The chapter is based on the content of the project's deliverable report on Belgium, which describes and evaluates the evolution of FRG in the country between 1995 and 2015 (Mees et al. 2016a). This period was selected for analysis because a number of significant flood events took place in it and important legislative and policy related initiatives were launched, both at EU and national/regional level. Due to the unique situation in the Brussels-Capital Region, we have limited our evaluation to the Flemish and Walloon Regions. More detailed information on the methods used in our research can be found in Mees et al. (2016a).

10.4 Flood risk governance in Belgium

Since the 1980s, the Belgian state has been subject to a series of constitutional reforms, which turned it from a centralist state into a federal country, with competences divided over four different governmental levels, i.e. federal, regional, provincial, and municipal. As a result, Belgian FRG is composed of five separate FRGAs.

10.4.1 The regional water system arrangements

The Water System Arrangements encapsulate the strategies of flood-risk prevention and protection.[1] Competences related to these strategies belong to the regional governments, which necessitates us to discern separate arrangements for the Flemish, the Walloon, and the Brussels-Capital Regions. As mentioned above, for the sake of analysis, the Brussels-Capital Region is not addressed in this analysis.

The regional Water System Arrangements are highly fragmented in terms of actors. Both in Flanders and Wallonia, the management of watercourses is split into five categories, with each a different set of water managers (see Table 10.1). The smallest watercourses (non-classified watercourses) are to be managed by landowners themselves, others are in hands of governmental bodies. The polders and wateringues form an exception to this rule.

Table 10.1 Watercourse managers for different watercourse categories.

	Flanders	Wallonia
Navigable watercourses	Waterwegen & Zeekanaal (W&Z) De Scheepvaart Department of Maritime Access	Direction Générale Opérationelle 2 (DGO2)
Non-navigable watercourses 1st category	Vlaamse Milieumaatschappij (VMM)	Direction Générale Opérationelle 3 (DGO3)
Non-navigable watercourses 2nd category	Provinces Polders and wateringues	Provinces
Non-navigable watercourses 3rd category	Municipalities Polders and wateringues	Municipalities Wateringues
Non-classified watercourses	Landowners	Landowners

They unite riparian landowners to collectively manage the watercourses of a specific area. Where they are still in place, these NGOs take over watercourse management from the government.

Apart from water management, spatial planning is also organized at three governmental echelons, i.e. regional, provincial, and municipal.

The principal legislative act on FRG in Flanders is the *Decree on Integrated Water Policy* (DIWP), which has been introduced in 2003 and substantially reformed in 2013.[2] In Wallonia, FRG is mainly steered by the *Plan PLUIES* (2003) and the subsequent *Water Code* (2004).[3]

10.4.2 The flood preparation and recovery arrangement

In contrast to water management and spatial planning, emergency planning and insurance policies are primarily regulated at the federal level. In addition, emergency plans are drafted at the provincial and municipal level. The main legislative act in terms of emergency planning is the Royal Decree of 16 February 2006, which harmonized emergency plans at different policy levels.[4]

In order to strengthen flood recovery, flood damage has been added to the general fire insurance policy with the 2005 Act on the Insurance of Natural Disasters.[5] Fire insurance packages are voluntary but 90–95% of the population is covered (Suykens et al. 2016). Flood damage not covered by insurance can in certain cases be compensated through the public disaster fund. This fund has been transferred to the regional governments through the 2014 Sixth State Reform.[6]

10.5 Comparing intra-state developments

Between 1995 and 2015, each of the Belgian FRGAs has undergone significant changes. Developments such as population growth, scarcely coordinated building expansion, state reforms, and financial constraints have severely challenged the Flemish and Walloon FRGAs. In response to these internal and external stressors, policymakers have taken considerable steps to adapt. Modifications took place under the form of spatial planning initiatives, improved coordination mechanisms, increased use of cost–benefit analysis, and altered state-citizen relationships.

10.5.1 Spatial planning initiatives

Extensive flood events in 1998 and 2002/03, pressure from European legislation, and international knowledge exchange gave rise to a new discourse to provide space for water (Mees et al. 2016a). This discourse stressed the importance of spatial planning instruments, apart from flood protection measures (embankments, flood control areas, and so on).

In Flanders, a series of innovative spatial planning instruments have been introduced since 2003. A first instrument has been the *water assessment* (2003), which obliges authorities to request advice from the relevant water manager on the impact of a permit, plan, or programme on the water system.[7] Second, the Flemish government has introduced the 'duty to inform' in 2013.[8] This instrument acknowledges the fact that inhabitants are often insufficiently aware of their flood risks. Therefore, it obliges property owners to provide widespread dissemination of information related to the vulnerability of floods of their property in the context of real estate transactions, in particular sale and rent exceeding nine years. The same year saw the introduction of 'signal areas'. These are undeveloped areas that could be valuable in mitigating floods due to their natural characteristics but which formally have a building destination (residential or industrial) (van Doorn-Hoekveld et al. 2016). As a result, there is an inconsistency between the spatial planning destination and the interests of the water system. For each of these areas, a deliberative trajectory should determine its further potential development. This process should be facilitated by another recently introduced instrument, namely 're-parcelling with destination exchange'.[9] Herewith, property rights in flood-prone areas can be exchanged with those in better located zones, which currently have a 'soft' destination (e.g. agriculture, nature). With the exchange, the first area's planning code changes into a soft destination and opposite.

In Wallonia, the flooding of 2002/03 was followed by the Plan PLUIES, the principal innovation of which is the introduction of flood cartography. Flood hazard and flood risk maps have been established, in accordance with the European Floods Directive (2007/60/EC). Furthermore, the relevance of spatial planning for a sound FRG has been acknowledged by the regional development scheme (SDER or *Schéma de Développement de l'Espace Regional*) for the Walloon Region of 8 November 2013. But the Walloon authorities' toolbox to take action in this matter is limited. The first SDER of 1999 stipulated as an objective that areas susceptible to flooding would be delineated.[10] Within the perimeters of these zones, restrictive provisions would guide further development. However, to this day, no planning regulation on this subject has been approved.

Similar to the water assessment in Flanders, the Walloon Spatial Planning Code (CWATUP) enables Walloon authorities to request the advice of water managers on spatial developments (article 136 CWATUP). In contrast to the Flemish Region, however, this request is not compulsory and the legal framework is less prescriptive on the thresholds for advice, the scope of the instrument, and the criteria applied.

10.5.2 Coordination mechanisms

Besides spatial planning tools, the DIWP and Plan PLUIES have created coordinating bodies in order to liaise developments in water management and spatial planning more effectively. In Flanders, this coordination takes place in the *Coördinatiecommissie voor Integraal Waterbeleid* (Coordination Commission on Integrated Policy, CIW), in Wallonia in the *Groupe Transversale Inondations* (Interdepartmental Flood Group, GTI).

The CIW assembles representatives from all Flemish water-management authorities and the departments of spatial planning, agriculture, and economy, next to representatives from the municipalities, provinces, and polders and wateringues (CIW 2015). The commission meets about five times a year to discuss ongoing developments and to establish a common approach toward integrated water management. The CIW hosts several working groups, of which one addresses water quantity management.

In Wallonia, the GTI has been established to coordinate the follow-up of the Plan PLUIES and to reflect on water quantity policy issues, including the consequences of climate change. The GTI is composed of representatives from regional and provincial administrations (spatial planning, environment, transport and infrastructure, agriculture, local authorities) and university experts. The group meets on a monthly basis and discusses all developments that take place within the different administrations in relation to flooding.

While the CIW and GTI were established after the 1998 and 2003 floods, in response to a persistent need for coordination at a regional level, experiments to coordinate water management at sub-basin scale go back further in time. Both in Wallonia and Flanders, coordinating bodies have been installed in several sub-basins since the beginning of the 1990s (Mees, Suykens, and Crabbé 2017).

- In Flanders, sub-basin committees were established in a number of sub-basins in 1990–1993. The DIWP (2003) replaced these organizations by the *sub-basin boards*, which cover all Flemish sub-basins, and provided them with a legal basis and uniform operational procedures. The Flemish sub-boards bring together representatives from relevant Flemish governmental departments and from local governments. The main task of these organizations is to develop the sub-basin specific parts of the River Basin Management Plans (RBMPs) for the Scheldt and the Meuse, which include both water quantity and quality issues. The RBMP is to be implemented by the partners of the sub-basin board.
- In Wallonia, river contracts (*contrat de rivière*) are active in most sub-basins. They are created through bottom-up initiatives but with support from the regional level. The first river contract was established in 1993. With the Order of the Walloon Government of 13 November 2008, the system of river contracts has been elaborated to the entire territory of Wallonia. The river contracts bring together a wide range of local and supra-local stakeholders dealing with water management. Together, these stakeholders draft an action programme, which is approved by the Walloon government. The coordination of the river contract is in hands of an independent non-profit organization, which is financially supported by the Walloon government and the other partners of the river contract.

Both in Wallonia and Flanders, the coordinating bodies at sub-basin scale are to guide and coordinate the development and execution of action programmes. In Flanders, this task is strongly subjected to formal requirements and counts for most of the provided staff's time. The planning procedures in the Walloon system are much less formal, leaving room to the coordinators of the river contracts to spend time and resources on the most relevant issues at stake in the basin. Another important difference between the Flemish and Walloon system is that the Flemish sub-basin coordinators belong to either VMM, W&Z, or De Scheepvaart, whereas the coordination of the river contracts is organized in an independent non-profit structure (Mees et al. 2017).

10.5.3 Citizen engagement

In accordance with EU legislation, water managers in Belgium are obliged to include certain forms of public participation in their decision-making, e.g. the six-month public consultation procedure for the Flood Risk Management Plans (Priest et al. 2016; Pettersson et al. 2017). Mostly, the participation takes place through passive forms of citizen consultation, i.e. by providing the possibility for (organized) citizens to react on a draft plan in a late stage of the decision-making. The consultation procedures require citizens to be familiar with the ongoing decision-making process and its terminology. In more and more cases, however, a trend of intensified citizen engagement is noticeable, whereby the input of citizens and their organizations is actively collected in hearings, information markets, discussion tables, and so on.

(Organized) citizens are involved both by the Flemish sub-basin boards and the Walloon river contracts but their engagement differs substantially (Mees et al. 2017).

- Generally speaking, the involvement of non-governmental actors in the Flemish sub-basins is limited to their representation in the formal sub-basin councils. In these councils, they can advise on the plans and reports developed within the sub-basin board and deliberate on more general issues of integrated water governance, e.g. how to deal with problems of erosion in the sub-basin, and the like. In specific cases, local stakeholders can become more actively involved in addressing local water issues.
- The river contracts, conversely, are the interface between civil society and the authorities in the Walloon water governance. Public participation is inherent to the structure of river contracts; every river contract includes a number of non-governmental stakeholder groups (e.g. nature organizations, fishing associations) among its partners. These stakeholders are thus actively involved in the draft of its action programme and can participate in the execution of the defined actions. In addition, the stakeholders can participate in working groups on specific projects within the river contract. Stakeholder involvement is thus strongly interwoven into the river contracts' management structure, whereas it is organized as a separate process in most of the Flemish sub-basin boards.

In general, it can be said that the involvement of citizens in Walloon and Flemish FRG is rather limited and is mainly restricted to passive public participation. This holds true for their inclusion in decision making but even more so in policy delivery

(Mees et al. 2016b). FRG is perceived as an exclusive governmental task and the protection of the population towards flooding consequently depends highly on governmental action and spending. Since 2013, however, the Flemish government advocates the so-called multi-layered water safety (MLWS), which stands for FRG based on prevention, protection, and preparation measures and for shared responsibilities between plural governmental actors, the private sector, and citizens. The first prudent steps towards increased citizen responsibilities have been the 'duty to inform' (2013), and a pilot study and awareness-raising brochure on individual protection measures. The fact that flood risks are obligatorily included in first-party property insurance in Belgium also adds to the responsibilities of private parties (Suykens et al. 2016). In Wallonia, the 3P-discourse (i.e. prevention, protection, preparation) has found its way to water managers as well but so far it has not been accompanied by a discourse to actively increase citizens' engagement in implementing flood-risk measures.

10.6 Evaluating resilience, efficiency, and legitimacy

Both the Walloon and Flemish FRG arrangement underwent changes in relation to spatial planning, policy coordination, and citizen engagement. In some cases, these developments were similar, in others very different approaches were applied. How do these developments affect the resilience, efficiency, and legitimacy of the arrangements?

10.6.1 Impact on resilience

In a densely populated country such as Belgium, risk-prevention measures are a valuable addition to classical flood protection to increase society's resilience to flooding. Urban sprawl is particularly outspoken in Flanders, which is likely one of the reasons why the Flemish government has initiated more ambitious initiatives related to spatial planning than its counterpart in Wallonia.

While the article 136 CWATUP in Wallonia offers spatial planning authorities the possibility but not an obligation to request water managers' advice, authorities in Flanders are, in appropriate cases, obliged to do so in accordance with the DIWP. The outcome of the Flemish water assessment is not binding for spatial planners, but in practice, few local authorities have deviated from it in recent years. Especially since the legal reform of the water assessment in 2011, the effectiveness of the instrument has increased. The media storm after the 2010 flood, which had caused considerable damage as a result of inadequate spatial planning, contributed to local awareness-raising on the subject. The water assessment does not aim to prohibit new constructions in flood-prone areas but imposes measures to reduce or cancel out their impact on the water system.

An evaluation of the impact of instruments such as the duty to inform, signal areas, and re-parcelling with destination exchange is premature considering their recent introduction. They have been, however, applauded as important tools to remediate spatial planning policies with the interests of the water system and FRG.

The Flemish and Walloon Regions have adopted divergent approaches to the challenge of spatial planning concerning flood risks. In Wallonia, actors aspire to an overruling legislative framework, stipulating uniform rules for flood vulnerable areas, which can be generally applied. It appears to be difficult, however, for such legislation to get through the necessary processes for final adoption. The Flemish government, in contrast, pursues a case-based approach, by which flood risks and their consequences are evaluated case by case, i.e. through the water assessment and the exercise of the signal areas. This strategy entails the risk that its intentions are watered down when meeting local political realities on the ground. So far, however, it seems to have led to more tangible results than does the general approach pursued in the Walloon Region.

But whether the initiatives taken in Flanders will suffice to tackle the challenges of flood risks today and in the future remains to be seen. Despite their best intentions, local authorities struggle to impose strict building regulations in flood-prone areas. In areas designated as building zones, local actors consider it politically and/or financially unfeasible to avoid further development despite their flood vulnerability. New buildings are in most cases still permitted, although on the condition of implementing compensation and damage reduction measures. But there is no systemic and robust mechanism to follow up on the implementation of these building and compensation measures on the ground. Moreover, even if these measures would manage to cancel out the impact of a building on the water system entirely, they do still put an increasing burden on crisis response services, thereby shifting the focus from flood risk prevention to flood preparation measures.

In addition, even if a further increase of new flood risks could be entirely avoided through the spatial planning instruments mentioned earlier, existing risks will remain and might even increase in the context of climate change. Both in Flanders and Wallonia, the existing spatial planning legislation is currently insufficiently adapted to also address existing flood risks, caused by spatial developments in the past.

Next to product indicators, the definition of resilience used in this chapter also focuses on the process that should lead to a resilient arrangement, namely by assessing its capacity to adapt. An important role hereby is played by bridging mechanisms. Bridging mechanisms have been defined as 'inter-linkages between sets of actors, aiming to intensify interactions in their pursuit of various flood risk management strategies in order to cope with the difficulties relating to fragmentation' (Gilissen et al. 2016). Both the Walloon and Flemish FRGA are highly fragmented and sufficient coordination is thus indispensable to achieve a resilient arrangement. In both regions, actors appear strongly aware of the need for better coordination of their actions. Consequently, coordinating bodies have been installed to exchange information and discuss uniform planning between the different actors involved, namely the CIW in Flanders and GTI in Wallonia. These actors do not only include water managers but also spatial planners and crisis managers. Hence, they facilitate a FRG that strengthens both the arrangement's capacity to resist, absorb and recover, and to adapt.

The CIW and GTI have improved coordination at the regional policy making level significantly, but their impact on FRG often proves insufficient at the level of policy implementation. A gap seems to exist between the policy makers at the regional level and the actors responsible for its implementation at local level. In Flanders particularly, FRG has been developed in a top-down manner, with limited participation possibilities for local authorities in the design of new policy initiatives.

Since local authorities are to a great extent responsible for the implementation of these actions, large-scale adoption is regularly lacking in the field. Municipalities are often unaware of the existence of new legislation and instruments or are incapable or unwilling to apply them.

Coordination bodies at sub-basin level, i.e. the sub-basin boards in Flanders and river contracts in Wallonia, are to facilitate coordination and knowledge exchange between local actors. Mees et al. (2017) found that the Walloon river contracts were, in general, evaluated remarkably more positively by the actors of the FRGA than the sub-basin boards in the Flemish FRGA. This difference could be due to the fact that the sub-basin boards have been imposed in a top-down manner by the Flemish government and have to fulfil a large range of formal duties, while the river contracts are developed through a bottom-up process and have a flexible governance structure. Of equal importance might be the fact that the river contracts are better equipped in terms of staff.

The coordination mechanisms installed at local level have been delineated per sub-basin, thereby crossing administrative boundaries such as municipalities and provinces. Interestingly, however, the coordinating bodies do not cross regional frontiers, despite the fact that several watercourses are located both in Wallonia and Flanders. At some places, interregional coordination exists on an ad-hoc basis, due to interpersonal contacts between local actors. In general, however, the information exchange between Flemish and Walloon actors is very limited. Coordination issues in basins crossing national borders can be found at several places in the EU (Suykens 2015). In the federal state of Belgium, where water management is governed in three separate Water System Arrangements, these issues exist as well within the state itself. Hence, increasing contacts between the two regions could potentially improve the resilience of both arrangements, particularly in terms of their capacity to adapt.

10.6.2 Impact on efficiency

In Flanders, the discourse to implement prevention, protection, and preparation measures aspires not only to increase the resilience of the arrangement resilience but also its efficiency. Through cost–benefit analysis, Flemish water managers aim to define a package of measures that generates the most optimal output to input ratio (VMM 2014). It is calculated that a well-considered mix of protection and spatial planning measures would benefit both the resilience and efficiency of the FRGA. In Wallonia, the linkage between the 3P-discourse and efficiency concerns is less outspoken.

In both regions, actor fragmentation forms a barrier to resource efficiency. Both at the regional and local scale, the high level of fragmentation often leads to overlaps in the implementation of projects by different actors. The installed coordination mechanisms are, therefore, an important step in improving the efficiency of the FRGAs. The coordination bodies have a direct impact through their activities, but even more important is that these bodies facilitate the establishment of an informal network between the actors involved, which leads to better information exchange and informal coordination. In both regions, interviewed actors indicated that their informal contact network has improved significantly since the development of local and regional coordination platforms.

10.6.3 Impact on legitimacy

The legitimacy of Belgian FRG is primarily focused on output rather than input legitimacy. Flood-risk managers in Belgium increasingly include public participation in the decision-making process, but these consultations mostly take place only in the final stages of decision making. Despite the limited citizen involvement, the acceptability of FRG appears to be high in general. Traditionally, FRG in Belgium is considered to be an exclusive governmental responsibility. Consequently, citizens' interest to participate actively in the decision-making process is limited.

In Flanders, however, authorities are increasingly advocating to share flood risk responsibilities between governmental actors, the private sector, and citizens (Mees et al. 2016b). But this appears to be a difficult message to bring in an FRGA that is strongly state-centred. Mees et al. (2016b) argue that if the government wants to share its flood-risk responsibilities with non-governmental actors in the future, it will also need to provide additional opportunities for them to participate in the decision-making process. If not, the new approach can severely hamper the FRGA's legitimacy and resilience.

Despite the fact that the river contracts are widely applauded by Walloon stakeholders, they are rarely used for public participation. In most cases, water managers prefer to organize participation processes outside the river contract structure. We consider this to be a missed opportunity since the river contracts unite the most important stakeholders and are well-connected to the local population. Also the Flemish sub-basin boards could strengthen the link between governmental flood-risk managers and civil society. Today, the input of non-governmental stakeholders in the activities of the sub-basin boards is to a large extent limited to a formal advisory role. By including (organized) citizens more actively in the development and implementation of sub-basin action plans, the sub-basin boards could offer a valuable contribution to the pursuit of integrated water governance.

10.7 Conclusion

Between 1995 and 2015, FRG in Belgium has undergone significant changes. Both in Flanders and Wallonia steps have been taken to increase the attention to flooding in the spatial planning domain, to improve the coordination between different actors and domains, and to involve non-governmental stakeholders in the FRGA.

We found that the Flemish legislative and policy initiatives in the spatial planning domain and the increased coordination have had a positive impact on the resilience and efficiency of the arrangement. Still, further efforts are needed to ensure adequate implementation on the ground and to deal with already built-up, flood-prone areas.

Also in the Walloon Region, additional action is needed to increase society's capacity to absorb and recover from flooding. The establishment of regional and local coordinating bodies, however, is an important step towards a more resilient and efficient FRGA.

The Walloon river contracts and Flemish sub-basin boards offer opportunities to more intensely involve (organized) citizens in FRG. Particularly in Wallonia, these sub-basin organizations constitute a valuable link between the governmental flood-risk managers and civil society. Considering the fact that water managers increasingly acknowledge the need to involve non-governmental actors in the approach of flooding, it would be advisable to employ the river contracts and sub-basin boards more intensely to facilitate citizen engagement both in the decision making and implementation of FRG. This way, an FRGA can be created that is co-produced by governments and citizens and scores well in terms of resilience, efficiency, and legitimacy.

In the past decades, Belgian policymakers have proven to be able to reform their FRGA in response to current and future challenges. On their own, however, these measures are insufficient for a resilient FRGA. Further efforts will be required, which can in some cases be costly, experimental, and/or controversial. In order to increase the chances of a successful FRG implementation, Belgian policymakers will need to pay sufficient attention to best practices from in- and outside the country and create room to experiment with different flood-risk solutions.

Notes

1. In this chapter, four flood risk strategies have been distinguished, whose definition is based on Hegger et al. (2014). Flood risk prevention aims to 'keep people away from water' and primarily includes measures of proactive spatial planning. Flood protection has the purpose to 'keep water away from people' through infrastructural measures such as embankments, flood retention basins, and so on. Flood preparation stimulates an adequate crisis response during flood events. Flood recovery facilitates rapid (financial) recovery following a flood event.
2. *Belgian Official Journal*, 14 November 2003.
3. *Belgian Official Journal*, 23 September 2004.
4. *Belgian Official Journal*, 15 March 2006.
5. *Belgian Official Journal*, 11 October 2005.
6. More specifically, through the Act of 6 January 2014 related to the Sixth State Reform, *Belgian Official Journal*, 6 January 2014.
7. Article 8 DIWP and further implemented by the Order of the Flemish Government of 20 July 2006 related to the water assessment.
8. Decree of 19 July 2013 modifying various provisions of the DIWP of 18 July 2003, *Belgian Official Journal*, 1 October 2013.
9. Article 1.1.4, Section 2 Decree on land organization of 28 March 2014, *Belgian Official Journal*, 22 August 2014.
10. Thematic Fiche 17 'Risque naturels et technologiques', SDER, adopted by the Walloon Government on 27 May 1999.

References

Alexander, M., Priest, S., and Mees, H. (2016). A framework for evaluating flood risk governance. *Environmental Science and Policy* 64: 38–47. https://doi.org/10.1016/j.envsci.2016.06.004.

Bruneau, M., Chang, S., Eguchi, R. et al. (2003). A framework to quantitatively assess and enhance the seismic resilience of communities. *Earthquake Spectra* 19: 733–752.

CIW (2015). *Over CIW*. http://www.integraalwaterbeleid.be/nl/over-ciw (accessed 13 February 2019).

Cosens, B.A. (2010). Transboundary river governance in the face of uncertainty: resilience theory and the Columbia River treaty. *Journal of Land, Resources & Environmental Law* 30 (2): 239–265.

Davoudi, S., Shaw, K., Haider, L.J. et al. (2012). Resilience: a bridging concept or a dead end? 'Reframing' resilience: challenges for planning theory and practice. *Planning Theory and Practice* 13 (2): 299–333.

de Bruijn, K.M. (2004). Resilience indicators for flood risk management systems of lowland rivers. *International Journal of River Basin Management* 2 (3): 199–210. https://doi.org/10.1080/15715124.2004.9635232.

Gilissen, H.K., Meghan, A., Beyers, J. et al. (2016). Bridges over troubled waters – an interdisciplinary framework for evaluating the interconnectedness within fragmented flood risk management systems. *Journal of Water Law* 25 (1): 12–26.

Hegger, D.L.T., Driessen, P.P.J., Dieperink, C. et al. (2014). Assessing stability and dynamics in flood risk governance: an empirically illustrated research approach. *Water Resources Management* 28 (12): 4127–4142. https://doi.org/10.1007/s11269-014-0732-x.

Huitema, D., Mostert, E., Egas, W. et al. (2009). Adaptive water governance: assessing the institutional prescriptions of adaptive (co-)management from a governance perspective and defining a research agenda. *Ecology and Society* 14 (1): https://doi.org/10.1111/j.1541-1338.2009.00421_2.x.

Larrue, C., Hegger, D., and Trémorin, J.B. (2013). Researching flood risk policies in Europe: a framework and methodology for assessing flood risk governance. (STAR-FLOOD report no D2.2.1) http://www.starflood.eu/documents/2014/07/d-2-2-1.pdf (accessed 13 February 2019).

Liefferink, D. (2006). The dynamics of policy arrangements: turning round the tetrahedron. In: *Institutional Dynamics in Environmental Governance* (ed. B. Arts and P. Leroy), 45–68. Dordrecht: Netherlands: Springer.

Lockwood, M., Davidson, J., Curtis, A. et al. (2010). Governance principles for natural resource management. *Society and Natural Resources* 23 (10): 986–1001. https://doi.org/10.1080/08941920802178214.

Mees, H., Suykens, C., Beyers, J.C. et al. (2016a). *Analysing and Evaluating Flood Risk Governance in Belgium. Dealing with Flood Risks in an Urbanised and Institutionally Complex Country*. UA, KULeuven, Belgium: STAR-FLOOD Consortium.

Mees, H., Tempels, B., Crabbé, A. et al. (2016b). Shifting public-private responsibilities in Flemish flood risk management. Towards a co-evolutionary approach. *Land Use Policy* 57: https://doi.org/10.1016/j.landusepol.2016.05.012.

Mees, H., Suykens, C., and Crabbé, A. (2017). Evaluating conditions for integrated water resource management at sub-basin scale: a comparison of the Flemish sub-basin boards and Walloon river contracts. *Environmental Policy and Governance* 27: 59–73. https://doi.org/10.1002/eet.1736.

Nelson, D.R., Adger, W.N., and Brown, K. (2007). Adaptation to environmental change: contributions of a resilience framework. *Annual Review of Environment and Resources* 32: 395–419. https://doi.org/10.1146/annurev.energy.32.051807.090348.

OECD (2015). *OECD Principles of Water Governance. Directorate for Public Governance and Territorial Development*. Paris: OECD.

Pettersson, M., Van Rijswick, M., Suykens, C. et al. (2017). Assessing the legitimacy of flood risk governance arrangements in Europe: insights from intra-country evaluations. *Water International* 42 (8): 929–944.

Priest, S.J., Suykens, C., Van Rijswick, H.F.M.W. et al. (2016). The European Union approach to flood risk management and improving societal resilience: lessons from the implementation of the floods directive in six European countries. *Ecology and Society* 21 (4): 50.

Scharpf, F.W. (1997). Economic integration, democracy and the welfare state. *Journal of European Public Policy* 4 (1): 18–36.

Schmidt, V.A. (2013). Democracy and legitimacy in the European Union revisited: input, output and 'throughput'. *Political Studies* 61 (1): 2–22. https://doi.org/10.1111/j.1467-9248.2012.00962.x.

Suykens, C. (2015). EU water quantity management in international River Basin districts: crystal clear? *European Energy and Environmental Law Review* 24: 134–143.

Suykens, C., Priest, S.J., van Doorn-Hoekveld, W.J. et al. (2016). Dealing with flood damages: will prevention, mitigation, and ex post compensation provide for a resilient triangle? *Ecology and Society* 21 (4): 1. https://doi.org/10.5751/ES-08592-210401.

United Nations Office for Disaster Risk Reduction (UNISDR) (2009). UNISDR Terminology on Disaster Risk Reduction. https://www.unisdr.org/we/inform/terminology (accessed 12 February 2019).

van Doorn-Hoekveld, W.J., Goytia, S.B., Suykens, C. et al. (2016). Distributional effects of flood risk management – a cross-country comparison of preflood compensation. *Ecology and Society* 21 (4): 26. https://doi.org/10.5751/ES-08648-210426.

Van Tatenhove, J., Arts, B., and Leroy, P. (eds.) (2000). *Political Modernisation and the Environment. The Renewal of Environmental Policy Arrangements*. Dordrecht/Boston/London: Kluwer Academic Publishers.

VMM (2014). *Onderbouwing van het Overstromingsrisicobeheerplan van de onbevaarbare waterlopen. ORBP-analyse Basisrapport*. Erembodegem: VMM.

VMM (2015). *MIRA Klimaatrapport 2015 over waargenomen en toekomstige klimaatveranderingen*. Aalst: VMM.

Walker, B., Holling, C.S., Carpenter, S.R. et al. (2004). Resilience, adaptability and transformability in social–ecological systems. *Ecology and Society* 9 (2): 5. http://www.ecologyandsociety.org/vol9/iss2/art5 (accessed 13 February 2019).

Part III
Droughts

III.1

Actors Involved in Drought Risk Management

11

European Actors and Institutions Involved in Water Scarcity and Drought Policy

Ulf Stein and Ruta Landgrebe
Ecologic Institute, Berlin, Germany

11.1 Introduction

The growing severity of economic, social, and environmental impacts during recent drought events has provided the grounds for increasing pressure on the EU to promote a common European drought policy (Brebbia and Kungolos 2007). Despite that, there is no one drought and scarcity policy at the EU level yet. The complexity of the droughts phenomenon involving many environmental and social elements is reflected in its governance system. Accordingly, the role of European-level actors in drought governance is shaped by very complex and highly inter-connected governance systems of water and climate change management. The actors in these core policy fields further touch upon many other fields and actors, such as land management, agricultural policy, and civil protection. Furthermore, policy implementation depends on actors in the economic sectors who use and depend on water such as those in agriculture, tourism, industry, energy, and transport. With the Communication in 2007 and the Policy Review in 2012, water scarcity and drought (WS&D) policy is relatively young and thus European-level actors and networks are still evolving. The role of European level actors in WS&D governance is poorly understood and not yet systematically analysed in the literature. Through the stakeholder mapping, we would like to close this gap and determine the possible additional value of European-level actors and their networks to national, regional, or local level drought risk and water scarcity management as well as the link to the international-level actors. The chapter discusses: (i) which actors or networks would most influence a positive change process, and (ii) what overlapping roles and/or

Facing Hydrometeorological Extreme Events: A Governance Issue, First Edition.
Edited by Isabelle La Jeunesse and Corinne Larrue.

gaps exist. This stakeholder mapping can support responsible actors in identifying the factors that support or limit effective development and implementation of WS&D policy.

11.2 Actors in the European Union related to WS&D policy

As shown in Table 11.1, the various types of actors can be grouped into four broad categories: European Union Institutions, European Technical Agencies, other pan-European actors, and international actors. The main policy actors involved in WS&D at

Table 11.1 Non-comprehensive overview of European Union Institutions and related actors in the field of water scarcity and drought (WS&D).

European Unions Institutions[a]	European Technical Agencies[b]	Other pan-European actors	International actors
The European Council; The Council of the European Union; The European Parliament; The European Commission; The Directorate-General for Environment (DG ENV); • The Directorate-General for Climate Action (DG CLIMA); • The Directorate-General for Agriculture and Rural Development (DG AGRI); • The Directorate-General for European Civil Protection and Humanitarian Aid Operations (DG ECHO); • Committee of the Regions (CoR)	The European Environment Agency (EEA) • European Water Data Centre • Biodiversity Data Centre • Climate Change Data Centre • Land Use Data Centre Joint Research Centre (JRC) • The European Drought Oberservatory (EDO) • The European Soil Data Centre	The European Water Partnership;[c] The International Office for Water (IOW);[d] The Mediterranean Working Group (MED-EU Water initiative);[e] WFD CIRCA Interest Group 'Implementing the; Water Framework Directive and the Floods Directive'; The European Drought Centre (EDC)[f]	The International Commission on Irrigation and Drainage (ICID);[g] Working Group on Managing Water Scarcity under Conflicting Demands (WG-MWSCD);[h] Integrated Drought Management Programme (IDMP)[i]

[a] https://europa.eu/european-union/about-eu/institutions-bodies_en
[b] https://europa.eu/european-union/about-eu/agencies_en
[c] http://www.ewp.eu.
[d] http://www.oieau.fr/?lang=en
[e] http://www.ypeka.gr/medeuwi.
[f] http://europeandroughtcentre.com.
[g] http://www.icid.org.
[h] https://www.icid.org/wg_mwscd.html
[i] http://www.droughtmanagement.info.

the European level are the European Union's Institutions. They include the institutions that set the policy agenda (the European Council), law-making institutions (the European Parliament, the Council of the European Union and the European Commission), other EU institutions that link regional and local authorities in the European decision-making process (e.g. the Committee of the Regions), technical agencies (e.g. the European Environment Agency (EEA), Joint Research Centre [JRC]), and advisory or expert groups (e.g. the WS&D Expert Group/Network). In addition, a number of interest platforms and influential groups support the implementation of WS&D policy (e.g. the European Drought Centre (EDC), the Integrated Drought Management Programme [IDMP]).

The scale of the multi-actor environment of drought governance at the European level owes its complexity to the WS&D issue itself, which touches upon many other fields such as water, climate change, and land use. The actors at the European level represent mainly the policy and science domains including the EU's institutions and their advisory or expert groups, technical and scientific agencies, and a number of interest platforms and influential groups supporting the implementation of WS&D policy. Whilst the EU institutions and technical and scientific agencies represent European public actors, the other pan-European actors include other groups, such as private actors and civil society. The international actors group though might be composed of representatives of international public bodies.

11.3 Roles and powers of European actors and institutions involved in WS&D policy

The roles of the European actors are determined by their level of authority, or power, and the various major responsibilities are clear. The European Council sets the policy agenda whilst the European Parliament asks for specific legislation. The European Commission actually proposes legislation and the Council of the European Union and the European Parliament amend and adopt proposed legislation. The Committee of the Regions advises the latter institutions when drawing up legislation on matters concerning local and regional governments. The European Commission is also responsible for executing policies by ensuring that all EU legislation is applied by all Member States, and that all Member States implement the budget. Lastly, the Council of the European Union and the European Parliament are responsible for approving the budget.

The Commission is collectively accountable to the Parliament, as it elects the President of the Commission and approves (or rejects) the appointment of the Commission as a whole. When comparing the decision-making institutions, they have power to address different levels. Whilst the Council of the European Union responds to proposed new legislation from the perspective of the representatives of the governments of the Member States and the European Parliament responds from the perspective of people's/voter's interests grouped by political affinity, the Committee of the Regions does it from the perspective of regional and local authorities, linking them in the European decision-making process. In this way, the link to national, regional, and local-level actors as well as governments and citizens is ensured.

When proposing new policies and measures, the Commission tries to find the best solution in the general interest of the EU and its Member States. Creating expert groups is one of the ways to gather the full range of non-binding views on an issue. For WS&D, the Commission's work is assisted by the expert groups on WS&D specifically and indirectly by the Steering Group on Adaptation to Climate Change. Generally, expert groups are set up for a limited period. Additionally, the Common Implementation Strategy (CIS) of the Water Framework Directive (WFD) recognized the need for working groups to provide guidance documents on key aspects of the WFD – including topics indirectly linked to WS&D – to support implementation. These permanent working groups are supported by temporary steering, drafting, and expert groups.

When implementing policies and making decisions, the EU institutions use the support of technical and scientific agencies, e.g. the EEA or the JRC. By performing technical and scientific tasks (mainly data gathering, analysis, and sharing, as well as technical advice), the technical and scientific agencies support the EU institutions with independent evidence throughout the whole policy cycle, thus contributing to the implementation of EU policies. The technical and scientific agencies also support cooperation between the EU and national governments by pooling technical and specialist expertise from both the EU institutions and national authorities (EC 2017a; EU 2017). Unlike expert groups, technical and scientific agencies are set up for an indefinite period.

In addition to the public actors, private actors and civil society also support WS&D policy at the European level. A number of influential groups and interest platforms have been formed mainly around the water and agriculture sectors. These include, for example, the European Water Partnership, the International Office for Water (IOW), the Mediterranean Working Group (MED-EU Water initiative) or the EDC. In addition, international actors with a specific focus on Europe also influence WS&D policy, for example WG-MWSCD, and the IDMP. The role of private actors and civil society focuses mainly on knowledge and experience exchange, gathering of empirical evidence, and policy advice (through projects).

In analysing the European actors and institutions involved in WS&D policy, we focus mainly on their role and power in (i) strategic policy making, (ii) financing – i.e. access to financial resources, (iii) data gathering and sharing – i.e. access to information resources, and (iv) technical and policy advice. Based on this methodology, Table 11.2 provides a simplified overview of the roles and powers of European actors and institutions involved in WS&D policy. Access to other resources, such as networks, is not shown in the table but will be discussed in the text, considering thereby strategic interest and preferences.

The roles and powers of the Council of the European Union are quite similar to the European Parliament though weaker. It has the same power to adopt the proposed legislation and approve the budget, but does not have supporting committees. Unlike the European Parliament, which represents the EU's citizens and is directly elected by them, the Council of the European Union is represented by the governments of the Member States.

In contrast to the European Commission and the European Parliament, the institutions such as the Committee of Regions have opinion-forming influence but lack the financial resources to provide financially backed-up instruments. The European Council sets the EU's overall political direction – but has no powers to pass laws or approve the budget.

Table 11.2 Roles and powers of European actors and institutions involved in water scarcity and drought (WS&D) policy.

Roles and powers	Policy agenda setting at EU level	Legislative power	Decision and advisory power	Implementation power
Strategic policy making	European Council	European Commission European Parliament	Council of the European Union European Parliament Committee of the Regions	European Commission
Financing (access to financial resources)	Council of the European Union European Parliament			European Commission
Data gathering and sharing (access to information resources)	Technical and scientific agencies	Technical and scientific agencies	Technical and scientific agencies	Technical and scientific agencies
Technical and policy advice	Influential groups and interest platforms	Expert groups		Technical and scientific agencies

The presented roles and powers of the main European actors and institutions is however not exceptional for WS&D policy and can be applied to many other policy sectors. A more in-depth analysis, including committees, commissions, working groups, experts, and interest groups provides better insights on the distribution of roles and powers in WS&D policy.

11.4 Mapping European actors and institutions involved in WS&D policy

With the various actors and institutions related to European WS&D policy described earlier and delineated according to their roles and powers, it is also beneficial to visually assess where and how actors may interrelate. Figure 11.1 provides a schematic overview of the key actors of WS&D policy in Europe. As mentioned above, the European Commission, especially DG Environment (DG ENV) and DG Agriculture (DG AGRI), is the most influential actor in terms of setting the topic on the political agenda by proposing specific legislation and can delegate responsibilities to other 'Main Actors'. However, the actions of DG ENV and DG AGRI are not coordinated with WS&D specifically in mind.

Whilst DG ENV is directly responsible for water management and addresses WS&D as a sub-theme, the interest of DG AGRI is related to the fact that the agriculture sector is recognized as one of the key vulnerable sectors to climate change, as well as a major water-demanding sector, especially in Southern Europe. Indeed, WS&D are

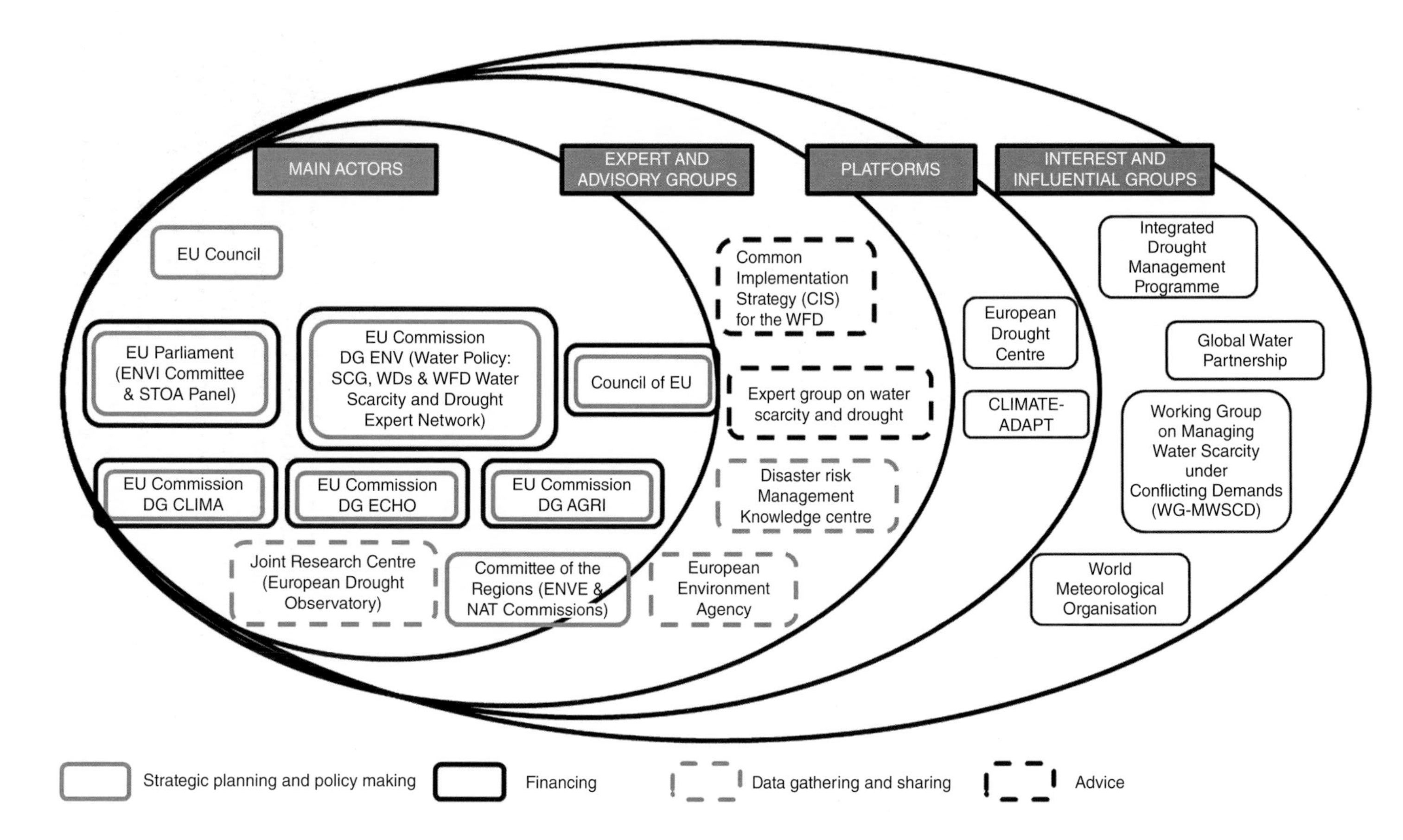

Figure 11.1 Institutional mapping of water scarcity and drought (WS&D) management at the EU level.

highly likely to increase in severity as a result of climate change. After DG ENV and DG Agri, both DG Clima and DG Echo are important as they tend to have a more indirect link to WS&D policy. The European Commission is responsible for the implementation of the policies in the Member States and correct spending of allocated budgets. In addition to the European Commission, the European Parliament also holds a very important role and strong power by adopting the proposed legislation, providing suggestions for a new legislation on the basis of a reviewed annual programme of the Commission, and approving the budget. The Parliamentary Committee ENVI (European Parliament 2017a) and the STOA Panel (European Parliament 2017b) have a direct influence on WS&D policies, as they provide opinion on these issues and focus on resource and water management; the STOA Panel in addition considers agricultural issues. Of no small importance is the fact that the European Commission is collectively accountable to the Parliament which acts to influence the power distribution. Nevertheless, the European Commission holds more expertise through the research of the JRC which provides independent scientific advice and support to EU policy makers through the whole policy cycle. The European Drought Observatory (EDO), established by the JRC, is responsible for drought detection, monitoring, and forecasting in Europe (EC 2017b). The EEA serves primarily the main group of the European-level policy actors, including the European Commission, the European Parliament, the Council and the Member States. In addition to this central group of European policy actors, it also serves other EU institutions such as the Economic and Social Committee and the Committee of the Regions.

11.5 Discussion

Taking into consideration the multi-level governance perspective, which suggests that policy-making responsibility is shared amongst a variety of actors at European, sub-national, and national levels, it is clear the EU institutions are not the only policy-making actors. Trnski (2004) argues that sub-national actors are sometimes as important to EU policy making as central governments and EU institutions. Each of these actors holds important resources, such as information, expertise, and political power and all are engaged in a bargaining relationship. Thus, even if inter-institutional decisions are taken on the basis of formalized procedures, the role of informal consultation, negotiation, and exchange of expertise remains significant (Popa 2010).

The influence of informal procedures is well visible in WS&D policy development. The most policy action and assessment occurred in 2007 and 2012. In recognition of the acuteness of WS&D challenges in Europe, the Commission undertook an in-depth assessment of the situation at the EU level, and issued a Communication on these issues in 2007. The WS&D policy was pushed at the EU level by the Portuguese presidency in 2007 (Kraemer 2007) and a group of mainly southern Member States informally lead by Portugal and Spain. These countries used their momentum as a counterweight to the floods policy initiated by the central and northern European countries (e.g. Germany).

The year 2012 was the European year of water and as part of that the European Commission published its 'Blueprint to safeguard European waters'. The Blueprint comprises reviews of Europe's policies on WS&D in addition to the reviews of the

WFD (EC 2000) and of the water related aspects of climate change adaptation and vulnerability. To accompany and inform the Blueprint, the EEA produced six reports throughout 2012 on the state of Europe's water. The reports were developed in coordination with the EEA's member countries and the European Commission DG ENV. The reports were also produced in coordination with three other review processes led by DG ENV: the review of Europe's WS&D policy, the review of water vulnerability and climate change adaptation policy, and the review of the River Basin Management Plans established under the WFD (CESR 2011; EC 2012a, 2012b; EEA 2012).

The European Parliament, which represents the EU's citizens and is directly elected by them, has a more national focus in comparison to the Council of the European Union which comprises representatives of the governments of the Member States, or the European Commission whose officials should be characterized by independence, impartiality, and exclusive pursuit of Community interests (Popa 2010). However, the experts appointed by Member States in the committees and working groups in different EU institutions may act as government officials representing national interests, as 'impartial' technocrats or as 'European experts, representing Community interests'. A clear delineation of these roles may be difficult to draw as one or the other is activated or becomes dominant depending on the context of the decision, the mandate of the working group or even the group dynamics (effects of the composition of intra-group interaction) (Popa 2010). Thus interest groups or lobby organizations aiming to influence outcomes in the European arena can use different EU institutions and shape the EU as a multilevel governance polity.

Another aspect influencing actor roles in WS&D policy is the definition of both terms themselves. Strosser et al. (2012) argues that despite the established concepts of WS&D, there is still debate about their precise definition, in particular regarding droughts. As both concepts relate to water quantity issues and can have similar effects, water managers, policy makers, public, and media often use both terms in an imprecise manner. This can lead to confusion in policy documents, in particular for defining adequate policy responses, and thus also confusion for defining responsibilities for taking policy actions to address both issues. Up to now, the majority of actors are responsible for both issues and this situation increases the chances of overlooking one of them, in particular droughts due to their complexity. Only a few actors, such as the Disaster Risk Management Knowledge Centre (DRMKC), the EDO and the European Drought Early-Warning System, focus solely on drought issue.

Due to the complexity of the issues and high inter-connectedness of water, land, and climate governance systems, the EU institutions address WS&D management issues from a multi-objective perspective, where each involved actor has its own mandate. The non-structural nature of drought as well as the lack of scientific data and approaches to predicting drought impacts comparing to water scarcity seems to create a political vacuum in which suitable interventions lose their position in comparison to other political priorities. This might explain why the WS&D issue is approached using water saving measures (which are also the majority of measures in drought management plans), thus addressing rather a water scarcity issue (the usual role of DG ENV – WFD) and not droughts. Responsibility on droughts is shared between the DG ENV, 'Civil Protection and Humanitarian Aid Work' (DG Echo) and the Budget Parliamentary Committee which allocates the money for the consequences of droughts. This explains partially why the drought issue remains a disaster-driven topic.

11.6 Conclusion

The growing severity of economic, social, and environmental impacts during recent drought events has provided the grounds for increasing pressure on the EU to promote a common European drought policy (Brebbia and Kungolos 2007). At the same time, WS&D have different drivers and thus require different policy responses. From the perspective of the governance regime, this means that the relevant actors and networks are highly dependent on the issues at stake. At the moment, WS&Ds policy at the European level focuses mainly on water scarcity issues based on the usual role of DG ENV to implement the WFD. Unlike water scarcity, the drought issue at the European level stays a rather disaster-driven topic and lacks a planning approach.

The main actors are the EU institutions (high power, interest shared with other issues). However, there is a general lack of interest in the topic, unless severe pan-European droughts occur. At the same time, lobby organizations are pushing the issue (increasing the visibility of the issue at the EU level, increasing the 'interest'). Due to the impacts of climate change to increase the severity of WS&D, climate change actors should show more interest for droughts in a more proactive policy.

None of the EU institutions have a decisive role when comparing their powers in the policy cycle, as they represent a part of a system of multilevel governance involving competition and interdependence amongst the European Commission, the European Parliament, and the Council, each with control of impressive resources in policy initiation. Hence, considering a multi-level governance perspective, it seems that the power of policy actions has increasingly become a shared and competitive competence amongst the European institutions.

Considering the facts that drought is a natural phenomenon, with driving forces relating mainly to meteorological aspects and with climate change often cited as a causing factor, actors responsible for climate change should take more responsibility in drought policy. Actors and policy makers should also try to ensure that measures to address WS&D are more equally focused on both WS&D as individual but related issues, rather than solely on water scarcity and by association drought. This habit of indirectly adapting to drought by adapting to water scarcity can be broken through specially designed policy instruments and strategies. Finally, case-study investigations provide an opportunity to evaluate the effectiveness of such instruments on drought adaptation.

References

Brebbia, C.A. and Kungolos, A. (eds.) (2007). *Water Resources Management IV*, vol. 103. Southampton, UK: WIT Press.

CESR (2011). Climate Adaptation – Modelling Water Scenarios and Sectoral Impacts (ClimWatAdapt), Second Stakeholder Workshop 30–31 March 2011, Ministry of Rural Development, Budapest, Hungary organized by Center for Environmental Systems Research. http://edepot.wur.nl/192357 (accessed 13 February 2019).

EC (2000) Water Framework Directive (2000/60/EC). Brussels: European Commission.

EC (2012a). Report from the Commission to the European Parliament and the Council on the Iiplementation of the Water Framework Directive (2000/60/EC) river basin management plans. Brussels: European Commission.

EC (2012b). Report on the Review of the European Water Scarcity and Drought Policy, Communication from the Commission to the European Parliament, the Council, the European Economic and Social Committee and the Committee of the Regions. Brussels: European Commission.

EC (2017a). JRC in brief – EU Science Hub – European Commission. Brussels: European Commission. https://ec.europa.eu/jrc/en/about/jrc-in-brief (accessed 8 March 2017).

EC (2017b). EU Policy – European Drought Observatory – JRC European Commission. http://edo.jrc.ec.europa.eu/edov2/php/index.php?id=1045 (accessed 8 March 2017).

EEA (2012). *European Waters – Current Status and Future Challenges: Synthesis (No. EEA Report No 9/2012)*. Copenhagen: European Environment Agency.

EU (2017). EUROPA – Agencies and other EU bodies [Text]. https://europa.eu/european-union/about-eu/agencies_en (accessed 13 February 2019).

European Parliament (2017a). Highlights | Home | ENVI | Committees | European Parliament. http://www.europarl.europa.eu/committees/en/envi/home.html (accessed 8 March 2017).

European Parliament (2017b). STOA | STOA Panel – 8th Legislature. http://www.europarl.europa.eu/stoa/cms/cache/offonce/home/panel;jsessionid=6C753ED9AC30EBD46D39E4310918A852 (accessed 8 March 2017).

Kraemer, A. (2007). Economic impact of droughts: challenges for water & environmental policies. In: *Water Scarcity and Drought – A Priority of the Portuguese Presidency* (ed. F.N. Correia), 62–68. Portugal: Ministério do ambiente, do ordenamento do território e do desenvolvimento regional.

Popa, F. (2010). Formal and informal decision-making at EU level. *COGITO Multidisciplinary Research Journal* 2 (4): http://cogito.ucdc.ro/n4e/FORMAL-AND-INFORMAL-DECISION-MAKING-AT-E-LEVEL.pdf (accessed 13 February 2019).

Strosser, P., Dworak, T., Garzon Delvaux, P.A. et al. (2012). Gap analysis of the Water Scarcity and Droughts policy in the EU (final report). http://ec.europa.eu/environment/water/quantity/pdf/WSDGapAnalysis.pdf (accessed 13 February 2019).

Trnski, M. (2004). Multilevel governance in the EU. Paper presented at the The Europe of Regions for the Regions of Europe. Regional Co-operation as Central European Perspective, 1st DRC Summer Summer School. http://drcsummerschool.eu/proceedings?order=getLinks&categoryId=3 (accessed 13 February 2019).

12

National and Local Actors of Drought Governance in Europe: A Comparative Review of Six Cases from North-West Europe

Gül Özerol
Department of Governance and Technology for Sustainability, University of Twente, Enschede, The Netherlands

12.1 Introduction

There is an intensifying pressure on the availability of freshwater resources in Europe (EEA 2010). This makes the equitable and sustainable governance of the limited water resources amongst the key sectors – domestic, industry, energy, agriculture, and the environment – a daunting challenge. In the past decades, the situation has become even more alarming due to the increasing number of water-related disasters such as floods and droughts (Adikari and Yoshitani 2009).

Most regions of Europe are typically known as water abundant and thus floods are seen as the main water-related disaster. However, since the 1970s large areas in Europe have been affected by drought, and the frequency of and damages from drought events have increased (EEA 2010). As a result, drought has received increasing attention of policy-makers at the European Union (EU) level (De Stefano et al. 2015; Stein et al. 2016). The main policy document that addressed drought, and the interrelated problem of water scarcity, is the communication of the European Commission (EC 2007). This communication both identifies the challenges regarding water scarcity and drought in the EU, as well as several policy options that are proposed to tackle those challenges. A key feature of the document is its emphasis on the need to prioritize demand-oriented solutions over supply-oriented ones, as also addressed in expert reports such as those

Facing Hydrometeorological Extreme Events: A Governance Issue, First Edition.
Edited by Isabelle La Jeunesse and Corinne Larrue.

by the European Environment Agency (EEA 2009) and the Water Scarcity and Droughts Expert Network (WSDEN 2007).

Three types of policy approaches can be discerned regarding drought: (i) the post-impact approach, which focuses on relief and assistance measures (ii) the pre-impact approach, which focuses on reducing vulnerability through mitigation measures, and (iii) the development and implementation of preparedness plans and policies that focus on improving the coordination and collaboration amongst all relevant actors at multiple levels (Sivakumar et al. 2014; Wilhite, Sivakumar, and Pulwarty 2014). From a public policy perspective, these three approaches can be seen as 'non-governmental', 'governmental', and 'governance' approaches to drought policy. The governance approach promotes the integration of multiple dimensions of water policy, such as actors, instruments, and responsibilities (Bressers et al. 2013). At the same time, other policy sectors, such as agriculture, energy, land use, and environmental protection, should feed into the development of the drought policy.

Drought policies in the EU countries have been largely shaped by the local circumstances and needs. In some countries, especially those in the south that are more familiar with drought events, there is a move from the crisis-oriented post-impact approach towards the preventive approaches (Kampragou et al. 2011; Stein et al. 2016). However, partially due to the historical influence of floods in most European countries, the risk perception and responsibilities regarding drought events are not integrated into broader water policies, neither at the EU nor at the member state levels (Kampragou et al. 2011).

Actors at multiple governance levels are amongst the key elements of any governance system. Accordingly, drought governance entails the involvement of various actors at multiple levels. These actors range from water users at the household and farm levels to the international organizations that are part of the decision-making processes at the global level. There is a need to align the ways with which these actors make decisions and take actions. The effectiveness of the EU drought policies depends on their operationalization through actions and initiatives at the national and local levels (Bressers et al. 2013; Stein et al. 2016). Therefore, the assessment and improvement of drought resilience should pay attention to the national and local levels. With the motivation to improve the understanding on the national and local levels of governance, this chapter focuses on the actors at these two levels, such as ministries, provincial and municipal authorities, and non-governmental organizations (NGOs). It specifically aims to create insights into the involvement of these actors in the design, implementation, and evaluation of drought resilience measures, and seeks to identify recommendations to improve this involvement.

12.2 Methodology

This chapter presents a comparative review of the six cases that were included within the scope of the 'DROP' (Benefits of governance in DROught adaPtation) project. The DROP project was implemented between January 2013 and September 2015 within the framework of the EU Interreg IVB programme, and it focused on six local cases from five countries in North-west Europe: Belgium, France, Germany, the Netherlands, and the United Kingdom (Figure 12.1). An overview of the cases is presented in Table 12.1.

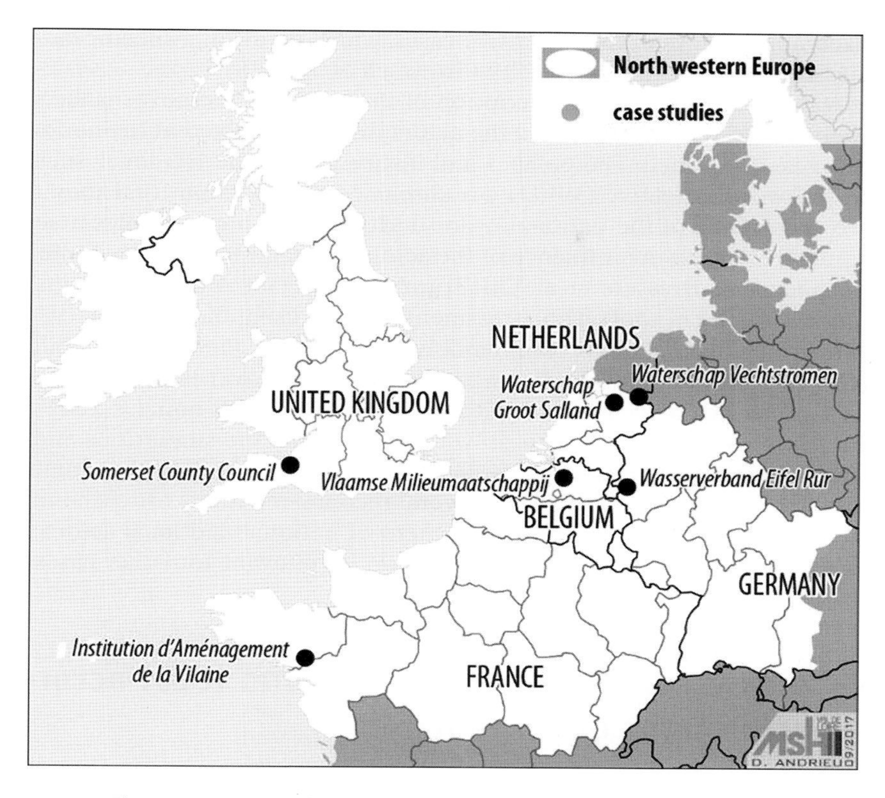

Figure 12.1 Location of the six case studies in North West Europe.

Table 12.1 Overview of the six cases included in the review.

Country	Location	Data source
Germany	Eifel-Rur Region	Vidaurre et al. (2016)
United Kingdom	Somerset County	Browne et al. (2016)
France	Vilaine Catchment	La Jeunesse et al. (2016)
Belgium	Flanders Region	Tröltzsch et al. (2016)
The Netherlands	Twente Region	Bressers et al. (2016a)
The Netherlands	Salland Region	Özerol et al. (2016)

Data sources are the findings from the individual case studies published within an edited volume (Bressers et al. 2016b) that came out as a result of the DROP project.

The geographical boundaries of the case studies included in the DROP project and compared in this chapter cover sub-national 'regions' that consists of several cities or towns. These regions do not necessarily form a jurisdictional level, yet provide rich data in terms of the interactions of actors at local, regional, and national levels. Amongst the six cases, the only case that covers a formally defined jurisdictional region is the Belgian case, where Flanders is examined as an entire federal region. This case reflects the importance of federalism in Belgium and the more centralized approach than the other five cases, which focus on a specific catchment or the jurisdiction of a water authority (Tröltzsch et al. 2016).

To provide a comprehensive description of the six cases, the governance systems of these countries are examined and the actors that are involved at the national and local levels are identified. The policies and institutions at the EU level, such as the Water Framework Directive (WFD), are discussed to the extent that they directly influence the actions of the national or local actors. Then a governance assessment tool is applied to assess the influence of the actors and their networks as a dimension of governance (see Bressers et al. [2013] for more details on the elements of the governance assessment tool). The assessment is made using the four quality criteria of the governance assessment tool, namely the extent, coherence, flexibility, and intensity, with a purpose to identify whether the given governance context is supportive or restrictive towards the implementation of drought policies. Assessments from the six cases are compared to draw conclusions on the similarities and differences amongst individual cases as well as overarching patterns that might exist across cases. Based on the conclusions, several recommendations for each individual case are identified to improve the involvement of national and local actors, as well as three overarching recommendations that can be relevant for other cases within and beyond Europe.

12.3 Assessment of the national and local actors of drought governance

12.3.1 The case of Eifel-Rur, Germany

Case study background In the 2000s, Germany adapted its national water policy to the WFD, mainly through a redesign of responsibilities and competences between the federal and state (*Länder*) levels. The Federal Water Act, the major water legislation at the national level, enacts that each state should adopt its water law, stipulating the distribution of responsibilities amongst cities, municipalities, and water authorities. In the state of North Rhine-Westphalia, where the Eifel-Rur region is located, various actors are involved in drought governance, through their roles in water management. The state ministry for Environment and Nature Protection, Agriculture and Consumer Protection is the highest-level water authority that is responsible for developing the river basin management plans (RBMPs), the catchment-wide plans required by the WFD. These plans are adopted in consultation with District Councils and the responsible committee of the state parliament (*Landtag*).

Local government, such as districts and cities, are responsible for implementation. Other actors such as nature organizations, water associations, and regional councils participate in the planning and implementation process. The District Councils are responsible for authorizing water withdrawals. In the Eifel-Rur region, the district government is responsible for implementing the WFD. Municipalities are the main actors responsible for drinking water and wastewater management. They may form associations for drinking water supply and wastewater disposal. To some extent, municipalities are members of water management associations (*Wasserverbände*), such as the Water Board Eifel-Rur (WVER) of the Eifel-Rur region.

Governance assessment ***Extent***: The two main actors are the district government and the water board, whereas municipalities are withdrawing from water management mainly due to resource constraints. The design of the water boards is supportive. The law of WVER allows the water users to be the members of the water board, implying that all major users can participate in river-basin management. However, farmers and nature organizations do not have the same voice as larger actors. Positively, there is a movement towards collaborative decision making, for instance through the participatory workshop and roundtable discussions, which are organized as part of the implementation of the WFD.

Coherence: Main actors at the state and local levels accept their legal share of tasks, responsibilities, and funding. The fact that most water management responsibilities are within WVER also improves the coherence. Since water users are WVER members, they participate in decision making and pay for its services. There is a sense of trust and mutual dependency amongst actors, who positively evaluate the participatory mechanisms. The actors believe in the value of the trust-based collaboration that has been built over the years and evolved positively over time. However, the consensual and voluntary approach often reaches its limits, as some negotiation processes on controversial topics have been at a standstill for several years.

Flexibility: The water management system has a rigid framework, shown in the priorities and responsibilities of WVER. Any decision that disregards the formal rules raises the question of legal responsibility. Thus, WVER has a strong incentive to follow the rules. The water management system is flexible at the local scale. There is a strong culture of discussion and collaboration, and openness to incorporate the interests of local stakeholders. The water board has to provide a certain level of flood protection and water supply. This results in an elaborate set of rules to manage the interaction of reservoirs and water bodies. There is little flexibility regarding implementation and crisis situations, since decisions are mostly taken by the highest authorities.

Intensity: The district government is the most relevant actor concerning drought decisions. The water board only has executive powers, and thus cannot implement drought measures on its own. Drought resilience has no priority on the national agenda and no resources are made available. The national and state-level strategies are too generic to result in concrete actions. Thus, the district government has no pressing obligations regarding climate change adaptation in general, and drought in particular. Individual actors initiate activities, such as the DROP project. The WVER emphasizes the technical projects to improve system robustness and risk management, and to develop solutions to extreme events. Since WVER is responsible for most water issues in the region, it is also the first organization to be affected by drought, which leads to its proactive approach.

12.3.2 The case of Somerset, the United Kingdom

Case study background This case focuses on England, because water resources are managed differently across country boundaries in the United Kingdom (UK). (DEFRA)Department of Environment, Food and Rural Affairs has the overall responsibility for water and sanitation policy. The Environment Agency (EA) is the executive body that is responsible for environmental regulation related to, amongst others, water provision, climate change adaptation, and WFD implementation. The Drinking Water Inspectorate monitors the drinking water quality. Irrigation Drainage Boards (IDBs) work in order to reduce flood risk. Whilst Defra is responsible for the IDBs, they work closely with the EA and lead local flood authorities. The drinking water and wastewater companies are private, but they are regulated by governmental bodies (Water UK 2019.).

Somerset County Council (SCC) is responsible for managing strategic local services. Their role in water management focuses on emergency planning, consumer protection, and local flood management. They also work to raise awareness on climate change adaptation. Finally, there are NGOs whose interests reflect the balance between nature and agriculture. Regarding climate change adaptation, the UK government has reporting power, which requires stakeholders and companies to report on the impacts of climate change on their organizations, and their proposals for adaptation. The water industry has also engaged with activities related to climate change forecasting and adaptation as part of their water management plans, which is also part of the WFD requirements.

Governance assessment ***Extent***: The main organizations responsible during a drought are the EA (through providing strategic oversight and coordinating the mitigation of the impact of drought), water companies (through managing and maintaining water supply, whilst minimizing environmental impact), and the government (through policies on water resources), whilst several other organizations play an important role, including local councils and representative bodies such as National Farmers' Union, UK Irrigation Association, and environmental charities (Environment Agency 2015). Actors' relationships are seen as positive and have further improved after the 2012 drought. Despite its overall negative impact, the 2013/2014 flood enhanced the range of actors involved in regional water management.

Coherence: There are effective relationships amongst actors, resulting both statutory requirements and voluntary efforts. The importance of this collaborative approach has been increasing, especially after the 2012 drought. The national drought framework further clarifies the role of different actors involved in drought management, especially for crisis periods. Although the 2013/2014 flood created risks regarding social and political divisions, constructive activities also took place. The activities carried out by the SCC and its partners in the DROP project, including their key role in the 20Year Flood Action Plan, indicates a proactive and collaborative approach to drought management.

Flexibility: A positive flexibility was observed in the sense that many actors could be involved in drought management. However, the 2013/2014 flood reduced the legitimacy of some actors, such as the EA, causing a shift to a slightly more restrictive flexibility. Nevertheless, a clear institutional approach exists to drought management and goals are flexible, too. However, some actors perceive a lack in the problem definition and goal setting processes, and their implementation into particular strategies and instruments for drought adaptation.

Intensity: Drought management is part of the core business for the lead actors. This results from both regulatory requirements and the ability of actors to go beyond regulatory defined agendas to promote the drought issues across various actors. The intensity is lower regarding the allocation of resources for the tasks of adaptation. After the 2013/2014 flood, the intensity with which certain actors were pushing for change and management reform for drought drastically decreased. Particularly, the experiences of actors after the 2013/2014 flood significantly changed their problem definition and goal ambitions in disfavour of drought. Whilst the catchment management approaches, which were promoted for flood mitigation and adaptation, could also include drought measures, the social and political environment limited the ability of actors to define drought as a problem.

12.3.3 The case of Vilaine, France

Case study background Since the 1970s, several droughts were experienced in France, which have led to changes in policies or adaptation measures. The drought of 1976 had strong impact on agriculture, and as a result the national government provided financial assistance to farmers. During the 1989 drought, drinking water could be supplied in most cases, however measures such as prohibiting or restricting certain water uses had to be adopted. Governance problems, such as the inability to quantify water abstractions and the absence of groundwater abstraction control, were also diagnosed. The drought of 2003 led to the preparation of a Drought Action Plan that aimed both to balance different water uses and to preserve the quality of aquatic ecosystems. A National Drought Committee was created to coordinate water uses during drought crises.

The six water agencies in France have their own river basin committee, involving representatives from local authorities, users, and the state. Based on the WFD and national water policies, the committee establishes guidelines that are driven by the RBMP. The national government is responsible for regulation, such as authorizing water abstractions and monitoring water use. The Water Management Direction and the Ministry of Ecology, Sustainable Development, and Energy define water management policies, including drought plans. At the regional level, water regulation and drought policies are the responsibility of prefects, who represent the national government and can issue bylaws on water use, particularly on drought control through water bans.

Whilst the water agencies manage large river basins, local issues are managed by groups of local authorities called 'EPTB' (*Etablissement Public Territorial de Bassin)* that work in hydraulics (e.g. low-flow management), environment (e.g. maintenance of banks), and local development (e.g. natural and cultural heritage). EPTBs are also usually the coordinators of the (SAGE) *Schéma d'Aménagement et de Gestion de l'Eau*, the water management plan at the sub-catchment level. SAGE is mainly developed by a Local Water Commission (LWC), which is formed by representatives from state, local authorities, and users. *Institution d'Aménagement de la Vilaine* (IAV) is the EPTB of the Vilaine catchment. Its primary role is to support water management at the basin scale and to ensure cohesion amongst users. The IAV is also responsible for the management of the Arzal Dam and the drinking water facility. These technical tasks give IAV an important role as an EPTB.

Governance assessment ***Extent***: A strong network, involving the relevant actors in the region has been built by the IAV. Frequent meetings organized by the IAV discuss shared issues create mutual trust. There are, however, no actor networks that specifically address drought at any level. The only exception to this is crisis management at the national level, which is the responsibility of the National Drought Committee.

Coherence: IAV facilitates dialogue amongst stakeholders. Some actions, such as connection between drinking water networks, are implemented coherently, but the integration of the climate change perspective remains mainly inside the IAV. Since the reform in the 1990s, the state has had little power in water management. As mentioned in the previous section, the decentralized services for managing water at the catchment level are under the authority of the prefect. However, there is a lack of consistency between these services. Whilst difficulties are experienced in coordinating interdepartmental relations, the actors are used to working together. All stakeholders representing the various water uses accept the priority given to drinking water. However, the perception of the drought risk and the potential impacts on freshwater availability is almost absent, which can be attributed to the frequent flooding issues in the area.

Flexibility: Due to the decentralization, most decisions on water management are made at the catchment and sub-catchment levels. The meetings held by the LWC are pragmatic and achieve their goals, whilst state services struggle to advance equally rapidly and anticipate the coordination needs of stakeholders. The LWC makes substantial efforts to integrate the different levels of decision making. All actors are represented in the LWC. Whilst it is possible to create committees that are devoted to special areas and topics, there is no drought committee. The priority for drinking water is not flexible, as the IAV has to deliver the amount of water dedicated to this water use.

Intensity: IAV has a strong role at the catchment level in preserving water supply and implementing specific measures. However, it lacks the capacity to propose solutions for other water uses during droughts. The low level of drought awareness amongst most of the actors also hinders the intensity with which levels are recognized and implemented in drought adaptation strategies. The intensity of IAV is very high in the lower Vilaine River area. It drives a network of all water uses within the catchment and succeeds in collaboration with water users and managers at the catchment level. Since the position of the IAV is too central, this can also be seen as a weakness of the network.

12.3.4 The case of Flanders, Belgium

Case study background The Belgian federal governmental system implies that responsibilities are allocated to both the regional and federal levels, which have exclusive and equivalent competences, with no hierarchy between their standards. The federal government is responsible for coastal and territorial waters, whereas the regions are responsible for environment and water policy, including technical regulations regarding drinking water quality, economic aspects of drinking water provision, land development, nature conservation and public works and transport. A regional approach to river basin planning is adopted. Thus, all water management plans are developed at the regional level, except the Federal Plan on Coastal Waters.

Three ministries are involved at the regional level: The Ministry of the Environment and Nature, the Ministry of Mobility, and the Ministry of Spatial Planning. The Flemish Environment Agency (VMM) is responsible for many tasks related to water management. Three hydrological levels are discerned: two international river basin districts (Scheldt and Meuse), four river basins at the Flemish region (Scheldt, Ijzer, Meuse, Polder of Bruges Polders) and 11 sub-basins. The Flemish government is responsible for the RBMPs of the Flemish parts of the international river basin districts.

The Coordination Committee on Integrated Water Policy (CIW) is the competent authority for the implementation of the WFD and the Floods Directive. It is chaired by VMM and its responsibilities include preparing the RBMPs for the Flemish region, reporting on WFD implementation, organizing the public consultation of the RBMPs, preparing the methodology and guidance for developing the RBMPs, and aligning them with the Flemish Water Policy Note.

In Flanders, there is a high pressure on water resources, partially due to intensive agriculture and the high population density. The main focus of the water managers has been guaranteeing water supply and quality, and mitigating the flood risk. Nevertheless, drought is a well-known problem in sectors that depend on a good water supply, such as agriculture. As have most European regions, Flanders has experienced several droughts, the main ones in 1996, 2006, and 2011, which were recognized as agricultural disasters. Farmers affected from those droughts were financially compensated. The drought policy is based on the implementation of the EU policy documents such as the WFD and the Communication on Drought and Water Scarcity (EC 2007).

Governance assessment ***Extent***: There is active involvement of the actors as well as strong interactions between them. The national level is retreating to a coordinating role, which is not seen as a problem. Municipalities are also disengaging from water management, for instance by handing over responsibilities for small watercourses to the provinces. This can lead to a situation that they miss the opportunities for water-related synergies, including drought preparedness. Extent is more limited in drought terms, since initiatives are restricted to the higher levels. The broad perspective of the EU water policy, which includes social and ecological criteria, increased the extent of actors involved. This inter-sectoral approach is also demonstrated in the CIW. However, local stakeholders are often not involved in policy design, such as in the selection and prioritization of measures to be implemented, but rather in the implementation phase.

Coherence: Whilst few actions address drought at the regional level, there is a 'culture of coordination' amongst the actors involved, who often work well together. Several committees, such as the CIW, are active at the national level. The coordination of local authorities with district and province levels is fruitful, with possibilities to move up and down between different governance levels. Partly due to the long tradition of collaboration in the region, there are good and productive relationships amongst actors. In the last decades, the involvement of less active actors, such as nature organizations, has also improved.

Flexibility: Informal coordination mechanisms complement the formal mechanisms. Many actors can be reached since Flanders is a relatively small region, with governance levels ranging from a central government to the local level. However, the flexibility regarding the type and timing of involvement is low, as the participation of some actors in the planning proves is limited.

Intensity: The regional level is responsible to bring drought aspects into water planning, making the actions dependent on this level. VMM is willing to develop such actions and is the main driver of drought-related initiatives, such as the DROP project. Whilst the importance and desirability of addressing droughts is recognized, no other actors have this topic on their agenda. Thus, change is driven by one actor, and the design and implementation of strategies and instruments are dependent on the regional level.

12.3.5 The case of Twente, the Netherlands

Case study background In the Netherlands, water resources management is a public responsibility. Four types of governmental organizations are involved in water management: central government, provinces, municipalities, and water authorities. Publicly owned drinking water companies often serve multiple provinces and cover their costs under private law, whilst being regulated by the central government. In the national administrative agreement on water, the state government, provinces, water authorities, and municipalities stipulate the division of their tasks and agree on the principles of cost sharing. The provinces and water authorities also agreed to elaborate on desired water levels in rural areas.

Being a country where 55% of the land is flood prone, the issue of drought has not received attention in the Netherlands until recently. In the 1980s, drought was recognized as a policy issue under 'desiccation', which threatened the vitality of wetlands and the fertility of soil. A policy document, which initiated the reassessment of national water management, addressed the problem of 'high sandy grounds', the areas in the east and south of the Netherlands that often cannot receive water from the national system and depend on rainwater and groundwater (Delta Commission 2008). The water authorities perceived that their problem was ignored and decided to cooperate at the regional level. In 2012, they organized a symposium, involving other actors, and called for attention to the water problems of the higher areas. After this symposium, drought was positioned on the national agenda. Until then, water shortage was mostly seen in relation to the freshwater needs of the lower areas that have to be flushed to prevent salt intrusion. The higher areas, including Twente, suffer from droughts and the situation is likely to worsen with climate change. To combat these challenges, guidelines were introduced that emphasize measures to increase buffering capacity and improve water use efficiency, whilst allocating responsibility to users for mitigating the damage from droughts.

In 2014, the actors in eastern part of the Netherlands, including Twente, signed the ZON (the Dutch acronym for freshwater supply for East Netherlands) Declaration, which acknowledges the shared responsibility of all actors for freshwater availability and the resilience of the water system for extreme weathers, including preparedness to contribute to programme costs. The ZON declaration constitutes a major step towards regional collaboration. Further, it is a message to the national political arena that the higher parts of the Netherlands, which are mainly in the south and east of the country, should be sufficiently represented in terms of attention and funding.

Governance assessment ***Extent***: Although freshwater availability is gaining attention, the national level is gradually withdrawing from the process. This is explained by the policy changes in 2011 that drastically reduced the budgets for nature-protection projects and as a result damaged the relationships between the environmental protection and water sectors. The drinking water company, general public, and nature organizations are the other actors whose involvement are perceived as less active than desired.

Coherence: In northeast Twente, four local authorities, the water board and the province prepared a regional vision, focusing mainly on tourism and recreation, whilst also including drought resilience. Thanks to these collaborative efforts, nature organizations and farmers are enthusiastic for participation. The drinking water company perceives a significant improvement in collaboration over the last years. Trust between farmers and government actors is also crucial, as it enables the alignment of agricultural interests with the interests of drought projects. Such a close collaboration provides coherence and facilitates implementation. It can also be seen as a successful adaptation to deal with a fragmented governance context, which is labelled as 'the fragmentation-coherence paradox' (Larrue et al. 2016). Fragmentation would normally lead to stalemates and ultimately disinterest, whilst in this context of positive experience with collaboration, it leads to a recognition that actors need each other and no single actor will become too dominant.

Flexibility: Two main indicators of flexibility are that the province and municipalities took over the role of the national level and the high degree of interconnectedness helps to receive political and financial support for projects. Some inflexibility results from the geophysical conditions of the Twente region or relate to the local land use planning system, which is known to have lengthy decision-making procedures. Stakeholder representation can also lead to inflexibilities. For example, it can be difficult for new actors to get recognition, such as the farmers that act independently from the farmer organization (*Land- en Tuinbouw Organisatie*, LTO). Further, the drought measures are restricted to voluntary approaches. The strategy to compensate this is involving farmer groups in projects in order to avoid obstacles.

Intensity: Political support exists only for voluntary, preventive measures. The withdrawal of the national level from the nature policy could lead to cuts in funds for projects. Whilst the province took over this role, it prefers to concentrate on Natura 2000 areas. Drought resilience is a priority for the water authority, making it a lead actor, even when other stakeholders are involved. The farmers' position has improved in the last few years. For them, the freedom in crop choice and increasing productivity are essential to support drought measures. The LTO intends to improve awareness on drought issues by training farmers and aims to collaborate with other stakeholders.

12.3.6 The case of Salland, The Netherlands

Case study background *The background description of the Twente case study also applies to this case.*

Governance assessment ***Extent***: The regional actors are prominent, whilst interactions between different levels occur for the implementation of relevant EU policies, such as the WFD. The national government enforces these policies. Regarding groundwater management, the province and the drinking water company play a role at the regional level. The province oversees groundwater use and protection through controlling the water authorities and issuing permits, whereas the drinking water company does groundwater abstraction.

Coherence: As a result of the multitude of the actors involved, some disagreements occur regarding the allocation of water resources to multiple uses and services. For instance, regarding the impacts of groundwater extractions on nature areas, the position of water authorities, farmer organizations, and drinking water companies differ from that of nature organizations. The 'Agriculture on Sight' (*Landbouw op Peil*) project is seen as a positive example. It was started by the water authorities and LTO and involved farmers to increase their awareness of soil and water sustainability. The project improved the communication and collaboration between farmers and the water authority.

Flexibility: There is a collaborative environment, involving actors from different levels to elaborate on problems and solutions without imposing a hierarchy, unless there is a law that regulates otherwise. For instance, the water boards developed an irrigation policy by upscaling the irrigation issue from the local level to the regional level. The water authorities successfully developed a common regulation at the regional level, without following a formal procedure, whilst the LTO and the umbrella organization of the nature conservation organizations (*Natuur en Milieu Overijssel*, NMO) were not able to participate effectively. In the example of ZON, the freshwater supply problem was downscaled from the national level to regional level to address the specific context of the high sandy grounds.

Intensity: The WFD requires that all the water authorities in the Rhine-East subbasin have to collaborate on water planning and management. The long-lasting good relations amongst the water authorities Are supportive towards realizing this regional collaboration, as is also reflected in the ZON declaration. Thus, drought is a priority issue for the water authorities to collaborate at the regional level, whilst other actors assign it a lower priority. NMO cannot participate in water projects due to a lack of financial and human resources, whereas the participation of LTO is restricted due to their limited technical knowledge. Recently, the province has changed its approach to water projects by integrating water into the spatial planning sector.

12.4 Conclusions and recommendations

12.4.1 Conclusions: What do we learn from the experiences in north-west Europe?

Two main conclusions are drawn regarding the actors and networks of drought governance in the six cases (Larrue et al. 2016). First, there is a low level of intensity. In most of the cases, many actors of water governance are unaware of the drought risks or do not see them as a priority. Mainly due to the historically wet situation of North-West Europe, the actors are mainly concerned about floods. This low level of drought awareness results in a low intensity, which is assessed as neutral in five cases and restrictive in one case. Second, there are high levels of extent and coherence. In contrast to the situation of low awareness and intensity, the governance context is much supportive regarding the extent and coherence of the 'actors and networks' dimension. In all the cases studied, the actors involved at different levels are mobilized, implying a context that is favourable to establishing a drought policy that integrates these levels. As summarized in Table 12.2, most of the qualities regarding the actors and networks dimension are assessed as supportive or neutral, whilst intensity remains as the least supportive quality.

In all cases, the overall water governance system is relatively effective, as a water policy culture exists since the 1960s. This policy has been ingrained by the implementation of the WFD since 2000, imposing a multilevel water policy through district management plans. Accordingly, in all the regions, coherence is assessed as supportive for all the dimensions, except the Salland case where it is neutral. This high level of coherence is mainly attributed to the trust-based interrelationships amongst the actors. Even if their positions can be contradictory, most of the actors share a mutual interest in water management. This mutual interest is likely to facilitate the formulation and implementation of future drought measures.

Table 12.2 Assessment of the 'actors and networks' dimension in the six cases.

Case/Criterion	Extent	Coherence	Flexibility	Intensity
Eifel-Rur	Supportive	Supportive	Neutral	Restrictive
Somerset	Supportive	Supportive	Neutral	Neutral
Vilaine	Supportive	Supportive	Neutral	Neutral
Flanders	Supportive	Supportive	Neutral	Neutral
Twente	Supportive	Supportive	Supportive	Neutral
Salland	Supportive	Neutral	Neutral	Neutral

Source: Adapted from Larrue et al. (2016)

Although multiple actors are involved both at the local and national levels, the interactions amongst the levels differ amongst the cases. Water authorities, or 'water boards', are the lead actor in most of the cases, whereas the involvement of other actors is often not equally representative. Particularly the environmental NGOs struggle to represent their interests, since the focus is often on agriculture and economic development. The priority on agriculture is, however, not a common practice. In the Netherlands, surface water is flushed to prevent salt intrusion in the low-lying western parts of the country, reducing the availability for agricultural water use in the higher eastern parts, and during dry periods, agricultural use has a lower priority than many other water uses.

12.4.2 Recommendations: How to improve the involvement of local and national actors in drought governance?

Based on the assessments provided for the individual cases, several recommendations are identified for the six regions. Table 12.3 presents an overview of the recommendations that are related to the involvement of national and local actors.

Based on the recommendations from the six individual cases, three overarching recommendations are identified that can also be useful in improving the involvement of national and local actors in drought governance in other European regions and elsewhere.

1. *Improve drought awareness*: The low level of drought awareness restricts the selection of interventions to increase drought resilience and obstructs the implementation of measures. Public awareness campaigns are amongst the drought measures existing in Europe (De Stefano et al. 2015). Improving public awareness is also seen as a key measure for developing and implementing drought policies consistent with national legislation, capabilities, and objectives (Sivakumar et al. 2014). Three strategies are proposed for strengthening the position of drought on public and political agendas:
   - Place drought on public and political agendas as an independent problem, for instance by providing continuous information to the public and addressing national water planners with a broad coalition of stakeholders.
   - Address drought by 'piggybacking', such as including drought-relevant measures in different planning initiatives and ensuring the coherence of other plans with drought objectives.

Table 12.3 Recommendations regarding the national and local actors.

Region and reference	Recommendations
Eifel-Rur, Germany Vidaurre et al. (2016)	*Improve authorities' review of decision-making processes*: Authorities should take a more active role in guiding the stakeholders on the process objectives and controlling the implementation of regulations. *Develop strategies to improve actors' involvement in the planning process*: The financing possibilities of measures addressing drought should be given thorough attention, particularly to improve the participation of municipalities. Other actors can be addressed by demonstrating the benefits of particular initiatives, such as local showcases on the implementation of certain measures. *Increase synergies with farmers*: Farmers are reluctant to align with water management objectives, when measures do not coincide with their aims. The relationship between the water board and farmers can be redesigned to improve drought preparedness. For instance, the agreements on water quantity can be revisited, with a special focus on preventing possible bottlenecks.
Somerset, UK Browne et al. (2016)	*Maintain collaboration to overcome broader governance failures*: The lead actors should continue collaborating with the widest possible range of stakeholders. This will ensure they represent a diverse set of stakeholder views, when adaptation policies are developed, and avoiding broader governance failures that ingrain fragmented views of resilience. At the national level, strategies should be developed to address such governance failures, which limit the resilience of water catchments.
Vilaine, France La Jeunesse et al. (2016)	*Improve drought awareness*: An outreach effort can raise the awareness of water users and managers, focusing on the impacts of climate change. The network can be mobilized around drought, especially with local actors. Geographic or thematic committees can facilitate the creation of knowledge on specific challenges. A task force dedicated to climate change impact can keep the attention on local needs. *Share low-flow forecasts with interested parties*: One of the aims of the DROP pilot project in this region was to develop a tool for forecasting low flows and enhancing water management. Sharing information is essential to ensure a transparent management. Numerical tools can help to confirm decisions or to provide evidence for unforeseen risks.
Flanders, Belgium Tröltzsch et al. (2016)	*Improve drought awareness*: It is essential to increase the awareness of actors, such as farmers and the energy industry, who do not incorporate the drought risk in their operations. The strategy of using scientific models as a basis for actions can be expanded with pilot and demonstration projects. Risk communication with different sectors should be improved, for instance through discussing impacts. *Mainstream drought risks and preparedness*: Drought management can be integrated in different policies and lead to a multi-objective planning. This can also be used for guiding the collaboration between the relevant ministries. Drought should be mainstreamed into private actors' activities. *Improve the engagement with public actors*: The relationships between regional and national actors can be improved through further dialogues. Decreasing involvement of municipalities is not seen as a problem given the good connection between regional actors. The impact of such disengagement cannot be completely foreseen; therefore, its consequences should be observed.

Table 12.3 (*Continued*)

Region and reference	Recommendations
Twente Region, The Netherlands Bressers et al. (2016a)	*Improve drought awareness and keep it on the national agenda*: Financial and political support for drought measures are limited and their priority is low. Thus, drought seems to be the problem of 'high sandy grounds', where dependency on rainwater and groundwater increases the risk of drought damage. The actors in these areas should make efforts keep the drought issue on the national agenda. *Maintain inter-collegial exchange and learning*: Due to the low saliency of drought, the drought resilience projects are voluntary and consensus-oriented. The expertise and experience gained from the drought projects in consensual management can be beneficial for managers in other water projects. Platforms should be formed for inter-collegial exchange and learning.
Salland, The Netherlands Özerol et al. (2016)	***Develop an integrated approach to droughts and floods*:** Water authorities should emphasize their vulnerable context to place drought in higher-level policies. Floods and droughts should be seen as interrelated policy issues. Assigning one actor to manage the system-level knowledge is infeasible, but with increasing competition amongst water users, such comprehensive mechanisms can become inevitable. ***Raise farmers' awareness to create ownership and drought-sensitive water demand*:** Information tools that deploy technical expertise and incorporate farmers' knowledge and needs can be developed. Clear rules should be established on water withdrawals, which farmers can understand. Such rules can contribute to demand management. Monitoring water withdrawals can also contribute to managing farmers' water demand. ***Enable the active involvement of NGOs*:** The NGOs are keen to contribute to improving the situation at the regional level, but they have limited resources. An active involvement can increase their willingness to share risks. Mutually agreeable mechanisms should be devised to assign fair and clear roles, whilst decoupling the level of responsibilities from that of financial contributions.

- Prepare a ready-to-implement strategy for when a drought event makes the topic climb the agenda and receive political attention, resulting in a call for action.

Through these strategies, a higher priority can be established for drought and it can be aligned with the recognized importance of flood risks. Tailoring these strategies to local needs can start by improving the knowledge base on the local impacts of climate change on water availability, and the possible drought scenarios at the national and sub-national levels.

2. *Focus on water demand management*: Most of the current drought measures involve distributing the water and decreasing water scarcity during dry periods to improve resilience through increased water buffering capacity, whereas demand-oriented measures have been less common. As addressed at the EU level (EC 2007), demand management is a crucial part of the drought policy. The repercussion of this policy

for the national and local levels is that data should be collected on water rights, and water withdrawals should be monitored (Urquijo et al. 2017). Furthermore, policy measures should involve incentives to reduce water use. Such incentives are often absent in regions such as North-west Europe, since water is not perceived as a scarce resource. Whilst the pricing of water is often promoted, this market-based measure can be insufficient to reduce water demand (Maggioni 2015). To improve water demand management, drought risk and preparedness should be mainstreamed into the decisions of private actors. Particularly, individual farmers and their organizations should be involved in the design and implementation of measures. Finally, expectations regarding the responsibilities of public vs. private actors should be managed more effectively. When water users perceive water as an abundant resource, they see the water authorities responsible for supplying water. The ownership and awareness of water users should be increased through challenging the limits of such public responsibility.

3. *Integrate flood and drought management*: As it has already been emerging in some regions, the decoupled approach to managing floods and drought should be replaced with an integrated approach for policy and practice (Rijke et al. 2014; Grobicki, MacLeod, and Pischke 2015). Such integrated approaches consider flood and drought as different points on the same spectrum, rather than independent water-related disasters. The possible occurrences of drought events should be taken into account, when designing flood resilience measures, and vice versa. This can imply for many national and local actors that they need to change the types of strategies and instruments that they have been using so far, for instance by introducing comprehensive, preventive measures rather than crisis-oriented measures. These changes would often imply governance changes that focus on the resilience of the entire water system. Despite the possible resistance to such changes, the result will be a governance framework that can support learning from the lessons of all relevant sub-sectors, and can effectively adapt to climate extremes.

References

Adikari, Y. and Yoshitani, J. (2009). Global trends in water-related disasters: an insight for policymakers. World Water Assessment Programme Side Publication Series, Insights. Paris: The United Nations, UNESCO. International Centre for Water Hazard and Risk Management.

Bressers, H., Bleumink, K., Bressers, N. et al. (2016a). The fragmentation-coherence paradox in Twente. In: *Governance for Drought Resilience* (ed. H. Bressers, N. Bressers and C. Larrue), 181–201. Cham: Springer.

Bressers, H., Bressers, N., and Larrue, C. (eds.) (2016b). *Governance for Drought Resilience*. Springer.

Bressers, H., de Boer, C., Lordkipanidze, M. et al. (2013). Water governance assessment tool. With an elaboration for drought resilience. Report to the DROP project. Enschede: CSTM University of Twente, Enschede.

Browne, A.L., Dury, S., de Boer, C. et al. (2016). Governing for drought and water scarcity in the context of flood disaster recovery: the curious case of Somerset, United Kingdom. In: *Governance for Drought Resilience* (ed. H. Bressers, N. Bressers and C. Larrue), 83–107. Cham: Springer.

De Stefano, L., Urquijo, J., Kampragou, E. et al. (2015). Lessons learnt from the analysis of past drought management practices in selected European regions: experience to guide future policies. *European Water* 49: 107–117.

Delta Commission (2008). *Working Together with Water – a Living Land Builds for its Future*. The Hague: Delta Commission http://www.deltacommissie.com/doc/deltareport_full.pdf (accessed 14 February 2019).

EC: European Commission (2007). Communication from the Commission to the European Parliament and the Council on addressing the challenge of water scarcity and droughts in the European Union. COM (2007) 414 final, Brussels: European Commission.

EEA: European Environment Agency (2009). Water resources across Europe - Confronting water scarcity and drought. EEA Report No 2/2009, Copenhagen: European Environment Agency. https://www.eea.europa.eu/publications/water-resources-across-europe (accessed 14 February 2019).

EEA: European Environment Agency (2010). The European Environment State and outlook 2010 - water resources: Quantity and flows. Copenhagen: European Environment Agency https://www.eea.europa.eu/soer/europe/water-resources-quantity-and-flows (accessed 14 February 2019).

Environment Agency (2015). *Drought Response: Our Framework for England*. London, UK: HM Government.

Grobicki, A., MacLeod, F., and Pischke, F. (2015). Integrated policies and practices for flood and drought risk management. *Water Policy* 17 (S1): 180–194.

Kampragou, E., Apostolaki, S., Manoli, E. et al. (2011). Towards the harmonization of water-related policies for managing drought risks across the EU. *Environmental Science and Policy* 14 (7): 815–824.

La Jeunesse, I., Larrue, C., Furusho, C. et al. (2016). The governance context of drought policy and pilot measures for the Arzal dam and reservoir, Vilaine catchment, Brittany, France. In: *Governance for Drought Resilience* (ed. H. Bressers, N. Bressers and C. Larrue), 109–138. Cham: Springer.

Larrue, C., Bressers, N., and Bressers, H. (2016). Towards a Drought Policy in North-West European Regions? In: *Governance for Drought Resilience* (ed. H. Bressers, N. Bressers and C. Larrue), 245–256. Cham: Springer.

Maggioni, E. (2015). Water demand management in times of drought: what matters for water conservation. *Water Resources Research* 51 (1): 125–139.

Özerol, G., Troeltzsch, J., Larrue, C. et al. (2016). Drought awareness through agricultural policy: multi-level action in Salland, the Netherlands. In: *Governance for Drought Resilience* (ed. H. Bressers, N. Bressers and C. Larrue), 159–179. Cham: Springer.

Rijke, J., Smith, J.V., Gersonius, B. et al. (2014). Operationalising resilience to drought: multi-layered safety for flooding applied to droughts. *Journal of Hydrology* 519: 2652–2659.

Sivakumar, M.V., Stefanski, R., Bazza, M. et al. (2014). High level meeting on national drought policy: summary and major outcomes. *Weather and Climate Extremes* 3: 126–132.

Stein, U., Özerol, G., Tröltzsch, J. et al. (2016). European drought and water scarcity policies. In: *Governance for Drought Resilience* (ed. H. Bressers, N. Bressers and C. Larrue), 17–43. Cham: Springer.

Tröltzsch, J., Vidaurre, R., Bressers, H. et al. (2016). Flanders: regional organization of water and drought and using data as driver for change. In: *Governance for Drought Resilience* (ed. H. Bressers, N. Bressers and C. Larrue), 139–158. Cham: Springer.

Urquijo, J., Pereira, D., Dias, S. et al. (2017). A methodology to assess drought management as applied to six European case studies. *International Journal of Water Resources Development* 33 (2): 246–269.

Vidaurre, R., Stein, U., Browne, A. et al. (2016). Eifel-Rur: old water rights and fixed frameworks for action. In: *Governance for Drought Resilience* (ed. H. Bressers, N. Bressers and C. Larrue), 67–82. Cham: Springer.

Water UK (2019). Regulation. www.water.org.uk/about-water-uk/regulation (accessed 14 February 2019).

Wilhite, D.A., Sivakumar, M.V., and Pulwarty, R. (2014). Managing drought risk in a changing climate: the role of national drought policy. *Weather and Climate Extremes* 3: 4–13.

WSDEN: Water Scarcity and Droughts Expert Network (2007). Drought management plan report: including agricultural, drought indicators and climate change aspects. Technical report 2008–023. http://ec.europa.eu/environment/water/quantity/pdf/dmp_report.pdf (accessed 14 February 2019).

III.2

Strategies, Instruments, and Resources Used to Face Droughts

13

Awareness of Drought Impacts in Europe: The Cause or the Consequence of the Level of Goal Ambitions?

Isabelle La Jeunesse
Laboratory CNRS 7324 Citeres, University of Tours, Tours, France

13.1 Introduction

Despite being governed by a common European water policy framework with the Water Framework Directive (WFD) and the development of a Common Implementation Strategy (CIS), drought and water scarcity management strategies seem to be highly variable across the European Union (EU). In fact, whilst Europe has already been confronted by heatwaves and severe droughts since the sixteenth century (Garnier 2015), as well as in 1976, 2003, 2008, and 2011, for instance (La Jeunesse et al. 2016a), strategies and instruments for drought and water scarcity management are not on the top of the agenda. Moreover, in addition to this long-term historical reconstruction of droughts, the IPCC report on extreme events (SREX, IPCC 2012) predicts an increase in droughts for all of Europe. Thus, without any instrument such as the Floods Directive for flood management (Stein et al. Chapter 11), how are European territories managing drought challenges? Do territories in most European areas have goal ambitions to anticipate drought issues planned for this century? Is there any link with the difficulty in being aware of drought impacts?

This chapter addresses the possible differences in the development of drought preparedness by analysing the intensity of drought impact perceptions in several case

Facing Hydrometeorological Extreme Events: A Governance Issue, First Edition.
Edited by Isabelle La Jeunesse and Corinne Larrue.

studies in Europe. In order to provide a broader image of drought perceptions in Europe, the chapter proposes an analysis of case study investigation outputs into two distinct European regions: the North-western Europe (NWE) region and the Mediterranean region, based on the outputs of two European projects: DROP (Benefits of governance in DROught adaPtation) and CLIMB (CLimate Induced Changes on the Hydrology of Mediterranean Basins). Whilst NWE is typically associated with a wet climate influenced by the Atlantic Ocean, this region is also projected to experience more and more severe periods of drought and water scarcity in the coming decades (Forzieri et al. 2014). However, the Mediterranean region, because of its semi-arid climate, is said to be used to drought and water scarcity management, and thus prepared for it. The assumption made in this chapter is that the Mediterranean region, since it is already confronted by droughts and water scarcity, is more prepared than the North-western region, which is usually confronted by floods. However, in the context of climate change, both regions will be confronted by more severe droughts and are supposed to be in the process of elaborating strategies to face new extreme situations (IPCC 2007).

After this introduction, the chapter describes the methodology used to study water and drought governance in the case studies for each region. Then, the six NWE case studies and the five Mediterranean case studies are briefly described. As the main part of the chapter, the third and fourth sections propose a comparison of drought perceptions and goal ambitions for water management in the context of climate change in both regions. The last section proposes conclusions.

13.2 Drought governance analysis based on two methodological approaches

This section exposes the methodology used to assess the state of drought governance and drought risk awareness in the case studies for the two regions investigated. In the NWE region, the governance assessment tool developed by Bressers et al. (2013) was implemented; whereas, in the Mediterranean region, the analysis of water rivalries proposed by La Jeunesse et al. (2010) and La Jeunesse and Cirelli (2012, 2014) was used.

13.2.1 The Governance Assessment Tool

The Governance Assessment Tool comprises a 'matrix' style model consisting of five elements, the dimensions of governance (levels and scales, actors and networks, perceptions of the problem and goal ambitions, strategies and instruments, and responsibilities and resources for implementation) and four criteria (extent, coherence, flexibility, and intensity), producing a matrix of 20 cells (Bressers et al. 2013, Bressers, Bressers, and Larrue 2016). In this matrix, evaluative questions (Table 13.1) are formulated that can then be asked to both local and regional stakeholders. Based on their answers and insights, a judgement can be reached on whether the governance circumstances investigated in the matrix box concerned are supportive, restrictive, or neutral for drought adaptation.

In this chapter, we focus on the four criteria characterizing the perceptions of drought-related issues and the goal ambitions to solve these: extent, coherence, flexibility, and intensity. The governance dimension 'problem perspectives and goal ambitions'

Table 13.1 Matrix form of the governance assessment tool.

Governance dimension	Quality of the governance regime			
	Extent	Coherence	Flexibility	Intensity
Levels and scales	How many levels are involved and dealing with an issue? Are there any important gaps or missing levels?	Do these levels work together and do they trust other between levels?	Is it possible to move up and down levels (upscaling and downscaling) given the issue at stake?	Is there a strong impact from a certain level to change behaviour?
Actors and networks	Are all relevant stakeholders involved? Who are excluded?	What is the strength of interactions between stakeholders? In what way are these interactions institutionalized in joint structures? What is the history of working together, is there a tradition of cooperation?	Is it practised that the lead shifts from one actor to another?	Is there a strong impact from an actor or actor coalition on water management?
Problem perspectives and goal ambitions	To what extent are the various problem perspectives taken care off?	To what extent do the various goals support each other, or are they in competition?	Are there opportunities to re-assess goals?	How different are the goal ambitions from the status quo?
Strategies and instruments	What types of instruments are included in the policy strategy?	To what extent is the resulting incentive system based on synergy?	Are there opportunities to combine or make use of different types of instruments? Is there a choice?	What is the implied behavioural deviation from current practice and How strongly do the instruments require and enforce this?
Responsibilities and resources	Are responsibilities clearly assigned and sufficiently facilitated with resources?	To what extent do the assigned responsibilities create competence struggles or cooperation within or across institutions?	What is the flexibility within the assigned responsibility to apply resources in order to do the right thing in an accountable and transparent way?	Is the amount of applied resources sufficient for the intended change?

is considered here as reflecting the level of awareness of drought and water scarcity issues in the case study concerned.

Interviews were performed in 2013 and 2014, in six case studies in NWE, as briefly described later. The material for the analysis was collected during two field visits, during which the Governance Team – composed of six researchers in social and hydrological sciences – met stakeholders, managers, and representatives of the relevant local action groups. Results of the investigations site by site are described in Bressers et al. (2016).

13.2.2 The study of water rivalries

In water-scarce regions, water scarcity can lead to conflicts. However, the conflict is a consequence of a succession of tensions and competition for water that water-allocation strategies did not succeed in avoiding. In this methodology for water-governance analysis, the concept of water rivalry (Aubin 2008) is considered as an intermediary stage in water conflict (La Jeunesse et al. 2013). The concept is used in order to identify both cases of competition and cooperation over water at the catchment scale and is supposed to reflect stakeholders' ability to develop individual and collective actions to face water scarcity (La Jeunesse et al. 2016b). Thus, this concept makes it possible to identify the efficiency of water governance in responding to scarce situations, and thus indirectly assess the drought-governance efficiency.

The assessment of water rivalry contexts and the awareness of threats induced by climate change was performed in two phases. First of all, a qualitative analysis of local water uses and rivalries was conducted on local case studies based on semi-directive interviews, mainly with water managers. Then, a questionnaire containing both closed and open questions was implemented with the main representatives of water users. The material for the analysis was collected during several field visits and exchanges with stakeholders during the whole duration of the project. The collection of information was managed by a team of social scientists (geography, social anthropology, and political science) between 2010 and 2014. In total, 146 interviews were performed and 129 filled in questionnaires were collected and analysed, amongst other interactions with stakeholders.

The analysis of one particular section of the questionnaire and interviews is presented in this chapter. It pertains to how stakeholders (water managers and users) described the main water-related issues and how tensions were either solved or not. A question in particular related to the possible changes in the near future that could affect water use. This question provided information on the awareness and representation of the local impacts of climate change on water use.

13.3 Case studies in NWE

Six catchments were investigated in NWE (Figure 13.1): Brittany (France), Twente and Salland (both in the Netherlands), Flanders (Belgium), Eifel-Rur (Germany), and Somerset (United Kingdom).

- The French case study concerns the Vilaine catchment in Brittany, and more specifically the Arzal dam and reservoir located at the outlet of the river. Accordingly, the analysis focuses on the lower part of the Vilaine. The Arzal dam is a freshwater

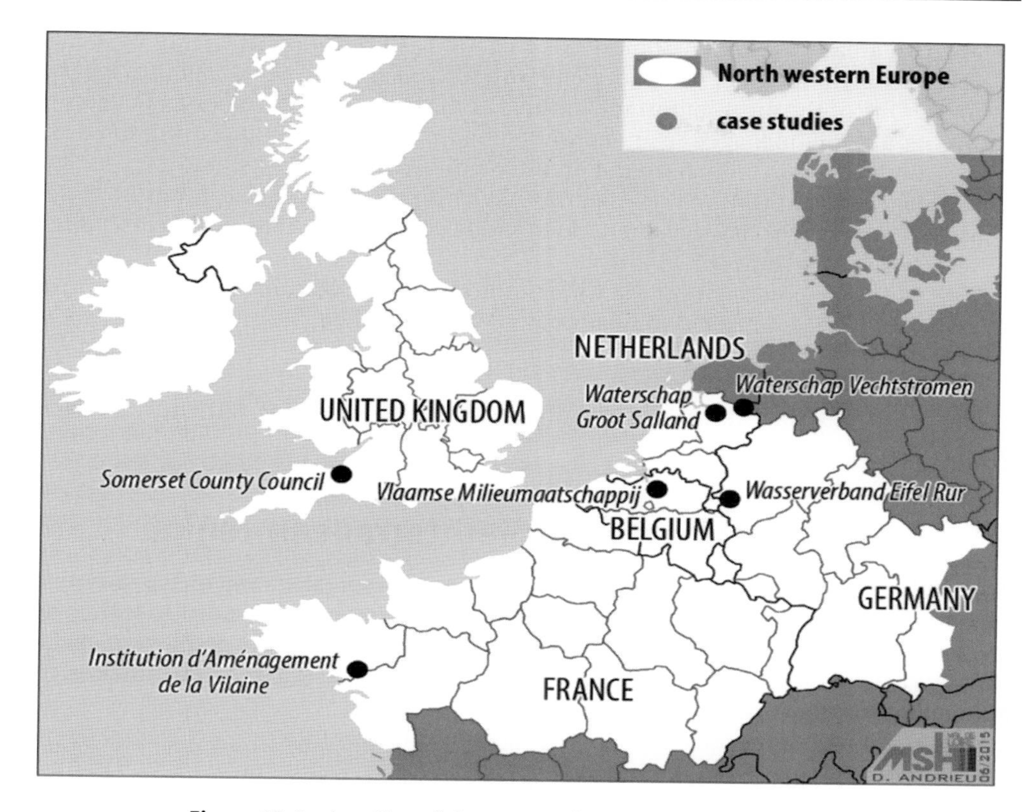

Figure 13.1 Location of the case studies in Northwestern Europe.

reservoir. Initially built in 1970 to protect the inland against salt water intrusion during spring tides, the water reservoir now plays a central role in providing drinking water for households in the region and supporting irrigation in the agricultural sector through regulating water levels. It also supports yachting activities.

- Water Authority Eifel-Rur (WVER) is a water authority in North Rhine-Westphalia covering the natural regions of Eifel and the Lower Rhine Embayment. The jurisdiction of WVER comprises approx. 2087 km^2 and c. 1.1 million inhabitants. With a capacity of approximately 300 million m^3, the large network of multifunctional water reservoirs of the upper part of the Rur plays an important role for many uses throughout the region. Originally built to control the effects of flooding, nowadays, it also serves tourism, drinking water provision, energy production, and recreation.
- The Salland region is included in the East-Rhine sub-basin. The case study conducted in this region focuses on the jurisdiction of the Groot Salland Water Authority (*Waterschap Groot Salland* WGS), a flat territory serving 360 000 inhabitants and numerous companies. The WGS manages a total of 4000 km of watercourse.
- Flanders covers 13 521 km^2 and has a population of 6.25 million inhabitants. The two WFD river basin districts are the Scheldt (466 inh/km^2) and the Meuse (258 inh/km^2). The high population density and intensive agriculture put the water resource under pressure and impact water quality.

- With an area of 4200 km², Somerset is England's seventh-largest county by area. It is a sparsely populated wetland area (with a population of about 508 000 inhabitants). One of the most important features of the region is the Somerset Levels and Moors, a highly-managed river and wetlands system which is artificially drained and irrigated in order to open up the area for productive settlement and uses such as farming. The peat soils of the Somerset Levels provide a multiple of ecosystem services. These peat soils are vulnerable to sudden and irreversible changes as a direct result of drought and dehydration.
- The Twente region is part of the water authority of Vechtstromen, which comprises 1 350 km² and about 630 000 inhabitants. Water availability is fully dependent on rainwater and groundwater.

The case studies are described in more detail in Bressers et al. (2016), available in open access.

13.4 Case studies in the Mediterranean region

The study was conducted on a set of five catchments representative of the variety of climates and social characteristics of the Mediterranean region (Figure 13.2). They are located along a latitudinal gradient from north to south, expressing a gradient of aridity, with one catchment in France, two in Italy (one located in the Alps and the other on the Mediterranean coast), one in Tunisia, and one in Palestine.

- The Noce River is a rural catchment located in the Dolomite range, in the Trentino-Alto Adige administrative region, in North-eastern Italy. This region supports an

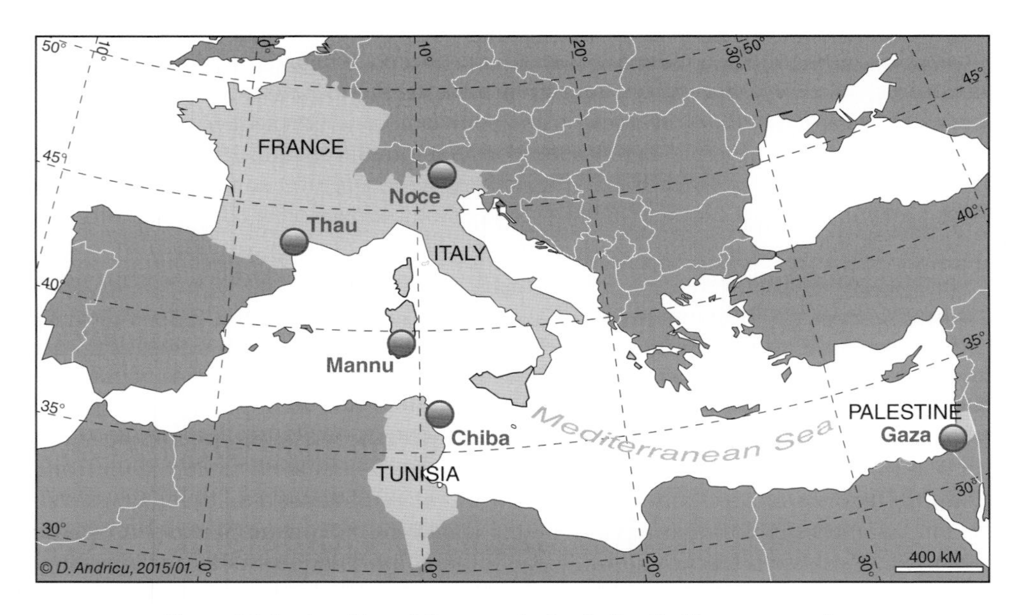

Figure 13.2 Location of the case studies in the Mediterranean region.

Table 13.2 Main characteristics of the five case studies.

Case studies/Characteristics	Noce	Thau	Mannu	Chiba	Gaza
Mean rainfall (mm/year)	1172	600	700	450	300
Catchment area (km^2)	1367	280	472	200	365
Population density (inh/km^2)	47	472	122	80	3836
Elevation min (m)	199	0	62	2	0
Elevation max (m)	3759	300	842	500	105
Urban area (% of the catchment)	3	14	4	5	23
Dams and reservoirs	Yes	No	Yes	Yes	Yes
Hydropower production	Yes	No	Yes	No	No
Coastal zone	No	Yes	No	Yes	Yes
Agriculture in the catchment	Yes	Yes	Yes	Yes	Yes
Tourism in the catchment	Yes	Yes	No	No	No
European Union member	Yes	Yes	Yes	No	No

intensive cultivation of apple trees for one valley (the Non Valley) and ski resorts for the other (the Sun Valley).

- The Thau lagoon is located in the Languedoc-Roussillon region, in the Hérault Department, in the south of France. This lagoon is an intensive area of shellfish production whilst the coast supports tourism and vineyards for wine production.
- The Rio Mannu di San Sperate catchment is a rural area in south Sardinia experiencing an intensification of agriculture. This stream is a tributary of the Flumini Mannu stream comprising dams that contribute to the Island's drinking water and electricity production.
- The Chiba catchment is a rural catchment located in the Nabeul Governorate territory, in the Cap Bon peninsula. Half of the catchment is controlled by a dam used for crop irrigation. The coastal aquifer is used to counterbalance the lack of water in this semi-arid climate and is now suffering from severe overexploitation and salinization.
- The Gaza Strip has a 42 km long coastline and is 6–12 km wide, covering an area of 374 km^2 with a high population density and a rapid increase in urbanization. All water resources are overexploited.

The main characteristics of the case studies are presented in Table 13.2. More details are available in La Jeunesse et al. (2016b).

13.5 Drought perceptions and goal ambitions in NWE

Based on the outcomes of the analysis of the drought-related water governance issues in NWE through the six case studies investigated, the first results show the low level of awareness regarding the drought issue. This creates a poor context for the development of drought governance and actions related to drought preparedness (Larrue et al. 2016). In fact, because of the wet situation, many actors involved in water governance are much more preoccupied with floods. It was found during the investigations that these regions have been flooded. Even in the Somerset region, where drought awareness was probably the highest at the beginning of the study, the particularly intense flooding event that occurred in the 2013–2014 winter – and thus, during the implementation of the GAT matrix – drought issues were probably forgotten for some period of time. This shows a small resilience of drought awareness and preparedness for this wet region.

As a matter of fact, in nearly all cases, the intensity of goal ambitions, and indeed drought perceptions, was the worst parameter in the drought governance context. In some cases, drought issues were even introduced during the interviews with the governance team. Thus, on the regional scale represented by these six case studies, the level of drought awareness can be considered as being very low, just as the intensity of the drought-governance context.

Moreover, the study highlighted that the anticipation capacity is very limited with a set of a few measures to be activated whenever a drought occurs. Finally, these measures can be considered more as a management of a drought crisis than the effective anticipation of potential drought hydrometeorological extreme events.

However, this low-level of drought governance context is not due to a lack of water governance development. In all cases, it was noticed that an effective water governance system is in place, with particularly intense networks, along with a significant involvement of actors on different levels and scales. In fact, in these European countries, water governance is highly developed thanks to a water policy already implemented since the 1960s and governed since the 2000s by the WFD (Larrue et al. 2016).

13.6 Drought perceptions and goal ambitions in the Mediterranean region

In the Mediterranean case studies investigated, water scarcity adaptation strategies caused by both climate variability and increasing water demand are mostly represented by fetching water from farther and farther afield. This shows a clear perception of water scarcity and the need to secure water availability with external sources. This strategy has indeed been quite an effective response to the increase in water demand.

However, this strategy does not solve local vulnerability as it neither proposes a long-term, sustainable, local strategy nor promotes the initiation of radical changes in how water is used. Moreover, it incites the development of economic activities based on external water resources.

This increased vulnerability on all sites can be illustrated by schematically looking at the adjustments already made in the case studies.

In northern Italy, by extending its agricultural frontier to the upper parts of the valleys, where a livestock industry along with tourism (hiking, etc.) are already present, or by producing artificial snow, which exerts greater pressure on local resources and requires investments in 'blankets' to slow down the melting of glaciers.

In Tunisia, more water is diverted from the north of the country at the same time as more water is allocated to tourism (outside the study area).

For the Thau coastal lagoon, water coming from the Rhône River – the main river of the French Mediterranean region – is used to support both domestic water uses and irrigate new crops. Introducing intensive irrigation in the catchment will impact the hydrological system and accelerate the transfer of contaminants to the coastal lagoon, whilst the shellfish farming activity, vital for the local economy, is particularly sensitive to any source of pollution.

The situation is a bit different in Sardinia. It is very likely that the limited access to water on an isolated island cannot allow for the expectation of external support. Reservoirs and dams are recognized here as a means of adaptation to increased water scarcity and drought frequency. The system has been optimized to face three consecutive dry years. However, there is no anticipation for a potential fourth drought year, as has occurred in another Mediterranean climate, in Cape Town, South Africa.

Now, we should admit that water use security is currently not an issue in the case studies described, with the exception of Gaza, where the cause is a particularly difficult geopolitical situation. However, this security is obtained by increasing the dependence on 'external' sources for local water uses.

In light of this, is creating external dependence a sustainable adaptation measure? Moreover, this increase of water supply as an answer to increasing water demand is not accompanied by the hoped-for radical changes in the dominant water consumption models for agriculture and tourism in the Mediterranean Region.

Furthermore, the study highlighted that all the analysed water management plans of catchments located on the Mediterranean coast mentioned desalination as an option. This is true both for the European case studies as well as the non-European ones. It seems this could be a next step to provide water supply in the Mediterranean region. Thus, fetching water farther and farther afield is still the main adaptation strategy to respond to an increase in water demand. This raises some questions, not only regarding the environmental and economic situation of the countries in question, since desalination requires a great amount of energy, whilst at the same time, countries are attempting to decrease their energy consumption. But also concerns the type of awareness of the threats induced by climate change whilst the complete integration of possible hydro-meteorological extreme event scenarios could initiate mutations in water uses. The situation seems to be more like a 'business as usual' strategy rather than the emergence of a process to face climate change threats with new adapted strategies for water use.

Finally, we can therefore understand that, even though the Mediterranean region is highly aware of drought impacts and water scarcity, information on the forecasted water scarcity seems still too slow to permeate into the decision-making process of the different water use sectors.

13.7 Conclusions

A comparison of drought-governance strategies in two different meteorological regions – respectively, North-western Europe and the Mediterranean Region – was performed in order to measure the role of drought perceptions in the development of goal ambitions in the context of climate change, a necessary step in the adaptation strategy development process. The assumption made for this chapter is that an arid region could be more prepared for an increase in the severity of droughts than a wet region.

Whilst drought awareness is very low in the case studies investigated in NWE, the actual awareness in the Mediterranean region did not support an emergence of revolutions in the way water is used. In fact, climate change in the Mediterranean is probably likely to generate tense situations amongst water resource users, similar to some examples of crises that occurred in the past for many catchment areas. The temptation is probably to believe that as it leads to comparable situations, with the successful management of these conflicts through integrated water management systems with local actors, there is no new challenge facing these situations. However, in the present state of knowledge on the impact of climate change, it is likely that the chains of causes and effects will lead to different situations which will be more difficult to address. Thus, the tools previously developed, although operational and valid today, might become obsolete if they are not updated continuously with the increased knowledge of all the criteria explaining the chain of causes and effects between climate change and water use. The main conclusion of this comparison is that in both regions, representation of drought threats is too low; this may be due to a difficulty in demonstrating drought impact mechanisms scientifically, compared to flood impacts that are obvious when shown a picture of the flooded area. What could be done to enhance drought awareness? Communicate on the Cape Town drama and the serial water 'Day zero' of this Mediterranean-climate city during the year 2018? As a matter of fact, the editorial of the *Lancet* global health Journal used this to denounce a failure in governance (Anonymous 2018), whilst the executive officer of the American water works association commented, in two editorials (La France, March and May 2018 – La France 2018a, 2018b) of the journal, that day zero finally did not occur due to the management of the crisis by local water managers. However, the severity of this drought, which was attributed to the effects of climate change, could have been anticipated. By considering a set of probabilities, a water-consumption strategy could have been tested and implemented instead of managing a potential Day zero crisis for 16 April 2018, with a water consumption 'as usual'. It seems as though there is an obligation to answer to the water demand as the main option, without any possible anticipated flexibility, except during a crisis.

This confirms the need to continue all efforts to disseminate the state of scientific knowledge on the impacts of climate change on water in both the Mediterranean and NWE regions, especially to local water managers, as initiated by various research programmes of the European Commission. Moreover, in both regions' context, solutions to increase awareness should be found. Models – integrating climate, hydrology, and water use – are an undeniably educational support tool in disseminating the scientific results, however drought indexes are still complex to assess and difficult to communicate. Also, drought modelling relies on the validity of the databases and moderating the dissemination by experts closest to local stakeholders.

To support such dissemination, the EU FP7 research programme CLIMB, the core objective of which was to decrease hydrological modelling uncertainties in the context of climate change in the Mediterranean Region, implemented a process of interactions with stakeholders to encourage local dissemination of scientific results so as to raise local awareness. The DROP project offered a handbook and a drought governance assessment guide communicated to stakeholders, teaching the stakeholders interviewed during the project in order for them to support a self-assessment of drought governance. Interactions with stakeholders after the end of the projects showed an increase in drought risk awareness, even though it this is probably not on the day-to-day agenda. Thus, in the context of the effective water governance in place in European catchments, along with a particularly intense and active network of actors involved on different levels and scales through decades of water-management policies, there is no reason to doubt that there will be a rapid development of strategies to anticipate drought events as soon as the awareness reaches a supportive level.

Acknowledgements

This chapter was funded by the European Commission Seventh Framework Programme through the CLIMB project, Grant Number 244151 and by the INTERREG IVB NWE Programme with the DROP project. The author would like to thank the coordinators respectively, Ralf Ludwig and Nanny Bressers, for their support throughout the project's duration. The author would like to gratefully acknowledge case study stakeholders for their participation in these studies.

References

Anonymous (2018). Water crisis in Cape Town: a failure in governance. *The Lancet Planetary Health* 2 (3): e95.

Aubin, D. (2008). Asserted rights; rule activation strategies in water user rivalries in Belgium and Switzerland. *Journal of Public Policies* 28 (2): 207–227.

Bressers, H., Bressers, N., and Larrue, C. (eds.) (2016). *Governance for Drought Resilience. Land and Water Drought Management in Europe*. Cham: Springer Open Science.

Bressers, H., Deboer, C., Lordkipanidze, M. et al. (2013). *Water Governance Assessment Tool with an Elaboration for Drought Resilience*. Almelo: DROP, 46p.

Forzieri, G., Feyen, L., Rojas, R. et al. (2014). Ensemble projections of future streamflow droughts in Europe. *Hydrology and Earth System Sciences* 18: 85–108.

Garnier, E. (2015). Strengthened resilience from historic experience. European societies confronted with hydrometeors in the sixteenth to twentieth centuries. In: *Hydrometeorological Hazards – Interfacing Science and Policy* (ed. P.H. Quevauviller), 4–25. Chichester, United Kingdom: Wiley Blackwell.

Intergovernmental Panel on Climate Change, IPCC (2007). *Climate Change 2007: Synthesis Report, Fourth Assessment Report*. Cambridge: Cambridge University Press.

Intergovernmental Panel on Climate Change, IPCC (2012). Managing the risks of extreme events and disasters to advance climate change adaptation. In: *A Special Report of Working Groups I and II of the Intergovernmental Panel on Climate Change* (ed. C.B. Field, V. Barros, T.F. Stocker, et al.), 582 p. Cambridge and New York: Cambridge University Press.

La France, D.B. (2018a). Day Zero, Defeat Day Zero, Open Channel. *American Water Works Association* 110 (3): https://doi.org/10.1002/awwa.1027.

La France, D.B. (2018b). Level 6B: Defeat Day Zero, part 2, Open Channel. *American Water Works Association* 110 (5): https://doi.org/10.1002/awwa.1076.

La Jeunesse, I. and Cirelli, C. (2012). Comparative analysis of water uses and rivalries in the case studies. Deliverable 7.2, CLIMB Project. Munich: CLIMB, 53p.

La Jeunesse, I. and Cirelli, C. (2014). Diagnosis of climate changes impacts on water uses and rivalries in the study sites. Deliverable 7.3, CLIMB Project. Munich: CLIMB, 65p.

La Jeunesse, I., Cirelli, C., Larrue, C. et al. (2016b). Is climate change a threat for water uses in the Mediterranean region? Results from a survey at local scale. *Science of the Total Environment* 543: 981–996. https://doi.org/10.1016/j.scitotenv.2015.04.062.

La Jeunesse, I., Gillet, V., Larrue, C. et al. (2010). Questionnaire for uses and rivalries identification in study sites. CLIMB, report. Deliverable7.1. Munich: CLIMB.

La Jeunesse, I., Gillet, V., Aubin, D. et al. (2013). Water rivalries for identification of both competitions and collaborations over water at catchment scale. In: *WATARID (vol.3.) Usages et Politiques de l'eau en zones arides et semi-arides* (ed. M.-F. Courel and M. Taleghani), 233–246. Paris: Eds Hermann, 564p.

La Jeunesse, I., Larrue, C., Furusho, C. et al. (2016a). The governance context of drought policy and pilot measures for the Arzal dam and reservoir, Vilaine catchment, Brittany, France. In: *Governance for Drought Resilience. Land and Water Drought Management in Europe* (ed. H. Bressers, N. Bressers and C. Larrue), 109–139. Cham: Springer Open Science.

Larrue, C., Bressers, N., and Bressers, H. (2016). Towards a Drought Policy in North-West European Regions? In: *Governance for Drought Resilience* (ed. H. Bressers, N. Bressers and C. Larrue), 245–256. Cham: Springer.

14
Strategies and Instruments to Face Drought and Water Scarcity

Hans Bressers[1], Nanny Bressers[2], and Stefan Kuks[3]
[1] *Department of Governance and Technology for Sustainability, University of Twente, Enschede, The Netherlands*
[2] *Former Project Leader of the European DROP project at the Water Authority of Vechtstromen, now a consultant at Vindsubsidies. Uthrecht, the Netherlands*
[3] *Department of governance and technology for sustainability, CSTM, University of Twente, Enschede, The Netherlands and Vechstromen Water Authority, The Netherlands*

14.1 Introduction

Extreme weather events are more easily related to too much water rather than to too little water. The reason is that droughts develop far less sudden than floods often do. However, even when the 'event' takes weeks to fully develop, droughts are no less damaging, nor are they less prone to become extreme. Due to climate change and increased demand, the availability of sufficient water of suitable quality for purposes such as agriculture, nature, and service and drinking water production has become less self-evident and will likely even become more uncertain in the future. Water-scarcity problems have long been experienced in the more arid Mediterranean countries within Europe. Natural conditions of the water basin play an important role in determining the frequency and amount of precipitation and storage in an area as well as the changing climate and rainfall patterns. As such, a number of issues related to water scarcity are also becoming important to many Northern European countries. This chapter discusses the strategies that are of importance in the development of policy responses to these water-scarcity issues.

Whilst differences of opinion exist with respect to the exact definition of water scarcity, it is often dealt with as a situation where the availability of water does not the meet the demands of people and nature (White 2012). Water utilization patterns are influenced by both natural factors that determine the resource status (water availability and chances for water utilization), and socio-economic factors that determine the resource use. Natural factors refer to the natural conditions of a water basin, conditions in terms

Facing Hydrometeorological Extreme Events: A Governance Issue, First Edition.
Edited by Isabelle La Jeunesse and Corinne Larrue.

of climate change and changing rainfall patterns. Socio-economic factors refer to traditional water demands that may be expanding and creating homogeneous rivalries (more same kind of users demanding use from a limited stock), new water demands or values (like natural, ecological, or recreational values of water) resulting in heterogeneous rivalries (between different use types); and demands for other natural resources creating rivalries (such as land use claims in floodplains). Natural factors and socio-economic factors together create a certain sense of urgency in terms of water scarcity.

To reduce the effects of water scarcity, a policy approach should focus on socio-economic factors related to the water utilization pattern. Adaptive responses should reconsider traditional and currently recognized natural and recreational water demands and take into account rival demands for rather vulnerable use functions in society. Traditional demands include mainly economic functions (e.g. guaranteeing the navigability of rivers, guaranteeing the use of rivers for cooling energy power stations, preventing salt water intrusion into aquifers for drinking water, and agricultural crop growing). Adaptive responses are crucial for managing these functions as well as for the more natural and recreational values of water systems. The result is often the development of minimum flow requirements (Kuks and De Boer 2013).

Different packages of measures are possible to increase resilience towards drought. Three general strategies can be discerned. These strategies are reactive, preventive, and adaptive. Each of these strategies can be coupled with types of measure. The *reactive strategy* consists of the more classic active interventions in the water system to bend it in the desirable direction. The *preventive strategy* largely concerns preventing water shortages from occurring in the first place, or at least limiting their consequences. The *adaptive strategy* works from the existing system and its limitations, in order to adapt to it in such a way that shortages occur less. In many situations combinations of these strategies will be applied in order to handle drought and water scarcity adequately. There is not a single preferred category, or, for that matter, categories that are wrong to apply.

Examples of *reactive measures* are the transport of water from other regions or from bigger rivers to the dry areas and the construction of dams and reservoirs. This is something that is not only costly, but can also have negative side effects, for instance on water quality in vulnerable nature areas. Amongst the *preventive measures* are all kinds of interventions to save and hold water available from wet periods, and to increase the buffering capacity of the soil and the water system. Examples of *adaptive measures* are to accept the limitations of the natural system and consequently adapt water use to the drought risks that the system generates. All strategies require appropriate water utilization, land use, and a robust water system. Whilst natural and weather conditions are a given fact that is generated by the climate, all of these strategies might be labelled forms of 'adaptation' in climate policy terms. However, real adaptive responses are more long-term oriented, based on learning from past experiences, and based on consultations of different opinions from rival stakeholders. Reactive responses could be more short term or satisfy one section of the population or a specific group of stakeholders, without sufficiently taking into account alternative options (Pahl-Wostl et al. 2007; Kuks and De Boer 2013).

To improve the evaluation of past experiences, for the development of a more long-term orientation, and the involvement of all stakeholders and their interests, we need to increase our consciousness of the natural conditions and socio-economic factors that affect water utilization patterns and the problems of drought and water scarcity that might result. This leads to a fourth necessary strategy that can be labelled as *supportive* measures, consisting mostly of research and communication measures.

In this chapter we will elaborate this taxonomy of drought measures further and give several examples. A major source of those examples are the many measures that have been studied in the European research project DROP (Benefit of governance in DROught adaPtation) on drought resilience in northwest Europe (DROP 2015; Bressers, Bressers, and Larrue 2016). These examples are from the regions of Somerset (UK), Twente and Salland (Netherlands), Vilaine (France), Flanders (Belgium), and Eifel-Rur (Germany). Other examples come from the European research project Aquadapt, which was part of a cluster of European research projects on coping with drought and water deficiency due to changing water utilization patterns in arid countries in the Mediterranean region (ARID 2005; Koundouri 2008; Kuks and De Boer 2013). These examples are from the Marina Baja catchment area around Alicante (Spain) and the Hérault river basin around Montpellier (France).

14.2 Reactive measures

In many cases of regular water shortages reactive measures have been the first form of public response. A typical example can be found the Marina Baja catchment area on Spain's eastern coast in the autonomous region of Valencia. Tourism along the coastline is very important for the Spanish economy, but also contributes to water stress. Fruit trees (especially orange and lemon trees), olive trees, and almond trees are cultivated here and the population has significantly increased over the last decades. The Marina Baja area suffers from water deficiency and salt-water intrusion due to over-exploitation. This is partly the result of natural conditions. It is a semi-arid area with a decreasing annual rainfall and frequent drought periods. In addition, in the autumn there are frequent floods and an irregular rainfall pattern. This is also the result of socio-economic conditions: overexploitation due to agriculture in inland areas (a decrease of rain-fed agriculture, an increase of irrigated agriculture), overexploitation due to urban growth and development and strong growing tourism in the coastal area. Since the 1960s there have been intra-basin transfers within the Jucar river basin district as well as the construction of new dams and wells in the Marina Baja area. Use limitations and recommendations have not been put in place to reduce consumption (demand management), although public bodies are aware since the 1960s that urban growth and tourism would create water problems for the area (Kuks and De Boer 2013).

The creation of dams and reservoirs and the transportation of large quantities of water from one basin to another are examples that long dominated drought management. One of the most controversial plans was the idea to transport enormous amounts of water from the Ebro basin in Spain to Andalusia in the south of the country where aquifers are being rapidly depleted. Whilst in Spain the National Hydrological Plan of 2001 on the one hand extended the eco-perspective, it promoted on the other hand more dams and interbasin transfers (Costèja et al. 2004). The opposition against this plan contributed to the fall of the Aznar government and lead to an international movement designing the 'European Declaration for a New Water Culture' (Arrojo et al. 2005). This declaration was partly in defence of the European Water Framework

Directive that explicitly turned against inter-basin water transports. Since 2004 the central government in Spain has developed a policy approach that aims more at addressing resource constraints and solutions in terms of desalinization and demand management. The Environment Ministry supervises the river basin authorities which develop water basin level policy (Costèja et al. 2004). However, one step below, the Jucar river basin authority did not have a lot of impact on the Autonomous Community of Valencia and the Municipalities in the Marina Baja catchment area. At regional and local levels the resource is not considered to be a limiting factor as they believe in and stick to supply management instead of demand management as a way of addressing water distribution issues (Kuks and De Boer 2013).

Having stated this, there is no reason to regard reactive measures as such purely negatively. The problem with the big schemes of the past is mainly that they were insensitive and thus incoherent with their side effects on other users and uses of the water systems that they affected, including the ecological aspects. In the Marina Baja catchment area a severe drought occurred in 1978, which created a strong sense of urgency. The prospect of growing demands and the threat of future water deficiencies that could occur if nothing were to be done triggered thinking about possibilities for water re-allocation. In 1978 the Marina Baja Water Consortium was created. This consortium is an arrangement between municipalities and farmers to re-allocate water in such a way that treated wastewater from urban and tourist areas could be reused in agriculture. Municipalities in the coastal area now cover most of the investment and exploitation costs, which keeps the price of irrigated water as low as possible. The required contribution from the general budget is justified by the importance of tourism for the area's economy. The consortium is an initiative by municipalities, but is not under the supervision of the river basin authority. The consortium is a closed community, protecting its strong economic interests. It is a corporative institution, not very open to rival interests (such as the interests of the ecosystem) or linking up with land use development. There is a low level of public participation and a lack of access to information (Kuks and De Boer 2013).

Examples of responsible reactive measures from the DROP-project can be found in the Dutch region of Salland (construction of a two-way water transport device), the river basin of the French Vilaine river (construction of a lock, preventing salt water intrusion in a fresh water reservoir) and the German region Eifel-Rur (adaptation of the management rules for the reservoir by adding a drought index). We now consider these examples in somewhat more detail.

In the Dutch region of Salland an area of some 18 000 hectares was prone to flooding due to an insufficient drainage system and to water shortage in periods of drought. This required a double-acting system that is able to drain and supply enough water in, respectively, wet and dry weather conditions, and also a management system that responds quickly and effectively to changing weather circumstances. Two weirs and two pumping stations were built to discharge water to the nearby Vecht River and to pump water from the Vecht River into the catchment area. Also an additional double-acting pumping station at the Vecht River will drain and discharge the area. A remote-controlled steering mechanism that is linked to weather forecasts will in the future be placed at the pumping stations to manage this system (Özerol et al. 2016).

In the region of Brittany in France the Arzal dam is located at the mouth of the Vilaine River, just above the outlet of the river into the Atlantic Ocean. It was built in 1970 to block the tidal flow that previously travelled far inland and aggrevated flooding when high tide coincides with high river water levels. It also creates a large freshwater reserve that became a major source of water for many purposes. Sailing and fishing is highly developed as a recreation industry. Sailing from the ocean to the reservoir and vice versa and fish passages create salt water intrusions. Especially in times of drought this affects water quality and its aptness for other uses. When water inflows tend to be the lowest, recreational activities, including sailing, are generally at their highest level (because it involves the summer period), and lead to a peak of salt water intrusions in the reservoir. To prevent salt intrusions, syphons have been installed upstream of the dam to pump the contaminated water from the reservoir back to the Atlantic Ocean. The syphons are not 100% efficient: some brackish water still enters the reservoir, and salinity peaks are regularly observed in late summer each year. Moreover, this system leads to losses of substantial quantities of freshwater, which, during prolonged periods of drought, may aggravate the problem of multiple uses of the reservoir and impact the freshwater supply. The water authority has worked on developing a new and innovative lock at the dam to prevent salt water intrusion when boats cross the dam. Significant efforts have been put into designing this new lock: preliminary and feasibility studies, 3D hydraulic modelling, and even a physical model at the scale of 1/12 (La Jeunesse et al. 2016).

In the region near Aachen in Germany in the upper catchment of the Rur, six reservoirs were built in the Eifel-hills mainly to control the effects of flooding and to maintain the flow during dry seasons. Five of them form an interconnected system around the main reservoir 'Rurtalsperre' with a stream/river flowing through the basin as a stream. With a total capacity of 300 m. m^3 their function for flood control is very important. The reservoir system serves additional important aims. Amongst the reservoir system's functions is that of providing good quality raw water for drinking water production. Whereas the different aims do not always go in line with each other, still all of them have to be served. For example, sometimes a controlled high discharge out of the reservoir is needed in order to prevent flooding, but this can only be carried out to such an extent that there is still enough water in the reservoirs to produce drinking water and maintain the flow in dry periods. Stillwater and falling water levels in reservoirs bear the risk of a decrease in water quality, which results in a higher amount of production work and possibly drinking water production problems. Stillwater in these reservoirs always bears the risk of eutrophication with effects such as algal blooms. The long dry periods in spring in recent years has already resulted in a loss of water quality in some of the reservoirs. Climate change with longer dry and sunny periods is expected to increase this problem. Consequently an adaptation to climate change is only possible by an adaptation of the management of the dams. The Waterboard Eifel-Rur analysed the inflow patterns in the different dams. Based on these results, a study was carried out on the management system of the dams with respect to water quantity and quality. Suggestions for the adaptation of the management plan emerged: one of the best results obtained is to add a drought index in the management plan, which will help prevent the release of too much discharge in an earlier stage compared to today's practice. This leads to a credit of water in dry periods. The project thus added flexibility to operational decisions to improve the performance of the management system, in that the different obligations of the system are now still met under a wider array of meteorological and flow conditions. Whereas certain dry conditions in the past would have made it impossible to meet all obligations, under the improved system this would now be possible (Vidaurre et al. 2016).

14.3 Preventive measures

Preventive measures are aiming at the re-use of water, like in the Spanish case, or at saving and holding water to be used for times of water shortage and to increase the buffering capacity of the soil and the water system. Preventive measures can be institutional arrangements for demand management or the redistribution of water resources. They can also be interventions in the water system by means of new constructions, just like the reactive measures. Their purpose can be both supply oriented and demand oriented and often combines both. An example of a supply-oriented preventive measure is the construction of a pool at a farm to store water that can be used in dry periods. An example of a demand oriented measure is the improvement of the water buffering capacity of the soil so that demand for external water supply in dry periods is lower.

The Spanish case of the Marina Baja Consortium demonstrates not only a reactive response, it is also an example of a mixture with preventive response in terms of constructing a new actor arrangement at the local level between municipalities and private stakeholders (the farming community and the tourism branch) for the purpose of redistributing resources amongst these actors. However, in terms of shared values we can see that the stakeholders still rely on water reuse, inter-basin transfers, and desalinization as their approaches to managing water supply. They believe these options are able to supply the increasing water demands for agriculture, domestic use, and tourism without demand restrictions. A strategy for demand management or integrated water resource management (IWRM) is missing, since the prevailing strategy is mostly supply oriented. In that way, strong property rights on agricultural water use do preserve existing modes of water management and water utilization, but compensation payments are introduced to better balance the rival uses of a limited stock of water resources. The Marina Baja Water Consortium now forms an important part of the governance regime relating to the trading of used water, and improving use efficiency (re-allocation amongst users). Water users are not yet fully bearing the costs of infrastructure, exploitation, and externalities. General budgetary funds of various administrations are still involved. In that sense, we consider it more as a mixture of reactive and preventive responses, than a real adaptive response (Kuks and De Boer 2013).

Also in Northern European cases we see examples of preventive measures in the form of institutional arrangements, like tailor-made management plans for specific areas/plots. This has for instance been done in the Dutch region of Twente (local water-management plans with farmers) and the UK region of Somerset (area water level management scheme and land management advisory work). These advices can lead to small construction works at the plot level. An important challenge is to upscale such very local improvements to whole areas instead of scattered individual farms.

Also the water authorities themselves use constructions in the water system to prevent effects from droughts. For example, in the Dutch region of Twente, creeks and also bigger water courses were reconstructed to elevate the ground water table. For the same purpose in some instances drainage tubes were removed. Also in the UK region of Somerset such measures were taken.

As a county, Somerset has the greatest variety of soils in England. Low lying inland areas of Somerset depend on water management to maintain their environmental features and agricultural interests. Areas with exposed peat soils are particularly vulnerable to drought, which can damage peat soils, affect agricultural production, and impact the natural and historic environment. Drought awareness was consequently relative high here. However, in the winter of 2013–2014 floods put a large part of the area under water for a substantial period. In response to the floods, a 20 Year Action Plan for water management in Somerset was developed that paid attention to both flood and drought risks. The Plan includes a range of interventions like woodland planting, increasing soil organic matter, run-off attenuation features, improving soil structure, and slowing watercourse flow that will improve drought resilience. Its implementation included working with farmers to investigate ways of storing, conserving, and recycling water for on-farm use. This has involved implementing water efficiency measures and water conservation techniques in land-based businesses. Four 'demonstration' farms showcase various in-situ soil protection measures, and open days have been held to encourage other farmers to implement these measures on their own land. Measures include (i) the reinstatement of ditches and drains to slow and elevate run-off on a historic rural estate in Somerset; (ii) use of temporary grassland to minimize soil erosion and installation of filter fences to prevent soil from washing into neighbouring properties; (iii) installation of a stone gabion and fencing feature to hold back fine soil that washes from the gently sloping arable field; (iv) soil bunds to prevent soil washing through a hedge and a newly installed filter fence, silt pond, drains, and established winter sown oats after maize at a local farm (Browne et al. 2016).

Several projects were realized in the north east of Twente in the Netherlands: drainage systems were removed, ditches were muted, streams were shoaled and water storage areas were constructed. Drainage systems are typically geared towards getting rid of the water as soon as possible. Deep ditches and creeks have a similar effect on nearby land: they extract groundwater from the shores with also some effect on ground water levels further away from the water streams. Water storage basins on the contrary give surface water ample time to make it into the ground water, and additionally provide service water in dry periods, creating a water buffer for dry periods. An example is the restructuring of the upper reaches of Snoeyinks brook. The project area comprised a number of small upper tributaries that flow into the Snoeyinks brook. The restoration of morphology and historical course of the river; the creation of new natural areas (hornbeam, oak woodland and poor-quality grassland); landscaping; recreational development, with access; and water management measures on farms and fields all contributed to making the area more drought resilient, whilst also improving the recreational and natural value of the area. Also water-management plans have been made together with farmers, tailored to their specific situation. The plans include tips and tricks on how to influence the water balance by storing water, resulting in a mutual gain for the farmer and the adjacent nature areas. The intensive communication established with the farmers created awareness, and motivated other stakeholders to work on drought adaptation as well. Especially the smaller traditional farmers need to see results like better crop growth at the neighbour's plot, to get interested (Bressers, Bleumink et al. 2016).

An interesting example of public debate about the need for more water buffering by farmers, to guarantee minimal flows, can be found in the Hérault river basin in the

South of France. This basin reaches the Mediterranean Sea near Agde, west of the city of Montpellier, and has a length of 148 km. The Hérault basin is situated in the region Languedoc-Roussillon. The Hérault river basin suffers from sanitation problems, as during times of low flow the amount of effluent from wastewater treatment plants has a very high impact on the water quality of the river (compared to periods of a high river flow). This is partly the result of natural conditions. Water shortage periods in summer constrain irrigation for a few weeks, but rarely result in a ban on crop irrigation. Supplies have not been low enough to affect drinking water supply, industrial use, and hydro-power in summer. Similarly, there is no pattern of increasing drought. However the awareness of issues related to water deficiency are increasing due to socio-economic conditions. There is a changing population due to new settlers from outside the area who now represent the majority of the inhabitants. The demand itself is not changing, but the sensitivity of new citizens to irrigation affecting recreational use options is. Tourism is an increasingly important source of income for the area and is thus playing a more important role in discussions on ways for determining water distribution.

In terms of mutual dependencies amongst stakeholders it is interesting to have a closer look at the role of property rights in the Hérault case. Rivalry exists between farmers (irrigation) and stakeholders making indirect use of water (fishing, swimming, and canoeing). Fishing associations argue that farmers do not comply with the 1994 Fishing Act, which requires minimum flow. They also argue that farmers could prevent water stress by changing irrigation practices (e.g. switching to drip irrigation). Farmers argue that they have historical legitimacy for water use in the basin and that they should not have to give up this position to new environmental concerns. Political pressure is often used to oppose cases that are taken to court. The social pressure on farmers has developed to a credible alternative threat and has led them to a more adaptive response of exploring the possibility of constructing small reservoirs. Financial institutions, such as the Rhône-Méditerranée-Corse Water Agency, support conflict resolution by subsidizing studies and investments to improve irrigation efficiency. In that way, political pressure, social pressure, and financial institutions function as institutional triggers that help to achieve negotiated compromises and to stay away from continuous litigation (Kuks and De Boer 2013).

14.4 Adaptive measures

Whilst reactive measures focus on redistributing water resources and preventive measures focus on improving the efficiency of water use, and together they form the perspective of supply optimization, we could say that adaptive measures focus more on demand management and IWRM. Adaptive responses are more long-term oriented, based on learning from past experiences, and based on consultations of different opinions from rival stakeholders.

Considering again the situation in the Hérault river basin, we see that as a reaction to increased water stress, government agencies in the Hérault area have promoted the principle of collaborative management and planning of water resources (Kuks and De Boer 2013). They used a planning instrument – the SAGE procedure (Schémas d'Aménagement et de Gestion des Eaux – in English: Integrated Water Management Plan at River Basin Level) – defined at the national level by the 1992 French Water Act.

SAGE creates a platform where stakeholders can debate and negotiate. Applying this model means a change from the traditional top-down model to a more flexible participatory model. This can be identified as an adaptive response. SAGE is one of the major innovations of the 1992 Water Act. The SAGE process intends to include all water issues and all relevant stakeholders in a river basin. The process results in active participation and a wider attention for river-related concerns as compared to urban and economic growth concerns (Sangaré and Larrue 2004).

The SAGE approach has been a game changer in the Hérault case, because it applies to the new paradigm of IWRM. Government agencies, traditionally working at the level of the department, have to cooperate at basin level. This strengthens the cooperation between the central and the department level. However, not all public agencies supported the SAGE planning procedure from the beginning. Another change has been that environmental associations, canoe rental companies, and other stakeholders, could contribute to policy making. However, not all stakeholders are well represented. Legitimacy issues arise since government agencies select the representatives and since only organized stakeholders are represented in the process. Consultation is very formal, and the rather technical language used by the expert community results in the general public not being involved. Traditional stakeholders refuse to accept the existence of water stress due to the inclusion of new values. As a result, public agencies avoid speaking about water conflicts and consider them as possible future tensions. Despite these legitimacy problems, the SAGE procedure plays a crucial role in providing an adaptive response to water deficiency in the Hérault case. The procedure restructures the cooperation between administrative levels, leads to the involvement of new actors, reformulates the policy problem, and results in more integrated management (Kuks and De Boer 2013).

Looking at the Northern European cases from the perspective of the adaptive strategy, we see that adaptive responses try to focus demand management on the acceptance of natural conditions and the limitations of water resources. Thus not 'water levels should be adjusted to follow chosen human land and water uses', but 'land and water uses should follow water levels as they result from natural circumstances'. An example of such adaptive measures is setting minimal flow requirements to protect aquatic biotopes.

Another example is to adapt crop choice to the drought risk profile. For example in Somerset this has been an issue. With the choice of crops also the division between private (entrepreneurial) and public (taxpayers) responsibilities is relevant. When water authorities clearly restrict their responsibility to a certain standard of water provision they can define the remaining risks to be private. Of course, farmers could still decide to grow high value but vulnerable crops on drought prone lands, but when yield fails every 10 years or so, they should view it as an entrepreneurial risk that they cannot have avoided or compensated by public authorities. Whilst water demand management measures like these are still unusual, it is wise to start preparing the support basis for them and develop ideas on their future implementation.

The most frequent measures in this strategy are however temporary water use bans. These can affect different stakeholders. One of the first measures taken is typically to prevent water use for irrigating private gardens and washing cars. But also economic interests are affected. In severe cases of drought, farmers are forbidden to take water from water courses to irrigate their field. This often hurts because it is precisely when they feel the crops are most vulnerable. Even electric power plants are in some cases restricted or even temporarily stopped because they cannot use river water as cooling water anymore. The problem is here not the quantity as such, but the fact that when water that is used for cooling forms a significant proportion of the flow, the water temperature rises to levels that create dangers for the ecosystem.

In some countries policies and plans are developed to guard the water authorities in dealing with such water use bans. For example The Netherlands has an official 'displacement chain' that elaborately prioritizes types of water uses and thus identifies the water uses in the sequence that they have to be given up for a whilst. New institutions (a national coordination committee) guide the distribution of water to areas and uses when necessary.

In the displacement chain guideline the first priority is to prevent irreparable damage to the water system, the soil (e.g. peat layers), or nature. Second in line are the utilities: drinking water and energy production. Third are high value agricultural and industrial production processes and last are the interests of shipping, general agriculture, nature with resilience, industry, recreation, and fishery. This displacement chain is in fact not often used, since limiting some of the last priority uses (like irrigation of agricultural fields and gardens and car washing) has been generally sufficient. Actually limiting the use of rivers for cooling water (thus restricting energy production from specific plants) has been another. There is a national coordination committee for water distribution. This consists of representatives of the ministry, including the public works agency, and the Union of Waterboards (UvW) and the Interprovincial Consultation (IPO). They meet when the water level in the transnational rivers gets lower than certain values or when even without this being the case there are drought problems in several regions. Apart from proposing measures (in principle using the displacement chain, but also including fine-tuning of the water system where it can be regulated) they also issue 'drought messages' to over 400 stakeholders whenever there are possible water shortage problems. In summer, April to September, this is generally done every fortnight.

14.5 Supportive measures

By supportive measures, we mean all those kinds of instruments that affect the water system indirectly, by enabling or improving measures from the first three strategies.

There are two main categories here. The first category consists of all kinds of *research*. This includes also modelling and forecasting. Examples are: Somerset (modelling and technology transfer in the Upper Parrett catchment on irrigation scheduling and water application management), Vilaine (forecast inflows to the reservoir during the low-flow season and help to anticipate critical situations to ensure better drought risk management), Eifel-Rur (inflow pattern analysis and drought index), Flanders (development and use of indicators for the monitoring and reporting of the drought situation and the modelling of drought impacts, on, e.g. agriculture). In Twente, two research projects have been testing level dependent drainage and the surface runoff to acquire better knowledge to improve adaptive drought management. In the Hérault case we have seen that financial institutions, such as the Rhône-Méditerranée-Corse Water Agency, support conflict resolution by subsidizing studies and investments to improve irrigation efficiency. On a more general level there has been a lot of attention given to the science-policy interface in the recent DROUGHT-R&SPI project (Andreu et al. 2015).

A second category of supportive measures are forms of *communication*. As the drought awareness amongst citizens (voters) and farmers (landowners) is still low and also relevant decision makers not always feel the urgency, awareness raising is very important, especially in the areas where droughts are less visible. Without a sufficient degree of awareness amongst relevant stakeholders preventive and adaptive measures get insufficient support. Some examples are: Flanders (a web portal with drought forecasts and drought impacts), Somerset (education at schools and colleges about drought alongside flooding as one of the future challenges for the Somerset Levels), and Salland (drought awareness campaign). Some of these examples will be explained more in detail below.

Combining research and communication, Flanders in Belgium focused on the improvement of drought-risk management and the improvement of knowledge and data collection, which were listed as policy options in the European Commission's Communication on drought and water scarcity (EC 2007). The Flemish Environment Agency operates a dense hydrological measurement network to monitor precipitation, evapotranspiration, river stage, and discharge. Real-time measurements from these stations are used for monitoring the water system, flood forecasting, and operating the water infrastructure in Flanders. Where possible, also a 10-day forecast for these indicators was developed to allow proactive management. To translate observations to impacts, modelling tools were developed. Through these models, the effect of soil and crop type, and drought severity and time of occurrence during the year on the impacts of the drought can be highlighted. A further key step towards effective drought management is turning these data into useful information and getting the information to the different actors and stakeholders. Therefore, the environment agency publishes a monthly bulletin on the state of the water system, focusing on drought or flood risks as required. Since 2014, five Flemish water-management services bundle their efforts to make real-time data and forecasts on the state of the water system available through the web portal www.waterinfo.be. Drought is included as one of the four main themes of the portal. Information on the drought situation can be used by actors such as decision makers or water managers, or individual stakeholders, such as farmers, to evaluate the drought situation and take action accordingly. At the same time, the environment agency participated actively in the creation of a coordination platform for drought to stimulate the cooperation between different actors and stakeholders on drought issues, such as the Flemish agricultural department, regional and national water managers, the provinces and municipalities. (Tröltzsch et al. 2016)

In France, the Vilaine river water authority has been working with the IRSTEA research centre to develop a modelling system to forecast inflows to the reservoir during the low-flow season and help anticipate critical situations to ensure better drought-risk management. The system incorporates weather information into a hydrological forecasting model and translates the results into a graphical representation of the drought risk. Future possible weather scenarios over the Vilaine catchment can thus be considered and transformed into modelled outcomes of river inflows all the way upstream to the dam. The graphical representation of the drought risk provides a visual assessment of the risk of being below given critical low-flow thresholds in the next weeks or months, both in terms of average flow and number of days below each critical threshold.

This risk assessment visualization tool aims to help the managers of the dam in deciding whether to release water from the reservoir and on how to operate the corresponding dam components. The tool is based on the development of a global forecasting chain, including the development of weather scenarios combining a short-term meteorological forecast (9 days), a long-term meteorological forecast (3 months) and an analysis of past events over the last 50 years (La Jeunesse et al. 2016).

A research project that was in fact testing a preventive measure that is clearly also aiming at gaining acceptance with farmers for drought resilience measures took place in the Dutch region of Twente. The project involved the testing of a level-dependent drainage system near a nature conservation area. Level-dependent drainage implies that the land owner or tenant can to some extent influence the water table by adjusting the drainage system. The advantage is that the table can be temporarily lowered when for instance the farmer wants to work on the land with machinery. This improves the farmer's preparedness to accept higher water tables than they would without a system they can influence themselves. Many water managers see level-dependent drainage as the primary means of preventing water depletion and of optimizing agricultural use of areas of land. However, this idea lacks scientific underpinning. There is only limited knowledge about the effects that level-dependent drainage has on nature. Therefore, the Vechtstromen water authority conducted a study in a nature area that is surrounded by water-depleted area of intensive agriculture. For that purpose Vechtstromen water authority constructed a system of level-dependent drainage in combination with raising the drainage basis of a small water-depleted nature conservation area to be able to test the effect on nature (Bressers, Bleumink et al. 2016).

A very special communication challenge occurred in England in the wake of the big Somerset Floods of 2013/2014. After some years of severe droughts, the flooding events created a public outcry and a 'blame game' and a 'militarization of emergency water management' (Browne et al. 2016, pp. 101–102). Whilst there had been well-balanced water-management plans already developed but not funded, suddenly money did not seem restricted anymore as politicians, including the prime minister, flew in and declared priority to improve discharge capacity. Also the media and many landowners blamed the nature protection organizations for delaying the necessary dredging and deepening of canals and rivers. For a while the call for one-sided measures was strong, that would have worsened subsequent impacts of new droughts. It is an example of how extreme weather events can disrupt networks of stakeholders in water management and frame the discussion in just one direction.

Not only in the UK, but across the areas studied in the DROP project the problem awareness is still low amongst land owners and the general public, and thus many politicians. This restricts the selection of forceful interventions to increase drought resilience and makes it sometimes more difficult to realize the measures chosen in practice. The project found three major strategies for pushing the position of the drought issue that is still experienced by many as a second-order issue.

1. Aiming to place drought and water scarcity on the public and political agenda on its own, as an independent problem. For instance, by providing continuous information to the public like in Flanders on the agency's website, or by directly addressing national water planners with a broad coalition of stakeholders like in the Netherlands' Delta programme process.

2. Addressing drought by 'piggy-backing' other issues, i.e. including drought-relevant measures in different planning initiatives and ensuring coherence of plans with drought objectives.
3. Using a 'plans in drawer' strategy by preparing a ready-to-implement strategy for when an extreme weather event makes the topic climb the agenda and receive political attention responding a call for action. The presence of elaborated ideas on how to combine flood and drought defence was likely what saved Somerset from an avalanche of one-sided measures that were called for when the big floods were upsetting the public, the media, and the politicians.

The careful application of a combination of these strategies leads to the best way to position drought issues and bring them more alongside the already recognized importance of dealing with flood risks.

14.6 Discussion and overview

Strategies to face drought and water scarcity have to take into account the natural circumstances, the socio-economic factors, and even the institutional circumstances in a specific area. Natural circumstances vary amongst the more arid areas in the south of Europe and the more humid areas in the northern parts of Europe. The weather conditions are different, and so is the natural availability of water resources. Demand restrictions are easier in humid areas than in arid areas. Socio-economic circumstances refer to existing water utilization patterns and possible rivalries that exist. We have seen cases with an increasing demand for agriculture, domestic use, or tourism and recreation. Such rivalries turn into a debate between those who consider water as an unlimited commodity versus those who think of water as a limited resource. This can be a rivalry between those who have and those who have not, between users with property rights and users without, between upstream users and downstream users, between present users and future users. Does gradual degradation of the resource (instead of sudden drought events) trigger policy makers to act in anticipation of a future threat of water scarcity? Overall, we have noticed that institutional factors like shared values, shared cognitions, and mutual dependencies affect the behaviour of the stakeholders and the chances for a more adaptive response.

In the previous sections we presented and illustrated four types of measures to promote drought resilience in the wake of more frequent extreme weather events. The various measures can be given a place in the overview table (Table 14.1). Their position regarding the water system was also used as the organizing device in this chapter. Neither the water use orientation nor the intervention type have a normative meaning. The fact that in the past there has been a too one-sided orientation towards reactive supply-oriented measures does not imply that none of these measures can have a positive contribution. This is mainly dependent on their coherence with the other functions of the water system.

Construction is used here to indicate changes in the water system itself. All other measures (including those in which for instance a model is 'constructed') are regarded as managerial in the Table 14.1.

In this overview we see that strategies to face drought and water scarcity traditionally are oriented on water-supply regulation, whilst nowadays more adaptive measures focus on demand regulation.

Table 14.1 Drought management instruments overview.

Water system position	Water use orientation	Intervention type
Reactive measures	Supply	Mostly construction
Preventive measures	Supply and demand	Mostly construction
Adaptive measures	Demand	Mostly management
Supportive measures		Mostly management
a. Research	Supply and demand	
b. Communication	Demand	

Water-supply regulation is often achieved by physical interventions to promote water transfers and water inflows into arid or drought prone areas. We labelled them as 'reactive measures'. Such transfers and inflows aim to keep water use functions going, like drinking water supply, irrigation for agricultural production, or creating a minimal water flow for ecological purposes. They could also be aimed to repulse salt water intrusion in coastal areas. Water transfers and inflows are costly measures of which the costs are often not recovered by the users that take profit. Water pricing of transfers and inflows could be used as an instrument to effectively regulate water demands, which on the contrary affects water supply. In times of water scarcity we see that use functions that depend on water supply can be prioritized by means of a 'displacement chain' in the Dutch example, by means of a 'drought index' in the German example.

Water-supply management need not only be a matter of engineered solutions to promote the transport or inflow of water. It can also be a matter of saving, storing or buffering, conserving or recycling water, which we labelled as 'preventive measures'. The sponge ability of natural areas or the soil could be utilized by elevating groundwater tables. This is a kind of utilizing nature ('building with nature') more than engineered solutions (reactive measures). In the Spanish case of the Marina Baja we have seen an example of setting up an institutional arrangement for compensation of access rights to a limited resource stock.

Water-demand regulation is about measures that aim to affect the behaviour of water users. We could think of command and control measures like temporary water use bans, or minimum flow requirements. Or we could think of financial incentives, like the pricing of water uses. And we could think of more communicative options to persuade water users to take into account the scarcity of the resource they want to make use of. When we have to deal with a limited water availability (no transfer or inflow is possible and no sufficient storage is available), then we need to adapt to natural circumstances as they are. Use functions should follow the drought-risk profile, like the example of adapting the choice of crops. We labelled such measures as 'adaptive measures'. In the French case of the Hérault we have seen an example of attempts to achieve IWRM.

Self-regulation of water use could also be an adaptive measure. We mentioned communication and research as supportive measures for adaptation. Self-regulation could be supported by cybernetic control tools, like a drainage system regulated by sensors interpreting weather forecasts: the system will reduce drainage and store water when dry and warm weather is predicted for the upcoming days. A related example we mentioned is the system of level-dependent drainage, which is a knowledge-based regulatory tool.

Although we distinguished reactive, preventive, and adaptive measures for analytical reasons, for strategic reasons in policy practice it might be wise to consider them as a triad, a set of three strategies that could work together quite well. A guiding perspective for policy practice could be to deliberate for each of the measure types if and to what extent they could be implemented. Because in case of severe drought and water deficiency every drop counts. However, reactive measures tend to take less care of rival users, whilst preventive and adaptive measures tend to better take into account the needs of rival users. So, with a scenario of climate change and thus more extreme weather events, it is not only more sustainable but even inevitable to invest sufficiently in the development of adaptation and self-regulation.

References

Andreu, J., Solera, A., Paredes-Arquiola, J. et al. (eds.) (2015). *Drought: Research and Science-Policy Interfacing*. Leiden/London: CRC Press, Taylor and Francis.

ARID (2005). ARID is a cluster of European research projects dealing with water resources use and management in arid and semi-arid regions, see for more information. http://arid.chemeng.ntua.gr (accessed 17 February 2019).

Arrojo, P., Dionysis, A., Barraque, B. et al. (2005). *European Declaration for a New Water Culture*. Madrid: Fundación Nueva Cultura del Aqua.

Bressers, H., Bressers, N., and Larrue, C. (eds.) (2016). *Governance for Drought Resilience: Land and Water Drought Management in Europe*. Dordrecht: Springer.

Bressers, H., Bleumink, K., Bressers, N. et al. (2016). The fragmentation-coherence paradox in Twente. In: *Governance for Drought Resilience: Land and Water Drought Management in Europe* (ed. H. Bressers, N. Bressers and C. Larrue), 181–202. Dordrecht: Springer.

Browne, A., Dury, S., de Boer, C. et al. (2016). Governing for drought and water scarcity adaptation and resilience in the context of flood disaster recovery: the curious case of Somerset, United Kingdom. In: *Governance for Drought Resilience: Land and Water Drought Management in Europe* (ed. H. Bressers, N. Bressers and C. Larrue), 83–108. Dordrecht: Springer.

Costèja, M., Font, N., Rigol, A. et al. (2004). The evolution of the water regime in Spain. In: *The Evolution of National Water Regimes in Europe: Transitions in Water Rights and Water Policies* (ed. I. Kissling-Näf and S. Kuks), 235–263. Dordrecht: Springer.

DROP (2015). Benefit of Governance in Drought Adaptation. For more information about this European INTERREG project http://www.nweurope.eu/about-the-programme/our-impact/challenge-5/the-drop-project (accessed 17 February 2019).

EC (2007). Addressing the challenge of water scarcity and droughts European Commission, Brussels.

Koundouri, P. (ed.) (2008). *Coping with Water Deficiency: From Research to Policymaking, Reports of the Results of the ARID Cluster of Projects*. Dordrecht: Springer.

Kuks, S. and de Boer, C. (2013). Adaptive repsonses to water scarcity: transfer of governance approaches across south and North Europe. In: *Water Governance, Policy and Knowledge Transfer: International Studies on Contextual Water Management* (ed. C. de Boer, J.V. de Kruijf, G. Özerol, et al.), 223–241. Abingdon, UK: Routledge.

La Jeunesse, I., Larrue, C., Furusho, C. et al. (2016). The governance context of drought policy and pilot measures for the Arzal dam and reservoir, Vilaine catchment, Brittany, France. In: *Governance for Drought Resilience: Land and Water Drought Management in Europe* (ed. H. Bressers, N. Bressers and C. Larrue), 109–138. Dordrecht: Springer.

Özerol, G., Troeltzsch, J., Larrue, C. et al. (2016). Drought awareness through agricultural policy: multi-level action in Salland, the Netherlands. In: *Governance for Drought Resilience: Land and Water Drought Management in Europe* (ed. H. Bressers, N. Bressers and C. Larrue), 159–180. Dordrecht: Springer.

Pahl-Wostl, C., Sendzimir, J., Jeffrey, P. et al. (2007). Managing change towards adaptive water management through social learning. *Ecology and Society* 12 (2): art 30.

Sangaré, I. and Larrue, C. (2004). The evolution of the water regime in France. In: *The Evolution of National Water Regimes in Europe: Transitions in Water Rights and Water Policies* (ed. I. Kissling-Näf and S. Kuks), 187–234. Dordrecht: Springer.

Tröltzsch, J., Vidaurre, R., Bressers, H. et al. (2016). Flanders: regional organization of water and drought and using data as driver for change. In: *Governance for Drought Resilience: Land and Water Drought Management in Europe* (ed. H. Bressers, N. Bressers and C. Larrue), 139–158. Dordrecht: Springer.

Vidaurre, R., Stein, U., Browne, A. et al. (2016). Eifel-Rur: old water rights and fixed frameworks for action. In: *Governance for Drought Resilience: Land and Water Drought Management in Europe* (ed. H. Bressers, N. Bressers and C. Larrue), 67–82. Dordrecht: Springer.

White, C. (2012). *Understanding water scarcity: Definitions and measurements*. GWF Discussion Paper 1217. Canberra, Australia: Global Water Forum. http://www.globalwaterforum.org/2012/05/07/understanding-water-scarcity-definitions-and-measurements (accessed 17 February 2019).

III.3

Lessons from Cases of Droughts Governance

15

Multilevel Governance for Drought Management in Flanders: Using a Centralized and Data Driven Approach

Jenny Tröltzsch
Ecologic Institute, Berlin, Germany

15.1 Introduction

This chapter presents a summary of the governance analysis of drought-related issues in the Flanders region of Belgium. In recent years several drought events happened in Flanders, with different impact severity. During some periods water extraction from the important Albertkanaal water channel needed to be restricted. Different drought events led to reduced yields in agriculture and farmers have been compensated. Also restrictions for navigation on the rivers have been applied. Climate projections show a clear trend towards higher average winter and summer temperature. The predictions show a decrease of summer precipitation but with strong variability and a moderate increase of winter precipitation. Impacts on groundwater level could balance out each other, but might have consequences for ecosystems saturated with water such as wetlands. Expected are impacts on agriculture and forestry due to drought caused by higher temperature and change of precipitation patterns.

The chapter is based on work that was part of the Interreg project: Benefit of governance in drought adaptation (DROP).[1] In the DROP-project, a governance assessment has been applied to six regions in North-West Europe: Twente and Salland in the Netherlands, Eifel-Rur in Germany, Brittany in France, Somerset in the United Kingdom, and Flanders in Belgium. These regions focus on drought aspects related to nature, agriculture, and freshwater. The Flanders analysis focused especially on agriculture.

Facing Hydrometeorological Extreme Events: A Governance Issue, First Edition.
Edited by Isabelle La Jeunesse and Corinne Larrue.

A research team from four universities and knowledge institutes visited Flanders twice to perform interviews with authorities and stakeholders (October 2013 and May 2014). The visits were supported by colleagues at the Flemish Environment Agency (Vlaamse Milieumaatschappij, VMM). The exchange was held in the form of individual and group interviews and workshops with stakeholders including representatives from different institutions and sectors, e.g. from the drinking water company, national and local nature protection organizations, local farmers and local and national farmers organizations, the VMM, different provinces, e.g. Province Vlaams-Brabant, and local municipalities, e.g. Kortemark Municipality. The analysis was guided by the drought-related Governance Assessment Tool (GAT) developed for the project.

This chapter presents the context of water management in Flanders, describes past drought events in Flanders and climate projections for this type of event. It explains the results of the analysis in terms of the GAT kit and presents possible conclusions and recommendations for improved drought governance.

15.2 Water management in Flanders

Belgium consists of two regions: Flanders and Wallonia. The federal and regional competences are exclusive and equivalent, with no hierarchy between the standards issued by each. The Belgium Federal Government has amongst other things environmental responsibilities for coastal and territorial waters (from the lowest low-waterline). Both regions are responsible in their territory for environment and water policy (including technical regulations regarding drinking water quality, responsibility for the economic aspects of drinking water provision, land development, nature conservation, and public works and transport). A mainly regional approach to river-basin planning is used in Belgium.[2]

At the regional level in Flanders, three Flemish ministries are involved in integrated water policy: the Ministry of the Environment and Nature, the Ministry of Mobility, and the Ministry of Spatial Planning. Many tasks related to integrated water management are assigned to the VMM.

For the organization and planning of integrated water management, the Decree on Integrated Water Policy (of 2003) distinguishes three levels:

- The two International River Basin Districts (Scheldt and Meuse).
- The Flemish region with its four river basins (Scheldt, Ijzer, Polders of Bruges, Meuse – of which IJzer and Bruges Polder are two comparatively small coastal catchments, added to the Scheldt river basin) (see Figure 15.1).
- The 11 sub-basins.

The Decree on Integrated Water Policy is the juridical implementation of the Water Framework Directive (WFD) (Coördinatiecommissie Integraal Waterbeleid, CIW 2003). River-basin management planning is an important part of the Decree. The Flemish Decree covers surface waters and groundwater, as well as infrastructure such as bridges, dikes, locks, and dams. Furthermore, the Decree contains regulations on water quality management as well as on water quantity management. In 2010, the implementation of the European Flood Directive was integrated in the Decree.

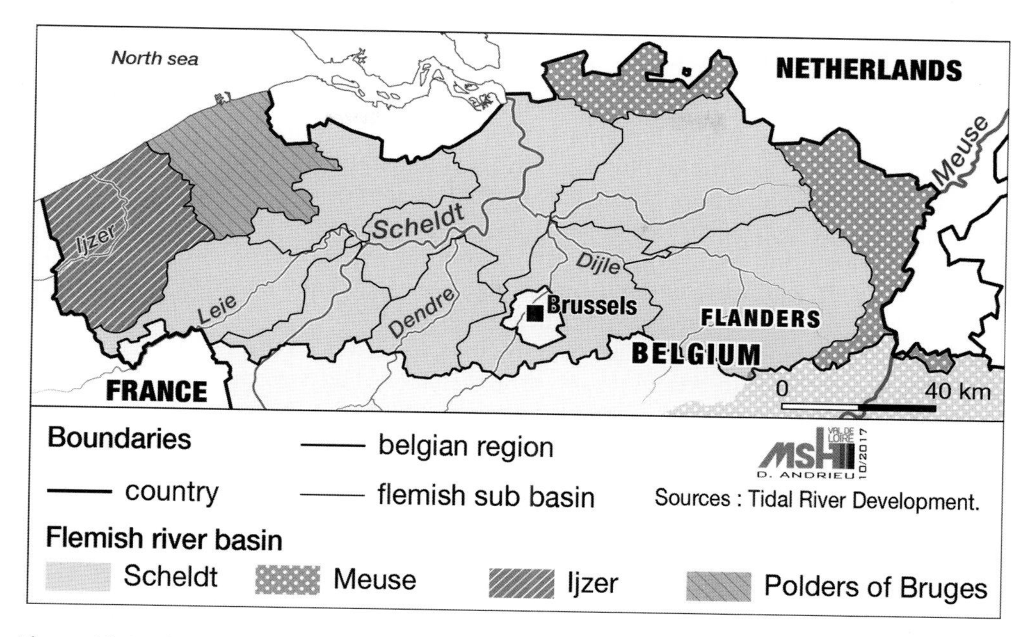

Figure 15.1 Overview of the four river basins in Flanders and location of the limits of the Flanders region.

In accordance with the Decree, one minister in the Flemish Government is appointed to be responsible for the coordination and organization of the integrated water policy. The Coordination Committee on Integrated Water Policy (CIW, chaired by VMM) assists the minister. CIW is designated as the competent authority for the implementation of the WFD, as well as for the Floods Directive. Amongst its responsibilities are the preparation of the River Basin Management Plans (RBMPs), including the sub-basin parts and groundwater specific parts, for the Flemish Region, and aligning the RBMPs with the Flemish Water Policy Note. This multidisciplinary commission unites different levels of water management and governance. It is responsible for the preparation, planning, and monitoring of integrated water policy, and it is responsible for the implementation of the Flemish government's decisions on integrated water policy. The CIW also oversees the uniform approach to the management of each basin.

The First Flemish Water Policy Note was prepared by CIW in 2005 and presents the Flemish Vision on water policy (CIW 2005). The vision was updated in 2013 with the Second Flemish Water Policy Note (CIW 2013). This document holds the goals of the Flemish government with regard to water management for the years 2014–2021. It names as its main goal the financing and implementation of integrated water management principles. Regarding drought and water scarcity it includes three guiding notions: Sustainable management of water resources and ensuring sustainable water supply, Integrated management of water scarcity and flooding, and Further stimulation of the multifunctional use of water.

15.3 Past and future drought events

In Flanders pressure on water resources is high, which is amongst others, due to the high population density. The Flemish region covers 13.521 km^2 and has a population of 6.35 million inhabitants. The Scheldt (466 inh./km^2) and the Meuse (258 inh./km^2) river basin districts show a very high population density compared to other European countries.

Water managers have historically paid much attention to guaranteeing water supply and water quality, and mitigating the risk of flooding. Nevertheless, drought extreme events and water scarcity resulting from drought, are well known problems in specific sectors relying on a good water supply, such as agriculture.

In the past, Flanders, as most European regions, experienced droughts in 1976, 1996, 2003, 2006, and 2011. In recent years, droughts have had several consequences in Flanders. On some occasions water extraction from the Albertkanaal has been restricted. The 1996, 2006, and 2011 droughts were recognized as agricultural disasters and affected farmers were financially compensated. Temporary restrictions have been placed on the draught of ships on the Meuse river and in 2003 also on the use of the locks. Furthermore, Flanders also experienced other problems related to droughts; due to the wildfire in the nature reserve Kalmthoutse Heide in 2011, about 600 ha of heathland were burned.

Climate predictions for Belgium posit a trend towards a warmer climate, with winter temperature increases ranging between 1.3 and 4.4 °C and summer increases of between 2.4 and 7.2 °C by the end of the century (Belgian National Climate Commission 2013). There are some indications of changes to the climate already occurring, the last two decades having seen very high annual average temperatures.

Since 1833 (the beginning of reords), there has been an increase in annual rainfall of approximately 7%. Projections for precipitation suggest a moderate increase in winter precipitation (between 5 and 22%), and a decrease for summer precipitation (projections have strong variability on this point, ranging from maintaining current conditions to a decrease of up to 25%) (Belgian National Climate Commission 2009, 2013).

Current observations already show evidence towards stronger extreme rainfall events and higher flood risk, but no such evidence is available for droughts. The longest periods without significant precipitation have not shown a major change since the early twentieth century. Whereas, due to higher climate variability and higher winter precipitation, a significant increase in risk of flooding is expected, the changes for droughts are again not as clear-cut (Belgian National Climate Commission 2013). In addition, since the 1980s, a trend towards increased evapotranspiration has been observed in every season; however, the most pronounced changes have been measured during winter.

The prediction for Belgium of hotter summers, with increased evapotranspiration and (possibly) reduced precipitation, will probably lead to a reduced groundwater recharge and a significantly reduced groundwater level during summer. However, increases in winter precipitation would possibly contribute to increases in recharge. Predicting the long-term impact of climate change on groundwater availability is thus difficult; a possible impact identified is a reduction of groundwater recharge and levels in the summer months, possibly affecting wetlands (UN Department of Economic and Social Affairs, Division of Sustainable Development 2004).

The projections for changes in mean river flow are either positive or negative, according to the different climate change scenarios used. However, a fall in summer precipitation in combination with a greater evaporation will lead to lower river flows, during summers at the end of the twenty-first century river flows could decrease by more than 50%. The reduced water flow in summer may increases the risk of water shortages and impact surface water quality (OECD 2013, Belgian National Climate Commission 2013). Lower river levels in summer and autumn could cause problems for river ecosystems and aquaculture (UN Department of Economic and Social Affairs, Division of Sustainable Development 2004).

The expected decrease in summer precipitation – coupled with a possible increase in summer water demand due to higher temperatures, in particular if irrigation becomes a widespread agricultural practice – can lead to a lack of water availability and problems for the agricultural sector (UN Department of Economic and Social Affairs, Division of Sustainable Development 2004). There are expected impacts on agriculture and forestry due to drought caused by higher temperature and change of precipitation patterns (Belgian National Climate Commission 2013).

15.4 Governance dimensions for Flemish drought management

This section describes the results of the governance assessment for Flanders. For the analyses the team of researchers developed a GAT,[3] through which the governance setting of a given region for planning and realizing drought adaptation measures could be assessed. Based on this assessment, recommendations can be provided to regional water authorities on how to operate most effectively towards increased drought and water scarcity resilience in this governance context. The GAT contains five governance dimensions (levels and scales, actors and networks, problem perceptions and goal ambitions, strategies and instruments, responsibilities and resources) and four governance criteria (extent, coherence, flexibility, and intensity). The following subsections are structured in line with the governance dimensions.

15.4.1 Administrative scales

The federalization process means that the Belgian national level has receded in water policy, with Flanders taking up most roles. Although the national level is thus not as predominant as the Flemish level, several coordination instances exist that align initiatives with the other Belgian regions, such as the CCIM (Inter Ministerial Conference for the Environment) and CCIEP (Coordination Commission International Environment Policy). The Belgian level also plays a coordinating role with the International River Basin Commissions of Meuse and Scheldt.

The Flemish level is the key agent in generating initiatives and policy. In principle all other relevant levels are present and included, with provinces, communities, and municipalities being involved in different aspects of water management. However, in what seems to be a reaction to the growing requirements and complexity of water management, there is a movement towards provinces taking over responsibility of small surface water bodies from municipalities. This could mean that municipalities end up 'out of the loop' regarding, for example, water quality and nature protection, and thus may be unaware of opportunities for synergies related to water management such as drought preparedness. This is problematic as they keep the responsibility for flood measures, and measures to dovetail drought prevention whilst addressing flood may not be incorporated as they could. This is also relevant when considering that groundwater permits are given by towns and municipalities. This being said, interviewed stakeholders tended not to see this change as a problem.

The coordination between different authorities at the local level with district and province levels seems to be both intense and fruitful, and planning moves up and down levels according to the size of the initiatives. The possible problems for drought preparedness at the lower levels are thus not related to the governance arrangements in themselves, but rather to questions of agenda-setting, i.e. what is the perception of drought. In general, there is a flexibility that is built into the system: problems are dealt with 'at the relevant level'. The possibility of issues moving up one level can be seen in cross-boundary management issues. On the topic of drought the lead is currently firmly in the hands of VMM, and no clear possibilities of issues moving downwards, to a level closer to local, on-the-ground implementation of actions, were observed. This is related to the agenda-setting point mentioned.

15.4.2 Inclusion of different actors and stakeholder groups

The drought governance arrangements in Flanders show a high extent of involvement of actors and networks. All relevant stakeholders seem to be involved in the management process, e.g. via CIW-Committee (Coordination Committee on Integrated Water Policy), Advisory Boards for CIW, and in sub-basins. There are frequent ministerial contacts with VMM (both formal and informal). Recent new regulations, e.g. from the European level, have a broader perspective, which includes social and ecological criteria. The actors and networks involved in water policy have recently expanded. The commitment to intersectoral involvement can be seen in the CIW, which includes leading officials of agriculture, nature, and planning.

Due to the long tradition of stakeholders working together with the Flemish region there are good and productive relationships in place, with regular exchanges. The drinking water companies and also other organizations (e.g. Boerenbond) for instance have a long history of working relationships with the Flemish region (VMM) on different topics, e.g. water-resource management planning. The established relationships provide channels for discussing drought risks and measures, and the first discussions of this type are taking place.

The last few decades have also seen great improvements in the dialogues with stakeholders traditionally less involved in decision making (e.g. nature organizations). Whereas planning processes in the 1980s were referred to as 'trench warfare', current-day planning would be based on dialogue and identifying possibilities for mutual cooperation and win-wins.

Whilst there is a high involvement of actors in processes, there seems to be a requirement for more flexibility in the kind of involvement, and at what stage of the process. For example, some actors highlight that they are not included in planning phase of policy, e.g. in the selection and prioritization of measures to be implemented, but only in the implementation phases.

15.4.3 Drought risk perceptions and goal ambitions

The issue of drought is already on the agenda, but it is still an upcoming topic. The awareness for water scarcity and drought problems is very low for some stakeholders. Drought resilience as a topic is weakly developed in current constellations of problems and goals defined by stakeholders in the region. Various problem perspectives, e.g. from farmers, nature organizations, drinking water availability, are being taken into account in discussions, but there is as yet not much work on prioritization of drought and water scarcity as an issue. For example, there is no 'hierarchy' or prioritization of different water uses/demand if a situation of water scarcity was faced in the region. The DROP-pilot of VMM, which developed drought indicators for the monitoring and reporting of the drought situation, serves to a large degree the function of getting the issue on the agenda. Since 2014, five Flemish water management services bundle their efforts to make real-time data and forecasts on the state of the water system available through the web portal www.waterinfo.be. Drought is included as one of the four main themes of the portal. Information on the drought situation can be used by actors such as decision makers or water managers, or individual stakeholders, such as farmers, to evaluate the drought situation and take action accordingly.

In the current discussion in Flanders many different perspectives are included: agriculture, nature conservation, drinking water, and groundwater issues. Indeed, groundwater problems are a well-developed topic in Flanders, due to problems with historical over-exploitation of the resource. However, environmental non-governmental organizations (NGOs) find it difficult to get their voice heard (and with this put forward the perspectives of nature conservation) because the focus is mainly on agriculture and economic development. The consequences of increasing water demand (domestic and industrial) are already being discussed on provincial level, but not on the local level.

Furthermore, at the moment the approach for flood prevention is about getting rid of the water as quickly as possible, so the water basins can be used immediately in case of a further flood. This goes against drought-preparedness approaches which aim to maintain water in the system. Also if sensitivity exists concerning (ground-)water saving in different sectors, an integrated flood and drought management has not, as yet, reached the agenda.

The Second Flemish Water Policy Note includes the sustainable management of water resources and the integrated management of water scarcity and flooding. At the moment, goals are not at the stage of being discussed within policy processes, and therefore are very far from being operational. Currently the emphasis is still on building up a knowledge base, e.g. via the model prepared by VMM. Therefore some actors see a further need for a more integral vision on the water conditions of sub-areas. The DROP pilot's modelling could create the means for this integral vision. The new RBMP with the integration of drought and water scarcity measure could support the discussion if the relations with other issues are seen. Willingness to find solutions can be seen, and water managers do meet and develop solutions when necessary on a case-by-case basis where local problems are clear. But even in these situations, gaining a long-term perspective versus just solving immediate problems is very problematic. This flexibility on local level is also the result of there not really being any formal mechanisms (e.g. strategies and instruments) in place.

15.4.4 Strategies and instruments for drought resilience

Regarding strategies and policy objectives, the first 'Flemish vision on water policy' from 2005 includes the sustainable use of water, and states that it is necessary to deal coherently with water shortage. It has been agreed that the second RBMPs in 2015 also include measures for water scarcity and droughts. In addition the Environmental Policy Plan 2011–2015 'MINA-Plan 4' mentions as objective: groundwater quantity and water use. Water scarcity and droughts are here seen as a new but important development. In the Coalition Agreement of the Government of Flanders for 2014–2019, a number of initiatives regarding drought and water scarcity are included, e.g. the promotion of water savings and the reuse of water (Government of Flanders 2013). The policy memorandum of minister Joke Schauvliege (Vlaamse Regering 2014) also contained a chapter on water scarcity and droughts, which included, e.g. developing a draft of a hierarchy of water uses in case of water scarcity and drought events.

Regarding instruments, different instruments are already implemented, such as groundwater taxes for business users (handpumps and households not included), groundwater permits, and restrictions for water extraction. Because the groundwater bodies are affected both in qualitative and quantitative terms, restoration programmes for groundwater bodies have been implemented. The reduction of groundwater permits is also in discussion. Different studies on water savings are published to promote Water Audits of companies. Further instruments are under development, e.g. insurance at regional level (currently regulated at federal level); this instrument is planned to be regionalized by 2020. Ministerial plans that subsidies will be given for insurances or weather impacts if there is compensation for yield damage over 30% have been tabled; events with damage below 30% will be borne by the farmer.

There have been problems with groundwater for a long time and a good monitoring scheme is in place. As a result a wide variety of measures were developed. However, problems exist with instruments such as source protection (quality), which requires land-use changes or change of agricultural practices. Here local authorities are resisting such measures. Advisory services for farmers seem to be quite developed and they work towards water saving. Water saving is also supported with actions at house scale as rainwater harvesting. At the moment no general approach for priorities or a hierarchy of water uses in conditions of water scarcity exist. Bans have been a possibility for a long time, but they have never been used. At the regional level the awareness for options also seems to exist.

For surface water partially the legal basis is lacking, e.g. water rights for surface water. A key instrument that seems to be missing is the agreement of flows over borders with France. Political will regarding a legal framework was seen as lacking, but there is also a problem of leverage with French government.

The permissions for water retention basins for farmers are complex, because farmers need to prove in the application process that the basin will not in reality end up capturing groundwater. This application procedure can take a long time and interviewed farmers voiced the feeling that they are hindered in their possibilities to adapt to water scarcity and droughts.

The main coherence problem regarding the instruments is that they are not developed via one strategic objective or strategy. The instruments were developed very independently and for each of the different relevant areas. Because only a limited number of instruments relevant for drought purposes are in place, in general the instruments do not overlap. Only smaller problems exist, e.g. between groundwater recharge and rainwater storage, which should be taken into consideration for new instruments.

VMM is using the same definition of water scarcity and droughts as the European Union (EU) and the European Environment Agency (EEA). Further activities for a coherent approach in water management in general can be seen in the synchronization of the planning period of sub-basin RBMPs and basin RBMPs. Also the integration of drought measures in the RBMPs from 2015 shows a tendency to a harmonized approach of developing and implementing drought measures with other water-management measures.

15.4.5 Drought management responsibilities and available resources

Different drought management instruments and initiatives exist, but every actor is driving activities based on the often local, individual water scarcity and drought problems. Resources and responsibilities are not clearly assigned; mainly there are only general responsibilities for water resources. For example, at the Flemish region level no sub-agency or department has the special responsibility for water scarcity and droughts; it is rather integrated in different water departments and therefore very fragmented. This fragmentation leads to individual discussions with the 'normally' involved institutions and levels. A coherent connection between the different discussions and approaches is not realized in the most involved organizations. Because the drought events issue is not that high on the agenda there are as yet no competence conflicts between the different actors.

In the case of some actors such as drinking water companies, there is an initiative on the side of the authorities to get them to assume responsibility via the implementation of 'Public Service Duties', which oblige them to plan to ensure, e.g. continuity of supply and provide alternative supplies. So a movement towards distributing responsibility to different actors is in place. The problem of low resources is voiced strongly by farmers who are interested in innovative measures to address water quality but have to deal with financial and staff limitations.

Because responsibilities and resources are not clearly assigned, a pooling of particular drought-related resources is not possible. But resources are available for corresponding issues such as land use or nature management. New funding has to be found for every planned project or process. So the lack of concrete drought resources leads to a quite flexible structure of finding resources on a case-by-case basis.

15.5 Summary and recommendations

This section presents a summary of the Flemish drought-management characteristics and possible recommendations to increase drought resilience by improving drought governance in Flanders.

The Flanders region researched in this chapter shows that VMM as the Flemish Environmental Agency is taking first steps toward an increased drought resilience together with some stakeholder groups such as individual farmers. Flooding is a by far more noticed problem in the region. During the last decade agricultural drought events have happened more frequently in Flanders. But some actors have still a limited awareness for the risk of drought events.

The approach by VMM is built around scientific modelling results to assess drought risks following the specificities of possible droughts (hydrological versus agricultural droughts or water scarcity). It can be described as a scientific-technocratic approach that is building up the knowledge base first. The modelling results are the basis for setting the policy agenda and development of drought-management actions.

The Flanders level is the highest scale policy for drought management, especially setting objectives, developing strategies as well as planning and coordination of implementation of drought management measures. The Flemish government and its institutions such as VMM is taking a centralized and less bottom-up approach for drought management as it is already used, e.g. for dealing with flood risks. The stakeholder involvement is improving but is still limited.

The existing governance settings can support the motivation, cognitions, and resources of actors for implementing drought activities if also in a limited manner. Most activities concerning drought resilience are initiated by VMM which has the drought topic on its agenda but is also limited, since a broad strategic vision for drought resilience still needs to be developed. The vision could lead to further coordination of different public and private actors for drought planning and management in Flanders and to integrated approaches including drought management as well as land use management, flood management, and nature protection. Different bottom-up activities are taken by individual stakeholders, e.g. farmers who are building water basins for collection of water. So, awareness is shown by some stakeholder groups or individuals. But a large number of actors still show a limited awareness and motivation to deal with drought risks. Drought management is based on limited resources. The resources for drought measures are organized on a case-by-case basis and depending on research money from different sources.

The governance assessment shows that Flanders is at an early stage of dealing with drought risks and implementing drought-management actions. Different actors have already started a limited number of initiatives. But for the implementation of further drought resilience activities the awareness for drought risks need to be increased. Intermediaries and multipliers would be a main target group as they can motivate stakeholders to incorporate drought risks in their operations. Awareness and additionally more involvement can also be increased if the flexibility to include stakeholders in the planning processes (especially selection, prioritization, and calibration of measure) can be increased. The further strengthening and support of bottom-up processes focusing on drought resilience could supplement the existing scientific modelling-based approach very well. Different activities at the local level could include the support of implementing pilot measures, showcasing actions, best-practice exchange and networking activities. Also dialogues for drought risks management within different economic sectors could initiate the implementation activities and increase of drought resilience.

To deal with future drought events an integrated central vision for Flanders could support integration of drought-related policies. A multi-objective planning between the different responsible ministries as well as agencies and other public institutions could be initiated. Private actors could be involved in drought resilience measures via voluntary agreements or the preparation of contingency plans for their operations. The governance setting can be improved by a further emphasis on transboundary drought management. A coordination with upstream regions such as Germany or France could be focused.

For the evaluation of drought resilience measures, their prioritization, and monitoring their effectiveness needs a robust data base. It would be useful to identify available data gaps, e.g. volume of surface water abstraction. The data can be the key for identification the potential and prioritization for drought management measures as well as enabling private actors to incorporate drought risk into their actions.

Notes

1. This chapter is based on Tröltzsch et al. (2015) and Tröltzsch et al. (2016).
2. Except the *Federal Plan on Coastal Waters*.
3. More information on the Governance Assessment Tool can be found in Bressers et al. (2015).

References

Belgian National Climate Commission (2009). Belgium's Fifth National Communication on Climate Change under the United Nations Framework Convention on Climate Change. http://unfccc.int/resource/docs/natc/bel_nc5_en_final.pdf (accessed 5 May 2017).

Belgian National Climate Commission (2013). Belgium's Sixth National Communication on Climate Change under the United Nations Framework Convention on Climate Change. http://unfccc.int/files/national_reports/annex_i_natcom/submitted_natcom/application/pdf/bel_nc6_rev_eng.pdf (accessed 5 May 2017).

Bressers, H., Bressers, N., Browne, A. et al. (2015). Benefit of governance in drought adaptation – governance assessment guide. Funded by the Interreg-project 'Benefit of Governance in Drought Adaptation' (DROP). https://ris.utwente.nl/ws/portalfiles/portal/5138088/DROP_Governance_Assesment_Guide_web.pdf (accessed 20 February 2019).

Coördinatiecommissie Integraal Waterbeleid, CIW (2003). Decreet betreffende het integraal waterbeleid. http://codex.vlaanderen.be/Portals/Codex/documenten/1011715.html (accessed 5 May 2017).

Coördinatiecommissie Integraal Waterbeleid, CIW (2005). De eerste waterbeleidsnota. www.vliz.be/imisdocs/publications/261558.pdf (accessed 20 February 2019).

Coördinatiecommissie Integraal Waterbeleid, CIW (2013). Tweede waterbeleidsnota – inclusief waterbeheerkwesties.https://www.vlaanderen.be/nl/publicaties/detail/tweede-waterbeleidsnota-inclusief-waterbeheerkwesties (accessed 5 May 2017).

Government of Flanders (2013). Coalition Agreement. 2014–2019. http://financeflanders.be/sites/default/files/atoms/files/coalition_agreement_2014-2019.pdf (accessed 21 February 2019).

OECD (2013). *Water and Climate Change Adaptation. Policies to Navigate Uncharted Waters.* Country profile Belgium. Paris: OECD.

Tröltzsch, J., Vidaurre, R., Bressers, H. et al. (2015). The governance context for drought policy and measures in Flanders. Almelo. Funded by the Interreg-project "Benefit of Governance in Drought Adaptation" (DROP). https://research.utwente.nl/en/publications/the-governance-context-for-drought-policy-and-measures-in-flander (accessed 20 February 2019).

Tröltzsch, J., Vidaurre, R., Bressers, H. et al. (2016). Flanders: regional organization of water and drought and using data as driver for change. In: *Governance for Drought Resilience* (ed. H. Bressers, N. Bressers and C. Larrue), 139–158. Dordrecht: Springer.

UN Department of Economic and Social Affairs, Division of Sustainable Development (2004). CSD-12/13 (2004–2005) Freshwater Country Profile Belgium, Section G. New York: DESA.

Vlaamse Regering (2014). Beleidsnota 2014–2019. Omgeving. ingediend door mevrouw Joke Schauvliege, Vlaams minister van Omgeving, Natuur en Landbouw. https://www.vlaanderen.be/nl/publicaties/detail/beleidsnota-2014-2019-omgeving (accessed 5 May 2017).

16

Drought Governance in the Eifel-Rur Region: The Interplay of Fixed Frameworks and Strong Working Relationships

Rodrigo Vidaurre
Ecologic Institute, Berlin, Germany

16.1 Introduction

This chapter presents an analysis of drought governance in the Eifel-Rur region of Germany. This research was performed within the framework of the INTERREG IV-B project 'DROP: Benefits of Governance in Drought Adaptation'. In the DROP-project, a governance assessment was applied to six regions in Northwest Europe: Twente and Salland in the Netherlands, Brittany in France, Somerset in the United Kingdom, Flanders in Belgium, and Eifel-Rur in Germany. A research team from five universities and knowledge institutes visited each of these regions twice to perform individual and group interviews and workshops with authorities and stakeholders, including representatives from different institutions and sectors.[1]

The Eifel-Rur region counts with a sophisticated water management system – which is however only starting to address the issue of droughts. This is due to the comparatively humid climate of the region, a result of its proximity to the Atlantic. In general, there is not much experience with drought episodes in the German state of North-Rhine Westphalia, in which the Eifel-Rur is located.

Our analysis shows that past and ongoing work on increasing drought resilience in the Eifel-Rur has produced a number of successful approaches that could be emulated in other regions – but it also highlights additional steps that could be taken to improve the region's preparedness for droughts. This chapter commences with two sections providing background information on the water system and its management

Facing Hydrometeorological Extreme Events: A Governance Issue, First Edition.
Edited by Isabelle La Jeunesse and Corinne Larrue.

in the Eifel-Rur region, moves on to deliver a succinct assessment of the system's governance from a drought perspective, and finalizes with a discussion of the successes and possible improvements of this system.

16.2 The water resources system in the Eifel-Rur region

North-Rhine Westphalia's water management is quite particular in the German context, as it relies on water boards to perform many of the duties of water management. This particular form of organizing water management has its origins in the nineteenth century, in response to the large-scale water-related challenges of North Rhine-Westphalian coal mining. The responsibilities of the water boards ('Wasserverbände') are established in a particular law for each single water board. They are put in charge of water management for a certain geographical region, usually following the boundaries of river basins. The region under analysis is that which is managed by the *Wasserverband Eifel-Rur* (WVER) (Eifel-Rur Water board), which is a public water corporation similar in nature to a water authority.

WVER's region lies in the district of Cologne, which covers the natural regions of Eifel and the Lower Rhine Embayment. The WVER region comprises mainly the catchment of the Rur and has approx. 2087 km^2 and ca. 1.1 million inhabitants. It stretches from Heinsberg in the North to Hellenthal in the South, and Aachen in the West to Düren in the East. The institutional structure of North-Rhine-Westphalian water boards is such that water users are also water board members, involved in decision making and paying for the services provided.

As a public body which is an operating organization, the WVER executes different tasks set by the aforementioned special North Rhine-Westphalian law.[2] WVER responsibilities comprise the full range of water services. By law the duties of WVER include control of water discharge in the catchment area, river maintenance, river restoration, supply of raw water for drinking water production, supply of production water, wastewater treatment, prevention of disadvantageous influences on river systems (in general looking at different issues), and hydrology. Groundwater is not included under WVER's duties, as only a very small area of WVER's region has significant groundwater bodies; their management has been entrusted to a neighbouring water board.

In addition to its legal obligations, WVER informally collaborates with further actors to achieve additional objectives including keeping reservoir levels high enough for water quality objectives, managing reservoir levels for sailing purposes and to ensure a pleasant landscape (tourism), managing reservoir levels in a way that minimizes disturbances of fish reproduction, and electricity production by the company RWE.

The area under WVER's responsibility is characterized by a network of connected reservoirs. WVER operates six reservoirs with a total capacity of 300 million m^3 in the northern part of the Eifel hills (the southern part of its service area, Figure 16.1). The reservoirs were mainly developed for flood control and flow maintenance during dry seasons.

The total length of flowing surface waters in the WVER service area is approx. 1900 km. WVER is responsible for the management of these waters, as well as for the operation of 50 flood retention basins and other flood control works. The average

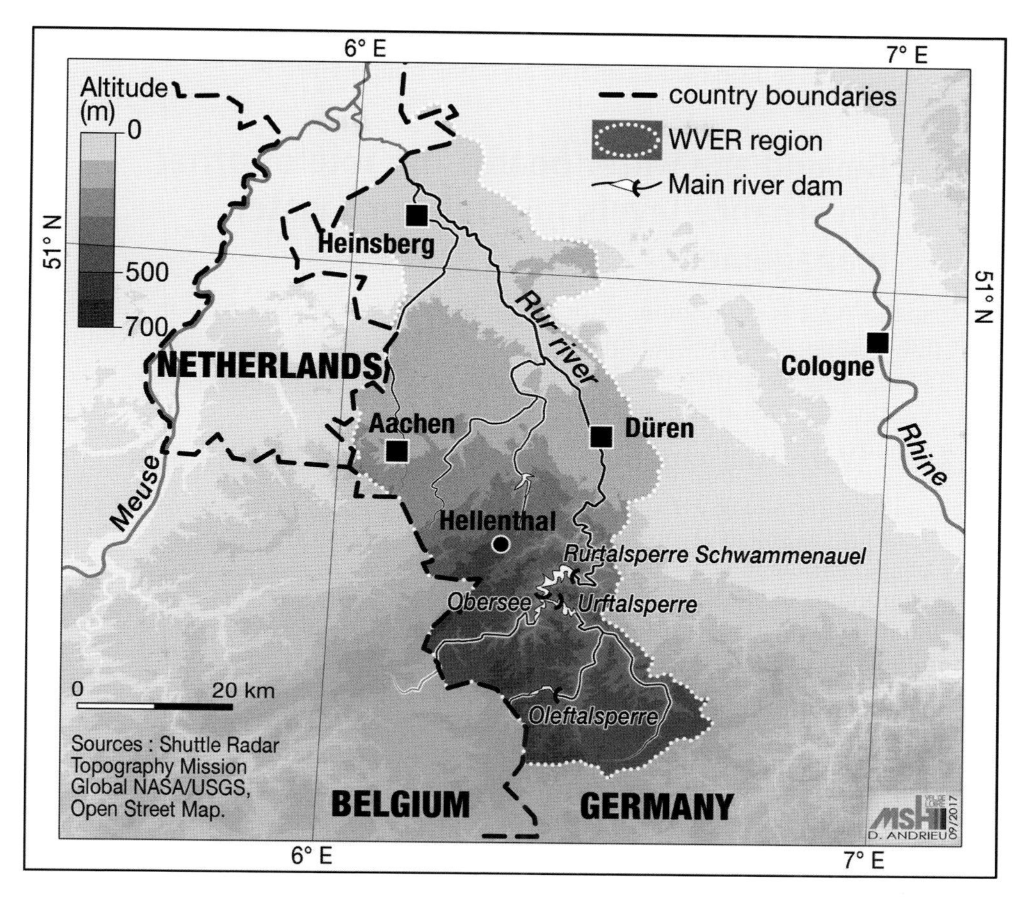

Figure 16.1 *Wasserverband Eifel-Rur* (Eifel-Rur Water board, WVER) region in the catchment of the Rur river and location of its dams and reservoirs.

annual output of the five power plants driven by water power is 61 MWh. The WVER reservoirs supply each year about 27 million m^3 of drinking water for the Aachen region. The Rur river provides annually some 5.5 million m^3 of drinking water for the Düren area.

Recently, the Eifel-Rur region has experienced somewhat dryer periods during the spring season, as a result of which the water flow through the reservoirs decreases. Still water and falling water levels in reservoirs bear the risk of a decrease in water quality, which results in a higher amount of production work and possibly drinking water production problems; still water in these reservoirs always bears the risk of eutrophication with effects such as algal blooms. Due to the topography and the limited capacity, the 'Oleftalsperre' and the 'Urfttalsperre' run the risk of more algal blooms during long dry periods. This can also have consequences on the reservoirs 'Obersee' and 'Rurtalsperre' downstream.

In the context of climate change – with longer dry and sunny periods – this problem is expected to increase. At present there is a lack of knowledge about the evolution of water quality within the dry scenario. The long dry periods in spring in recent years has already resulted in reduced water quality in some of the reservoirs.

16.3 Beyond the water board: The role of other governance levels in Eifel-Rur's water management

Whereas the WVER is the main actor in water-resources management in the region, other public authorities are also involved in the region's water management, both at a higher and at a lower geographical scale.

16.3.1 Water management in North-Rhine Westphalia

In Germany, the European Union's (EU's) Water Framework Directive (WFD) was transposed into national law via the seventh amendment to the Federal Water Act (*Wasserhaushaltsgesetz*, WHG) in June 2002. Due to a major restructuring of responsibilities and competencies between the Federal Government and the German Länder in 2006, the German water legislation was modified in 2009; the modified Federal Water Act entered into force in 2010.

According to this act, in their implementation of the WFD the German federal states must adapt their state water laws to encompass water protection and to formulate the roles for cities, municipalities, and water authorities, who bear the concrete responsibility for implementing measures. In the case of North Rhine-Westphalia, the responsibility for developing the river-basin management plans lie with the Highest Level Water Authority which is the North Rhine-Westphalian Ministry for Environment and Nature Protection, Agriculture and Consumer Protection. Plans are adopted in consultation with the High Water Level Authorities (District Councils) and the responsible committee of the North Rhine Westphalia regional parliament (*Landtag*). Responsibility for implementation lies with lower level public administration, such as districts and cities. Further actors such as nature protection organizations, water associations, and regional councils should participate in the planning and particularly in the implementation process. Regarding water abstractions, it is the District Councils who are responsible for authorization of water abstraction for surface water and groundwater.

In the particular case of the Eifel-Rur river basin, the district government in Cologne (Aachen) is responsible for implementation of the WFD. The measures are financed 80% from the state government and 20% from own contribution (e.g. from the municipalities where they are responsible).

16.3.2 The role of municipalities and lower water authorities in water management

The German Basic Law (Article 28 [2]) and most constitutions of the German Länder ensure the local self-government of districts, towns, and municipalities. Self-government comprises all matters concerning the local community. Municipal regulations and the water laws of the different German federal states stipulate that drinking water supply is usually, and wastewater disposal is always, an obligation of the local authorities. On that basis, municipalities decide on the local implementation and organization of water supply and wastewater disposal.

With a view to effectively realizing drinking water supply and wastewater disposal, municipalities may form associations for voluntary cooperation. To some extent, municipalities (such as in North Rhine-Westphalia) are members of water-management associations (*Wasserverbände*, such as the WVER), subject to special laws. In addition to these compulsory tasks, municipalities have to fulfil partial tasks regarding the implementation of environmental laws issued by the government and the German Federal States.

Amongst other responsibilities, the lower water authorities, as supervisory/executive authorities, approve flooding areas, wastewater systems, wastewater treatment plants, small sewage works, wastewater and rainwater discharges, water supply facilities, the use of water bodies, such as abstraction from surface water, and exceptional approvals for water and medicinal spring protection areas.

16.4 The drought perspective on Eifel-Rur's water governance

This section summarizes the water governance system of the Eifel-Rur region, as evaluated from a drought perspective. The analysis follows the four qualities of the 'Water Governance Assessment Tool with an Elaboration for Drought Resilience' (2013), developed in the framework of the DROP project. Along these four qualities – extent, coherence, flexibility, and intensity – the dimensions of the governance system are mapped, to identify the elements or aspects of the governance system that could be modified so as to achieve improvements from a drought perspective. As an overall statement, the higher the system ranks according to these qualities, the better suited it is for dealing adequately with drought episodes.

16.4.1 Extent

The *extent* aspects of the governance context can mostly be regarded as somewhat positive, covering all levels and scales of the system. Many administrative levels are directly involved in the water-management system. The two main actors are the district government (second authority level) and the water board itself. When it comes to droughts, the national level (German Ministry of Environment) is still somewhat decoupled, mainly providing first studies and visions. The EU level is quite relevant for its directives. However, a negative point is that municipalities are seen to be withdrawing from their water-management responsibilities, mainly due to serious resource issues. Although municipalities are only required to co-fund 20% of the cost of water-management measures, many municipalities are unable to provide these resources.

When focusing on the actors, the same positive extent can be appreciated. This is a result of the design of the North-Rhine Westphalian water boards: according to the law regulating WVER, users with a water right of a certain size are automatically members of the water board, whether they like it or not. This means that all major users participate – also economically – in the management of the water basin. There are, however, some restrictions to this positive extent regarding actor involvement. Smaller (and thus non-paying) actors, such as farmers and nature organizations, do not have the same voice as larger actors. This being said, there is a strong movement towards collaborative and inclusive decision-making processes on the side of the water authorities, as well as on relationship building on the side of WVER. The implementation of the WFD and the Floods Directive for instance were based on a huge number of participatory workshops and roundtable discussions, and there is a strong emphasis on voluntary implementation of measures. Interviewed stakeholders repeatedly mentioned that the developments over the last decades had been very positive in this sense.

The implementation of WFD and Floods Directive has provided the region with a set of new instruments and experience in consultation processes with stakeholders, and all in all, there is a broad extent of strategies and instruments in place. However, from a *drought* perspective there are very significant elements still missing, e.g. water demand management, drought contingency planning, drought communication. In this context there is still room for improvement, e.g. via knowledge transfer of experiences from other pilot regions. Another point affecting drought management in particular is the fact that the district government does not currently see itself in a position to actively push the topic of droughts – whereas they welcome the water board's actions on the topic, they are currently suffering due to overstretched resources. This means that those actors that could implement measures on the ground, if they were required to, face no external pressure to act on the topic.

16.4.2 Coherence

Passing to the governance system's *coherence*, the evaluation is also quite positive. The main actors, such as the state level, the district, the municipalities, the water authorities, and the drinking water companies, mutually accept their share of the tasks, responsibilities, and funding given by law. The dependence amongst these levels is well recognized by the interviewed individuals. EU environmental policies seem to have played an important role in introducing a more holistic and synergistic approach to the management of the reservoirs. To some degree the coordination of the lower competent authorities appears to be more coherent than that at a higher level.

Amongst the factors determining a positive degree of coherence is that the WVER is in charge of practically all relevant water-management tasks in the region. All these responsibilities being within one organization rather than distributed between different actors is probably helpful in establishing a coherent framework of action. The institutional structure of WVER also helps: with water users also being the water board members, involved in decision-making and paying for the services provided, this structure provides a framework conducive to good coherence of, e.g. perceptions, goal ambitions, strategies, and instruments. In addition, different stakeholders have goals that match quite well. For instance, the fishermen associations are interested in large fish populations in the reservoirs, which are also of interest to the national park authorities and to the water board in its role of drinking water supplier (because of fish population's positive effects on water quality) and in its role as responsible for WFD implementation (which measures success via ecosystem quality).

In the WVER region, the interviews show a sense of trust and mutual dependency between the actors, expressed for instance in their positive evaluation of participatory approaches being used in water management. All actors interviewed were quite satisfied by the way the water board is working with them and how actors' perspectives are considered when proposing measures, for instance for the implementation of the WFD. All in all, the stakeholders interviewed expressed their belief in the extremely high value of the trust-based collaboration that has been built over the years, and that has evolved positively over time. However, the consensual and voluntary approach towards measure implementation seems in some cases to be reaching its limits, with some negotiation processes on contentious topics being practically at a standstill for a number of years.

This notwithstanding, we can identify potential conflicts of interest that could worsen in case of water scarcity. The existence of very old water rights (with strong legal precedence) seems to create opposing goals between some users from a drought perspective. Industry users with a certain water right do not, at present, have incentives to reduce their water use or to partly reduce their water rights. A further point is that the strategy for flood prevention implies keeping the water level in the reservoir sufficiently low until the spring, to ensure enough storage capacity in case of exceptional precipitation events which may be associated with intense rainfall or snow melt. However, if there is not enough precipitation or snow melt during the spring period when water is collected, there is not enough water in the reservoirs to ensure adequate water quality for some drinking water providers (e.g. water temperature below 10 °C and oxygen above 4 mg/l). Furthermore, there is a lack of coherence when it comes to resources; in particular, there are a lot of issues with municipalities being extremely cash-strapped at the moment, as well as in the foreseeable future.

That said, it is also true that drought can be considered a second-order problem in the region, and the extent to which conflicts related to drought and water scarcity have emerged is really quite limited – with the exception of isolated issues of water supply and water quality between WVER and a water supply company.

16.4.3 Flexibility

The overall evaluation of the governance system's *flexibility* is only intermediate, with however some positive developments over time. This evaluation is based on the fact that the water-management system has a quite rigid large-scale framework, shown fundamentally in the priorities and responsibilities of WVER (established by law) and its operational procedures. The management of the water system in the Eifel valley follows a clearly established set of complex management rules which WVER helps elaborate and which are authorized by the district authority of Cologne. Any management decision that disregards these rules can bring with it the question of legal responsibility – for instance for flood damages ensuing due to incorrect flood protection. This means that WVER and its personnel have a strong incentive not to stray from this set of rules.

The framework is both difficult and slow to change, and some actors see in this a problem for the system to take on board new responsibilities. However, the water-management system shows very significant flexibility at the small scale, within the room provided by this overall fixed framework. There is a strong culture of discussion and collaboration between actors, and interviewed stakeholders were broadly of the opinion that their interests are considered and taken on board as much as possible.

The legal obligation of the water board to provide a certain established level of protection (floods) and of supply (deliver water for drinking water production) and the responsibilities associated with it have resulted in an elaborate and sophisticated set of rules that manage the interaction of reservoirs and water bodies. However, these same legal obligations imply that there is no short-term possibility of officially incorporating additional risks (e.g. droughts) into the set of principles that govern the system. Even smaller changes (e.g. to operational procedures) have to be extremely well-founded and well-argued, based on thorough evidence and modelling of historic data, which means that the overall framework is more reactive than proactive, and that the timeframe for these reactions is a long one. The management of secondary objectives or of other unconsidered aspects can only be improved if it can be shown that primary objectives are not affected. This means that the adaptation of dam management rules (e.g. so that they incorporate drought considerations) is a lengthy procedure.

This said, there is significant capacity, responsibilities, and resources to address different issues in a way that does not interfere with the overall framework; there is also the will amongst actors to address new risks and topics. The district authorities and WVER's approach to the implementation of EU directives foresees amicable agreements/cooperation with affected parties, showing a high degree of flexibility in on-the-ground implementation. It seems possible to reassign responsibilities in the definition of water resource problems related to flooding, and possibly nature. The question of resources, however, seem a different issue altogether, with the system quite fixed. The issue of available resources seems very important in the final implementation, particularly where municipalities are involved.

Flexibility is also shown in the way that topics pushed by stakeholders have been taken up by the relevant authorities. The question of enabling the return of salmon to the region's rivers was initially pushed by fishermen, who managed to convince authorities to take up these objectives. Regarding implementation and crisis situations, there does not seem to be much flexibility in moving up and down levels, as main decisions are mostly taken by the highest authorities in realizing certain plans. Depending on the issue at stake, the decision is often brought up automatically to the superior levels, e.g. the district government.

The ability to include new actors into formal structures of responsibility seems questionable as the structures within WVER and its 'assembly' seem fixed. However, informal relationships are seen to be a way forward in this regard, with new actors being addressed in participation processes, and adjustments to the distribution of responsibility seeming possible.

16.4.4 Intensity

Currently, the relatively weakest point of the governance context for drought resilience policies and measures is its *intensity*. (However, this also holds true in other DROP pilot regions in Northwest Europe, due to the region being overall quite water-abundant and not yet often affected by droughts.)

The district government seems to constitute the most relevant level in the decision chain concerning water issues related to drought. (It should be remembered that the water board only has executive powers, and thus cannot implement on its own accord measures for a new issue such as drought.) At the national and at the *Länder* level, initiatives addressing climate change adaptations have been launched, but are as yet only limited to knowledge exchange and studies. Improving drought resilience has no priority on the political agenda (or not yet at least) and no resources are made available for this topic. According to the district government, the German and North Rhine-Westphalian Adaptation Strategies do not have implications for their daily work, because they are too unspecific to result in concrete requirements and actions. As a consequence, there seems to be a lack of plans or other instruments regarding drought adaptation, as well as a lack of long-term vision for this issue.

This means that the district government is under no pressure due to obligations on the topic of Climate Change adaptation in general or drought in particular, and nor have the relevant resources been made available. Although they recognize the importance of the issue, they do not see themselves in a position to take it up, and so the district government is currently not driving any process (e.g. establishing its own guidelines, programmes, or implementing adaptation initiatives). It is individual actors that are initiating interesting activities – the DROP project having been one of them. The WVER, a drinking water producer and a hydroelectricity producer interviewed all emphasized the existence of technical projects to enhance the system's robustness, improve risk management, and develop backup solutions in case of extreme events. Drought prevention is thus being addressed in the context of general risk management strategies that use precipitation patterns as inputs.

WVER can thus be described as the driving force of change in the region. As the authority responsible for most things water in the Eifel, they are also the first in line to be affected by drought issues, which explains their taking a proactive approach. The overall assessment of the intensity is thus low, but with increasing strength.

16.5 Conclusions: Factors for current and future success

Currently, and in spite of the Eifel-Rur region not having been frequently affected by droughts in the last few decades, the water management system is incorporating drought-relevant elements in its decision making. This proactive engagement and positive evolution of the management system seem related to the following factors, which could be seen in this and similar contexts as 'success factors':

1. **The culture of strong communication and consensual decision making within the state of North-Rhine Westphalia**
   The approach of the North-Rhine Westphalian authorities on, e.g. the implementation of WFD's river-basin management plans or the Flood Directive's flood-risk management plans has emphasized collaboration and consensual implementation. This means that there is a strong culture of communication and exchange between the different actors. Actors may not always see eye to eye (indeed, some of the processes are currently bogged down because of disagreements between actors)

but these forums provide space for understanding other actors' requirements and identifying solutions that are more palatable to the different groups.

2. **The organizational conception and structure of WVER**
The WVER, comprising as it does many different actors who by law automatically become (paying) members, is in itself a forum for balancing out different water management interests. Its organizational structure itself reflects the need to consider and level out the interests of the different water actors. This leads to intense exchanges between actors, as well as to a thorough understanding both of actor groups' interests and implementation possibilities within the water-management system.
Similarly, the level of collaboration with the district authority ensures that the available flexibility of the system – admittedly somewhat limited, due to the legal requirements (which do not include drought preparedness) and its rule-based approach – is used to its full extent. As a result of modelling efforts, the WVER and the district authorities agreed on operational decisions that were flexibilized so as to improve the performance of the management system, in that the different obligations of the system are now still met under a wider array of meteorological and flow conditions (including drought-relevant ones). This improvement was reached without requiring a (very lengthy) legal modification of the WVER's obligations.

3. **The broad mandate of the responsible organization (WVER)**
The water board's mandate ensures ownership for and a proactive approach to emerging topics such as drought resilience.

However, the system can improve its drought resilience on a number of counts. Operational decisions have been successfully flexibilized, to the point that whereas certain dry conditions in the past would have made it impossible to meet all (legal) water management obligations, under the improved system this would now be possible. But there would seem to be an open issue in the flexibilization of the obligations, in our opinion not yet satisfactorily addressed. Particularly the water rights regime ensures constant supply to water users, and provides no incentives to reduce these water rights where there could be potential for such reductions.

Although the water system is managed comprehensively and sophisticatedly in the Eifel-Rur, there seems to be a mismatch between the instruments in use for floods, water quality, and groundwater,[3] and those addressing quantitative aspects of surface-water management (including those relevant for drought purposes). Whereas the former have profited from recent EU regulations that have driven comprehensive updates of planning objectives and tools, the latter is rather the result of the historical development of regional water regulations. For this reason, numerous elements seem to some degree incompatible with each other and with modern water-resources management. For instance, there seems to be no real incentive structure in place to manage water demand – which would have significant overall benefits from a drought perspective. The options we have identified are:

1. **Develop a strategy that addresses current inefficiencies**
From a climate-adaptation perspective, but also from a broader governance objective of reducing resource use conflicts and thus enhancing planning security for economic actors, a number of possibilities are currently being missed. These inefficiencies could

be reduced if a better use is made of existing instruments that could reduce unused water rights to bring them in line with actual use – including realistic development potential for the local industry in the future. Whereas some instruments to this purpose exist, updating historic water rights in the Eifel-Rur may be resisted by affected users, which means that authorities need to count with political will behind their initiative. They would probably also require an improved resource base to address this extra task over several years, as resources already now seem stretched quite thin.

2. **Review water rights and water pricing strategies**
   New, additional instruments which provide adequate steering mechanisms for managing water demand could also be implemented. For instance, current water charges in the Eifel-Rur region are linked to water use, and not to water rights. Including a link to the size of a water right in the charges, for instance by making charges both water rights and water use (e.g. weighting them in an average) could help address current inefficiencies and missed opportunities.

   Interviewees highlighted that owners of water rights would hang on to existing surplus rights for possible future expansion of operations. 'Old' rights often provide more legal guarantees than newer ones, which creates unwillingness to trade in old rights for new rights.
3. **Create incentives to explore alternative water supply options**
   Incentives for increased water efficiency (e.g. water recycling) are not felt everywhere, as water recycling comes at a cost (of energy). There seem to be no initiatives in place exploring alternative water supplies, e.g. rainwater harvesting, significant process water recycling. An impulse to increase process water recycling could be given by making an economic case (e.g. making it financially beneficial) for the private companies that are the largest water users in the Eifel-Rur region.
4. **Develop a comprehensive and up-to-date database on water rights and water uses**
   Related to the previous points are the significant data issues affecting surface water. Up-to-date information would not always be available, both for the different types of surface water rights, as well as for the different types of actual uses of water. Options such as systematic water metering do not seem to be in discussion. The lack of data would be related to the lack of updated legal requirements mentioned in the previous point.

   An adequate management of water resources requires comprehensive and up-to-date data on these points. This is a necessary basis for understanding the system and evaluating the potential for increasing system resilience, e.g. by water-demand management. Again, a push for data improvement would probably require both political will and to some extent additional resources. The benefits of increasing the water-management system's resilience would in all probability far outweigh the expenditures.

Notes

1. For a more lengthy discussion of background and findings please consult the project outcomes (Bressers et al. 2015; Governance Assessment Team DROP 2015; Vidaurre et al. 2015, 2016).
2. Compared to a Dutch water board, the WVER is limited to executing powers, without any rights of an authority (e.g. it cannot give permissions).

3. Strictly speaking, these instruments are not in use in the area covered by the WVER, as it does not have any relevant groundwater bodies. However, the instruments exist and are in use in the neighbouring water boards' regions.

References

Bressers, H., Bressers, N., Browne, A. et al. (2015). Benefit of governance in drought adaptation – governance assessment guide. Funded by the Interreg-project "Benefit of Governance in Drought Adaptation" (DROP). https://ris.utwente.nl/ws/portalfiles/portal/5138088/DROP_Governance_Assesment_Guide_web.pdf (accessed 21 February 2019).

Governance Assessment Team DROP (2015). Roadmaps to improved drought resilience measures' implementation. Collection of recommendations to the six practice partners in the DROP project. https://research.utwente.nl/en/publications/roadmaps-to-improved-drought-resilience-measures-implementation-c (accessed 21 February 2019).

Vidaurre, R., Stein, U., Browne, A. et al. (2015). Observations on the governance context of drought policy and pilot measures for Waterboard Eifel-Rur (WVER). Final report. (unpublished).

Vidaurre, R., Stein, U., Browne, A. et al. (2016). Eifel-Rur: old water rights and fixed frameworks for action. In: *Governance for Drought Resilience. Land and Water Drought Management in Europe* (ed. H. Bressers, N. Bressers and C. Larrue), 67–82. Heidelberg: SpringerOpen.

17

Adaptation of Water Management to Face Drought and Water Scarcity: Lessons Learned from Two Italian Case Studies

Claudia Cirelli and Isabelle La Jeunesse
Laboratory CNRS 7324 Citeres, University of Tours, Tours, France

17.1 Introduction

Climate change is expected to cause alterations of streamflow regimes in the Mediterranean region, with possible relevant consequences for several socio-economic sectors including hydropower production. Italy is concerned by these changes. In one location of the Sardinian island, in the Nuoro province, 334 consecutive days without any rainfall were recorded. Even the Alpine area of Italy is concerned as the volume of northern Italian glaciers has drastically decreased in the last decades (Gobiet et al. 2014). According to the 12th report on climate indicators in Italy issued by the ISPRA (Istituto Superiore per la Protezione e la Ricerca Ambientale), after a very hot year 2015 and a year 2016 considered as the sixth hottest since 1961, the year 2017 was also an abnormal year with a rainfall deficit of −41% (between January and August), and the second hottest year since 1961 (+2.48 °C), after 2003. For the third consecutive year, the annual Italian average temperature has increased: + 1.35 °C compared with the 1961–1990 reference period.

The negative effects of these extreme events on economic sectors and in particular agriculture soon manifested themselves. Between 2003 and 2012, droughts and downpours have damaged agricultural production, facilities, and infrastructures, for a total cost of €14 billion. In 2017, at least two-thirds of cultivated land was not irrigated and the damage for farming and livestock rearing activities is estimated at over €2 billion (Coldiretti[1] 2013, 2018).

Facing Hydrometeorological Extreme Events: A Governance Issue, First Edition.
Edited by Isabelle La Jeunesse and Corinne Larrue.

In this context of water threats in the Mediterranean region, the purpose of this chapter is to conduct a comparative analysis of the governance of water uses in two catchments of the same European country, Italy. The analysis was conducted in the Rio Mannu basin, in Sardinia, and in the Noce basin in the Trentino-Alto Adige region, two autonomous regions of Italy. How are these two different regions of the same country preparing themselves for drought? Is there any difference in the awareness of climate change impacts? What are the main criteria explaining the differences?

To answer these questions, the study first identified the water uses, the water users, and the water managers. Then, questionnaires were administered to at least one person representing each water use and water use management system. Respectively in the Noce and in the Rio Mannu case studies, 37/43 interviews were conducted and questionnaires were filled in by 10/14 water managers and 9/19 water users. Organized in five sections, the questionnaires comprised 32 questions for water users and 40 questions for water managers. For more information on the content of the questionnaires or the investigations at each site, readers can refer to the CLIMB reports on the EU project website (La Jeunesse et al. 2010; La Jeunesse and Cirelli 2012, 2014) as well as the analysis of water uses (La Jeunesse et al. 2016).

The chapter is divided into four sections. After the introductory section where the methodology used to collect information is presented, the following section describes the water management in Italy and the specificities of the two autonomous regions of the case studies: Trentino Alto-Adige and Sardinia. The next section is dedicated to the description of the two case studies, the Rio Mannu catchment in Sardinia and the Noce catchment in Trentino. The last section provides a comparative analysis of the two sites with four criteria underlined to compare the state of water and drought governance: water uses, water management, water savings, and risks assessment.

17.2 Water management in Italy and the autonomous regime

17.2.1 Changes in water legislation and hydraulic management since 2006

In 2006 (decree 152/2006), Italy set up a single law for environmental rules, part of which is dedicated to the management and protection of water resources. This law is in fact the transposition of the Water Framework Directive (EC/WFD2000/60 [WFD 2006]). According to this, Italy is subdivided into eight hydrographic districts. Sardinia represents an entire district, whereas the Trentino Region is part of the Oriental Alps district.

Since the end of the 1980s, Italy had started to reorganize water management on its territory at the catchment scale (both regionally and inter-regionally). A new entity was created: '*l'Autorità di Bacino*', the authority of the basin. The basin authorities are water-management entities with mixed Regional and National responsibilities. They govern hydrographic districts as a single management entity. Their objectives are to (i) protect hydrological and hydrographic networks, (ii) preserve water quality,

(iii) ration water resource uses, and (iv) regulate soil uses. The Basin Authority proposes the integrated management system at the catchment scale, drafts a plan, and supervises its implementation. Thus, the water-management plan constitutes the rules and regulations for general water management. It is valid for three years and is the official document used for sectorial planning.

Parallel to this national context, the two case studies of this chapter belong to two Italian regions with the special status of autonomous regime.

17.2.2 The specificities of the autonomous regime

Although both regions studied in this chapter have an autonomous status, significant operating differences within these regions have an impact on water-management systems. First of all, the provinces in the Trentino-Alto Adige region retain 90% of local taxes, whereas the Sardinia region retains 70%.

For the Trentino region, most legislative and administrative responsibilities as well as the financial resources were transferred from the Trentino-Alto Adige region to the two provinces of Trento and Bolzano. It is therefore the province that plays the crucial role in decision-making processes concerning water running over its territory. The Adige river basin authority, which govern the hydrographic basin – including the Noce basin, does not intervene directly for the territory concerned by the Trento province. Planning is made at province level and drinking water management at municipal level.

The main instruments deployed to manage the resource along with water services and uses are the *Piano di Tutela delle Acque* (water protection plan) of 2015 and the *Piano Generale di utilizzazione delle acque pubbliche* (general plan for the use of public water, PGUAP) of 2006 (Provincia Autonoma di Trento 2006) which, together, comprise the reference framework for the integrated management of water as regards quantity, quality, and territorial protection. In 2015 the Provincial agency for water and energy resources was created in order to set up a water policy consistent with that of energy production on the scale of the province, and create a link with the national basin authorities and hydrographic districts.

In Sardinia, the regional centralization process means that the actors involved in water management over the territory are now more restricted in number. The region plays the main role, through the Regional Basin Authority in charge of planning and scheduling, and above all the ENAS (Ente Acque Sardegna) manager. The region also strongly conditions the actions and strategies implemented by local actors.

Water management is based on the multi-sectoral regional water system which controls the whole network. The infrastructures of this complex system (reservoirs, aqueducts, canals, water diversion and pumping systems, hydro-electrical power stations) belong to the region. Thus, the region owns all the public water concessions.

As regards planning, the approved instruments currently deployed in Sardinia are the *Secondo Piano di gestione del distretto idrografico della Sardegna* of 2016 (second plan for the management of the hydrographic basin of Sardinia, the first dating from 2010; Regione Autonoma Sardegna 2010, 2016) and the *Piano Stralcio di Bacino regionale per l'utilizzo delle risorse idriche* (sectoral water use plan, PSURI) of 2006 (Regione Autonoma Sardegna 2006).

17.3 The Rio Mannu catchment

The Rio Mannu di San Sperate stream, tributary of the Flumini Mannu stream, spreads over a catchment area of 504 km^2. The stream flows for 42 km before reaching the Campidano plain, one of the areas with the highest agricultural productivity on the island (Figure 17.1).

The climate is typically Mediterranean (Perra et al. 2018). The Rio Mannu river is ephemeral more than eight months per year (Mascaro, Deidda, and Hellies 2013).

In the regional sectorization of hydrosystems, the Rio Mannu di San Sperate stream is part of the hydrological zone of Flumendosa-Cixerri. The smaller unit to represent to Rio Mannu water uses is the information available for the Fluminimannu sector were irrigation represents 86% of the water use. The water is stored in a series of reservoirs which are now connected. The total capacity of the system is around 650 Mm^3 (ENAS 2011), mainly shared between the Flumendosa and Mulargia dams. In the catchment of the Rio Mannu only one reservoir exists in Monastir (Figure 17.2).

Concerning water allocation, the Rio Mannu catchment is part of the *Consorzio di Bonifica de la Sardegna Meridionale* (CBSM).[2] The latter spreads over 263 000 ha, 20% of which is irrigated.

The basin is mainly covered by croplands and grasslands. A small part of the area is still covered by Mediterranean forest. In the non-irrigated areas, wheat production

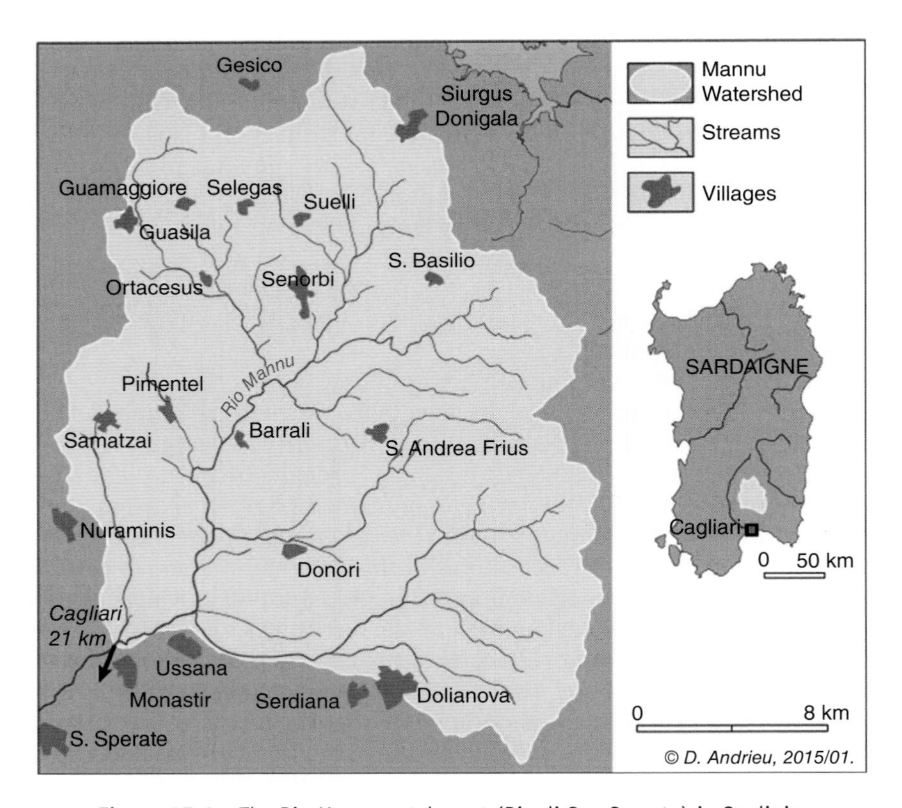

Figure 17.1 The Rio Mannu catchment (Rio di San Sperate) in Sardinia.

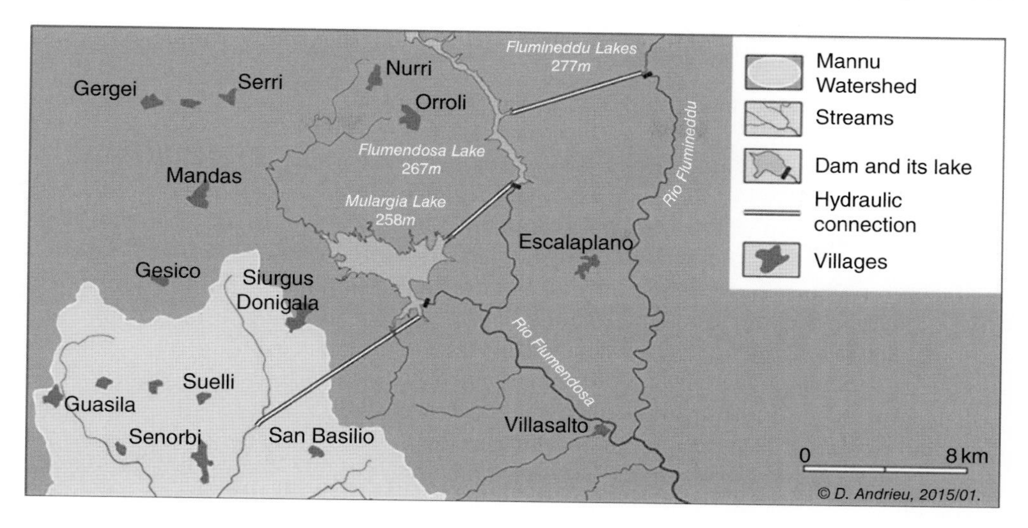

Figure 17.2 The Flumendosa-Campidano-Cixerri system. *Source*: Adapted from Botti et al. (2001, p. 212).

represents 33% of the agricultural area. The rest is dedicated to vineyards (18.5%), orchards (57.5%) and olive groves (24%). These are mainly located in the steep slope areas. In irrigated districts, the areas are divided as following: 30% artichokes, 16% different species of market gardening such as tomatoes, 13% citrus fruits trees, 8% other fruit trees, and about 12% grassland and alfalfa.

17.3.1 Projections of the impacts of climate change

Studies of the impacts of climate change on the hydrology of the Rio Mannu catchment show a future reduction in all hydrological quantities occurring in the spring season. Furthermore, simulation results reveal an earlier onset of dry conditions in the catchment (Meyer et al. 2016). The projections assess a future reduction in mean annual precipitation (total average of –12.70%) and a raise in temperature (+2.18 °C). These generate a significant decrease of mean annual runoff (–32%) and a drop in the groundwater table, with spatial variations related to terrain and soil properties (Piras et al. 2014). A general reduction of soil water content and actual evapotranspiration is also projected (Piras et al. 2016).

17.4 The Noce catchment

The Noce River is located in the Dolomite range, in the Trentino-Alto Adige administrative region, in North-eastern Italy (Figure 17.3). It drains a catchment area of 1,367 km^2 receiving the contribution of the two glaciers Ortles-Cevedale and Adamello-Presanella.

Since the past century, the Trentino region has intensified the construction of new reservoirs on its territory. These reservoirs were built in the 1950s as part of a large

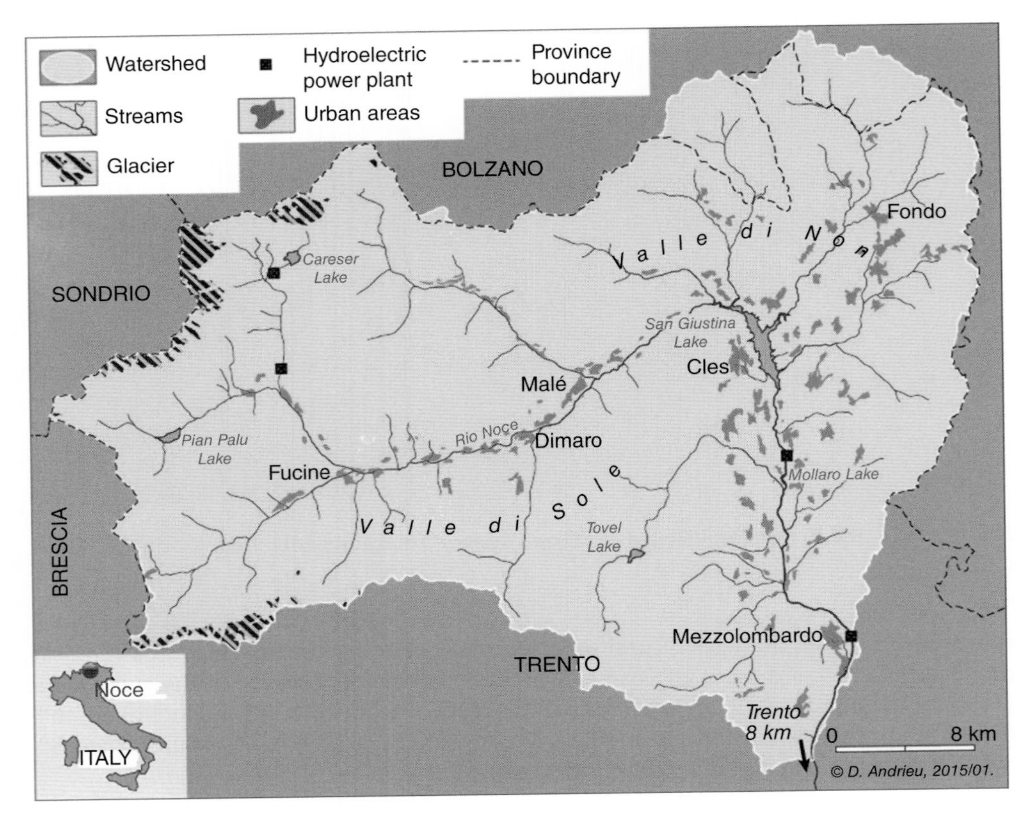

Figure 17.3 Rio Noce catchment, watercourse, and reservoirs.

national endeavour to increase the production of electricity intended to sustain and foster the industrialization of the country. The main purpose of these reservoirs was to store water in spring and through the summer, and use it in winter, when the energy demand for industry and domestic uses is at its peak. Since the mid-1990s, this interest has been reinforced because of incentives to increase the share of energy produced from renewable resources. The main water use in the catchment is hydropower which is regulated by four reservoirs (Figure 17.3): Pian Palù and Careser in the headwaters and Santa Giustina and Mollaro in the middle course of the river. The second water use is dedicated to irrigation for agriculture (6.9%). Then comes domestic water (2.6%) (Provincia Autonoma di Trento 2006). The river also provides services for recreational activities such as fishing and rafting.

The alpine's climatic characteristics, cold in winter and fresh in summer (20 °C as mean temperature in July) are varying depending on the microclimates in valleys, the exposure and orientation of slopes and the thermic inertia of Lakes. The yearly mean precipitation is of 1170 mm and the yearly mean river flow is of 34 m^3/s (Provincia Autonoma di Trento 2006).

Administratively, the catchment is divided into two districts: the Valle di Sole, close to the border with Lombardia, and the Valle di Non, located in the South-eastern part. In the Valle di Sole, tourism is the main activity, conducted in a few ski resorts and hot springs, followed by agriculture, often in connection with tourism. The lower part of the river is dominated by hydropower. The Valle di Non is mainly agricultural, with the intensive cultivation of apple trees. The total population of 53 000 inhabitants is spread amongst 60 small villages. The two major towns are Malè (2,200 inh.) and Cles (6,500 inh.).

17.4.1 Projections of the impacts of climate change

Climate change is projected to have significant impacts on water resources and hydropower production in the Noce catchment. Projections indicate an increase of the mean temperature in the catchment, in the range of 2–4 °K, depending on the climate model used (Majone et al. 2016). This can induce an increase in water demand for irrigation. However, hydrological simulations show an increase of water yield during the 2040–2070 period with respect to the 1970–2000 period, accompanied by early spring melt, which leads to a shift from a glacio-nival to a nival regime for the catchment. The magnitude of changes in both runoff and hydropower potential is location-dependent and in general larger at higher elevations (Majone et al. 2016).

17.5 Comparative analysis and discussion

17.5.1 Changes in water uses

In the Mannu basin, agriculture is the main water use, with several types of crops (vine, olive trees, fruit trees, alfalfa and cereals for fodder, artichokes) yet in the Flumendosa district, the city of Cagliari must be considered with more domestic needs. In the Noce basin, without taking into account the hydroelectric use, the main volume of which returns to the river, agriculture is also the main use where apples are the most developed crop, and pastureland to a lesser extent. However, there has been a particularly high increase in urban demand in Sardinia, notably characterized by a greater pressure in the summer, in the metropolitan area of Cagliari with both inhabitants and tourists. Water for urban-domestic use mainly comes from the Flumendosa dam. This use competes with farming uses, particularly during low water periods, especially as there is a hierarchy of uses, with priority given to urban use.

In the Noce catchment, there has also been an increase in the urban-domestic demand for water but not at the same period. It concerns holiday periods, both in summer and winter, where the fluctuating population of tourists creates greater pressure on the resource. However, many other sources of water can supply the basin's municipalities with drinking water. This factor makes the system more autonomous than the Rio Mannu basin, and therefore less subject to supply restrictions during water shortage periods.

The study highlighted that in both Italian sites, the same debate exists as to the river's minimum ecological flow, although the level of development and significance is not the same in Sardinia and Noce. In the Noce basin, the question of ecological flow creates genuine controversy due to the diversity of uses on the river. In the Rio Mannu basin, which is an intermittent river, this matter is hardly brought up by the manager ENAS, whose main concern is to introduce an anticipation programme and capacity in relation to short and mid-term low-water phenomena. The lack of environmental groups concerned by this matter in the Mannu basin also indicates the secondary importance of this subject on the local level. However, in the Noce basin, nature protection associations, notably the *Comitato di salvaguardia del fiume Noce* for the protection of the river Noce, are campaigning for the river's environmental use, arguing that the economic benefits of quality tourism in natural and protected areas requires a river in excellent ecological health.

As regards rivalries and the risk of conflict, in the Noce basin, the current risk is very clearly represented by tensions between upholding economic activities (production of hydroelectricity and apples), the benefits of which are essential for the Province and its municipalities, and the development of a quality area where water is an essential development factor. Thus, interviewees' awareness of the impact of domestic wastewater and diffuse pollution on the water resource shows how difficult it is to combine an excellent ecological status with the maintenance of intensive apple farming, as currently practiced. In the Rio Mannu, the rivalries are explained by the water deficit that affected the island, along with the rest of the countries in the Mediterranean basin (Cadoni, Silvano, and Vacca 2010). For about 30 years, Sardinia has seen volumes of water available for different uses diminish fairly drastically. The crisis in 2000 mainly came about due to this reduced water availability, associated with other political and socio-technical factors that characterized management methods until very recently (fragmentation of decision-making bodies, diversity of players in charge of local and regional management) that created a fairly sensitive water context because it was unable to meet water emergency situations.

17.5.2 Saving water versus developing uses

In the Sardinian case, certain farmers have adapted to the lack of water and responded to the incentives to save the resource by anticipating the sowing period, using autumn and winter crops or early crops which require less water. Even if after the hydraulic crisis of 2002, two types of water-intensive crops (artichokes and market garden) have begun to occupy half the irrigated surface area of the CBSM, as farmers seek to maximize profits, over the past 10 years, a trend towards reducing the amount of water used per cultivated hectare has been observed, with today almost 20% less water used (CBSM 2012). The implementation of more efficient irrigation systems, such as drip systems, instead of spray irrigation and the deployment of automated distribution systems with water meters have undoubtedly played a role in reducing consumption.

It is interesting to note that after a first period during which farmers led resistance actions, and even sabotaged these new water-saving processes, the interviews showed that the system has now been accepted. The farmers met said they learned to regulate their water allocations according to crop needs through these new techniques. Indeed, it is understandable that farmers feared not having enough water allocated for the upcoming farming cycle, which began before the managers knew how much water they had in the dams, as crops are sowed in the spring, when the dams are still filling up. Moreover, the new pricing conditions based on the volumes used instead of the irrigable surface led farmers to use water sparingly. Some farmers do not even use the whole volume of water allocated to them and their farm each year. Thus, all seems to indicate that the water-saving system implemented has been integrated in the local farming and social irrigation practices. In parallel, research aiming to select local seeds requiring less water are conducted by the regional agronomic research bodies (Dettori, Cesaraccio, and Duce 2017). However, in the strongest crisis periods, several users interviewed declared having dug wells, these remaining little-controlled by authorities. Other users said they acquired pumps to draw water directly from watercourses. Indeed, these two resources are not subject to a purchase cost per cubic metre of water used, as is the case for water from the dams.[3]

Water-saving strategies were also adopted over the past few years in the Noce basin. Drip irrigation for apple trees, the creation of small reservoirs for agriculture (rainwater and artificial filling), as well as the establishment of agreements between different interest groups to exchange water rights (in particular between agricultural and hydro-electrical uses)[4] at key hydrological periods of the year were deployed throughout the territory. However, the emergence of the opportunity to access a new quantity of available water though these water-saving techniques in the main dams has actually led to the expansion of the agricultural border towards ever higher altitude zones, which were previously used as pasture land and are now also sough-after for recreational purposes. Therefore, water was not truly saved by the agricultural activity, but the use of the resource was maximized which ultimately led to the development of the activity.

As an effect of European policies, both sites grant a lot of attention and investments in so-called renewable energies through hydro-electrical production. This leads to a paradox. Whereas the production of renewable energy is promoted with a wish expressed by policies to limit large dams, the increasing number of small hydro-electrical power stations, in the Noce territory in particular, will have an impact on the ecological continuity of the watercourses concerned. Over 1000 installation applications were filed in the Trento Province alone. Moreover, these structures can only operate through releasing water from large dams, which leads to a loss of income for the latter. In contrast, for small municipalities and agricultural consortiums, these new structures are a new source of income (sale of energy) and economy (production and use of energy).

17.5.3 Adapting water management

In the area of resource management, a fundamental difference between the two Italian sites lies in the pressure exercised by the climatic context on the decision-making process. Hydrological variability conditions differently the choices as regards public water actions. In the Sardinian case, the natural water conditions marked by recurring drought cycles, and especially after the water crises over the past 20 years, are not only etched durably in the local hydraulic experience (all actors interviewed referred to this), but now seem to weigh in offer and demand management.

After several cycles of drought (1988–1995) and water crises (2000–2004), which led national government to appoint an extraordinary commissioner (with powers that derogated ordinary management) to take charge of water administration, water management in Sardinia entered into a regional centralization process acted in 2006. The past years of crisis have justified the current policy for centralized management, storage, and allocation of water. This, supported by the re-creation of ENAS entity, in order to rationalize the water use at regional scale and guarantee water availability for at least three consecutive droughts. Today, whilst management of the ENAS is sometimes criticized for having centralized decisions in a single body, farmers and managers of irrigation consortiums finally acknowledge that maintaining artificial reservoirs and carrying out interconnection work between basins has resulted in multi-annual management which should reduce the effects of extreme hydrometeorological events.

On the Noce site, although it also in summer that water demand is at its highest in the basin, potential conflicts seemed to be linked to the multiplication of water uses in the basin, as well as the hydro-electrical production use, which dominates all the other uses. If in Sardinia, the question of water rivalries includes water shortage situations, in Noce, the question of rivalries is associated with the current intensification of already existing uses: hydro-electrical (mini hydro-electrical powers stations along the river); agricultural (intensification and expansion of the activity towards higher altitudes); tourism (even though there was a significant increase in the winter tourism offer over the past 10 years with new ski stations, the activity has tended to shift towards the summer season due to the variability of the snow cover and the melting glaciers). The advent of new uses, such as the environmental use (ecological flow) or recreational use (water sports) further complicates the issue. Although melting glaciers seems to be a condition that may affect the local availability of water in the long run, in the mid- and short-terms, the former may be at the centre of rivalries for the management of the water volumes allocated.

17.5.4 The divergence in the risk assessment by users and managers

In a context where the variability of the resource is increasing, which managers are asked to address so as to guarantee use continuity but have no control over weather events, what seems to gain importance in the decision-making process is the tension between managers and users induced by a difference in their risk perception. Hydraulic management authorities aim to manage weather events by setting up a sparing water offer policy that rationalizes the resource based on a multi-year management system taking into account the probability of an event. Thus, the risk is reassessed each spring during the assessment of the water stocks accumulated during the autumn and winter, and thus before the irrigation season. Here, risk management corresponds to deciphering potential future shortage signals in order to avoid the risk of any water crisis occurring. Farmers assess risks yearly in relation to the assurance of a good harvest. The risk corresponds to a potential production loss if the crops do not receive the water needed during the vegetative or crucial phase. Yet, excluding multi-year plantations, crops are selected at the end of the winter, long before the managers assess the quantity of water stored. It is therefore impossible for farmers to then adapt to a low hydrological assessment by selecting less water-intensive crops. Indeed, it is asked of farmers to declare their crops beforehand, then they will be informed of the quantity of water they will benefit from after having taken into account the other requests and the assessment of the water available compared with a potential drought risk over three consecutive years. Whereas the managers bank on limited damage in the present to eliminate larger risks in the future, farmers gamble on present productions, even meagre, by refusing immediate damage and potentially taking the risk of not sowing the next year. It is therefore very difficult for a farmer losing their harvest to understand the importance of keeping water in the dams should another drought occur the following year, whilst little efforts are made to save water for other uses, including domestic, considered as priority.

In the case of the Noce site, the risk is now represented by the multiplicity of uses and their development between maintaining hydro-electrical production, agricultural production – and especially AOP apples – and the development of nature tourism both in winter and summer (ski, national park, watercourses, lakes, biodiversity). Yet, the economic impacts of all the activities are vital for the territory, with in particular the generous compensation granted by the small hydro-electrical companies to the municipalities located in the scope of these small dams. It is therefore easy to understand that the maintenance of an ecological flow is at the origin of many tensions as it corresponds directly to losses and profits for all the activities over the territory.

17.5.5 Low awareness of climate change

In both sites, excluding a few environmentalists and managers, the issue of climate change impacts on the resource is not directly evoked by the actors. However, the Sardinian site appears to be a lot more familiar with the risks of less water availability than the Noce site, with a drought risk really taken into account.

In both basins, forecasts are able to assess the hydrological vulnerability of the territory (lack of water/too much water) through risk management linked with the operation of the infrastructures. However, in the Sardinian case, the risk of three consecutive drought years was assessed based on past crises and not on the forecast potential impacts of climate change. Naturally, this is an improvement for the management of water scarcity, but not in a climate change context. This management change led to consultations for the deployment of actions aiming to save water in order to face scarcity phenomena. We may think that these changes may also facilitate the development of water-saving techniques for other water uses. Moreover, in Sardinia, we observe a regionalization of the basin scope and decision-making process that can speed up the implementation of an action plan related to climate change as soon as this is established. For Noce, in a context of more water physically available compared with Sardinia, a reduction of the water resources could in all likelihood give rise to more arrangements between the different users present in the basin. We may also think that the resource-saving strategy will continue to be deployed and uses will be hierarchized. The recreational and sporting uses would certainly be the most impacted, however this remains to be confirmed by future studies.

The recourse or not to substitute water resources such as seawater desalination or recycling waste water is a major difference between the two sites. This difference shows the historical difference concerning water scarcity in Noce. Moreover, in the Trento Province, the mid-term situation of water abundance will probably make it a lot more difficult to anticipate climate change adaptation strategies. Maybe once the glacier melting process is very advanced, this will have an effect on taking into account of an 'after glacier' phase in climate change impacts. Therefore, in this case, the risk time scale is the most difficult variable to consider for local actors.

Finally, for Sardinia, it is certain that the context of insularity, which does not permit bringing water from a neighbouring region, has forced authorities, unlike many other Mediterranean sites studied, to face periods of drought on the basis of the existing water resource. The pooling of efforts, the saving of water, and the rationalization of the resource are more the result of past droughts than the emergence of awareness of the current challenges of climate change.

17.6 Conclusions

In this chapter, the objective was to compare the awareness of climate change impacts on water availability in two different case studies of the same country, Italy, through a local analysis of water uses and water-use rivalries.

In the two sites, both water managers and water users see that new water resources will be required in the near future to support all water uses throughout their territory and this, with or without any climate change impacts. Considering the forecasted impacts of climate change on the hydrology at the level of the entire Mediterranean region or even regionally, along with an increase in water demand, we may ask how stakeholders can believe in an increase in water availability. The fact that most stakeholders interviewed were feeling confident in the capacity of their territory to have more water available is a significant result as it reflects the state of knowledge on local climate change threats. In the case of Noce, this can be explained by the current abundance of water and by the projected first stage (2040–2070) of climate change impacts inducing the melting of glaciers which will thus release more water into the water system. For the Rio Mannu case study, this misperception of climate change threats that will enhance an already existing water scarcity and drought context, is probably due to the high performance of the water management system, which has increased the water availability since 2006.

This observation is also probably a logical continuation of the trust placed in the hydraulic technique of storing water to solve water scarcity and increase the water supply through the development of technologies, to almost always meet the demand. In fact, in both regions, the technology for collecting, storing, and allocating water has been particularly developed and efficient. In Sardinia, the water resource has been politically regionalized to allow for a single water authority and the different basins and reservoirs have been linked together, up to technological limits. Thus, the situation can be compared to the Noce case. In Noce, the different water availability conditions (large reservoirs and the presence of the river) clearly mark the decision-making context. As in Sardinia, water management is centralized by one actor, the Province. In terms of water management, the implementation of a coordinated policy for an entire region, along with basin interconnection and resource transfer from one basin to the next, has permitted managers to reduce the number of decision makers and therefore shortened the time frame for responding to crisis situations.

However, whilst in Noce 89% of the water volume is allocated to hydroelectric energy production, the biggest percentage of water consumption in Sardinia is for agricultural irrigation, which cannot count on abundant local sources. Storing water is the solution to cope with variable rainfall in order to meet both domestic and farming needs. In Noce, small reservoirs were built to cope with water requirements for farming in the event of drought. This project competes with requests for the installation of small hydroelectric power stations, the development of which is encouraged by the European Commission as part of its renewable energy directive. This will probably increase water rivalries along the river in the very near future.

In both cases, there is significant recourse to water stored in dams which is used for the different uses present on the territory. Whereas, previously, low water conditions could only be handled using resources available in the basins, the new interconnection system increases the basin's water supply through transfers from areas with more plentiful reserves. Whilst basins with shortages are thus able to access additional water volumes, which serves to reduce internal rivalries, the regional scale becomes the stage of other possible water-related conflicts amongst actors who had not been in competition previously, since they had not been sharing the resources existing on their territory. Consequently, an interconnection-based management strategy has indeed expanded the water supply. Yet, at the same time, the 'multi-sector' system composed of hydraulic facilities imposes a multiyear scheduling on authorities able to prevent crisis situations through 'rationalizing' (managers' parlance) uses. This option was simply infeasible when these facilities were supervised either by distinct managers or by managers of individual river basins. The crises experienced have underscored the limitations in the former system, which was highly fragmented and organized along sector-specific lines. However, as part of the reorganization, basin-wide hydrological criteria no longer influence water management policies. Regional authorities have adopted a water-management scope that covers the entire regional territory. The exchanges between water industry actors no longer take place within the confines of a catchment, but can extend beyond, crossing the hydrographic boundaries. Such exchanges are no longer tied to a physical space and loose their territorial rationale to focus solely on the political arena and the ability of individual actors to impose their project. The efficiency of water management in Sardinia is also probably due to the fact that Sardinia is an 'isolated' island without the possibility of looking at an externalization of water resource from other regions. Last but not least, whilst this situation optimizes the economic use of water on the regional scale, it does not yet take into account climate change threats but will probably facilitate the implementation of new strategies as soon as these will be on the agenda.

Acknowledgements

The studies presented in this chapter were funded by the European Commission through the Seventh Framework Programme for the CLIMB project. Authors would like to thank the coordinator of CLIMB Ralf Ludwig for his support all along the project duration. The authors would like to gratefully acknowledge local stakeholders for their active contribution and the time allocated during field investigations.

Notes

1. Confederazione Nazionale Coltivatori Diretti, the largest organization representing and supporting Italian farmers, and in particular small producers and cooperatives.
2. The Reclamation Consortia were established in 1933. The Consortium is responsible for the management and the maintenance of existing reclamation works. The realization of new reclamation works is the responsibility of the State and the Region but which can ask the Consortium, with an appropriate public funding, to execute works.
3. Although the single price applied by ENAS to all agricultural consortiums is rather low (€0.007 cent per litre in 2012), the water price applied by the CBSM to associates is higher due to management costs. In the CBSM, the cost per hectare can reach €260 (which corresponds to 6000 m^3/ha/year of water) and sometimes even more if additional sums are required to balance the consortium's accounts or for extraordinary expenses.
4. This is what happens in particular between large hydro-electrical companies and agricultural consortiums: an agricultural consortium cedes its rights to pump from a source to the hydro-electrical company at a time of the year when the farmers do not need it. In exchange, the agricultural consortium has access to the Santa Giustina reservoir water during the irrigation period, when the pressure on the resource is at its highest.

References

Botti, P., Dessena, M.A., Fiori, M. et al. (2001). The Medio Flumendosa reservoir. *Rendiconti Seminario Facoltà Scienze* 71 (2): 209–220.

Cadoni D., R. Silvano, and S. Vacca (2010). Drought watch system in Sardinia to reduce vulnerability and improve management during periods of drought. ICEMT 2010 - The First International Conference on Environmental Management & Technologies, 1–3 November 2010, Amman, Jordan.

Coldiretti (2013). Statistical reports. https://www.coldiretti.it (accessed 21 February 2019).

Coldiretti (2018). Statistical reports. https://www.coldiretti.it (accessed 21 February 2019).

CBSM (2012). Consorzio di Bonifica de la Sardegna Meridionale. https://www.cbsm.it/it/index.php?pg=7&op=0 (accessed 21 February 2019).

Dettori, M., Cesaraccio, C., and Duce, P. (2017). Simulation of climate change impacts on production and phenology of durum wheat in Mediterranean environments using CERES-wheat model. *Field Crops Research* 206: 43–53. https://doi.org/10.1016/j.fcr.2017.02.013.

ENAS, Ente Acque de la Sardegna (2011). Resoconto di fine gestione 2007–2011. Sardegna: ENAS.

Gobiet, A., Kotlarski, S., Beniston, M. et al. (2014). 21st century climate change in the European Alps – a review. *Science of the Total Environment* 493: 1138–1151. https://doi.org/10.1016/j.scitotenv.2013.07.050.

La Jeunesse, I., Gillet, V., Larrue, C. et al. (2010). Questionnaire for uses and rivalries identification in study sites. CLIMB report. Deliverable7.1, 2010, 27p. Munich: CLIMB.

La Jeunesse, I. and Cirelli, C. (2012). Comparative analysis of water uses and rivalries in the case studies. Deliverable 7.2, CLIMB Project, 2012, 53p. Munich: CLIMB.

La Jeunesse, I. and Cirelli, C. (2014). Diagnosis of climate changes impacts on water uses and rivalries in the study sites. Deliverable 7.3, CLIMB Project, 2014, 65p. Munich CLIMB.

La Jeunesse, I., Larrue, C., Furusho, C. et al. (2016). The governance context of drought policy and pilot measures for the Arzal Dam and reservoir, Vilaine Catchment, Brittany, France. In: *Governance for Drought Resilience - Land and Water Drought Management in Europe* (ed. H. Bressers, N. Bressers and C. Larrue), 109–138. Cham: Springer.

Majone, B., Villa, F., Deidda, R. et al. (2016). Impact of climate change and water use policies on hydropower potential in the south-eastern alpine region. *Science of the Total Environment* 543: 965–980.

Mascaro, G., Deidda, R., and Hellies, M. (2013). On the nature of rainfall intermittency as revealed by different metrics and sampling approaches. *Hydrology and Earth System Sciences* 17: 355–369. https://doi.org/10.5194/hess-17-355-2013.

Meyer, S., Blaschek, M., Duttmann, R. et al. (2016). Improved hydrological model parametrization for climate change impact assessment under data scarcity – the potential of field monitoring techniques and geostatistics. *Science of the Total Environment* 543: 906–923. https://doi.org/10.1016/j.scitotenv.2015.07.116.

Perra, E., Piras, M., Deidda, R. et al. (2018). Multimodel assessment of climate change-induced hydrologic impacts for a Mediterranean catchment. *Hydrology and Earth System Sciences* 22: 4125–4143. https://doi.org/10.5194/hess-22-4125-2018.

Piras, M., Mascaro, G., Deidda, R. et al. (2014). Quantification of hydrologic impacts of climate change in a Mediterranean basin in Sardinia, Italy, through high resolution simulations. *Hydrology and Earth System Sciences* 18: 5201–5217. https://doi.org/10.5194/hess-18-5201-2014.

Piras, M., Mascaro, G., Deidda, R. et al. (2016). Impacts of climate change on precipitation and discharge extremes through the use of statistical downscaling approaches in a Mediterranean basin. *Science of the Total Environment* 543: 952–964.

Provincia Autonoma di Trento (2006). Piano Generale di Utilizzazione delle Acque Pubbliche (PGUAP). *Gazzetta Ufficiale* (119): 1.

Regione Autonoma Sardegna (2006). Piano Stralcio di Bacino regionale per l'utilizzo delle risorse idriche Sardegna, Agenzia del Distretto Idrografico.

Regione Autonoma Sardegna (2010). *Piano di gestione delle acque, Regione Autonoma*. Sardegna: Agenzia del Distretto Idrografico.

Regione Autonoma Sardegna (2016). *Secondo piano di gestione delle acque*. Sardegna: Agenzia del Distretto Idrografico.

WFD (2006). Water Framework Directive 2000/60, EC. Brussels: European Commission. https://eur-lex.europa.eu/legal-content/en/TXT/?uri=CELEX:32000L0060 (accessed 21 February 2019).

18

Power Asymmetries, Migrant Agricultural Labour, and Adaptation Governance in Turkey: A Political Ecology of Double Exposures

Ethemcan Turhan[1], Giorgos Kallis[2], and Christos Zografos[3]
[1]*Laboratory of Environmental Humanities, KTH Royal Institute of Technology, Stockholm, Sweden*
[2]*Institut de Ciència i Tecnologia Ambientals (ICTA), Universitat Autonoma de Barcelona and Institució Catalana de Recerca i Estudis Avançats (ICREA), Barcelona, Spain*
[3]*Environmental science and technology Institute ICTA, Autonomous University of Barcelona, Barcelona, Spain*

18.1 Introduction

The politics of agriculture and climate change are likely to remain high on the global agenda in the decades to come. The IPCC Fifth Assessment Report (IPCC 2014) anticipates high risks in global agriculture due to the impacts of climate change. The Mediterranean region is particularly a hotspot for climate impacts where the warming is much stronger in summer than in winter and the number of warm days and nights are expected to increase sharply (Lelieveld et al. 2016). Those risks manifest themselves as concerns for future water availability, food security, agricultural incomes, and shifts in production zones of food and non-food crops, all of which become discernible at local and regional scales. In the absence of ambitious and concerted mitigation efforts, climate impacts are likely to 'alter ecosystems in the Mediterranean in a way that is without precedent during the past 10 millennia' (Guiot and Cramer 2016, p. 468). Located at the eastern Mediterranean, Turkey's agricultural sector is seriously threatened by the impacts of climate change (Demircan et al. 2017). In fact, recent analyses show that whilst the cultivable land area may increase in Turkey under climate change, negative

Facing Hydrometeorological Extreme Events: A Governance Issue, First Edition.
Edited by Isabelle La Jeunesse and Corinne Larrue.

impacts will mostly be felt in terms of declining yields (Le Mouel et al. 2016). This is particularly important since the share of gross value-added contribution of agriculture to the Turkish economy was 6.1% in 2017 with 21.22 billion USD worth of exports (World Bank, 2019). Although the share of agriculture in GDP retreated from 11% in 2000 to 6.1% in 2017 (ibid.), Turkey still ranks seventh in terms of total agricultural production (OECD 2017) and has more than 20% of its formal labour force in the agricultural sector (MOFAL 2015). It is within this context that the multiple and overlapping interactions of global economic and environmental changes are posing threats to the Turkish agriculture and those who live off it (Yano, Aydın, and Harguchi 2007; Aydın 2010; Keyder and Yenal 2011).

In order to overcome the 'adaptation paradox',[1] it is important to develop coherent explanations of how global changes are experienced locally and how responses to those changes in turn shape and are shaped by other global processes. A key approach for the analytical study of such interaction can be found in Leichenko and O'Brien's (2008) double exposure (DE) framework. This framework suggests that regions, sectors, ecosystems, or social groups, which are confronting mutually reinforcing and interacting pressures from both global environmental change and neoliberal globalization, are subject to DE. The DE framework helps us to analyse how 'global processes occurring both simultaneously and sequentially [create] positive and negative outcomes for individuals, households, communities and social groups' (Leichenko and O'Brien 2008, p. 33). Nonetheless, the questions of how the complex interactions of 'neoliberal rules and institutions produce or exacerbate vulnerability' (Fieldman 2011, p. 160) as well as how the power asymmetries operate in particular cases of DE still remain as incomplete tasks.

In an effort to explore the importance of local power asymmetries as determinants of multiple and overlapping vulnerabilities, we present a case study on labour-intensive agriculture. Understanding multiscalar, multidimensional vulnerabilities and mechanisms giving way to them beyond the biophysical and livelihood aspects are essential to address 'more fundamental concerns that shape vulnerability' (Nightingale 2017) in the context of adaptation governance. Hence through this chapter, we aim at unravelling the role of power asymmetries in creating and maintaining vulnerable conditions. We employed an in-depth case study method to bring forth social power relations, which are crucial for understanding 'causality in local vulnerability and adaptive potential' (Birkenholtz 2012, p. 296). Moreover, our approach to local power constellations is rooted in the political ecology tradition with a particular focus on a relatively ignored social group: migrant seasonal agricultural workers. We argue that powerful groups actively and knowingly shift the costs and risks of ecological and economic uncertainty to less powerful groups in order to avoid economic and/or political losses, particularly in times of climate stresses such as droughts. State-led adaptation policies often facilitate this cost and risk-shifting through their framing of the agricultural sector as a homogeneous unit as well as their oversimplification of the forces that structure the vulnerabilities.

The next section presents the DE framework, which informs the analytical approach of this case study. Section 18.3 then introduces the reader to the case, a watermelon producing community in southern Turkey, and outlines the research methods used for the collection of empirical data. Section 18.4, then, shows the outcomes of DE upon political and economically the most marginalized community in the region, migrant seasonal agricultural workers (hereafter *seasonal workers*). In this section, we provide an explanation of how the adaptive responses of landowners increase vulnerabilities and shift risks and costs to seasonal workers. Section 18.5 discusses the policy implications of this analysis and Section 18.6 concludes with some thoughts for future research.

18.2 Double Exposures and political ecology of vulnerability

Despite its consolidation as a crosscutting concept for the study of the human dimensions of global environmental change, until recently vulnerability to environmental change has been studied separately from other stressors, such as those propagated by neoliberal globalization (O'Brien and Leichenko 2000). Nonetheless handling these two large-scale changes together is an initial step to avoid, at least conceptually, compartmentalization of adaptation policymaking (Leichenko and O'Brien 2008, p. 5). In this regard, the DE framework (Leichenko and O'Brien 2008) provides us with an analytical frame to explain how two major global processes; neoliberal globalization and global environmental change simultaneously impact communities, sectors, or regions. In doing so, it also accounts for the feedbacks arising from their interaction as well as paying attention to the ways in which these processes spread vulnerability over space and time (O'Brien and Leichenko 2000; Leichenko and O'Brien 2008).

The DE framework comprises five main components: (i) processes of global change, (ii) exposure unit, (iii) contextual environment, (iv) responses, and finally, (v) outcomes (Leichenko and O'Brien 2008, p. 39). Whilst all five components are subject to change simultaneously, alterations in contextual conditions can impact exposure and responses to future global change processes, hence resulting in new patterns of vulnerability and new challenges for socio-ecological systems. Responses to changes, on the other hand, can vary in accordance with political and economic priorities and shape outcomes accordingly. Probably the most enticing element in this framework is its explicit focus on feedbacks arising from different configurations of its components. O'Brien and Leichenko (2000) call these configurations 'double exposure pathways'. There are three main pathways of interaction.

The first pathway is outcome DE, referring to discrete events or gradual changes from which the exposure units (i.e. an agrarian community) are affected as a result of the interaction between major processes of global change. An example is the loss of livelihood due to the combined effects of a drought and the removal of subsidies and trade protections on a product that a certain community depends on. In their analysis of the DE in New Zealand's sheep and dairy industry, Burton and Peoples (2014) draw a picture of overlapping conditions of consequent droughts and neoliberalization of the sector as demonstrated by gradual removal of subsidies and tightening of drought assistance. Consequent to an overlap of environmental change and economic restructuring which left many indebted, a significant number of farmers set off to seek off-farm work. According to Leichenko and O'Brien (2008, p.45), this outcome pathway raises a major, equity-related question: 'Who is most likely to experience negative outcomes related to both processes?'

The second pathway identified in the DE framework is context-related. Emerging conditions associated with global changes may change the contexts (i.e. existing socio-ecological environments) in which livelihoods are embedded (Leichenko and O'Brien 2008, p. 47). Contextual changes affect the adaptive responses of people or weaken their traditional coping mechanisms thereby exacerbating vulnerability. For example, Milman and Arsano (2014) demonstrate that a state-led adaptation policy in Ethiopia, aiming at sedentarizing formerly dispersed communities into larger and fixed villages, was supported by an agricultural development-oriented industrialization scheme. This scheme promoted a shift towards cultivating market-oriented crops with an aim to mobilize 'underutilized and unproductive rural labour' (Milman and Arsano 2014, p. 6). Nonetheless, by restricting the mobility of formerly dispersed populations, these schemes increased their exposure to drought and tied their livelihood options to global commodity prices. In essence, the scheme limited viable livelihood options (i.e. seasonal migration) of these communities by changing contextual conditions.

The final DE pathway, identified by Leichencko and O'Brien (2008), is via feedbacks. This pathway acknowledges that responses to one or both of the global changes (environmental or socio-economic) may change the basic structures of the system. Such feedbacks may initiate from outcomes as much as from the exposure unit itself or the contextual environment. Consider for example, the shift experienced by large wineries and small producers of Okanagan Valley, Canada. Belliveau, Smit, and Bradshaw (2006) demonstrate that with entry into force of the North American Free Trade Agreement (NAFTA), market conditions drove wine producers towards a higher quality product which required further capital investment. Hence small producers were encouraged to get in the agro-tourism sector to make up for the losses incurred. Nonetheless, deteriorating climatic conditions with increasingly uncertain weather patterns also impeded tourism and once the relevant investments were made, 'there were few options to cope with reductions in tourism' (Belliveau et al. 2006. p. 371). As shown in this example, adaptation policy produces feedbacks, which further create complexity due to increasing climatic and market uncertainties. As demonstrated in this latter case, differential vulnerabilities between large and small-scale farmers are not only shaped by changing climatic conditions but also by differential access to new resources and technologies.

The DE framework identifies unequal power relations as a key driver that gives rise to uneven vulnerabilities. Yet, we need further analytical tools to decipher the underlying mechanisms of these three different pathways. This is precisely the point where political ecology provides us with a toolbox that is fit to purpose. Political ecology helps with a thorough questioning of uneven power relations and 'all [the] struggle hidden in the quiet vista' (Robbins 2004, p. xvi). On one hand, power in political ecology, as Peet, Robbins, and Watts (2011, p. 31) observe, 'is most crudely and commonly understood as the capacity of a polity or state to control the actions of people in its jurisdiction'. On the other hand, it also refers to the 'concrete power which every individual holds and whose partial or total cession enables political power or sovereignty to be established' (Foucault 1980 , p. 88). Consequently, power operates in a capillary manner where it 'reaches into the very grain of individuals, touches their bodies and inserts itself into their actions and attitudes, their discourses, learning processes and everyday lives' (Foucault 1980, p. 39). This capillary operation eventually shapes the everyday interactions between different social groups and their surroundings. A political ecology of vulnerability, in this sense, attempts to 'denaturalize' the given conditions and vulnerabilities of social groups by making the topics of contestation visible (Robbins 2004, p. 12). This approach also reveals uneven geographical development with unequal distribution of socio-ecological costs and benefits.

As Gertel and Sippel (2014, p. 4) observe, labour-intensive modes of agriculture often rely on exploitation of seasonal (and frequently migrant) workers despite providing a vital income resources at the same time. Yet pressing challenges posed by climatic and socio-economic changes give rise to new insecurities for the seasonal agricultural labour force. Nevertheless, the migrant labour dimension of market-driven agriculture under a changing climate still remains understudied. It is precisely this gap that this study aims to fill with an eye on power asymmetries in adaptation to climate change manifested as environmental cost shifting.

18.3 Case study and methods

Çukurova ('Low Plain' in Turkish) is the Mediterranean delta system that lies in the downstream of Seyhan and Ceyhan river basins with Adana, the fifth most populous city of Turkey, lying at its heart. At the onset of the nineteenth century, this region was hardly more than a 'badly drained, fever-ridden, thinly populated piece of land' (Kıray 1974). Çukurova assumed a new economic role during 'Ottoman modernization' around 1870s to facilitate integration with global markets (Toksöz 2010). In the following decades, the region successfully connected with the global agricultural markets mostly through its large-scale cotton plantations (Toksöz 2010). However the real transformation arrived after the Second World War with a gradual shift from agrarian to agro-industrial capitalism in Çukurova (Gürel 2011). Still a young republic, Turkey willingly accepted an agricultural exporter role in the emerging postwar global economic order also in order to join the key international institutions such as the International Monetary Fund (IMF) and the Organisation for Economic Co-operation and Development (OECD) (Tören 2007). Conforming to the US Marshall Plan's main objective to tackle the problems of overaccumulation in the core economies whilst redesigning the periphery with development aid, Çukurova became one of the hotspots of agricultural modernization and intensification drawing ever-increasing masses of agricultural workers. The region was further pounded with two decades of neoliberal restructuring with severe socio-economic impacts on its agricultural population (Keyder and Yenal 2013; İslamoğlu 2017).

In this chapter, we focus on a very local scale within Çukurova, namely the case of Kapı village in Karataş district of Adana, forming part of the lower Seyhan river basin (Figure 18.1). The Seyhan river has the second biggest drainage basin area in the Eastern Mediterranean after the Nile, with an area of approximately 25 000 km^2. The lower flat area of the basin (where Kapı is located) is characterized by irrigated agriculture, cultivating maize, wheat, fruit, and other cash crops (Watanabe 2007). As Birkenholtz (2012, p. 301) maintains, empirical generalizability remains a key challenge for intensive case study research on vulnerability. We chose Kapı precisely because it makes for a typical case for analysing and analytically generalizing responses to DE, given its high reliance on a single economic activity (agriculture) and the expected severe implications of climate change.

Kapı is typically representative of the production conditions in the rest of the region. Practising mainly early grown horticulture with manual labour, Kapı residents rely on migrant labour to meet their needs. Karataş, the administrative district where Kapı is located, has the highest share of agricultural production and the highest rural agricultural

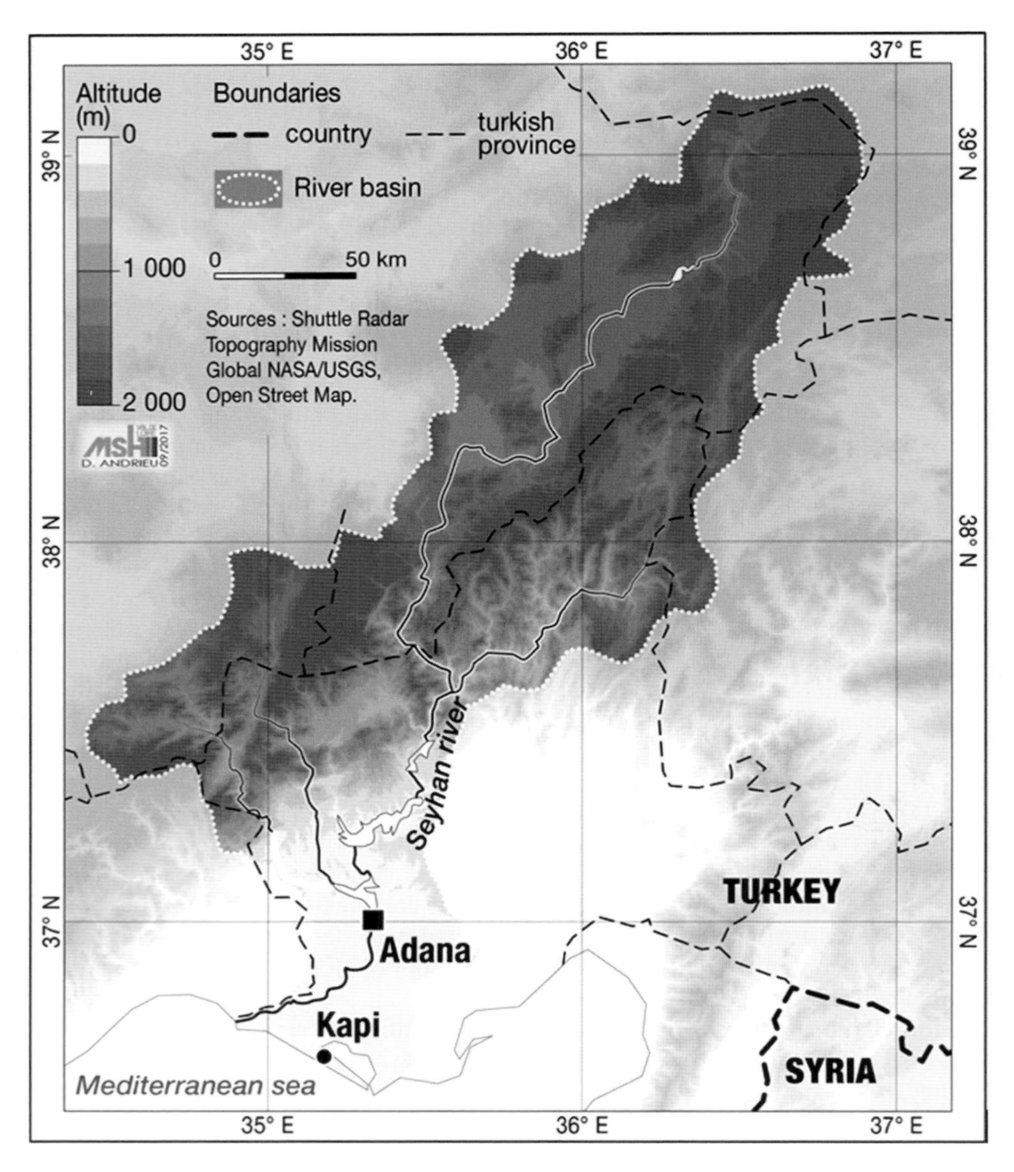

Figure 18.1 Map of Seyhan river basin.

employment (DPT 2004; UNDP Turkey 2009) in the lower Seyhan basin. Moreover, this district scores 21st out of 872 in the whole of Turkey in terms of gross agricultural production (DPT 2004). Watermelon, which is the focus of this study, is one of the principal products of the district and definitely a key product for Kapı. Turkey is a global runner-up in watermelon production, where it ranks in second place after China amongst largest producers by accounting for 3.8% of the global market (FAO 2014). Watermelon cultivation in Adana increased 3.18% between 2008 and 2013 whereas agricultural land dedicated to watermelons was reduced by 2.57% at the provincial level with the cultivation in this region corresponding to 20% of the total national production (AGV 2013).

As in many developing country contexts, the Turkish countryside has been transformed by capitalist development through 'the commercialization of production leading inexorably to the elimination of peasant family farming, a process of depeasantization linked in turn to the growth of landless workers' (Jacoby 2008, p. 260). The shift towards commercial fresh vegetable and fruit production, which accounted for half of national agricultural exports between 1996 and 2004 testify this (Yercan and Işıklı 2006 as cited in Keyder and Yenal 2011). Whilst fresh fruit (13.68%) and fresh vegetable (1.96%) production generally expanded in Turkey, this expansion has been at a much larger scale in lower Seyhan region with a 96.8% expansion in fresh fruits and 3.18% in vegetables in the period 2008–2012 (AGV 2013). Yet, such rapid expansion often comes at a cost: the social cost of labour-intensive agriculture (Holmes 2013; Gertel and Sippel 2014).

This said, a focus on the political economy of agriculture fails to account for the whole picture in the region. Climatic predictions for the Seyhan basin foresee mean surface air temperature increases in the range of 2–3.5 °C accompanied by a 25% reduction in winter precipitation by 2070 (Watanabe 2007). Regional climate models report an even more serious 6–7 °C increase in mean temperatures by the end of this century (Şen et al. 2011). Biophysical risks associated with these changes include but are not limited to decrease in groundwater recharge in the range of 24.7–27.4% as well as possible decreases in precipitation up to 25% for the period 2070–2080 (Tezcan et al. 2007; Watanabe 2007). Recent assessments suggest that water deficit in the basin may increase from 15% by the 2040s to more than 30% by the end of the century (Selek and Tunçok 2014). It is also estimated that with a 50% increase in groundwater abstraction, which is the main means of irrigation in Kapı village, seawater intrusion in the lower Seyhan basin may reach as far as 10 km inland by 2080 (Tezcan et al. 2007). Consequently, all these impacts of climate change are expected to have very significant adverse effects on the yield, crop prices, and employment in the region (Özkan and Akçaöz 2003).

Given the importance of agriculture in the region and the anticipated impacts of climate change, there is a surprising research gap on vulnerabilities of different social groups. To gain a better understanding of the local context of vulnerability and the ways in which power relations shape DE, we held 20 semi-structured in-depth interviews, eight of which were with seasonal workers and four with agricultural intermediaries.[2] On top of this, we interviewed three landowners, three representatives of different local government authorities, and two academics who had previously worked on agriculture in Kapı. Additionally, two focus group meetings were organized, the first with 12 landowners discussing changes in watermelon production in Kapı, and the second with 11 different actors from the region including but not limited to watermelon producers. The data stemming from these qualitative inquiries were triangulated with direct observation with seasonal workers in the field.

The research took place in two stages, first a short two-week scoping phase (February 2009), and then a return period of two months (March–April 2011). There were some limitations in fieldwork particularly in communicating with the Kurdish seasonal workers on a one-to-one basis, especially during the politically charged general election campaign, which coincided with the second fieldwork period. There was a particular suspicion by local authorities, rural police, and the landowners on the presence of a male researcher of Turkish origin speaking to Kurdish seasonal workers.[3] Such political and practical constraints may have created a certain bias. Nonetheless, we believe that the original, empirical material here allows us to sustain the basic thesis of this chapter, that exposure in Kapı is unevenly distributed, and that landowners often successfully shift the costs of DE to seasonal workers intentionally, with climate change adaptation policies facilitating this shift. We present this in what follows.

18.4 A political ecology of Double Exposure in Kapı village

18.4.1 Outcome Double Exposure in watermelon production

Seasonal agricultural labour migration constitutes approximately 65% of the agricultural labour in the lower Seyhan region (Erkan 2000). An unofficial estimate suggests that 100000 seasonal workers arrive in the lower Seyhan region every year between January and September (Gümüş 2005). Kapı, whose population is 300, hosts approximately 400 seasonal agricultural workers every year. These workers, mostly ethnic Kurds from southeast Turkey, arrive in Kapı for sowing, setting up greenhouses, hoeing, and harvesting watermelons and groundnuts between January and September.

The roots of seasonal agricultural labour migration in Turkey lie in processes of dispossession. Dispossession, in this context, is often the result of multiple, overlapping factors (Gürel 2011). Constant demand for cheap agricultural labour in the western, central, and northern parts of Turkey also induced a pattern of seasonal migration for those who were not qualified or not willing to join the informal urban labour force (Harris 2009; Kadirbeyoğlu 2010, see also Grineski et al. 2013). On top of these, a key factor in the prominence of seasonal migration in the past three decades has been the ethnic strife between Kurdish insurgents and the Turkish army in the eastern and south-eastern Turkey. This violent conflict internally displaced approximately a million people with more than 75% of them being from rural zones (Hacettepe University 2006). Adding further complexity to the landscape today, refugees fleeing from the nearby Syrian border are by now contributing significantly to the seasonal agricultural labour demand in the region (Çetin 2016).

In Kapı, a heightened exposure of seasonal workers to climate risks, which are most commonly manifested as climatic extremes such as prolonged drought and increased frequency of hail and frost events, can be observed. Two main negative outcomes of this exposure prevail. First, there are increased incidences of contagious diseases amongst the seasonal workers linked to increased climatic variations. Since their work is mediated by an informal verbal contract between the agricultural intermediaries and the landowners, seasonal workers remain out of the formal social security and health coverage. Malaria is one of the vector-borne diseases with well-known linkages of climate and living condition related to its transmission. Ergönül (2007) links the increased historical trend of malaria incidences with increasing mean surface temperatures in Adana for the periods 1977–1987 and 1993–1998. Out of the 77 malaria cases seen in Kapı in 2001, 57 of them were observed amongst seasonal workers (Özbek 2007). In 2002, these figures were 25 out of 31 cases. Moreover, out of 1399 cases of seasonal workers registered at the regional clinic, 342 (24,4%) of them suffered from respiratory diseases during March–October 2002 (Sütoluk et al. 2004). A recent study on 1662 Syrian seasonal workers in Adana also found out that 68.8% of the respondents suffered health problems linked to their daily work practices (Kalkınma Atölyesi 2016). The majority of these observed health problems are related to high temperatures (notably heat stroke), lack of potable water, contamination, malnutrition, and increase in vectors. Due to an absence of social security coverage, seasonal workers either end up having to pay for such occupational hazards themselves or rely on charities for health service.

Second, worker housing conditions in the region pose a significant threat under heightened climatic uncertainties. Workers in Kapı reside in makeshift tents in encampments, mostly next to the plots that they work in, under extremely poor sanitary conditions often without access to clean water and toilets. Tent fires due to wood stoves inside the dwellings pose a significant problem for seasonal workers particularly in early spring. Inappropriate housing conditions in these encampments exacerbate vulnerability to climate-induced health risks, specifically during very cold and very warm periods. Climate-induced health risks also manifest themselves in sudden on-set events. During our initial interviews in early 2009, landowners in Kapı stated their satisfaction with the weather conditions following three consecutive dry years. However, by May of the same year, approximately 6000 ha of watermelon plantation in the district were inundated due to heavy rainfall. Such events have deleterious effects on seasonal workers not only due to their precarious housing conditions but also in terms of non-payment of their wages. The superimposition of climatic and economic stressors manifest themselves as loss or delay of wages (often in the scale of months), severely worsened housing conditions and deteriorating individual health. In the words of a seasonal worker:

> *Landowners come [here] and tell us that 'there was a storm last night' and we all go to the field. We spent days and nights working there. We go there at night to see but it [the watermelon in the field] dies at night, what can he do? Eventually if he cannot sell it, you cannot get your money. These [watermelons] also belong to us. (Seasonal worker, 29.03.2011)*

Watermelon farming is increasingly popular amongst many small and medium-sized producers. Yet despite the risk of anticipated impacts, this practice is not only plagued with climatic uncertainties. It is also affected by global and national agricultural market fluctuations. 'Gamble' is the typical metaphor for watermelon farming due to its lucrative yet risky character, as we are told by our respondents.

> *Watermelon is like a gamble. You never know what you will get back or if you will be able to pay back your debts.* (Landowner, 22.04.2011)
>
> *It is all about the weather conditions. It needs to be chilly in here and warmer in bigger cities like Ankara or Istanbul so that there is higher consumption. Look, what happened last year? We started sending out [watermelons] in late May and there was no sun in these cities for 20 days. No one wanted watermelons and prices fell. If there is heavy wind [indicating frost] in here, producers will harvest 10 days before the normal, prices will fall and no one benefits. (Landowner, 11.04.2011)*

The gamble metaphor resonates well with Keyder and Yenal's (2011, p. 63) observation that there is a 'growing feeling of insecurity and indeterminacy on the part of the farmers about the prospects of their production and marketing decisions'. Eventually simultaneous occurrence of climate shocks and landowners' gamble with agricultural market prices manifest themselves as an outcome DE for seasonal workers. For example, marketable yield of watermelons in the region are estimated to decrease significantly in the likely case of reduction in irrigation under a scenario where evapotranspiration of the plants increase by 50% and water availability decreases (Kuşçu et al. 2015). This loss, however, is often passed onto seasonal workers. Absence of formal labour contracts, lack of access to social security, and social and spatial exclusion of their encampments all contribute to consolidate a context where seasonal workers bear the majority of the burdens. It is with this vision that we now turn to context-related DE of seasonal workers.

18.4.2 Contextual Double Exposure and agricultural insurance

The increased frequency of climatic and market shocks damages both landowners and seasonal workers alike in Kapı. However, climate change adaptation policies as in the case of state-subsidized agricultural insurance (enacted with Turkish law no. 5363, dated 14.06.2005) bring about opportunities for risk reduction for landowners. The Agricultural Insurance Pool Enterprise (TARSIM), a private-public partnership including 23 insurance companies, was established in 2006 to offer 50% subsidized agricultural insurance. TARSIM covers risks from hail, storm, fire, cyclone, landslide, earthquake, floods, and flash floods. One of the key bottlenecks associated with this insurance scheme, however, is its indifference to drought. Despite increasing calls from the landowners to cover for droughts, TARSIM officials are reluctant if not outright against including drought in the cover; stating that the company is not financially strong enough to insure drought risks yet (Hürriyet 2010).[4]

Despite its shortcomings, TARSIM still provides a fall-back measure to compensate for the economic loss after other strategies fail. Nonetheless for seasonal workers, such schemes repeatedly fail to deliver. Whilst the Turkish national climate change adaptation strategy pays rhetorical lip service to seasonal workers being amongst the most vulnerable groups (MOEU 2011, p.104), these national policies provide no specific measure whatsoever to address the living and working conditions of 'hands and feet' of the agricultural production in Turkey. For example, inclusion of seasonal workers in the formal social security scheme with subsidized premiums would go a long way into reducing their vulnerabilities, yet such proposals never make it to the tables where adaptation policies are made. Whilst landowners can purchase insurance with state-funded back up against financial losses, seasonal workers remain with no security against the losses of their income source. DE of seasonal workers to a multiplicity of stressors at the outcome level is further exacerbated with the lack of social safety nets and income support. Landowners are still able to change their contexts by enrolling in subsidized insurance schemes despite increasing protest on their part, for instance, due to nonpayment of compensation in due time as in the case of hail-stricken watermelons in Kapı in 2014 (Evrensel 2014). Yet, as our respondents repeatedly mentioned to us, farmers receiving insurance payments on-time still delay or do not even dare to pay wages to seasonal workers. Seasonal workers become entrapped, waiting for their payments from the previous growing season whilst trying to make sure they have a job in the next year. This is captured well in the words of a seasonal worker in Kapı:

> *If market prices are low and watermelons do not cost a dime that year, what are you going to do? Are you going to kill the landowner? [Agricultural] intermediaries get their money anyway but when the harvest do not get a good return, they usually delay our money. (Seasonal worker, 09.03.2011)*

Agrawal and Perrin (2009) remind us that resorting solely to market-based climate change adaptation (i.e. insurance) policies should be treated with extreme caution, given the highly unequal access to these financial tools, especially for those who are in marginalized situations. Moreover fleshing out insurance as a success story seems to be rather driven by an 'ideological commitment to market-driven risk mitigation through financial extension' (Taylor 2017). As we elaborate later, this unequal access contributes to the larger picture in which costs of DE are successfully shifted to seasonal workers in labour-intensive agriculture.

18.4.3 Feedback Double Exposures and cost-shifting successes

Joan Martinez-Alier (2002, p. 30) suggests that 'the poor sell cheap, not out of choice but out of lack of power' when referring to 'cost-shifting successes'; a term he uses to highlight the uneven distribution of socio-ecological cost to the already vulnerable segments of society and appropriation of the benefits by the dominant social groups particularly in reference to ecological conflicts. In line with the DE framework, we observe that two processes come out strongly when we shift the focus towards landowners instead of seasonal workers. First, there is a negative feedback insofar as actions taken by landowners to adapt to climatic and economic changes end up producing new vulnerabilities. Second, to the extent possible, landowners shift the costs and risks of their adaptations (or maladaptation for that matter, see also D'Alisa and Kallis 2016) onto the seasonal workers. Such cost shifting eventually causes a relative safety for powerful segments of the population in Kapı at the cost of relational insecurity of other groups.[5]

Landowners face exposure to both changing climatic and market conditions, to which they respond in ways that create different feedbacks. The mass shift to watermelon farming in Kapı was mainly due to watermelon's high productivity compared to traditional crops like maize and barley, which are failing in efficiency with the rising temperatures. Tsujii and Erkan (2007, p. 10) suggest that, in this region, watermelon farming may increase to cover 54.4% of the irrigated lands by 2070s. However, watermelon production is labour-intensive and hence augments the demand for stable seasonal labour inflow. In one of our focus group discussions, landowners reported an increase in the yield due to aggressive introduction of fertilizers and grafted seedlings in the last 10 years. By raising the yield from 30 to 40–45 ton/ha in watermelon and expanding the sown area, more seasonal workers were attracted to the region. The negative feedback from this intensification of production, nonetheless, was the rise of *fusarium oxysporum* disease in the crops; a problem our respondents repeatedly linked to bad agricultural practices such as soil overexploitation, intense salinization due to excess groundwater extraction, and the monoculture of watermelons. Derviş et al. (2009) suggest a link between *fusarium* and the excessive exploitation of land for watermelon without crop rotation. It is important to underline that such intensification is undertaken in order to meet the domestic and international market demands and to deliver fruits around-the-clock.

In an attempt to adapt to changing conditions, producers in Kapı responded by introducing zucchini-grafted watermelon seedlings. These grafted seedlings are promoted for their resistance against climatic extremes (Habertürk 2017). Moreover, they are also resistant both to *fusarium* and longer periods of drought. Nevertheless, as in most technical fixes, these seedlings created unexpected problems. On the one hand, the shift to grafted seedlings created a market dependency since these seedlings cannot be sown for a second time hence forcing farmers to purchase new seedlings each year. Our respondents in Kapı repeatedly lamented their dependence on seedlings that come from large agri-business companies from the Netherlands and Israel. On the other hand, grafted seedlings not only produced new dependencies but also adversely affected the taste of the product and augment the use of chemical fertilizers; at times leading to refusal of watermelons by international markets (e.g. as in the Russian market in 2008, Yeni Şafak 2008).[6]

Despite these points, agri-business representatives continue to argue that falling prices and taste issues are mainly due to early harvest of the fruit. Competition in the national and global markets drive landowners to harvest as early as possible. Kapı's comparative advantage is that it can grow watermelons early in the season hence giving producers a chance to serve domestic markets first when the prices are at their highest. Therefore, producers here harvest their watermelons between mid-May to mid-June, placing them ahead of other producers elsewhere and thus giving them advantage to set the initial prices. Nevertheless, these circumstances changed drastically with trade liberalization and the re-opening of border trade with Iran. Turkish law no. 4910 (enacted on 1.7.2003) changes the border trade regime and authorizes the Council of Ministers to make exceptions on tariffs and quotas for border trade with Turkey's neighbours, at times eliminating trade barriers to allow for free circulation of goods. The introduction of this law coincided with the implementation of ARIP (Agricultural Reform Implementation Project) spearheaded by the World Bank in the aftermath of the 2001 economic crisis. Aiming at overcoming trade protectionism, these reforms redesigned Turkish agriculture towards an agricultural basin model, which prioritizes particular crops by removing subsidies and introducing direct income support. Hence an overlap of an increased interest towards watermelon farming, expanding border trade, and removal of subsidies in agriculture led to a convoluted stress for producers in Kapı.[7]

The problems related to grafted seedlings, we argue, point at a negative feedback DE. Grafted seedlings were introduced as more climate- and *fusarium*-resistant varieties, but in turn decreased marketability of watermelons in Kapı and increased the costs for the producers by exposing them to further new markets. Landowners themselves in Kapı are acutely aware of these risks and their exposure to both climatic and global market forces. A landowner reflects on this as follows:

> *We started cultivating grafted seedlings back in 2006 because everything was getting more and more uncertain. Grafted seedlings are immune to diseases such as* fusarium. *Imagine,* fusarium: *You work to sow [the land], watermelons become ripe, time for harvest arrives and all of a sudden you see that* fusarium *hits all your work. There is no cure to it. [...] Grafted seedlings [on the other hand] are a bit hard-bitten, they grow under any weather conditions be it heavy rain, extreme hot or extreme cold. Moreover, they are resistant to* fusarium. *(Landowner, 22.04.2011)*

Landowners therefore knowingly *gamble* in the face of climatic and economic uncertainty, and know that their adaptation might backfire. Climate change only worsens this gamble by increasing the bets and risks alike, therefore creating a terrain fit for shifting the costs of potential failures to those at the bottom of the social ladder. Precisely for this reason, landowners make sure that a considerable part of this burden and risk falls on the shoulders of the seasonal workers in Kapı. For instance, the landowner cited above, told us that with the watermelon market turning sour (as a result of competition with Iran, the backfiring from *fusarium*, and the grafted seedlings) landowners now favour a shift from watermelons to mechanized wheat and maize production despite decreasing returns. The reason is that these traditional crops require much less wage labour due to increased mechanization and therefore reduce labour costs. In other words, the costs of the failed gamble with watermelon are now being shifted to seasonal workers, who kept coming to this region for work in the past two decades.

Apart from such structural changes, socio-ecological cost-shifting also takes place on years with bad returns, when landowners delay or even refrain from paying the promised wages to seasonal workers. Producers in Kapı argue that they incurred serious losses in the period 2007–2009 given the decline of prices in the domestic market due to the competition with imports from Iran. Many of them did not even harvest their watermelons, as it was less costly to leave the watermelon on the ground. To our surprise, this period also coincided with prolonged droughts, where many landowners moved to other products due to the excessive cost of irrigation. Related delay of payments due to this DE led to a massive strike of 50 000 seasonal workers in the region in 2007, the year when drought most severely struck Turkish agriculture (Sol 2007). In what follows, we provide some critical thoughts on what these mean in the context of broader adaptation governance efforts.

18.5 Discussion

In his preface to *The Development Dictionary*, Sachs (2010, p. ix) argues that 'the competitive struggle of the global middle classes for a greater share of income and power is often carried out at the expense of the fundamental rights of the poor and powerless'. We argue that this is also the case for adaptation governance in seasonal labour-intensive agriculture in the lower Seyhan basin and beyond. Power asymmetries, shaped by divisions in ethnic, class, and gender lines, often contribute to a successful socio-ecological cost-shifting in adapting to climate and market contingencies alike. These uneven relations allow powerful groups to shift the costs and risks of future uncertainties to those who are at the lower end of the agricultural seesaw. Adaptation policies, embedded in 'market-first' ideologies, arguably facilitate this shift. We argue that a certain blame here goes to national climate change adaptation planning with its perception of the agricultural sector as a homogeneous unit and oversimplifying the forces that structure the vulnerabilities of diverse social groups within it.

In the light of these findings, we argue that local power dynamics and uneven hierarchical relations embedded in rural relations underscore cost shifting in the name of climate change adaptation to seasonal workers. Understanding the inherent uneven relations in agrarian settings requires more than a mere practice of mapping vulnerability to climate change impacts (Taylor 2013). This practice also involves an exploration of the structural conditions relating to the means of production, class relations, gender hierarchies, and ethnic divisions with an eye for transforming those conditions. In Kapı, a rift between Turkish landowner and Kurdish seasonal workers marks these conditions. Social and spatial exclusion of Kurdish seasonal workers that temporarily settle in the outskirts of the village topped with a relation of dependency towards intermediaries contribute to workers' vulnerability in the face of multiple stressors. Whilst landowners are increasingly facing more uncertainty, they also seek to maintain the inflow of cheap and docile labour. This requires a continuation of the status quo, implying that in order to avoid any economic burden for landowners, wages and living conditions of seasonal workers must be kept at the minimum. In this context, we argue that a justice-oriented climate change adaptation policy needs to move beyond interventions to partially amend biophysical vulnerabilities and problematize production relations ridden with ethnic and class divisions in the region (see also Holland 2017).

One of the key tenets of the DE framework is the acknowledgement of winners and losers within the context of global changes (O'Brien and Leichenko 2003). Nonetheless, the empirical literature on DE to multiple stressors often focuses on rural communities as homogeneous groups of losers. Our analysis suggests that there are significant differences between social groups at the local scale, which are brushed aside. These differences are shaped by power asymmetries and further facilitate successful socio-ecological cost shifting from the relatively more powerful to the relatively weaker segments of the population, not least in adaptation. Therefore, our case study contributes to an understanding in which winner and losers are not universal categories but shaped by ethnicity, class, and gender amongst other factors. This becomes even more evident with the arrival of Syrian refugees to join the ranks of reserve army of free labour in the region and hence pushing the bar of vulnerability even lower (Pelek 2018). Recent accounts of encounters of Syrian agricultural workers with other migrant seasonal workers in the region suggest a case of 'adverse incorporation' (Kavak 2016) of these communities in an already fragile socio-ecological landscape.

We maintain that a key notion that helps us to understand the conditions comes from the meticulous focus of political ecology on power asymmetries. William Kapp (1950) once argued that externalities are not so much market failures as cost-shifting successes. In a similar vein, we find that adaptation provides space for cost shifting to be performed by powerful-at-risk groups in order to reap the benefits (i.e. in case of agricultural insurance) and outsource socio-economic costs (i.e. in the case of wage payment delays for seasonal workers) of DEs. Similar to Don Mitchell's (2012) account of Mexican farmworkers in Californian agriculture, the DEs of seasonal workers in Turkish agriculture are instrumentalized to save more than the crops. As such they serve to 'save the system in all its iniquitous glory' (Don Mitchell 2012, p. 419). In trying to implement incremental adaptation to climate change, current policies to save the system rather facilitate risk-shifting than pursuing human security for all by generating and maintaining particular social vulnerabilities (Zografos 2017).

The inclusion of affected groups in adaptation policies needs to go beyond safeguarding the existing patterns of production and calls for rethinking the components of the agricultural sector (i.e. labour, land, markets, production relations, etc.) holistically anew (Zografos, Goulden, and Kallis 2014). Monodimensional policies, which only target at compensating landowners, will remain indifferent to the losses of more marginalized, albeit highly vulnerable social groups. Although Tunçok (2016) suggests that a 'combination of tools including optimized crop cycles/patterns and technology-driven solutions' will do the job in the Seyhan basin, a holistic adaptation governance that will do justice to different vulnerabilities in the region require more than that. If 'adaptation is a social development issue as much as (if not more than) an environmental and technological issue' as Eriksen and O'Brien (2007, p. 348) suggest, then it means that responses to DE need to address 'the political and economic structures and frameworks within which people adapt'. As we have shown elsewhere, a thorough rethinking of adaptation governance needs to go beyond increasing individual resilience and consider collective transformation of the political-economic assumptions (Turhan, Zografos, and Kallis 2015).

Highlighting social justice and environmental integrity as key concerns, the concept of transformative adaptation diverges both from reformist visions as well as from understandings of adaptation as a fine-tuning of business-as-usual development practices (Eriksen et al. 2011). However insofar as multiple and intertwined exposures are concerned, the development practice itself predominantly delivers a 'palliative care' for climate change adaptation (Boyd et al. 2008, p. 391). Such palliative solutions offer few perspectives to transform the uneven power relations and avoid cost shifting to vulnerable segments of the population. Changing this is possible, if uneven vulnerabilities within the agricultural sector are to be addressed thoroughly. For example, McMichael's (2009, p. 147) call for a shift from 'Food from Nowhere' to 'Food from Somewhere' in the global food regimes is a case in this direction as it involves establishing strategies that make explicit the differential vulnerabilities and power asymmetries embedded amongst the actors of the labour-intensive agricultural system. We contend that only such rethinking of the seasonal labour-intensive agriculture could address the different pathways of DE by recognizing that labour-intensive agriculture in the Mediterranean relies heavily on a flexible and mobile workforce (often dubbed as 'undesirably desired', e.g. Gertel and Sippel 2014, p. 249). Therefore, to reach a just adaptation, a move to 'the place-based forms of agro-ecology' (McMichael 2009) is imperative. This requires us to be 'attentive to the way [...] adaptation fashions a closely-knit body of ideas around the causes and impacts of climate change that are seamlessly integrated into prevailing institutional rationalities' (Taylor 2015, p. 189). The added value of a political ecology analysis here is that it makes evident that responding meaningfully to socio-environmental injustice and overwhelming power asymmetries involves advancing adaptation through a more equitable distribution of power, a genuine consideration of alternatives and a deepened democracy beyond the existing institutional rationalities (Kallis and Zografos 2014).

18.6 Conclusion

This chapter advanced the DE framework by making explicit a dimension that was implicit in it, that is the fact that within exposure units there are population groups that face uneven exposures, with powerful groups shifting risks and exposures to less powerful groups. In the empirical case analysed here, power differentials ran across lines of ethnicity and citizenship, with seasonal workers left out of the provisions of the welfare state, and hence rendered extremely vulnerable to the whims of climate and economic forces that influence intensive agricultural production in southern Turkey. It is conceivable that in other cases such lines of uneven exposure may run across divisions of class, wealth, gender, or race. This calls for more empirical studies of DE that are more aware and explicit of such divisions and shed light on the ways in which economic and political power shape exposure not only at the macro but also at the micro levels.

Agriculture in Turkey, like many other parts of the world, is increasingly exposed both to the impacts of climate change, via more intense and frequent droughts, floods, and high temperatures, as well as global economic challenges, mainly experienced as declining prices due to market integration and intensified global competition. Landowners respond to such exposures with adaptation measures that intensify inputs and production, often taking a gamble that makes matters worse in the long term. A significant part of the costs of this gamble however is shifted to workers with much less power that lack basic citizenship rights, in this case, both Turkey's Kurdish and Syrian seasonal workers. This is done via wage cuts as well as by shifts to new crops where possible. Whilst national policies, preoccupied with the continued profitability of the agricultural sector, develop mechanisms to insure and compensate landowners against losses, they tend to ignore those who are deeply vulnerable, and whose vulnerability can be easily reduced by the provision of access to the basic services of a social security system. We contend that the massive numbers of Syrian refugees who joined the ranks of already vulnerable Kurdish seasonal workers in Turkish agriculture will face the same – if not the worse – dire conditions, if adaptation policies continue to be oblivious to those whose hands and bodies shape the agricultural landscapes. Our political ecology lens on unevenly distributed and power-determined DEs allows us to understand why climate policies that aim at the adaptation of the agricultural sector as a whole, are bound to end up increasing rather than reducing, the vulnerability of those who are most vulnerable.

Acknowledgements

We would like to thank the editors and anonymous reviewers for their comments. Ethemcan Turhan acknowledges FI-DGR 2011 fellowship of Generalitat de Catalunya and two visiting research periods in Durham University and Middle East Technical University. We are most grateful to all seasonal workers and landowners, who contributed to this research, for their time and hospitality.

Notes

1. Ayers (2011) defines 'adaptation paradox' as coming to terms with the fact that climate risks are instigated globally but negative impacts experienced locally.
2. Agricultural intermediary refers to labour-brokers, who are the middleman between the producer and the workers. They often both act as an exploiting agent as well as playing a facilitative role for workers to find farm placement. Intermediaries act like an informal buffer for the workers in the absence of access to a formal social security system. Some of the intermediaries have grocery markets back in workers' hometowns and provide food and coal to the workers during the winter in exchange of their labour during the summer months. Such practices inevitably lock-in the worker to the mercy of the intermediaries.

3. Doing fieldwork-based research on seasonal workers in Turkey has always been politically charged since it often attempts to undo the long-established power relations. This is even more valid today in a time when academics who call for a peaceful end to the Kurdish conflict in Turkey are ostracized. See also Küçükkırca (2012) and Uzun (2015).
4. Hence it is not fully groundless when some commentators suggest 'TARSIM produces gains only for the insurance industry, not for the farmers' (Turhan, 2011, p. 9).
5. Within the context of seasonal labour, Laurent (2013, p. 164) calls this phenomenon 'social dumping' or put otherwise 'having the work [done] without [the hassle of] the worker'.
6. A similar but even more severe blow from Russia came in 2016 when the Russian Federation banned vegetables, fruit, and other agricultural products from Turkey in the aftermath of the downing of a Russian jet in Syrian territory by Turkish jets. Although the terms of this import ban were eased later on, the socio-economic aftershocks are still being borne by the producers and workers alike.
7. When asked about the worrying state of inflation in the country, Iran's Khomeini once famously said, '[the] revolution was not about the price of watermelons'. However, the price of watermelons from Iran does matter in Turkey. Due to its favourable weather conditions, Iran can supply both the Turkish domestic market (up to 50 tons/day) and global markets starting from April, thus earlier than Kapı. This inevitably leads to decreased prices for producers in Kapı both in local and global markets.

References

Agrawal, A. and Perrin, N. (2009). Climate adaptation, local institutions and rural livelihoods. In: *Adapting to Climate Change: Thresholds, Values, Governance* (ed. W.N. Adger, I. Lorenzoni and K.L. O'Brien), 350–367. Cambridge: Cambridge University Press.

AGV (2013). *Adana Vizyon 2023: Adana İli Tarımsal Üretim Durum Raporu*. https://web.archive.org/web/20190301092737/ http://www.agv.org.tr/upload/2013/08/adana-ili-tarimsal-uretim-raporu.pdf (accessed 1 March 2019).

Atölyesi, K. (2016). Fertile Lands, bitter lives: the situation analysis report on Syrian seasonal agricultural workers in the Adana Plain. https://www.academia.edu/30811191/Fertile_Lands_Bitter_Lives_THE_SITUATION_ANALYSIS_REPORT_ON_SYRIAN_SEASONAL_AGRICULTURAL_WORKERS_IN_THE_ADANA_PLAIN (accessed 25 February 2019).

Aydın, Z. (2010). Neo-Liberal transformation of Turkish agriculture. *Journal of Agrarian Change* 10 (2): 149–187.

Ayers, J. (2011). Resolving the adaptation paradox: exploring the potential for deliberative adaptation policy-making in Bangladesh. *Global Environmental Politics* 11 (1): 62–88.

Belliveau, S., Smit, B., and Bradshaw, B. (2006). Multiple exposures and dynamic vulnerability: evidence from the grape industry in the Okanagan Valley, Canada. *Global Environmental Change* 16 (4): 364–378.

Birkenholtz, T. (2012). Network political ecology: method and theory in climate change vulnerability and adaptation research. *Progress in Human Geography* 36 (3): 295–315.

Boyd, E., Osbahr, H., Ericksen, P.J. et al. (2008). Resilience and 'climatizing' development: examples and policy implications. *Development* 51 (3): 390–396.

Burton, R.J. and Peoples, S. (2014). Market liberalisation and drought in New Zealand: a case of 'double exposure'for dryland sheep farmers? *Journal of Rural Studies* 33: 82–94.

Çetin, İ. (2016). Labor force participation of Syrian refugees and integration: case of Adana and Mersin cities. *Gaziantep University Journal of Social Sciences* 15 (4): 1001–1016. https://doi.org/10.21547/jss.265320.

D'Alisa, G. and Kallis, G. (2016). A political ecology of maladaptation: insights from a Gramscian theory of the state. *Global Environmental Change* 38: 230–242.

Demircan, M., Gürkan, H., Eskioğlu, O. et al. (2017). Climate change projections for Turkey: three models and two scenarios. *Turkish Journal of Water Science and Management* 1: 22–43.

Derviş, S., Yetişir, H., Tok, F.M. et al. (2009). Vegetative compatibility groups and pathogenicity of verticillium dahliae isolates from watermelon in Turkey. *African Journal of Agricultural Research* 4 (11): 1268–1275.

DPT (State Planning Organization) (2004). *İlçelerin Sosyo-Ekonomik Gelişmişlik Sıralaması Araştırması (Research on Socio-Economic Development Ranking of Districts)*. Ankara: DPT.

Ergönül, Ö. (2007). Correlation between temperature, rainfall and malaria incidence in Turkey. In: *Climate Change and Turkey: Impacts, Sectoral Analyses and Socio-Economic Dimensions* (ed. Ç. Güven), 28–29. Ankara: UNDP Turkey.

Eriksen, S., Aldunce, P., Bahinipati, C.S. et al. (2011). When not every response to climate change is a good one: identifying principles for sustainable adaptation. *Climate and Development* 3 (1): 7–20.

Eriksen, S.H. and O'Brien, K. (2007). Vulnerability, poverty and the need for sustainable adaptation measures. *Climate Policy* 7 (4): 337–352.

Erkan, O. (2000). *Tarım-İş: Adana Sosyo-Ekonomik Rapor (Adana Socio-Economic Report)*. Adana: Adana Chamber of Commerce.

Evrensel (2014). TARSIM zararı karşılamıyor (TARSIM is not paying compensation for the damage). http://www.evrensel.net/haber/82222/tarsim-zarari-karsilamiyor.html (accessed 27 April 2014).

FAO (2014). Food and agricultural commodities production. http://faostat.fao.org/site/339/default.aspx (accessed 24 February 2017).

Fieldman, G. (2011). Neoliberalism, the production of vulnerability and the hobbled state: systemic barriers to climate adaptation. *Climate and Development* 3 (2): 159–174.

Foucault, M. (1980). *Power/Knowledge: Selected Interviews and Other Writings, 1972–1977*. New York: Pantheon Books.

Gertel, J. and Sippel, S.R. (eds.) (2014). *Seasonal Workers in Mediterranean Agriculture: The Social Costs of Eating Fresh*. New York: Routledge.

Grineski, S.E., Collins, T.W., McDonald, Y.J. et al. (2013). Double exposure and the climate gap: changing demographics and extreme heat in Ciudad Juárez, Mexico, local environment. *The International Journal of Justice and Sustainability* https://doi.org/10.1080/13549839.2013.839644.

Guiot, J. and Cramer, W. (2016). Climate change: the 2015 Paris agreement thresholds and Mediterranean basin ecosystems. *Science* 354 (6311): 465–468.

Gümüş, A. (2005). Çukurova'nın Ötekileri (The others of Çukurova). *Tiroj Magazine* 22–26. April–June.

Gürel, B. (2011). Agrarian change and labour supply in Turkey, 1950–1980. *Journal of Agrarian Change* 11 (2): 195–219.

Habertürk. (2017). Altı kabak üstü karpuz. http://www.haberturk.com/ekonomi/is-yasam/haber/1540659-karpuzun-tadi-neden-kabak-tadi-vermeye-basladi (accessed 4 July 2017).

Hacettepe University (2006). *Turkey: Migration and Internally Displaced Population Survey*. Ankara: Hacettepe University Institute of Population Studies.

Harris, L. (2009). Contested sustainabilities: assessing narratives of environmental change in Southeastern Turkey. *Local Environment* 14 (8): 699–720.

Holland, B. (2017). Procedural justice in local climate adaptation: political capabilities and transformational change. *Environmental Politics* 1–22. https://doi.org/10.1080/09644016.2017.1287625.

Holmes, S. (2013). *Fresh Fruit, Broken Bodies: Migrant Farmworkers in the United States*. Berkeley: University of California Press.

Hürriyet (2010). Çiftçi kuraklık teminatı için bastırıyor, sigortacılar 'risk yüksek' diye direniyor (Farmers push for drought coverage, insurers resist saying 'the risk is high'). www.hurriyet.com.tr/ekonomi/14948884.asp (accessed 24 February 2017).

IPCC (2014). Climate Change 2014: Impacts, adaptation, and vulnerability. https://www.ipcc.ch/report/ar5/wg2/ (accessed 24 February 2017).

İslamoğlu, H. (2017). The politics of agricultural production in Turkey. In: *Neoliberal Turkey and its Discontents: Economic Policy and the Environment under Erdoğan* (ed. F. Adaman, B. Akbulut and M. Arsel), 75–102. London: IB Tauris.

Jacoby, T. (2008). The development of Turkish agriculture: debates, legacies and dynamics. *The Journal of Peasant Studies* 35 (2): 249–267.

Kadirbeyoğlu, Z. (2010). In the land of ostriches: developmentalism, environmental degradation, and forced migration in Turkey. In: *Environment, Forced Migration and Social Vulnerability* (ed. T. Afifi and J. Jager), 223–234. Heidelberg: Springer.

Kallis, G. and Zografos, C. (2014). Hydro-climatic change, conflict and security. *Climatic Change* 123 (1): 69–82.

Kapp, K.W. (1950). *The Social Costs of Private Enterprise*. Cambridge, MA: Harvard University Press.

Kavak, S. (2016). Syrian refugees in seasonal agricultural work: a case of adverse incorporation in Turkey. *New Perspectives on Turkey* 54: 33–53.

Keyder, Ç. and Yenal, Z. (2011). Agrarian change under globalization: markets and insecurity in Turkish agriculture. *Journal of Agrarian Change* 11 (1): 60–86.

Keyder, Ç. and Yenal, Z. (2013). *Bildiğimiz Tarımın Sonu: Küresel İktidar ve Köylülük*. İstanbul: İletişim Yayınları.

Kıray, M. (1974). Social change in Çukurova: a comparison of four villages. In: *Turkey: Geographic and Social Perspectives* (ed. P. Benedict, E. Tümertekin and F. Mansur), 179–203. Leiden: Brill.

Küçükkırca, İ.A. (2012). Etnisite, Toplumsal Cinsiyet ve Sınıf Ekseninde Mevsimlik Kürt Tarım İşçileri. *Toplum ve Kuram* 6: 1–16.

Kuşçu, H., Turhan, A., Özmen, N. et al. (2015). Deficit irrigation effects on watermelon (Citrullus Vulgaris) in a sub humid environment. *JAPS: Journal of Animal & Plant Sciences* 25 (6): 1652–1659.

Laurent, C. (2013). The ambiguities of French Mediterranean agriculture: images of the multifunctional agriculture to mask social dumping? *Research in Rural Sociology and Development* 19: 149–171.

Le Mouel, C., Forslun, D.A., Marty, P. et al. (2016). Climate change and dependence on agricultural imports in the MENA region. In: *The Mediterranean Region under Climate Change: A Scientific Update* (ed. S. Thiébault and J.P. Moatti), 511–517. Marseille: IRD Editions.

Leichenko, R. and O'Brien, K. (2008). *Environmental Change and Globalization: Double Exposures*. Oxford: Oxford University Press.

Lelieveld, J., Proestos, Y., Hadjinicolaou, P. et al. (2016). Strongly increasing heat extremes in the Middle East and North Africa (MENA) in the 21st century. *Climatic Change* 137 (1–2): 245–260.

Martinez-Alier, J. (2002). *The Environmentalism of the Poor: A Study of Ecological Conflicts and Valuation*. Cheltenham: Edward Elgar Publishing.

McMichael, P. (2009). A food regime genealogy. *The Journal of Peasant Studies* 36 (1): 139–169.

Milman, A. and Arsano, Y. (2014). Climate adaptation and development: contradictions for human security in Gambella, Ethiopia. *Global Environmental Change* 29: 349–359.

Mitchell, D. (2012). *They Saved the Crops: Labor, Landscape, and the Struggle over Industrial Farming in Bracero-Era California*. Athens: University of Georgia Press.

MOEU (Ministry of Environment and Urbanization) (2011). Turkey's National Climate Change Adaptation Strategy and Action Plan (Draft). Ankara MOEU. http://www.forclimadapt.eu/sites/default/files/TURQUIE.pdf (accessed 24 February 2017).

MOFAL (Ministry of Food, Agriculture and Livestock) (2015). Tarımsal Veriler. Ankara: MOFAL. www.tarim.gov.tr/Belgeler/SagMenuVeriler/Tarimsal_Veriler.pdf (accessed 24 February 2017).

Nightingale, A.J. (2017). Power and politics in climate change adaptation efforts: struggles over authority and recognition in the context of political instability. *Geoforum* 84: 11–20.

O'Brien, K.L. and Leichenko, R.M. (2000). Double exposure: assessing the impacts of climate change within the context of economic globalization. *Global Environmental Change* 10 (3): 221–232.

O'Brien, K.L. and Leichenko, R.M. (2003). Winners and losers in the context of global change. *Annals of the Association of American Geographers* 93 (1): 89–103.

OECD (2017). Crop production (indicator). doi: 10.1787/49a4e677-en (accessed 24 February 2017).

Özbek, A. (2007). New Actors of New Poverty: The "Other" Children of Çukurova. Master's Thesis, Department of Sociology, Middle East Technical University: Ankara, Turkey.

Özkan, B. and Akçaöz, H. (2003). Impacts of climate factors on yields for selected crops in Southern Turkey. *Mitigation and Adaptation Strategies for Global Change* 7: 367–380.

Peet, R., Robbins, P., and Watts, M. (eds.) (2011). *Global Political Ecology*. London: Taylor & Francis.

Pelek, D. (2018). Syrian Refugees as Seasonal Migrant Workers: Re-Construction of Unequal Power Relations in Turkish Agriculture. Journal of Refugee Studies. DOI: 10.1093/jrs/fey050.

Robbins, P. (2004). *Political Ecology: A Critical Introduction*. Malden, MA: Wiley.

Sachs, W. (ed.) (2010). *The Development Dictionary: A Guide to Knowledge as Power*, 2e. London: Zed Books.

Şafak, Y. (2008). Rusya'nın takıntısı karpuzu bitirdi (Russia's obsession destroyed watermelon). http://yenisafak.com.tr/ekonomi-haber/rusyanin-takintisi-karpuzu-bitirdi-8.6.2008%200-121955 (accessed 24 February 2017).

Selek, B. and Tuncok, I.K. (2014). Effects of climate change on surface water management of Seyhan Basin, Turkey. *Environmental and Ecological Statistics* 21 (3): 391–409.

Şen, Ö.L., Önol, B., Bozkurt, D. et al. (2011). Seyhan Havzası için İklim Değişikliği Projeksiyonları (Climate Change Projections for Seyhan Basin). Unpublished project report prepared for UNDP Turkey under MDG-F 1680: Enhancing the Capacity of Turkey to Adapt to Climate Change Project.

Sol (2007). Çukurova'da kazanan tarım işçileri oldu (Agricultural Workers Won in Çukurova). http://arsiv.sol.org.tr/index.php?yazino=26317 (accessed 24 February 2017).

Sütoluk, Z., Tanır, F., Savaş, N. et al. (2004). Assessment of health status of seasonal agricultural workers. *TTB Mesleki Sağlık ve Güvenlik Dergisi* 17: 34–38.

Taylor, M. (2013). Climate change, relational vulnerability and human security: rethinking sustainable adaptation in agrarian environments. *Climate and Development* 5 (4): 318–327.

Taylor, M. (2015). *The Political Ecology of Climate Change Adaptation: Livelihoods, Agrarian Change and the Conflicts of Development*. London: Routledge.

Taylor, M. (2017). Climate-smart agriculture: what is it good for? *Journal of Peasant Studies* https://doi.org/10.1080/03066150.2017.1312355.

Tezcan, L., Ekmekçi, M., Atilla, Ö. et al. (2007). Assessment of climate change impacts on water resources of Seyhan river basin. ICCAP Report. www.chikyu.ac.jp/iccap/finalreport.htm (accessed 24 February 2017).

Toksöz, M. (2010). *Nomads, Migrants and Cotton in the Eastern Mediterranean: The Making of the Adana-Mersin Region 1850–1908*. Leiden: Brill.

Tören, T. (2007). *Yeniden Yapılanan Dünya Ekonomisinde Marshall Planı ve Türkiye Uygulaması*. İstanbul: Sosyal Araştırmalar Vakfı Yayınları.

Tsujii, H. and Erkan, O. (2007). The final report of the socio-economic sub-group of the ICCAP Project: ICCAP Project Report. www.chikyu.ac.jp/iccap/finalreport.htm (accessed 24 February 2017).

Tuncok, I.K. (2016). Drought planning and management: experience in the Seyhan River Basin, Turkey. *Water Policy* 18 (S2): 177–209.
Turhan, E. (2011). *Policy Analysis of Turkish Agricultural Insurance and National Strategy and Action Plan on Combating Agricultural Drought*. Barcelona: CLICO Project Deliverable, Universitat Autònoma de Barcelona.
Turhan, E., Zografos, C., and Kallis, G. (2015). Adaptation as biopolitics: why state policies in Turkey do not reduce the vulnerability of seasonal agricultural workers to climate change. *Global Environmental Change* 31: 296–306.
UNDP Turkey (2009). Livelihood analysis in Seyhan river basin: Draft report for MDG-F 1680 UN joint programme (unpublished report).
Uzun, E. (2015). Kürt Fındık İşçileri: Bir Karşılaşma Mekânı Olarak Akçakoca. *Moment Dergi* 2 (1): 100–132. https://doi.org/10.17572/mj2015.1.100132.
Watanabe, T. (2007). Summary of ICCAP: Framework, outcomes and implications of the project. www.chikyu.ac.jp/iccap/ICCAP_Final_Report/1/7-summary.pdf (accessed 24 February 2017).
World Bank (2019). Agriculture, forestry, and fishing, value added (% of GDP), https://data.worldbank.org/indicator/NV.AGR.TOTL.ZS?locations=TR (Accessed on 17.05.2019).
Yano, T., Aydın, M., and Haraguchi, T. (2007). Impact of climate change on irrigation demand and crop growth in a Mediterranean environment of Turkey. *Sensors* 7 (10): 2297–2315.
Yercan, M. and Işıklı, E. (2006). International competitiveness of Turkish agriculture: A case for horticultural products. Conference paper presented in the 98th EAAE Seminar, Marketing Dynamics within the Global Trading System: New Perspectives, Crete, 29 June–2 July.
Zografos, C. (2017). Flows of sediment, flows of insecurity: climate change adaptation and the social contract in the Ebro Delta, Catalonia. *Geoforum* 80: 49–60.
Zografos, C., Goulden, M.C., and Kallis, G. (2014). Sources of human insecurity in the face of hydro-climatic change. *Global Environmental Change* 29: 327–336.

19

Drought Governance in Catalonia: Lessons Learnt?

Alba Ballester[1] and Abel La Calle[2]

[1] *Autonomous University of Barcelona-Institute of Government and Public Policies, Barcelona, Spain*

[2] *Autonomous University of Barcelona, Institute of Government and Public Policy, Barcelona, Spain*

19.1 Introduction

The drought crisis in Catalonia in 2008 is a paradigmatic case of drought management in Spain, not for its effectiveness and success but due to the lack of effective advance planning, institutional coordination and transparency, as well as the use of a critical situation in order to adopt measures which could not outweigh ordinary social, economic, and environmental controls.

Catalonia is one of the 17 Autonomous Communities, regional authorities with a decentralization of power similar to that of the Federal States, in Spain. Nevertheless, the map of these regional authorities, made up of political-administrative divisions, does not coincide with the map of water competencies corresponding thereto or with that of river basins. Therefore, water and drought management in Catalonia is divided hydrological wise into two management areas: inland river basins of Catalonia (IBC) and Catalan river basins of the Ebro (CBE), whose management responsibilities lie with the Agència Catalana de l'Aigua (ACA, Catalan Water Agency-Autonomous Community) and the Confederación Hidrográfica del Ebro (CHE, Ebro River Basin Organization-State) respectively. Drought management in Catalonia is both regulated by State regulations, relating to the entire Spanish State, and specific autonomous legislation related to the IBC.

Water management in Catalonia, and in particular the supply to the Regió Metropolitana de Barcelona (Metropolitan Area of Barcelona, MAB), is embroiled in a historical social conflict, motivated to a great extent by the utilization of external flows to the MAB in order to guarantee the supply of the over 3.3 million inhabitants who live in this region. Two main social movements take part in the social conflict, one within the hydrographical area of the IBCs and another in the CBEs. The first movement stems from the deterioration of the quality of the river ecosystems and the perception of territorial imbalance derivative from the Ter river basin transfer, the

Facing Hydrometeorological Extreme Events: A Governance Issue, First Edition.
Edited by Isabelle La Jeunesse and Corinne Larrue.

main supply system of the MAB. Elsewhere, there is the constant threat of alternative supply proposals which, without having been implemented, have a major social reaction, such as the transfer of the Ebro or Segre rivers to the MAB. Whenever these proposals are publicly posed and considered, in particular that in relation to the Ebro, the debate and conflict stands on a wider scale, first of all given that it is a transfer between basins of different public (regional and state) competence and jurisdiction, and secondly given that the Ebro transfer is a reference benchmark in the history of water conflicts in Spain that activated the debate as regards hydraulic mega-projects designed in the 1930s to stock Spain and the underlying conflicts to this type of proposals such as: the territorial imbalance, the economic growth model as well as the concept of general interest or 'surplus resources', the constraints faced by governments in the management of complexity and uncertainty, as well as the absence of scientific certainties, and simultaneously the existence of social demands for the construction of a common project with a broader community of players (Hernández-Mora et al. 2014).

The critical analysis of the drought crisis in 2007–2008 in Catalonia sheds light on drought governance in Spain, and as a result of the importance of certain decisions, who should determine the decisions and measures to be taken in which situations, what decision-making method is the most appropriate for each type of decision, the manner in which participatory management can be guaranteed, or how a drought situation should be notified. Taking this analysis as a starting point we will seek to explore to what extent the analysed crisis has entailed a change in drought management in Catalonia, and what are the main characteristics of management 10 years after that drought period.

19.2 Drought management in Spain

The Mediterranean hydro-climatic reality is characterized by low annual measured rainfall (300–450 mm). Climatic normality is defined as the irregular nature of the rainfall distribution, both in terms of space and time, resulting in the dry years to be lower than half the average rainfall. It is likewise usual to find a period of two or three consecutive years with below-average rainfall. Another characteristic is the presence of a significantly dry season, in which periods of many consecutive months without rain, or insignificant rainfall for the purposes of the replenishment of soil moisture and the reactivation of the river flows are normal. Moreover, the dry years are generally accompanied by high temperatures, which contribute to an increase of the effects of pluviometric drought and to decrease the hydrological efficiency of the rains (Martínez 2006).

In the Spanish contextual reality, the most severe droughts are those ensuing from the increased stress that has been introduced into the ecosystems, by a demand and consumption positioned over and above that which the ecosystem could normally cope.

Drought situations have often been used as a political weapon, and it is common to find urgent decision-making with approach deficiencies, generally accompanied by the creation of ad hoc expert committees. Crisis scenarios facilitate the adoption of emergency measures on behalf of the general interest, which are not sufficiently evaluated and in some cases have been implemented despite the fact that the exceptional nature of the circumstances have ceased with the arrival of the first rains (Urquijo, De Stefano, and La Calle 2015).

19.2.1 Legal framework

In Spain the first legal regulation of a general nature on drought accompanies the birth of a liberal State and is reflected in the laws of 1866 and 1879. This legislation perceived drought as a situation of a temporary water shortage, susceptible to varying degrees of intensity and whose manifestation authorized the Administration to take exceptional measures such as the expropriation of the waters necessary for water supplies.

The 1985 Water Law continued with a similar concept, qualifying that 'extraordinary drought' is a state of necessity, emergency, urgency, or anomalous or exceptional situation, and broadened the authority of the Administration so as to adopt 'the measures which are required in relation to the use of the public hydraulic domain'.

The incorporation of Spain to the European Community in 1986 did not produce any amendment of the drought regime in Spanish legislation.

Following the drought of the early 1990s, Law 9/1996 amended the 1985 Water Law and introduced two amendments which made it possible to demand efficiency in the use of water for water supplies and irrigation and to tighten penalties for cases of non-compliance of the measures adopted in cases of extraordinary drought. On the grounds of the substantiation of the aforementioned drought, Law 46/1999 amended the 1985 Water Law in the 'pursuit for alternative solutions' that make it possible to increase water production with desalination or reutilization and to enhance efficiency through the easing of the concessional regime.

In tandem with the adoption of the Water Framework Directive (Directive 2000/60/EC) in Spain, the Law on National Hydrological Plan (NHP) was adopted. Said Law 10/2001 stipulates in its explanatory statement that 'drought management' is one of the measures that underscores the interest to ensure a rational and sustainable use of water resources. A mandate for the development of guidelines and instruments for drought management was established in its body of articles.

Law 10/2001 mandated the Government to establish, within two years, the coordination criteria for the revision of river-basin management plans, amongst other matters in drought management. The Government adopted these criteria in the Hydrological Planning Regulations (Royal Decree 907/2007). The aforementioned law likewise established: a 'global system of hydrological indicators' for the prediction, forecast, and reference thereof in the formal declaration of 'alert situations and temporary drought', the special plans of action in situations of alert and temporary drought in the aspects of the water basin management plans, and emergency plans for drought situations to be carried out in the water supply of populations greater or equal to than twenty thousand inhabitants.

The consolidated text of the 2001 Water Law likewise introduced an amendment in the over-exploited aquifers or at risk of being over-exploited regime by permitting the authorization of further extractions in drought situations.

Law 62/2003 intended to 'transpose Directive 2000/60/EC into Spanish law' but in fact this bringing into line was insufficient as it was subsequently proven (see Court of Justice case nos. C-516/07 and C-151/12), nevertheless, a new reference as regards drought by broadening the objectives of water protection, including, inter alia, 'mitigating the effects of floods and droughts' was specified therein.

With the onset of the 2004–2005 drought and the plans prescribed in Law 10/2001 had still not yet been approved, nor had the regulation on permissible environmental deterioration limits in the event of drought set forth in the Water Framework Directive (La Calle 2005). The National Drought Observatory was set up in an accelerated manner, both action protocols in the event of drought in river basin districts, and exceptional legislative and regulatory measures were adopted.

In 2007, the special action plans in situations of alert and temporary drought (Ministerial Order MAM/698/2007) and the Hydrological Planning Regulations (Royal Decree 907/2007) were approved which partially supplemented the deficiencies in the transposition of the Water Framework Directive and compiled the regulation of the temporary deterioration of the state of water bodies caused by prolonged droughts (Article 38).

So far, a review has been made of the evolution of the drought legal framework in Spanish law until the drought period which is the subject matter of this chapter occurred. But this analysis must not be concluded without setting forth certain critical conclusions which serve to comprise said framework.

The first conclusion is the misuse by different Spanish governments that occurred upon the enactment in an urgent and exceptional manner of certain hydraulic works alleging grounds of guarantee of supply. This political rhetoric has been analysed by Julia Urquijo (Urquijo et al. 2015) who highlight the often unjustified nature of the recourse to exceptionality.

The second is that there is a deliberate confusion between natural indicators (rainfall, soil moisture, and so on) with management indicators (reservoir filling levels), which on the one hand prevents obtaining objective knowledge of the natural origin of the drought and on the other enabling a voluntary modification of those facts that determine the situation of drought. This problem was already evident prior to the establishment of the indicators in the Special Drought Plan (SDP) but was overlooked (La Calle 2005).

The third is that there is a blatant noncompliance of the general obligation of active and real public participation in the implementation of the Water Framework Directive (Article 14) due to failure to stipulate the manner of implementing that participation in drought-related measures.

19.2.2 Administrative organization

In order to understand water policy in Spain it is necessary to take into account the existence of the three tiers of decentralization of power: the General State Administration, the Autonomous Communities, and the Local Entities.

The 1978 Spanish Constitution created a system of division of competences shared between the General State Administration and the Autonomous Communities. This implies, on the one hand, that the state sets forth basic regulations on water that the Autonomous Communities may implement. And on the other, as far as river-basin management and planning is concerned, entails the conferral of competences according to river basins, irrespective as to whether or not these basins included more than one or several Autonomous Communities. The river-basin districts which comprise more than one Autonomous Community and therefore are under the jurisdiction of the General State Administration are Guadalquivir, Segura, Jucar, and Western Cantabrian (located in Spain); and Miño-Limia, Eastern Cantabrian, Douro, Tagus, Guadiana, Ebro, Ceuta, and Melilla (shared with other countries).[1]

The Autonomous Communities may assume competences in water-related matters to enact implementing regulations of the basic state legislation and for the planning and management of the river basins which are entirely in its territory. Thus, there are Autonomous Communities that have enacted their own water laws and have assumed the competence of watersheds that are entirely in their territory, especially in the case of Andalusia, Balearic Islands, Canary Islands, Catalonia, Galicia, and the Basque Country. The case of the Cantabrian Eastern River Basin District is a particular case given that it is a jointly managed by the General State Administration and the Basque Country.

The 1978 Constitution (Spanish Constitution 1978) likewise prescribes an autonomous autonomy, albeit does not stipulate which competences as regards water correspond thereto. The legislation has specified the competences in this area that focus on the services of supply and sanitation of urban agglomerations. Notwithstanding the competences of the Autonomous Communities which have been used to a certain extent as regards sanitation.

19.3 Drought management in Catalonia

Catalonia is characterized by a Mediterranean climate having a high inter-annual variability and very irregular rainfall throughout its territory, which tend to increase further north and to decrease further south. The Autonomous Community is divided into two hydrographical areas, the IBC,[2] which represent 52% of the Catalan territory, and CBE,[3] corresponding to 48% of the remaining territory (ACA 2008) (Figure 19.1). Between both hydrographical areas there are different management responsibilities, the Catalan Water Agency is the competent regional authority in the IBCs, and the Ebro River Basin Organization is the competent state authority in the CBEs.

Of the total population of Catalonia, 92% live in the IBC and are concentrated mainly in two Besòs and Baix Llobregat river basins, where domestic demand represents 70% of the total domestic demand of the IBCs. In the CBEs the highest demand derives from irrigation in the Ebro and Segre basins, accounting for 89.1% of total demand (ACA 2008).

One of the greatest historical hydrological pressures in Catalonia has been the lack of guarantee of water supply for the MAB, located in the IBC. The volume of reserves in the entire IBC is very close to the demand volume. This situation is particularly paradigmatic in the case of the Ter-Llobregat system, responsible for supplying water to the MAB, in which a 'structural deficit' of 80 hm^3/year is estimated (ACA 2008). This means that even if the utilization of the system may be exhaustive (for example, saving, optimization, regulation) it will still remain a system with hydric stress in relation to available resources and existing demands. It is estimated that these demands may exceed the reserves in the event of two hydrological dry years occurring (ACA 2008). Moreover, consumption per inhabitant per day is at 120l/inhab/day, a reason why demand elasticity in the MAB is low.

These circumstances have led to tensions and water conflicts throughout the supply history of the MAB, where the search for new supply sources to meet the continuous growth of demand has been constant. These conflicts are generated to a large extent as regards to three transfer projects: The Ter river transfer, currently

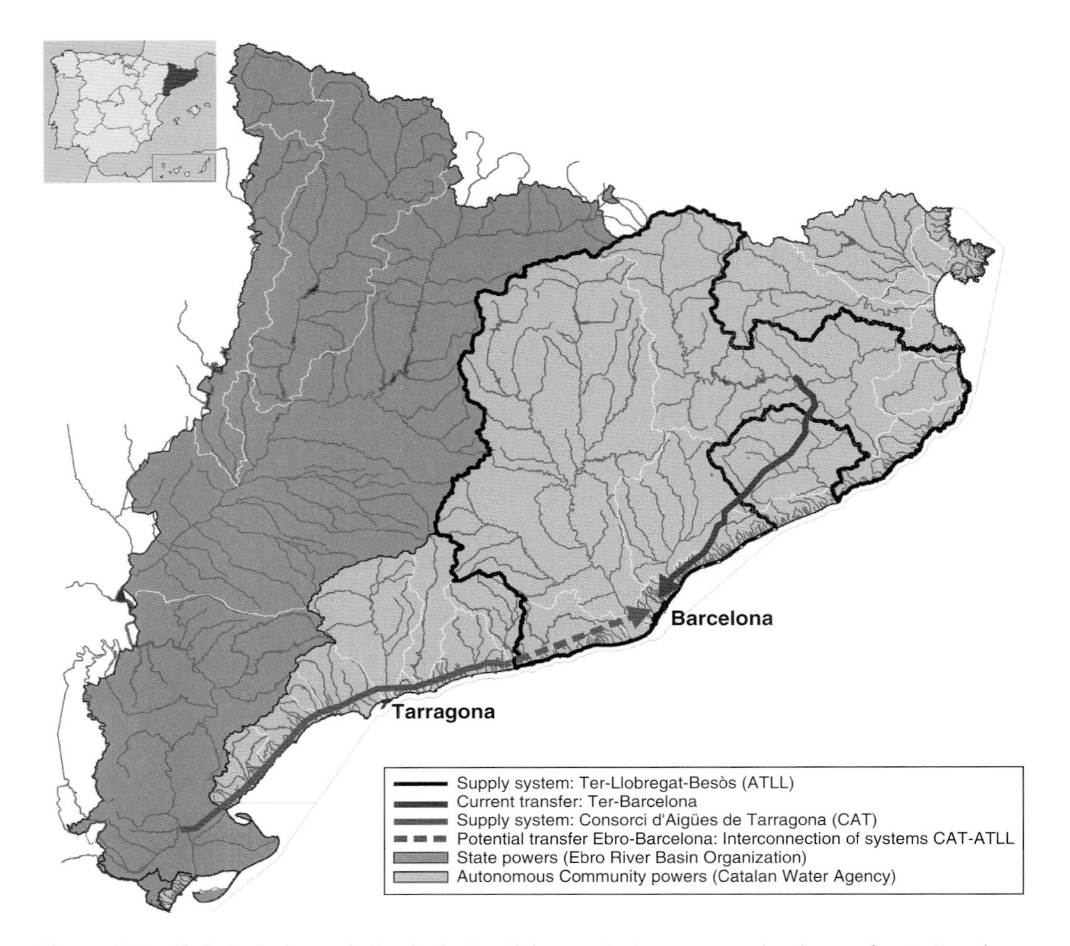

Figure 19.1 Hydrological areas in Catalonia, Ter-Llobregat-Besòs system, and main transfers to Barcelona.

underway, and the transfer projects of the Segre and Ebro rivers (Figure 19.1), the latter through the interconnection of networks of the Tarragona Water Consortium, which uses waters from the Ebro River, with the Ter-Llobregat-Besòs system. In all three cases the social response has been widespread and particularly confrontational during drought periods, moments in which said projects are subjected again to renewed debate.

The administrative management of these conflicts by the autonomous government is especially complex due to its disparity of competences in the hydrological and administrative area. On the one hand, the government must be able to manage a conflict in its territory, but nevertheless does not have any hydrological management competences in the specified area of its territory within the basin of the Ebro. This fact implies an institutional coordination challenge between the different competent water administrations in the Catalan territory, which to date has not yet taken place and which has caused tension in the last periods of drought in this territory.

19.4 Drought crisis in Catalonia 2007–2008

An insight into the drought crisis in 2007–2008 necessarily implies an understanding of the socio-institutional conflict ensuing from crisis management. To that end, work has been carried out as regards a selection of 1000 press reports during the drought period, in 10 media: *el País*, *Periódico, la Vanguardia*, *Público* and *Mundo* at state level, and *Avui, el Punt, Segre* at regional-Catalan level, as well as two online publications: Europa Press and iAgua.

The following sections specify the salient chronological aspects of the drought period, the public debates held, and the stakeholders concerned with the objective of understanding the evolution of the drought period and the conflict, as well as the dimension taken by them on a state level. Understanding this conflict enables the identification of key aspects to be taken into account for drought governance in Catalonia, which can be reproduced to a greater or lesser extent in other Spanish territories. Furthermore we review the state of drought management in Catalonia 10 years after the crisis and explore the changes that have taken place.

19.4.1 Chronological milestones

As regards the drought crisis in Catalonia between 2007 and in particular 2008, four key salient chronological aspects can be distinguished: (i) The identification of the potential problem of lack of supply in January 2007 with the 51.6% of reserves, together with the enactment of the Drought Decree 84/2007 by the Catalan government, and the creation of the Permanent Drought Committee (PDC) and Drought Management Committee (DMC); (ii) the adoption of corresponding measures pursuant to D84/ 2007 and extension of the PDC, the onset of exceptional situation level 1 in all IBCs in August 2007, with reserves at 41% of capacity; (iii) the onset of exceptional situation level 2, with reserves of 23.6% in February 2008, together with the proposal for emergency measures alternatives, this period is marked by political and socio-territorial conflict, and the authorization of the Ebro transfer in the Royal Decree Law (RDL) 3/2008 by the state government; (iv) the cessation of exceptional situation level 2 in June 2008, having values of 62.3% of its reserves, and slow advancement in rendering null and void and subsequent repeal of RDL 3/2008 (Figure 19.2).

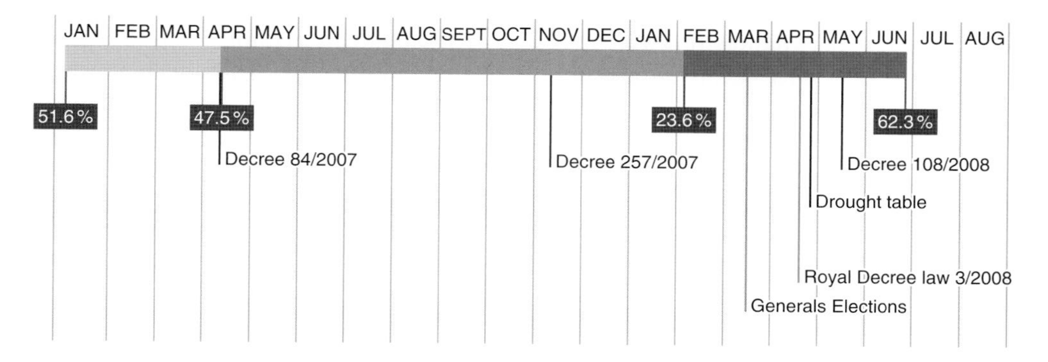

Figure 19.2 Chronological milestones of drought crisis in Catalonia 2007–2008.

According to the values prescribed in Decree 84/2007 a hydrological emergency situation, which is currently set at 20% of the total supply capacity in the IBC, never occurred nevertheless the measures did respond to an emergency situation, possibly not only as a response to the water stress situation but also to the social conflict derivative from the crisis management and alternatives proposals to avoid supply shortages in the MAB.

19.4.2 Public debates, adoption of measures, and actors involved

The public debate on alternatives to drought-risk management was concentrated during the Exceptionality 2 period. The first public proposals for alternative measures to the transfers to guarantee the supply to the MAB were made in early February 2008, when the ACA proposed the creation of an exchange centre in the Segre (CBE) to mitigate the water deficit in Barcelona. This proposal sparked the first reactions from various stakeholders in the Segre, in particular the Urgell irrigation association who rejected the transfer of water from the headwaters, but who in turn demonstrated their willingness to stop irrigation if they were guaranteed financial compensation.

During February this situation was marked by the public manifestations of those holding political offices involved in the problem where the first signs of conflict appeared. Elsewhere, the Partit Polular de Catalunya (Popular Party of Catalonia) submitted a motion to the Parliament of Catalonia in favour of the interconnection of 'Consorci d'Aigües de Tarragona and Aigües Ter-Llobregat' (CAT-ATLL) water systems, to which the remaining Catalan political parties voted against. In relation to this proposal, the president of the CAT publicly stated his rejection of the interconnection of the CAT-ATLL systems, at the same time that the Tarragona City Council offered surplus resources from several of its wells to supply Barcelona. In parallel, entrepreneurs from the municipality of Girona and the newly created Plataforma del Ter (Ter social Platform) joined the debate. The former demanding the transfer of the Segre to free up the Ter, and the latter demanding the partial return of the Ter water flows.

Meanwhile, reserves fell to 22.1%, and mitigating measures and savings campaigns continued, but no public demonstrations of solution proposals were made until after the 9 March general election in Spain, a date which became a turning point in the management undertaking of the crisis.

Five days after the general election, a news item was released notifying the appearance of certain pegs delineating a circular water-catchment area connected to the Segre, supposedly related to a transfer project of the Segre (within CBE) from its headwaters towards the MAB (within IBC), although said fact was denied in the Catalan Parliament and by the head for the Environment of the Government of Catalonia. It was he who publicly presented two weeks later this project as the solution to the emergency situation: the transfer would supply water in the amount of 4 hm^3/month and would entail a total cost of €45 M, and have a works execution deadline of five months in order to be ready before October, at which time when the need for residential water cut-offs was foreseen.

Two days after the Segre transfer proposal and before the first wave of criticisms as regards the project, the Spanish Minister who was still then acting Minister for the Environment rejected the proposal alleging that the competences to be adopted for that decision were that of the state, an opinion to which first vice president of the state joined in on categorically stating that there was no transfer nor would there ever be a transfer; and finally the president of the Spanish government stated that the transfer was not the alternative nor the right solution.

Amongst the voices sounding their opposition to the transfer of the Segre from the headwaters were: Ecologistas en Acció de Catalunya (Ecologists in Action of Catalonia), who publicly manifested their opposition to all transfers; Convergencia y Unió (political party) which in turn demanded a transfer from the Rhône river; social movements and irrigation association communities of the Terres de l'Ebre, who not only opposed the transfer of the Segre but also the contribution of water from the aquifers of Tarragona, claiming that if Tarragona had water to offer Barcelona, why was the current transfer to Tarragona (CAT) necessary; and the irrigation association of Lleida, who declared that they did not have enough water for themselves let alone others.

Elsewhere, the reassertion of a proposal made in 2004, certain members of the scientific group linked to the Fundación Nueva Cultura del Agua (Foundation for a New Water Culture) (FNCA), proposed an exchange centre in the Segre as one of the most efficient solutions and similarly considered that it had less impact than the proposed transfer through the Cadí Tunnel. This proposal was firmly rejected by the Plataforma en Defensa de l'Ebre (Platform in Defence of the Ebro), which considered such a proposal as a backstop to the transfer of the Segre.

In light of the Spanish government's refusal to the Segre project, the head for the Environment proposed the contribution of water via rail a supplementary measure to the contribution via boat as already contemplated. Meanwhile, the President of the Catalan Government demanded solutions from the Spanish Government, which ended by the latter announcing at the beginning of April the proposal for the interconnection of CAT-ATLL networks, which connected a water transfer to Tarragona (CAT) with Barcelona. Prior to this proposal there was an initial confrontation between the autonomous government and the state government, but the former rejected the Segre proposal to finally endorse the Ebro proposal. The Ebro transfer would entail a cost of €178 M, a maximum monthly contribution of 4.16 hm^3/month, and a works execution deadline of 21 months according to the ACA, and six months according to the Ministry.

The Spanish government, aware of the negative reactions which would imply its transfer proposal of the Ebro to Barcelona, and after having set aside the proposal with the derogation of the Ebro transfer envisaged in the National Hydrological Plan, diligently stated that the interconnection proposal was not any type of transfer. First it was argued that it was not intended to supply water permanently, but was intended as an emergency and reversible water capture-collection. This argument failed to convince the social movements. It was subsequently called again into question prior to the approval of the RDL where it was stated that what would be used would be those water savings ensuing from the upgrading of the irrigation in the Ebro delta (works for which financial allocations were foreseen), but finally, after the legal report of the State's legal counsel wherein a new transfer was rejected, it was stipulated in the RDL that the water that was wished to be transferred would ensue from the 40 hm^3 not used by the CAT (from the concession of 120 hm^3/year that CAT annually purchases from irrigation association of the Ebro, the average annual consumption is 80 hm^3/year). Therefore, according to the Government, there would be no additional drainage of water from the Ebro, but simply an authorization to carry water from the current transfer further.

Table 19.1 Level of government and position as regards the Ebro transfer.

Level	Typology of groups	Groups	F	A
State	Public authorities	General State Administration	X	
	Social groups	Ecologists in Action		X
	Scientific community	Foundation for a New Water Culture (FNCA)		X
Region	Public authorities (autonomous communities)	Cataluña	X	
		Aragón		X
		Valencia	X	
		Murcia	X	
		Castilla la Mancha	X	
	Social groups	Platform in defence of the Ebro river		X
		Catalan network for a new water culture		X
		Committee of people affected by large dams and transfers		X
	Economic sector	Irrigators of the Ebro	X	X
		Irrigators of the El Egido		X
		Irrigators of the Tajo-Segura	X	
		Irrigators of the Rhone: Offering their water.		
Local	Public authorities (municipalities)	Girona	X	
		Barcelona	X	
		Terres de l'Ebre		X
		Platform of mayors from Murcia and Valencia		X
	Social groups	Metropolitan platform against transfers (Barcelona)		X
	Economic sector	Some important entities of the business sector of Barcelona	X	
		Organized business sector of Valencia	X	

F = In favour, A = Against.

Notwithstanding the substantiation and explanation attempts by the Spanish Ministry and the Government of Catalonia, this proposal generated a major social mobilization in the Les Terres de l'Ebre. Likewise reactions ensued at the autonomous level by environmental groups and platforms, and the conflict was extended to other autonomous territories, such as Valencia, Murcia, Aragon, or Castile-La Mancha, and the stakeholders in these territories.

As a summary, the conflict during the 2008 drought crisis focused on four stakeholder groups: social groups, environmental organizations, and trade unions; economic sectors; public administrations and political parties; and the scientific community. Contrary positions could be found within the different groups which each group comprised, as well as between different groups and collectives. (Table 19.1) summarizes the main groups and positions to the Ebro transfer proposal.

19.4.3 Reflections on governance

An analysis of the crisis enables the undertaking of a reaffirmation of the need to define new management models that are capable of integrating all dimensions of a problem and to proffer a more coherent response to the current reality. A certain number of the most significant debates on drought governance in Spain are reflected in the drought crisis management in Catalonia in 2007–2008.

19.4.3.1 Multilevel management, with no institutional coordination or public participation The legal framework, the administrative organization of drought in Spain, and the distribution of water management competences in Catalonia, places the debate on drought in Catalonia at different levels of those governments involved in the management of a public, local, autonomous, and state policy, and in each of the policies in various departmental areas (i.e. water, agriculture, territorial policy, and so on). Coordination between public authorities when proposing supply problem solutions was very limited. This lack of coordination contributed to the generation of conflicts between them, in particular between the autonomous and state governments, as diverse political interests became evident in the alternative proposals.

The fact that different public authorities come into play when making decisions requires coordination between them, and we are of the opinion that this occurs when there previously has been a continued cooperation with a technical and administrative apparatus which prompts it, as when everything depends on ad hoc policy decisions that lobby groups have managed to impose on a particular solution. As in the case at hand, we are of the opinion that the existence of participatory action protocols designed and agreed upon beforehand to the drought period would have been effective.

Table 19.1 likewise demonstrates how different stakeholders are involved in the conflict at various levels as well as in diverse Spanish territories. It has been seen how a proposed transfer of the Ebro can arouse mistrust, rejection, and an intense political debate on a state level, which introduces other hydrological debates, such as the Tagus-Segura transfer. Although in the case of the 2008 drought, a solution to an emergency situation was sought, to propose a transfer of the Ebro, in Spain implied initiating a much broader debate, which transcends the autonomous ambit and the water policy, in which the interests of different autonomous communities, political parties, and stakeholders come into play. These stakeholders joined the debate in a reactive manner in light of the different alternative proposals formulated by the governments during the crisis, provoking the need for a previous and extensive public debate on the alternative proposals.

19.4.3.2 Lack of transparency and political and technical distrust If the political management carried out in the different transfer projects prior to the 2008 drought had fuelled a mistrust of the political decisions and the lack of credibility of them, the confused management of the drought period encouraged those stakeholders. In the political sphere, there was a lack of information transparency as to what was happening and as regards the proposals which were being discussed; an example of this was the nondisclosure of information pending the holding of general elections. A lack of coherence of certain political parties with their ideas was likewise perceived, for example, the very same political parties that were proposing the Segre and Ebro transfers were publicly declared as anti-transfer advocates during the Ebro transfer debate in the NHP.

With the historical background of mistrust acquired by the different transfer proposals, and given the lack of transparency and coherence of policy makers in crisis management, the stakeholders' voices of disbelief grew in light of the proposals for reversible, temporary, or detachable pipelines. There were likewise statements which predicted that such an infrastructure would entail opening the floodgates to permanent water transfer to Barcelona, but similarly to the south (Spanish East Coast). Another sign of this mistrust, coupled with the great difficulty of disclosing the facts clearly and justifying the measures taken, was the questioning, by certain social movements, of whether or not the drought was an exceptional drought and whether the proposed infrastructure was justified or not. This mistrust was reinforced when the President of the Catalan government declared that the construction the works did not include an obligation for its use, at a time when the accumulated rainfall during May had rendered null and void the reasons stated in the RDL for construction of the interconnection works. Perhaps this was a good time to positively reinforce the government's credibility if it had been undertaken in due time that neither the works nor the transfer were necessary.

Particularly in the face of a conflict situation, it would seem reasonable to explain clearly to the general public the grounds that lead us to a crisis situation, to report on what thesis underpinned the decision-making, on the process of delimiting indicators for the determination of different drought scenarios, and when a drought can be considered as prolonged, infrequent, or exceptional. It would likewise seem reasonable to have explained all possible measures, and to analyse and discuss in detail the possible impacts, costs, and benefits of each. And at the time when exceptional measures were taken, the criteria for determining said exceptionality should have been explained in a transparent and thorough manner and demonstrated that the measures were not an impediment to the compliance of the environmental objectives to which we were bound by the new EU legislative framework.

A period of crisis may sometimes entail an opportunity for change. The 2008 drought experience should invite reflection as regards the reaction of the general public in light of political irresponsibility and the amount and diversity of information disseminated. But it would be more accurate to think that an opportunity to better manage this period of crisis was lost and for which complex answers will be needed to be initiated and found for complex challenges and problems.

Beyond that of political mistrust, the perception of delegitimization and in light of the lack of confidence in the technical staff of the Catalan water administration was similarly been a fact worth emphasizing in the development of the conflict (Borràs 2009). Despite the fact that the water policy developed by the ACA before and during the crisis was marked by an anti-transfer advocate line, the acceptance of the transfer proposals managed to push into the background the actions undertaken so far in this vein. The result was an accusation of an absence of technical independence as regards political will and an increasing distrust of the actual interests of the Catalan water administration in the proposals made. Not only did the question arise as to why the desalination works were not timely, but there was similarly an accusation of having withheld the Segre transfer proposal during the pre-election period.

Elsewhere, at the moment when the decision making moved onto a political plane, the ACA lost its proposal capacity, and with that, publicly diminished its authority. As a result, the perception of incapacity in finding a solution to the supply problem was accentuated and furthermore, the lack of confidence in the technical objectivity as regards the possible environmental effects which may be occasioned by the different proposed options. Another ensuing result of this situation was the increase in the climate of uncertainty as regards which was the best option to take. There was likewise a growing diversity of views, criticisms, and a variety of proposals made by different stakeholders, and even the general public, on the decisions to be taken, to the extent that the hydrological situation worsened.

One could ask whether it would have been possible to incorporate spaces for dialogue with the territory in the discussion on the different possible alternatives to the problem and whether this dialogue would have contributed to a shared responsibility in the decisions taken in light of a situation of uncertainty such as that of the 2008 Catalan drought.

The drought is a further example through which the inability of the policy to respond to complex problems and situations becomes evident, as well as the difficulties of scientific-technical knowledge for the purposes of providing a comprehensive response to the problems, and their inability to control the uncertainty of a complex reality. In recent years, many authors have studied the origin of this phenomenon and how it influences society, politics, and decision making, concluding that it is necessary to revise and rethink the relationship model between science and politics (Funtowicz and Strand 2007). In light of the failure of scientific certainties, these authors suggest incorporating dialogue with an increased community of experts in identifying problems and solutions.

19.4.3.3 Background of debates, the importance of information, and social communication The debate on the alternatives to supply the MAB in order to prevent restrictions and residential water cuts offs represents the most visible aspect of the conflict, nevertheless, the understanding of why these alternatives result in a social conflict of such dimensions are derived from an analysis of the rhetoric of the different players during the drought, whose main positions can been seen in Table 19.1.

In overall terms it is an examination of the distributive conflicts, values, identities, related to spatial planning, and in which the ideas built historically around the perception of territorial imbalance have been used. The term 'territorial imbalance' refers to the perception of inequality or economic discrimination of the regions of Lleida, Girona, and Terres de l'Ebre (TTEE) compared to the MAB. These regions, especially Girona with the transfer of the Ter, and the TTEEs, with the transfer to Tarragona and the thermal power, nuclear, and wind power plants for the production of energy, consider that the MAB benefits from its resources so as to obtain wealth which subsequently is not redistributed or passed on to its territory. Furthermore these regions perceive that there is not even analysis of the economic, social, and environmental damage caused by these actions on the transferring resource districts.

In relation to the territorial imbalance and the distribution of wealth and risks, the questioning of the concepts of 'general interest' and 'surplus resource' arises. On the one hand, the resource-granting regions consider that those projects specified as actions of 'general interest' by the state are actually the result of an interest of the regions of Barcelona, which are usually the only beneficiaries of the works. And on the other, different social groups and environmentalists question the possibility that a river has surplus resources, and the question is posed as to whether the surplus resources can only be found in the regions located outside the MAB. To question these concepts evidences the lack of a collective definition thereof; and secondly to note the growing social, technical, and political disagreement in the determination and definition thereof, as well as the different perceptions as regards the goods and services that a river can provide. It would be a positive point to reasonably explain the criteria used to determine that a work is of general interest, and to carry out a study of economic, technical, environmental, and social feasibility prior to the execution of the work, as required by prevailing regulations. It must be explained how the calculations are made so as to determine the surplus resources of a river, as well as to think about how the definition of these concepts can be addressed, and whether it can be the result of a collective discussion and definition.

Another significant debate on crisis management is the role played by the media during the drought crisis, its political affinity and questionable objectivity in the handling of information, as well as its potential for social and political impact, and the ability to create opinion and change the perception of a particular subject matter or situation. The media has become a key player during the drought crisis as it was the means through which the general public received information, but the management of information through said media was ill-advised, creating confusion rather than clarity as regards the ensuing situation.

19.5 Drought planning in Catalonia after the crisis

In response to the 2007–2008 crisis, in 2009 the ACA drew up an Special Drought Plan, which was put to public consultation prior to the regional elections in November 2009. With the change of government in Catalonia the plan was not approved and it was not until July 2016 that a new drought plan was put to public consultation, and whose comment period thereof ended in October 2016. It seems to be no coincidence that this plan saw the light of day at that precise moment after 2015 having been the driest year in the last 100 years, and for which there were forecasts of a new drought period in 2016. This plan seeks to address and respond to drought periods in Catalonia in the last 20 years and meets the mandate of the NHP (10/2001) which stipulates the obligation for all river-basin bodies to draw up a drought plan. The IBC SDP includes the definition of drought indexes for each unit of exploitation of the territory, thus allowing a different management depending on the hydrological situation; the establishment of exploitation regulations that enable the optimal use of non-conventional resources (desalination and reutilization) and the coordinated harnessing of surface waters and groundwater; and the measures to be applied in each period of drought.

According to consultations with the competent body, 40 substantive comments were received by the end of the comment period, which implied the need of an in-depth review of the plan. To this end, a work group on the drought plan was set up in December 2016 through the 'Taula del Ter', a forum for dialogue with the main water management stakeholders in the Ter basin. It had the objective to find meeting points on what measures should be taken in order to mitigate risks in the event of drought, opting for an independent self-management of external contributions to the IBC (Figure 19.3). The autonomous government stated publicly that the plan would not be approved until an agreement of the working group was reached. That public agreement arrived on July 2017 after several meetings, setting up the steps to be taken for the reduction of the withdrawal of flows of the Ter river (Gencat 2017). This agreement is a milestone in water management in Catalonia, and it will be interesting to follow-up its implementation.

The doubt which arises in this scenario is whether or not the measures adopted in the SDP of the IBC will be sufficient in event of a new drought period. And in the case that they are not, the manner in which the debate will be addressed as regards potential transfers from the Catalan Basins of the Ebro. In this regard, the ACA did not actively participate in the SDP of the Ebro Basin, although being a direct stakeholder. Elsewhere,

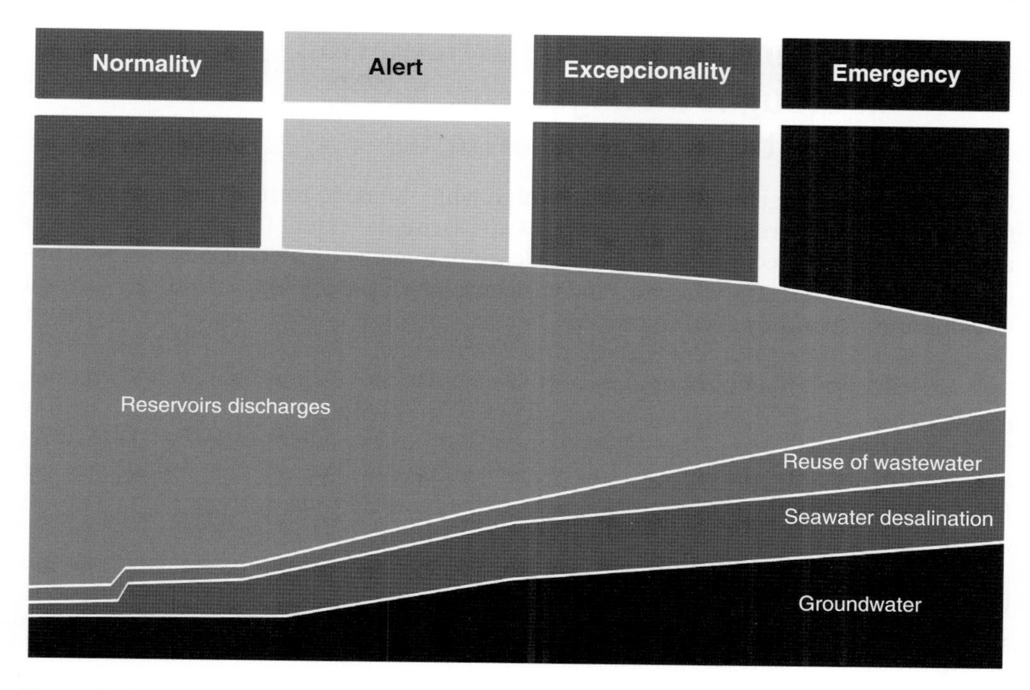

Figure 19.3 Measures contemplated in the Special Drought Plan (SDP) of the inland river basins of Catalonia (IBCs). *Source*: ACA 2016.

neither has a debate been opened with the social movements active in the CBE. It has a certain raison d'être, whereas the measures contemplated in the SDP of the IBC do not envisage that the management of future droughts require a debate beyond this hydrographical area, where elsewhere it is the only area where they have competence, and therefore the only area for debate as regards SDP has been created with IBC stakeholders. Nevertheless, taking into consideration the trajectory of existing conflicts in relation to possible water transfers between hydrographical areas, and with many options that the transfer proposal is once again at some point in the future back on the table, perhaps it would be most expedient to open a debate beyond that of the IBC, which includes the stakeholders of the CBE, and to propose this as a strategic territorial debate, incorporating both the manner of economic growth in Catalonia as well as the territorial balance, and the environmental sustainability of the measures proposed. In fact, certain measures currently being adopted are perceived from the social movement in TTEE as potential gateways to future transfers, so that the set of small measures would facilitate in the future the conveyance of water to the MAB and transfers to other parts of Spain. Institutional and political mistrust has a long history and even more so after a lack of information transparency during the 2008 drought, which we think it can only be amended by opening up a wide-ranging public debate with all parties involved, and the construction of a common project, which entails the entire Catalan territory.

19.6 Deliberative public participation in drought management: Need, obligation, and opportunity

Drought crisis management in Catalonia highlights the need for the opening of a participatory forum in drought management as an essential element for conflict resolution and for reaching collective agreements, to learn to manage that uncertainty, to contribute to social learning, transparency, and greater joint responsibility in the follow-up and implementation of the decisions taken. Far beyond that need, drought management plans must be integrated into hydrological planning, and therefore there is an obligation to promote public participation in the elaboration and implementation thereof, pursuant to Article 14 of the Water Framework Directive and the 1998 Aarhus Convention and the implementing regulations thereof. We make certain recommendations for the inclusion of public participation in drought planning for the purposes of the drafting, implementation, and follow-up of the Drought Management Plans (DMPs) which likewise must be integrated into the river-basin management plans.

First of all, it would seem appropriate that the participatory forums coincide with the exploitation systems or committees, areas where the occurrence probability of droughts can be established, and then subsequently the undertaking of a wider debate at the autonomous regional level.

Both the general public and stakeholders should have the opportunity to take part in drought management. First of all, channels of communication and dissemination of information to the general public must be sought and found, and the contribution to their understanding of the problems and solutions proposed, in particular when discussing periods of crisis, wherein the general public may feel directly adversely affected. As far as stakeholders are concerned, a 'drought committee' or similar could be set up, which was for a continuous period of time and that was directly implicated in the drafting and revision of the DMPs. This committee should likewise play an important role during drought periods, wherein said committee must be actively involved in the process of implementing the measures contemplated in the DMP.

The contents of the deliberative public participatory forums may include a debate on analysis and measures, both for the drafting process as well as the implementation and follow-up process, whilst the introductory and outgoing thresholds to the different management scenarios would be defined through technical studies and widely disseminated to society. Doing so would reduce the arbitrariness with which sometimes the decision is made to move from one scenario to another without, on certain occasions, notifying why this decision is taken and what implications this would entail; will increase the credibility of the decision maker, and contribute to the understanding of decision making (and decisions taken) by the general public. In the analysis phase information related to the different scenarios of demands and related guarantees, as well as the priorities of water allocation must be available. The results of the analysis phase would bring us to a debate on proposed measures for each scenario, which responds to what we are willing to assume, what objectives do we have, and how to obtain them. To this end, a risk analysis may be carried out, wherein each measure is related to a level of risk, its costs (environmental, social, and economic), its effects on the different exploitation systems, and the time required for its activation, as well as the legal measures necessary to adopt the proposed measures.

The entire process should incorporate up-to-date and quality information and thoroughly analyse the role that the media may play, in particular as regards dissemination during a drought period.

19.7 Conclusions

The analysis of the drought crisis in Catalonia brings us another step closer to the complexity of drought management in Spain, characterized by: multilevel decision making, with weak institutional coordination; a general public comprising of a very broad set of players with diverse positions and interests, who are organized and able to exert pressure, particularly sensitive to certain territorial implications such as transfers between river basins, but without channels or forums for a public and transparent dialogue; and a drought management traditionally focused on emergency management, and with limited planning capacity.

Nevertheless, in recent years there has been a required and necessary shift towards drought planning, as a means of improving the management thereof and the adoption of more effective preventive measures. In this shift of focus, public participation is a key element that is not always adequately taken into account. It is our opinion that dialogue with the territory should not only be an isolated process for deciding one or another public policy, but should be integrated into the manner of governance as a different way of doing politics. Therefore, this is not about the compliance of a legislative framework or of the solution of problems as they ensue, but of a transformation process in the manner of governance where the dialogue must be part of the entire water planning and management cycle (including droughts), and therefore must occur not only during crisis management but likewise before and after the event, ensuring transparency and quality information at all times. Elsewhere, the implementation of public participation in drought management, as well as the scope of effective drought management, requires the identification of finding effective technical and political institutional coordination mechanisms, and which are capable of integrating the river basin management criteria with the political-territorial management rationale.

Notes

1. Royal Decree 125/2007, of 2 February, which sets forth the territorial scope of the river basin districts.
2. Which shall be referred to herein by its initials in Spanish, that is, IBC.
3. Which shall be referred to herein by its initials in Spanish, that is, CBE.

References

Agència Catalana de l'Aigua (2008). L'Aigua a Catalunya. Diagnosi i propostes d'actuació. Esquema provisional de temes més importants que es plantegen en la redacció del Pla de Gestió del Districte de Conca Fluvial de Catalunya. Barcelona: Generalitat de Catalunya-Departament de Medi Ambient i Habitatge.

Agència Catalana de l'Aigua (2016). Pla especial d'actuació en situació d'alerta i eventual sequera (PES). DOGC 7163 (15/07/2016). Generalitat de Catalunya Barcelona.

Borràs, G. (2009). *Els guardians de l'aigua*. Barcelona: Cipmedia edicions.
Decree 84/2007 (2007). Decree 84/2007 d'adopció de mesures excepcionals i d'emergència en relació amb la utilització dels recurso hídrics. DOGC 4860 (12/04/2007). Barcelona: Government.
Directive 2000/60/EC (2000). Directive 2000/60/EC of the European Parliament and of the Council establishing a framework for the Community action in the field of water policy (EU Water Framework Directive). Official Journal (L 327) (EC). Brussels: European Commission.
Funtowicz, S. and Strand, R. (2007). De la demostración experta al diálogo participativo. *Revista iberoamericana de ciencia tecnología y sociedad* 8: 97–103.
Generalitat de Catalunya (Gencat) (2017). Acord de la Taula del Ter, 14/07/2017. Girona: Gencat.
Hernández-Mora, N., Ituarte, L.d., La-Roca, F. et al. (2014). Interbasin water transfers in Spain: interregional conflicts and governance responses. In: *Globalized Water: A Question of Governance* (ed. G. Schneier-Madanes), 175–194. Dordrecht: Springer Science+Business Media.
La Calle, A. (2005). Sequía y adaptación de la Directiva Marco del Agua. In: *La sequía en España. Directrices para minimizar su impacto* (ed. Comité de Expertos en Sequía del Ministerio de Medio Ambiente). Mdrid, Spain: Ministerio de Medio Ambiente.
Law 62/2003 (2003). Law 62/2003 de medidas fiscales, administrativas y de orden social. BOE 313 (31/12/2003). Madrid.
Martínez, J. (2006.) Las sequías en España, un fenómeno recurrente. In: Congreso homenaje al Duero (ed. Fundación Nueva Cultura del Agua). https://goo.gl/L5UdGS (accessed 1 March 2019).
Ministerial Order MAM/698/2007 (2007). Ministerial Order MAM/698/2007 por la que se aprueban los planes especiales de actuación en situaciones de alerta y eventual sequía en los ámbitos de los planes hidrológicos de cuencas intercomunitarias. BOE 71 (23/03/2007). Madrid.
Royal Decree 907/2007 (2007). Royal Decree 907/2007 por el que se aprueba el Reglamento de la planificación hidrológica. BOE 162 (07/07/2007). Madrid.
Royal Decree Law 3/2008 (2008). Royal Decree Law 3/ de medidas excepcionales y urgentes para garantizar el abastecimiento de poblaciones afectadas por la sequía en la provincia de Barcelona. BOE 97 (22/04/2008). Madrid.
Spanish Constitution (1978). BOE 311 (29/12/1978). Madrid.
Urquijo, J., De Stefano, L., and La Calle, A. (2015). Drought and exceptional laws in Spain: the official water discourse. *International Environmental Agreements: Politics, Law and Economics* 15 (3): 273–292.
Royal Decree 125/2007, por el que se fija el ámbito territorial de las demarcaciones hidrográficas. BOE 30 (03/02/2007). Madrid.

20

What Could Change Drought Governance in Europe?: A Comparative Analysis between Two Case Studies in France and the UK

Isabelle La Jeunesse[1], Hans Bressers[2], and Alison Browne[3]
[1] *Laboratory CNRS 7324 Citeres, University of Tours, Tours, France*
[2] *Department of Governance and Technology for Sustainability, University of Twente, Enschede, The Netherlands*
[3] *Department of Geography, Sustainable Consumption Institute, University of Manchester, Manchester, United Kingdom*

20.1 Introduction

According to the latest Intergovernmental Panel on Climate Change (IPCC) reports (IPCC 2007, 2012), even in North West Europe (NWE) the balance between water demand and availability could be rapidly under pressure of severe drought periods. Adapting to such changes requires governance contexts that are readying themselves for such extremes whilst flooding is still very frequent. In fact, weather conditions are and will likely to be often more extreme than they used to be and water management will have to cope with this increase in water variability, through increasing the resilience of water systems (Kuks 2016). However, one singularity of NWE is that severe droughts have been and are still rare events in the region and thus water scarcity has hardly been experienced in its recent history. Although the lack of a drought history in wet areas can explain why drought and water scarcity are not necessarily the focus of the agenda of water managers and water users, the increase of water demand for both environmental, economic, and domestic needs in wet regions can provide situations of water scarcity

Facing Hydrometeorological Extreme Events: A Governance Issue, First Edition.
Edited by Isabelle La Jeunesse and Corinne Larrue.

in case of an important variability of the water resource. Thus, the NWE, an actual particularly wet region, could face new drought and water scarcity situations due to climate changes.

During the severe European drought that occurred in 1976, the worst and most persistent streamflow droughts occurred in southern England and northern France (Zaidman, Rees and Young 2002). As a matter of fact, due to its gradual emergence, droughts tend to show strong spatial correlation over a wide region. For instance, some of the severest droughts that occurred in Europe over the last 40 years, such as in 1972, 1976, 1989, were widespread (Bradford 2000). Thus, the way a territory prepares for such an event can have some instructional benefit for other territories. In this purpose, the present chapter proposes an analysis of the water and drought governance contexts in case studies of two countries of NWE that have been jointly impacted during the past severe droughts, France and the United Kingdom with respectively two case studies: the Vilaine catchment in French Brittany, and the Somerset Levels and Moors in south west UK.

This chapter presents a comparative analysis of the outputs of the implementation of the Governance Assessment Tool (GAT) developed by Bressers et al. (2015) with the Contextual Interaction Theory, the theory at the origins of the GAT. The outputs of the GAT in the French and the English case study have already been described in details respectively in La Jeunesse et al. (2016) and Browne et al. (2016).

After a brief description of the two case studies, the methodology used to study drought governance, both the GAT as refined in the European DROP-project for drought-governance analysis and the Contextual Interaction Theory, is presented. We then present the main results and discussion. The last section is dedicated to go up in general conclusions for drought-governance resilience in NWE.

20.2 Vilaine catchment and Arzal dam

The French case study concerns the Vilaine catchment in the Brittany region in France and, more specifically, of the Arzal dam and reservoir located at the outlet of the river (Figure 20.1). Accordingly, the analysis focuses on the lower part of the Vilaine. The Arzal dam is a freshwater reservoir. Initially built in 1970 to protect the inland against salt water intrusion during spring tides, the water reservoir today plays a central role in providing drinking water for households and the agricultural sector through water level regulations. Next to that, the reservoir is used for recreational purposes, such as sailing and fishing (La Jeunesse et al. 2016).

The multi-functionality of this reservoir is depending on the amount of water. Sailing is now highly developed in the area with an average of 18 000 boats crossing the dam per year, with approximately 85% of boat crossings occurring during May to October thus during the low-flow period. Because of an intrusion of salt water when the lock is operated for navigation, sailing has an influence on the quality of the water in the reservoir. Because the most significant inputs of salt water to the reservoir occur when boats pass through the existing lock of the dam, navigation can be restricted during the summer. Management rules with restrictions on the time schedule and number of lock openings can thus generate conflicts, especially during the summer high season for recreational sailing.

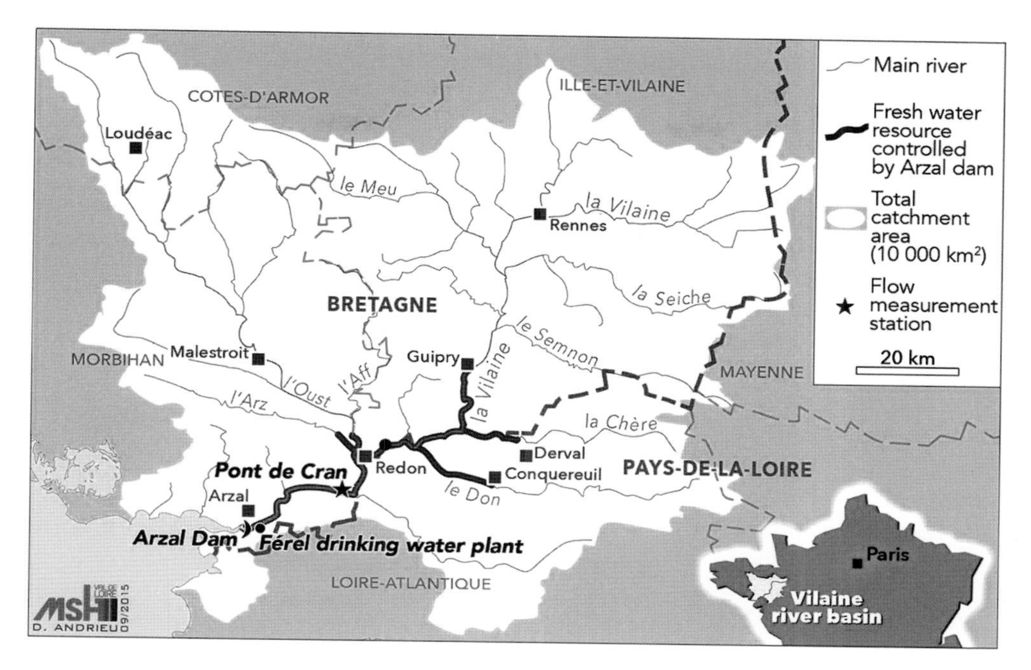

Figure 20.1 Location of the Vilaine catchment, the Arzal dam, and the zone of influence of the dam in the upper part of the riverbed.

20.3 Somerset Levels and moors

The county of Somerset is a sparsely populated wetland area in the south-west of the UK. Soils consist of marine clay levels along the coast and often peat-based moors inland. One of the most important features of the region is the Somerset Levels and Moors, a highly managed river and wetlands system which is artificially drained and irrigated in order to open the area for productive settlement and uses such as farming (Figure 20.2). The peat soils of the Somerset Levels provide a multiple set of ecosystem services including, amongst others, food production, nature, and protection of historic environment. However, these peat soils are vulnerable to sudden and irreversible changes as a direct result of drought and dehydration (Browne et al. 2016).

20.4 Methodology

20.4.1 Governance Assessment Tool

The GAT (Bressers et al. 2015), which has been developed and adapted to study drought governance during the DROP project, consists of five elements and four criteria. Here, governance is the combination of the relevant multiplicity of responsibilities and resources, instruments and strategies, problem perceptions and goal ambitions, actors and networks, and scales. These factors form a context that, to some degree, restricts and, to some degree, enables actions and interactions. Through a variety of

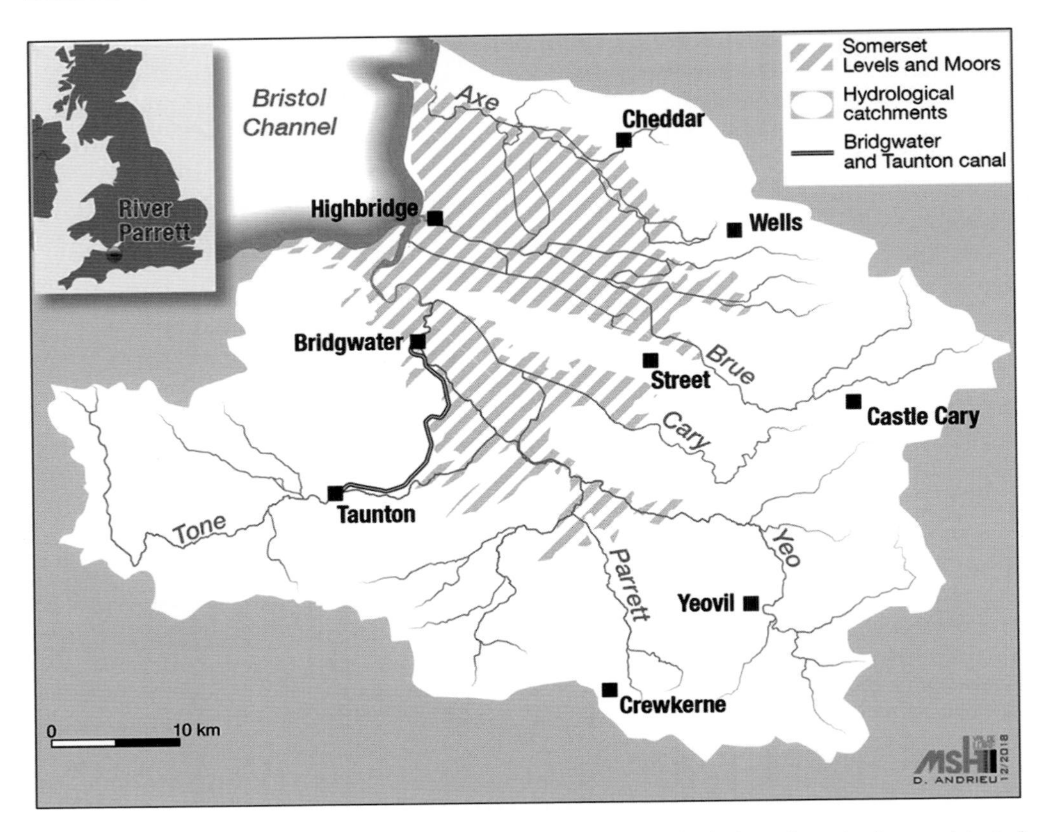

Figure 20.2 Location of the Somerset Levels and Moors and hydrological catchments. *Source*: Adapted from a map of the Enviromnent Agency, Browne et al., 2016.

interview questions with a range of relevant regional stakeholders (water managers, water users, and so on), each of these elements is assessed for its extent, coherence, flexibility, and intensity. This concept finds its origin in the contextual interaction theory developed by Bressers et al. and explained later.

20.4.2 Contextual Interaction Theory

The GAT is indeed rooted in a theory of policy implementation that is labelled Contextual Interaction Theory (Bressers and Kuks 2004; Bressers 2009; de Boer and Bressers 2011). It views implementation processes not top down, as just the application of policy decisions, but as multi-actor interaction processes that are ultimately driven by the actors involved. Thus it makes sense to explain the course and results of the process from that simple starting point and to place these actors and their main characteristics central stage in any analytical model.

The theory is based on a conceptual framework that aligns the motivations, cognitions, and resources of people to their contexts. It uses this to assess the likelihood of the governance instruments and structures in their ability to support implementation of given policies and achieve certain goals.

Normally this framework (Contextual Interaction Theory) is used to produce academic studies of policy implementation whilst the GAT is aligned to the needs of practitioners who are interested in understanding their own contexts.

Since interaction processes are human activities, all influences flow via the key characteristics of the actors involved (Bressers and Klok 1988). Thus, it is possible to explain the course and effects of implementation processes with a set of three core factors per actor (Figure 20.3). All other factors, including governance conditions, are regarded as belonging to the context that may influence this set of core factors. In Figure 20.3 we include these factors: their motivations that may spur the actors into action, their cognitions, information held to be true, and their resources, providing them with the capacity to act individually and their power in relation to other actors. Amongst the actors involved in the process there need to be a sufficiently strong combination of motivations, cognitions, and resources to enable the process to succeed (Bressers and Kuks 2004; Owens 2008).

The basic assumptions of Contextual Interaction Theory are quite simple and straightforward. The first assumption is that policy processes – that could conduct drought governance in our cases – are not mechanisms but human and social interaction processes between a set of actors. This includes policy implementation management but also project realization. The dynamic interaction between the key actor-characteristics that drive social-interaction processes can also be reshaped by the process. The actor characteristics are here considered as the ultimate process setting. The characteristics of actors are (i) their motivations which drive their actions, (ii) their cognitions with which the situation is interpreted and (iii) the resources that provide the capacity and the power to realize the project.

The origins of motivations for behaviour, including for the positions taken in interaction processes, first of all lay in the actors' own goals and values. Self-interests, as in many economic theories, of course play a strong role here.

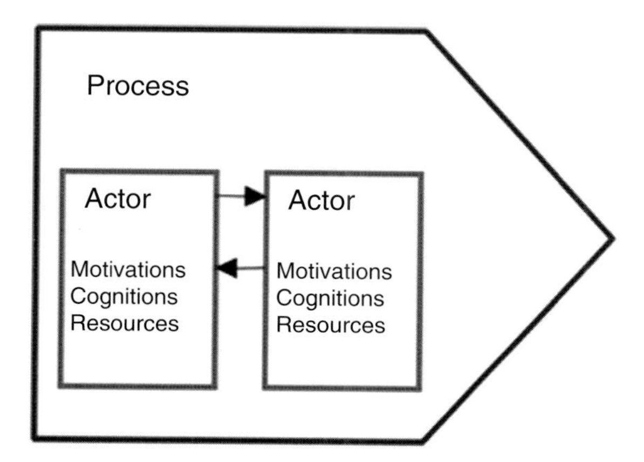

Figure 20.3 Process model with the actor characteristics used in contextual interaction theory (Bressers 2009).

The cognitions, that is to say the cognitions of actors (interpretations of reality held to be true), are not only a matter of observations and information processing capacity, though these aspects are important, and with the information technology revolution can change quickly.

Whilst the resources of an actor are important characteristics to provide capacity to act, in the relational setting of an interaction process they are also relevant as a source of power. Resources are here meant to be any asset that public and private actors can use to support their actions. This implies that the relevance of resources is dependent on the actions an actor wants to perform. Next to the resources of the actors, motivations and cognitions also play an important role in creating productive or non-productive settings for the process. Thus, resources only get meaning in the context of cognitions and motivations.

20.4.3 Conducting the governance analysis

In the French case study, the material for the analysis was collected during two field visits. The first visit occurred from 16 to 18 September 2013, the second from 16 to 18 June 2014, during which the Governance Team (GT) met stakeholders, managers, and representatives of the relevant local action groups. The analysis is also based on several documents provided by the *Institution d'Aménagement de la Vilaine* (IAV), one practice partner of the DROP project responsible for the management of the Arzal dam and its reservoir. IAV also plays a key role in supporting water management at the catchment scale of the Vilaine River (La Jeunesse et al. 2016).

In the Somerset case study, the governance conditions for drought have been assessed during two visits to the region – one in September 2013 following the period of drought, and another again in October 2014 after a period of flooding recovery (Browne et al. 2016).

20.5 Results and discussion

In the Vilaine, except for emergency measures, there is no global plan set up to manage drought vulnerabilities induced by climate change. The current situation of low drought risk perception, compared to a more significant flood risk perception, is explained by a lack of drought risk awareness, due to the absence of critical drought events in the past years in the region, and the lack of a culture of drought forecasting and risk communication. However, it is expected that as drought perceptions are raised, drought adaptation measures can rapidly be designed and implemented by the efficient, existing water governance for freshwater in the basin, which is supported by a dense stakeholder network driven by IAV (the river water managers – Institution d'Aménagement de la Vilaine).

In the Vilaine, a main issue is that the acceptance of climate change (*cognitions*) as a reality or at least as a relevant issue for the stakeholders involved is very weak. This is identified as a major problem and a root cause for the low degree of openness towards adaptation (*motivation*). However, also plain interests play a role in this low motivation of some of the users. With their legal rights (*resources*) they are also in the position to block the development of the process, at least until a higher level of awareness (*cognitions*) has been developed (Figure 20.4).

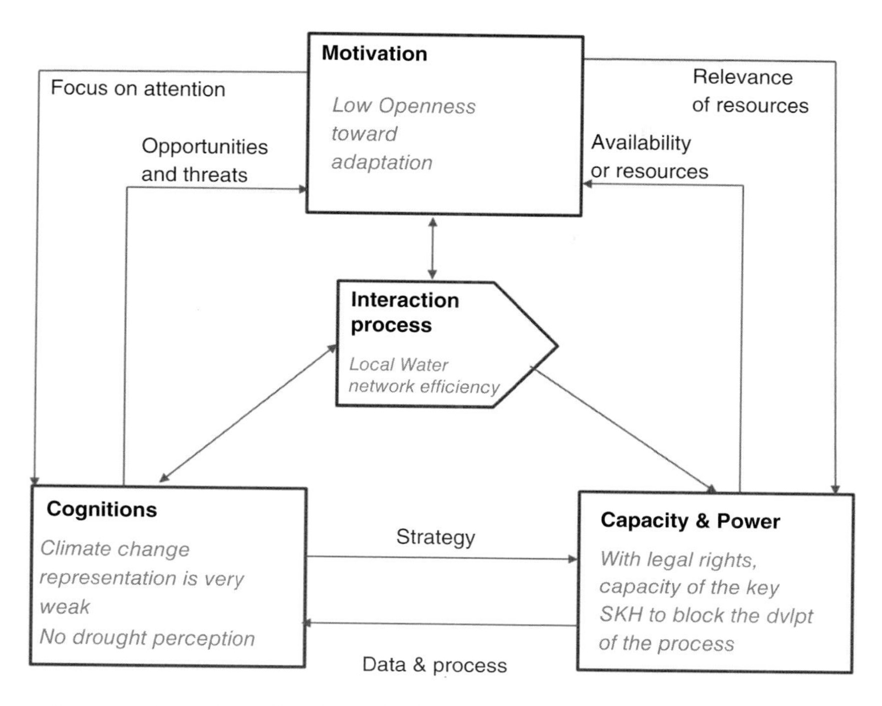

Figure 20.4 Contextual interaction theory in the Vilaine catchment during interviews (2013–2014).

As the matter of fact, there has been no seismic shock at the moment to raise awareness and the drought governance could be characterized as not actually supportive. However, whilst strong resources to block the development could also be regarded a very negative, that is just a matter of interpretation. If you relate to strength as such that the answer is 'strong', but if you relate to the influence on the dependent variable then the score is very negative. Maybe the story for Vilaine as also formulated in words is that all three main factors now have a negative influence on the progress of adaptation, but that when change occurs (likely from cognitions [more awareness] to motivation), the same strong resource positions that are now capable of blocking then could be getting productive to the development of adaptation measures. As a consequence relatively quick change is not impossible and even likely to happen, as proposed in Figure 20.5. A strong drought could maybe be sufficient to bring about a drastic change in the situation.

In the Somerset Levels and Moors whilst, historically, flooding has dominated the physical and political landscape there is in fact a greater awareness of drought and water scarcity impacts. The UK was the first European country to clearly integrate climate change in its water management strategy. In 2008 the UK government ratified the Climate Change Act. One aspect of this piece of legislation ensures relevant public bodies put plans in place to adapt to climate change. Since then there are a range of planning activities to increase the resilience of the English water system to short- and longer-term changes as a result of climate change and increasing water demand, including actions specifically related to exceptional drought events.

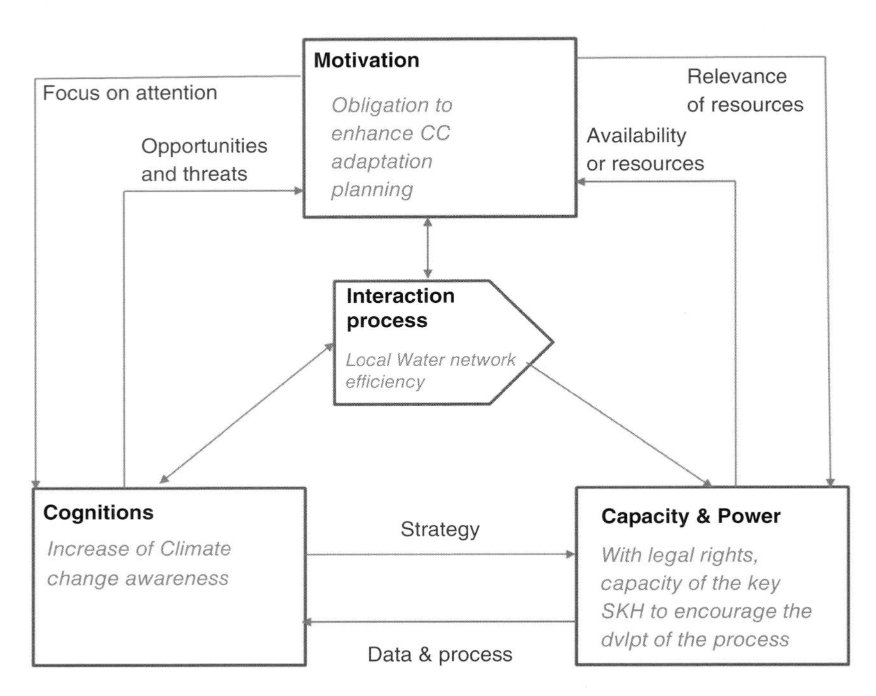

Figure 20.5 Contextual interaction theory in the Vilaine catchment with climate change awareness.

Thus, contrary to the French case study, in Somerset there was much more acceptance of climate change and its double effect on water levels (*cognitions*), with stakeholders engaged in adaptation projects. But already at that time the dependency on external funds (*resources*) was preventing the climate adaptation measures being put into practice. Then an external 'seismic shock' of the 2013/2014 flooding modified the picture. The politicization of flooding in the region (including high-pressure media exposure) led to a re-interpretation of water management (*cognitions*) that was far more one-sided and focused largely on mitigating flood events. Actors that were before prepared to cooperate were suddenly not only aroused (*motivation*), but also feeling themselves much stronger (*resources*) in this new constellation (Figure 20.6).

However, the severe floods of winter 2013/2014 that particularly affected Somerset, have provided a seismic shock with a particular increased politicization of the issues of flooding for the region, leaving a residue of risk of maladaptation of measures to deal with climate change as flooding and drought are currently governed in silos.

We can see here that a 'game-changer' that could only be imagined in the Vilaine case study, has actually taken place with the big 2013/2014 flooding. It is even hard to picture the situation in just one diagram (Figure 20.7) as the 2013/2014 floods have really modified the picture. Thus you have a before and an after situation. In this case the after situation is not really an improvement in relation to the dependent variable of increased drought adaptation. Before, a situation with high openness in

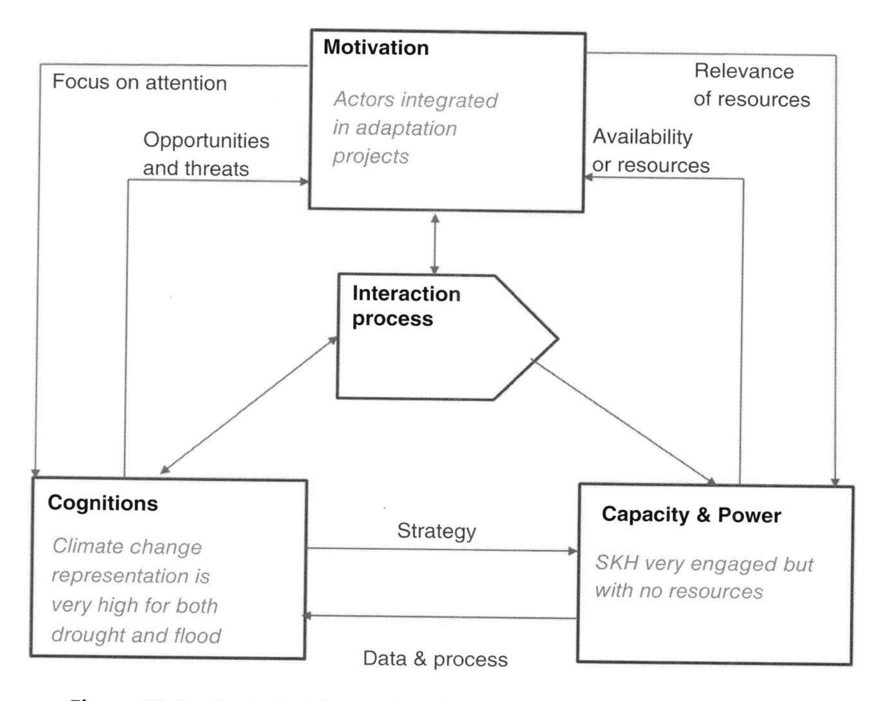

Figure 20.6 Contextual interaction theory in Somerset before 2014 floods.

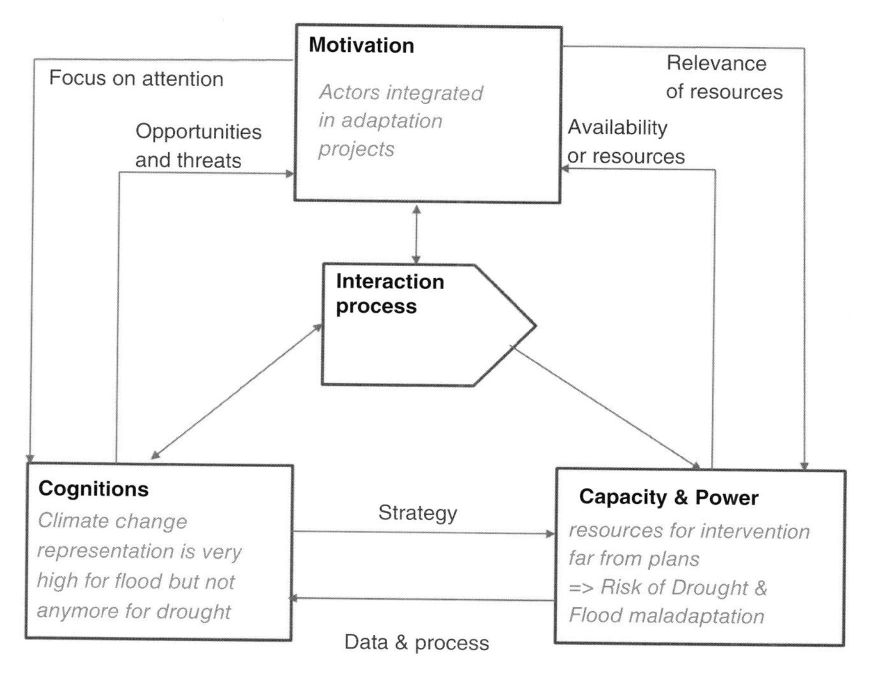

Figure 20.7 Contextual interaction theory in Somerset after 2014 floods.

the cognitive sphere, reasonable joint motivation amongst stakeholders and thus low resources to realize the measures was effective. After the flooding, the situation relates polarized cognitions (Figure 20.7). Those of 'flood-fighters' being in the advance have predominantly low motivation for drought measures with stakeholders and funders. But they have in fact high motivation for more one-sided flood protection measures and high resources.

20.6 Conclusions

The results of the drought governance analysis on two case studies, the Vilaine catchment in French Brittany and the Somerset Levels and Moors in the United Kingdom are re-analysed through the Contextual Interaction Theory in order to support a comparative analysis of key governance factors in these two cases. The governance context influences these processes through its impact on these actor characteristics: the drivers of processes are ultimately people, sometimes representing themselves, but often as part of organizations or groups and themselves driven by their *motivations*, *cognitions,* and *resources*.

In the Vilaine, a main issue is that the acceptance of climate change (*cognitions*) as a reality or at least as a relevant issue for the stakeholders involved is very weak. This is identified as a major problem and a root cause for the low degree of openness towards adaptation (*motivation*). However, also self-interests play a role in this low motivation of some of the users. With their legal rights (*resources*) they are also in the position to block the development of the process, at least until a higher level of awareness (*cognitions*) has been developed.

In contrast, in Somerset there was much more acceptance of climate change and its double effect on water levels (*cognitions*), with stakeholders engaged in adaptation projects. But already at that time the dependency on external funds (*resources*) was preventing the climate adaptation measures being put into practice. Then an external 'seismic shock' of the 2013/2014 flooding modified the picture. The politicization of flooding in the region (including high-pressure media exposure) led to a re-interpretation of water management (*cognitions*) that was far more one-sided and focused largely on mitigating flood events. Actors that were before prepared to cooperate were suddenly not only aroused (*motivation*), but also feeling themselves much stronger (*resources*) in this new constellation. Policy and implementation silos exist between drought and flood in the definition of the target of adaptation efforts for a future of climate change. These silos need to be addressed in on-going water-management policy, and on the ground adaptation actions in the Somerset region but more generally in Europe also, if resiliency to future events is to be increased.

Recognizing the need to address the impact of floods, whilst still acknowledging that there is also a very real threat for water scarcity in the NWE region, changes the range of strategies and instruments that could be used to effectively mitigate variability and extremes. The more joined-up approach of different forms of water management that is needed in these two case study examples draws together a range of lessons for more effective governance for climate change adaptation across the whole of NWE; that is governance approaches focused on adaptation and resilience of the whole water system rather than crisis management of extreme events separately.

Acknowledgements

Based on the INTERREG IVB funded DROP project this chapter presents an assessment of two of the six case studies of governance contexts across NWE. The authors would like to thank the DROP coordination, DROP practice partners, and all stakeholders interviewed.

References

de Boer, C. and Bressers, H. (2011). *Complex and Dynamic Implementation Processes. Analyzing the Renaturalization of the Dutch Regge River*. Enschede: University of Twente and Water Governance Centre.

Bradford, D. (2000). Drought events in Europe. In: *Drought and Drought Mitigation in Europe* (ed. J.V. Vogt and F. Somma), 7–22. Dordrecht, The Netherlands: Kluwer.

Bressers, H. (2009). From public administration to policy networks. Contextual interaction analysis. In: *Rediscovering Public Law and Public Administration in Comparative Policy Analysis. A Tribute to Peter Knoepfel* (ed. N. Stéphane and F. Varone), 123–142. Lausanne: Presses Polytechniques.

Bressers, H., Bressers, N., Browne, A. et al. (2015). Governance assessment guide. Adapt to drought and water scarcity now. 24 p. https://www.utwente.nl/en/bms/cstm/research/drop-governance-assesment-guide-web.pdf (accessed 3 March 2019).

Bressers, H. and Klok, P.J. (1988). Fundamental for a theory of policy instruments. *International Journal of Social Economics* 15 (3/4): 22–41.

Bressers, H. and Kuks, S. (eds.) (2004). Integrated governance and water basin management. In: *Integrated Governance and Water Basin Management*, Environment & Policy, vol 41, 247–265. Dordrecht: Springer.

Browne, A.L., Dury, S., De Boer, C. et al. (2016). Governing for drought and water scarcity adaptation in the context of flooding recovery: the curious case of Somerset, United Kingdom. In: *Governance for Drought Resilience. Land and Water Drought Management in Europe* (ed. H. Bressers, N. Bressers and C. Larrue), 83–109. Cham: Springer Open Science.

IPCC (2007). *Climate Change 2007: Synthesis Report, Fourth Assessment Report*. Cambridge, UK: Inter Governmental Panel on Climate Change, Cambridge University Press.

IPCC (2012). *Managing the Risks of Extreme Events and Disasters to Advance Climate Change Adaptation. A Special Report of Working Groups I and II of the Intergovernmental Panel on Climate Change*, 582p. (ed. C.B. Field, V. Barros, T.F. Stocker, et al.). Cambridge and New York: Cambridge University Press.

Kuks, S. (2016). Foreword: overcoming drought and water shortages with good governance. In: *Governance for drought resilience. Land and Water Drought Management in Europe* (ed. H. Bressers, N. Bressers and C. Larrue), v–vii. Cham: Springer Open Science.

La Jeunesse, I., Larrue, C., Furusho, C. et al. (2016). The governance context of drought policy and pilot measures for the Arzal dam and reservoir, Vilaine catchment, Brittany, France In: *Governance for Drought Resilience. Land and Water Drought Management in Europe*, Cham: Springer Open Science, Bressers H., N. Bressers, and C. Larrue (ed.), pp. 109–139.

Owens, K. (2008). *Understanding how Actors Influence Policy Implementation. A Comparative Study of Wetland Restorations in New Jersey, Oregon, the Netherlands and Finland*. Enschede: University of Twente.

Zaidman, M.D., Rees, H.G., and Young, A.R. (2002). Spatio-temporal development of streamflow droughts in north-west Europe. *Hydrology and Earth System Sciences* 6: 733–751. https://doi.org/10.5194/hess-6-733-2002.

Part IV
Coastal and Wind Storms

IV.1

Actors Involved in Coastal Risks Prevention and Management

21

Sustainable Communities and Multilevel Governance in the Age of Coastal Storms

Yves Henocque

Maritime Policy and Governance, French Research Institute for the Development of the Sea, IFREMER, Paris, France

21.1 Introduction: Addressing a social-ecological system

Scientific research relates more and more climate change to the fact that hurricanes, typhoons, and coastal storms in general are becoming more intense, lasting longer, unleashing stronger winds, and causing more damage to coastal ecosystems and communities. Moreover, the addition of increased storm activity and flooding superimposed on sea-level rise projections is getting more and more problematic, more particularly in deltas; estuaries, and low-lying areas. Increasingly, large-scale linkages are being recognized between atmospheric phenomena (e.g. the El Niño-Southern Oscillation, ENSO, or North Atlantic Oscillation, NAO) and the variation of intensity of regional weather patterns under the name of 'teleconnections' (Wallace and Gutzler 1981).

Actually, climate change and its consequences mainly come into play by exacerbating existing threats and problems in the coastal area whilst coastal management issues stem first from the impacts of development policies and processes that progressively become unsustainable. In Europe, the main pressures result from fishing, seafloor damage, pollution by nutrient enrichment and contaminants, and the spreading of non-indigenous species whilst marine litter and underwater noise are of growing concern (EEA 2014). The key is then to understand the cumulative pressures and impacts affecting the coastal and marine environment including climate change.

This understanding is crucial to the building up of an appropriate response about the sustainability of human activities and the resilience to change of underlying ecosystems. Towards such a goal, multilevel strategies (in space and time) are increasingly formulated as in the case of the European Union (EU): (i) short-term targets for individual

Facing Hydrometeorological Extreme Events: A Governance Issue, First Edition.
Edited by Isabelle La Jeunesse and Corinne Larrue.

sectoral policies; (ii) mid-term goals linking policy ambitions in more comprehensive policies like the Seventh Environment Action Programme (7th EAP); (iii) towards a vision of societal transition, informed by the concepts of planetary boundaries, green economy, and resilience of society and ecosystems.

Hence, disaster risk reduction and adaptation to climate change programmes, as well as integrated coastal management (ICM), poverty reduction, and good governance, have mutually supportive objectives. They all acknowledge both the social and bio-physical dimensions of the risks they address, call for a stronger political commitment and place significant importance upon changing human behaviour and attitudes. Strategies are all geared towards management of the bio-physical environment (e.g. zoning, habitat restoration), promotion of sustainable livelihoods, poverty alleviation, and promotion of public participation. In a nutshell, they all contribute to improving sustainability through the implementation of an integrated ecosystem-based approach to management.

Amongst one of the most recent and famous examples of large-scale impact of coastal storms is Hurricane Katrina in August 2005, one of the most devastating hazard events in US history. In New Orleans, historical development was concentrated on the higher ground, but with the economic expansion and protective works, residential development spread out into low-lying areas. The extend of flooding was certainly a product of topography but Katrina also became a 'man-made disaster' of catastrophic proportions because of the social vulnerability of New Orleans and the failure of public institutions to protect citizens, evacuate them, provide a timely and effective emergency response, and facilitate quick and safe rebuilding (Glavovic 2008). The underlying drivers of such a vulnerability are not in the coastal defence infrastructures but in long-time accumulating socio-economic issues showing that besides vital measures (coastal defence, emergency relief, reconstruction), there is a need to address pre-event vulnerabilities to build more resilient communities, i.e. the capacity of coupled socio-ecological systems to absorb hazard impacts and retain vital structures, processes, and feedback functions implying self-organizing, learning, and adaptive capabilities (Adger et al. 2005).

21.2 Harmonizing coastal management, disaster risk reduction, and climate change adaptation goals through meaningful public participation

21.2.1 Paradigm shifting

For more than two decades, there has been significant conceptual development in defining the elements of ICM and ecosystem-based management (EBM), which are both contributing to improving the governance of 'social-ecological' systems, emphasizing that, (i) the focus is upon both the environment and humans as interconnected and interdependent systems, and (ii) the term 'governance' calls for a re-examination of the formal and informal arrangements and institutions pertaining to the marketplace, the governments, and the civil society (Olsen 2009) (Box 21.1).

With this paradigm shift, the integrated management of coastal areas is closely associated with the implementation of the ecosystem approach, whose principles are central

Box 21.1 EBM as a paradigm shift

From	To
• Individual species	• Ecosystems
• Small spatial scale	• Multiple scales
• Short-term perspective	• Long-term perspective
• Humans independent of ecosystems	• Humans are integral parts of ecosystems
• Management divorced from research	• Adaptive management
• Managing commodities	• Sustained production potential for ecosystem goods and services

to the 1992 Biodiversity Convention (Box 21.2). As shown by the first four principles, the main challenge related to the ecosystem approach, together with the integrated management of coastal areas, is found in the social system and its complex interactions. Holling (1986) showed that ecological surprises are more likely to occur where large-scale systems become highly interconnected. This is also the case for social systems, which become more prone to instability the bigger and more interconnected they become. Under such conditions, it is not surprising that, when two unstable and highly interconnected systems meet, their overall stability becomes very uncertain and the question goes back to the capacity of local communities to deal with it.

Therefore, in the so-called 'Anthropocene' era which is ours, there is an urgent need for more integrated policies as regards the oceans and their coastal areas. This should be done using ICM and EBM in a mutually supportive, 'polycentric' and adaptive way where the local actors should be more in a position to play their part, use their experience and responsibility in working with public institutions (Henocque and Kalaora 2015) (Box 21.3). Under these conditions, coastal areas and the ocean may be seen as catalysts of shared values, creating a dynamic where science, ethics, and engagement represent major resources in the face of social ecological systems' complexity and uncertainty.

21.2.2 International framework

The frameworks put forth by the United Nations International Strategy for Disaster Reduction (UN-ISDR) and the United Nations Framework Convention on Climate Change (UNFCC) both adopt sustainable development as their overarching principle. In the aftermath of the 2004 tsunami in the Indian Ocean, the United Nations Environment Programme/Global Programme of Action (UNEP/GPA) (Global Programme of Action for the Protection of the Marine Environment from Land-based Activities) convened a meeting in February 2005 in Cairo (Egypt) to discuss post-tsunami reconstruction and coastal zone rehabilitation and management in affected countries (Box 21.4). The meeting adopted another 12 guiding principles which were initially drafted by ICM practitioners and, to a large extent, encapsulate the approaches that seek to reduce coastal areas' vulnerability to both man-made and natural hazards. In the end, these principles were endorsed by senior government

Box 21.2 The twelve principles of the ecosystem approach (Biodiversity Convention, United Nations 1992)

1. The objectives of management of land, water, and living resources are a matter of societal choices.
2. Management should be decentralized to the lowest appropriate level.
3. Ecosystem managers should consider the effects (actual or potential) of their activities on adjacent and other ecosystems.
4. Recognizing potential gains from management, there is usually a need to understand and manage the ecosystem in an economic context. Any such ecosystem-management programme should: reduce those market distortions that adversely affect biological diversity; align incentives to promote biodiversity conservation and sustainable use; internalize costs and benefits in the given ecosystem to the extent feasible.
5. Conservation of ecosystem structure and functioning, in order to maintain ecosystem services, should be a priority target of the ecosystem approach.
6. Ecosystems must be managed within the limits of their functioning.
7. The ecosystem approach should be undertaken at the appropriate spatial and temporal scales.
8. Recognizing the varying temporal scales and lag-effects that characterize ecosystem processes, objectives for ecosystem management should be set for the long term.
9. Management must recognize the change is inevitable.
10. The ecosystem approach should seek the appropriate balance between, and integration of, conservation and use of biological diversity.
11. The ecosystem approach should consider all forms of relevant information, including scientific and indigenous and local knowledge, innovations and practices.
12. The ecosystem approach should involve all relevant sectors of society and scientific disciplines.

Box 21.3 ICM and EBM principles of action are mutually supportive

ICM and EBM lead to a strategic approach towards sustainable development that recognizes the need for:

- Ecosystem approach – integrity and functioning of socio-ecosystems.
- Threats caused by climate change.
- Appropriate and ecologically responsible coastal protection.
- Sustainable economy and employment.
- Functioning social and cultural system.
- Land accessibility.
- Maintenance/promotion of remote communities.
- Improved coordination between implementing agencies.

Box 21.4 The Cairo Principles

1. Overarching principle: Reduce the vulnerability of coastal communities to natural hazards by establishing a regional early warning system, applying construction setbacks, greenbelts, and other no-build areas in each nation, founded on a scientifically mapped 'reference line'.

Using concepts of ICM, including public engagement in local decision-making, rapid assessment zoning and planning processes will be used to:

2. Promote early resettlement with provision for safe housing; debris clearance; potable water; sanitation and drainage services; and access to sustainable livelihood options.
3. Enhance the ability of the natural system to act as a 'bioshield' to protect people and their livelihoods by conserving, managing, and restoring wetlands, mangroves, spawning areas, seagrass beds and coral reefs; and by seeking alternative sustainable sources of building materials, with the aim of keeping coastal sand, coral, mangroves, and rock in place.
4. Promote design that is cost-effective, appropriate, and consistent with best practise and placement of infrastructure away from hazard and resource areas, favouring innovative and soft engineering solutions to coastal erosion control.
5. Respect traditional public access and uses of the shoreline, and protect religious and cultural sites.
6. Adopt EBM measures; promote sustainable fisheries management in overfished areas, and encourage low-impact aquaculture.
7. Promote sustainable tourism that respects setback lines and carrying capacity, benefits local communities and applies adequate management practices.

How things are done is as important, sometimes more important, than *what* is done. Local knowledge and insights are critically important to successful planning and decision-making, and local citizens must be engaged in the rehabilitation and reconstruction process at every stage. It is essential that the application of the construction setback line and the boundaries of bioshields are defined in consultation with the local communities, coastal reach by coastal reach.

8. Secure commitments from governments and international organizations to abide to these *Principles* and build on and strengthen existing institutional arrangements where possible.
9. Ensure public participation through capacity building and the effective utilization of all means of communication to achieve outcomes that meet the needs and realities of each situation.
10. Make full use of tools such as strategic environmental assessment, spatial planning and environmental impact assessment, to identify tradeoffs and options for a sustainable future.
11. Develop mechanisms and tools to monitor and periodically communicate the outcomes of the reconstruction through indicators that reflect socio-economic change and ecosystem health.
12. Widely disseminate good practices and lessons learned as they emerge.

officials from tsunami-affected countries, representatives from UNEP Regional Seas Programmes, and other UN agencies including the Food and Agriculture Organization (FAO) and the Intergovernmental Oceanographic Commission of the United Nations Educational, Scientific and Cultural Organization (IOC-UNESCO), as well as by international institutions such as the International Union for Conservation of Nature (IUCN), the World Bank and the World Wide Fund for Nature (WWF) (Chua 2006).

The likely and potential impacts of sea level rise, increased frequency of storm events, acidification of seawater, desertification of arable land, and the associated declines in ecosystem function should be considered on the short- (10 years), mid- (30 years), and long-term (100 years) periods.

Some adaptation handbooks already exist like the USAID (2009) one, which offers a comprehensive overview of the impacts of climate change on coastlines and the tools that may be applied to the mitigation of its impacts. The guidebook 'proposes an approach for assessing vulnerability to climate change and climate variability, developing and implementing adaptation options, and integrating options into programs, development plans, and projects at the national and local levels'. It highly recommends stepping up efforts for guiding proactive adaptation actions from assessing vulnerability (Step 1) to implementing adaptation (Step 4) and evaluating for adaptive management (Step 5), which is actually exactly the same steps recommended in any integrated management strategy putting particular emphasis on the following adaptation measures: (i) functioning and healthy coastal ecosystems, (ii) built environment less exposed, (iii) diversified livelihoods, (iv) human health and safety enhanced, (v) overarching planning and governance which include coastal watershed management, ICM, and special area management planning (including Marne Protected Areas).

Confirming this close interconnection between integrated management and climate change adaptation, the UK *Charting Progress* report (Defra 2005) for an integrated assessment of the state of UK seas, underlines the importance of fully integrating the assessment of possible climate change impacts in future marine and coastal management strategies, considering that 'in the long term, the greatest threat to the planet, including the marine environment, could be the impacts of climate change'. Therefore, climate change adaptation measures should be fully integrated into integrated management strategies and implementation.

21.3 As a response, are national climate change strategies efficient enough?

National climate change strategies have now been drafted in quite a number of countries, both developed and developing, all around the world. But if they reflect preliminary adaptation planning efforts they still require elaboration or supplementation, preferably in sectoral or regional planning documents, to achieve greater specificity, in particular regarding adaptation measures in the coastal area. A comparative analysis of seven countries' national climate change strategies (India, Brazil, China, Mexico, South Africa on one hand – Table 21.1; France and UK on the other hand – Table 21.2) illustrates the fact.

The seven plans do provide directions of national adaptation needs and priorities. They articulate the potential effects of climate impacts on livelihoods, economies, and natural systems like the coast, but stop short of providing concrete procedures and

Table 21.1 Comparative analysis of five national climate change strategies.

	INDIA National Action Plan on Climate Change	BRAZIL National Plan on Climate Change	CHINA National Climate Change Programme	MEXICO Special Programme on Climate Change	SOUTH AFRICA Long Term Mitigation Scenarios
Issuing entity	Prime Minister's Council on Climate Change, July 2008	Inter-Ministerial Committee on Climate Change, December 2008	National Development and Reform Commission, June 2007	Inter-Secretarial Commission, March 2009 (review draught)	Cabinet of South Africa, July 2008
Overview and scope	Address mitigation, adaptation Defines eight national missions: solar, energy efficiency, sustainable habitat, water, Himalayan ecosystem, green India, sustainable agriculture, and strategic knowledge	Address mitigation, adaptation Covers energy, forests and agriculture, and other sectors like industry, waste, transport, and health.	Address mitigation, adaptation Covers energy production and transformation, energy efficiency, industrial processes, agriculture, forestry, and waste.	Address mitigation, adaptation Covers energy generation, energy use, agriculture, forests, and other land uses, waste, private sector.	Identifies measures to reduce emissions and adapt: immediate activities and measures to scale these actions up with additional resources, tax, and incentive packages.

(*Continued*)

Table 21.1 *(Continued)*

	INDIA National Action Plan on Climate Change	BRAZIL National Plan on Climate Change	CHINA National Climate Change Programme	MEXICO Special Programme on Climate Change	SOUTH AFRICA Long Term Mitigation Scenarios
Examples of proposed adaptation interventions related to the coast	• Improve land use and development planning • Improve coastal protection through infrastructure and forest/mangrove restoration • Wetlands conservation • Desalination technology development • Support climate change research and modelling	• Improve regional modelling of climate change impacts • Vulnerability mapping for coastal zones, biodiversity, water resources… • Identify most vulnerable groups and address socio-economic factors	• Incorporate climate change into laws and regulations on forests and wetlands • Expand ecosystem monitoring system • Slope and shore protection through engineering and biological measures	• Preserve, widen, and connect protected areas • Build ecosystem resilience • Avoid and control spread of invasive species • Evaluate current national capacities and seek sectoral integration • Deepen understanding of impacts of climate change	The scenarios do not specifically address adaptation. South Africa is developing a National Climate Change Response Policy that touches on vulnerability and adaptation

Source: Adapted from: National Climate Change Strategies: comparative analysis of developing country plans. World Resources Institute, June 2009.

Table 21.2 Comparative analysis of two national climate change strategies within the EU.

	FRANCE National Climate Change Impact Adaptation Plan	UNITED KINGDOM National Adaptation Programme
Issuing entity	Ministry of Environment, Energy and the Sea, 2011	Secretary of State for Environment, Food and Rural Affairs
Overview and scope	Address adaptation as a pre-emptive approach by stakeholders to future problems and to society's perception of identified changes. Define four main goals (Protecting people and property; taking social aspects into account; limiting costs and exploiting benefits; preserving natural heritage applied to 20 fields including amongst others, Health, Water, Biodiversity, Natural hazards, Fisheries and aquaculture, Coastlines, Transport infrastructures, Tourism	Address adaptation by dealing with the risks and making the most of the opportunities. Define policies and actions in seven priority areas: Built environment; Infrastructure; Healthy and Resilient community; Agriculture and Forestry; Natural environment, Business, Local government
Examples of proposed adaptation interventions related to the coast	Pursue and promote existing approaches in protected areas networks relating to the study of current and potential future consequences of climate change on biodiversity Suggest adaptation measures for existing coastal defences Adapt the French shellfish sector to climate change issues Develop coastal observation networks Improve understanding of the coastline Adapt regulations and forms of governance Assess coastal population and housing issues Support the development of regional climate change adaptation strategies	Develop local flood-risk management strategies that set out the approach to managing local flood risk and consider the effect of future climate change and increasing severity of weather events Make national coastal erosion risk mapping Communities and civil society group taking action to build resilience to extreme weather events and impacts of climate change Develop a strategic plan for coastal realignment Embedding understanding of climate change into payment for ecosystem services and the business case for investing in natural solutions

strategies for meeting adaptation needs. As a matter of fact, all seven documents are meant as national overall frameworks requiring further adaptation planning efforts through regional and local planning documents as in the case of any national action plan to become practical and concrete. The argument could be extended to other established frameworks, particularly those that have to do with biodiversity conservation, land-based sources of pollution, or fisheries.

Any of those national strategies should be considered as overarching national frameworks within which the characteristics of provinces, regions, and localities are fully addressed through and with the concerned stakeholders.

From a global perspective, though poorly applied, the 1992 Earth Summit's Agenda 21 challenge is about envisioning the entire spectrum, encompassing both the land and water sides, emphasizing as well the need for proper management of any sector in ocean areas under national control and the importance of the connection between land and sea, i.e. covering the inland areas that affect the oceans mainly via rivers and non-point sources of pollution, coastal lands (wetlands, marshes, and so on) where human activity is concentrated and directly affects the adjacent waters, coastal waters (estuaries, lagoons, and shallow waters generally) where the effects of land-based activities are dominant. With the Agenda 21, Chapter 17, the concept of ICM was born. Five years after issuing its ICM Recommendation (2002), the evaluation report from the European Commission to the European Parliament and the Council (CEC, 2007) emphasized the following: 'When launching its strategy to implement the EU ICM Recommendation, the Commission indicated that coastal areas are particularly in need of an integrated territorial approach, but that, in essence, such good territorial governance is relevant for other areas facing multiple pressures and conflicting interests. This is increasingly the case for the seas and oceans. Notwithstanding the continued need for ICM on-shore, further emphasis will need to be placed on the implementation of ICM across the land-sea boundary and in a regional seas context'.

Nevertheless, ICM does not aim to replace sectoral policies: it does not delegitimize nor make them obsolete. Sectoral socio-economic activities and interventions (such as fisheries, mining, tourism, protected areas…) are the key drivers of change in coastal areas. Strong sectoral policies and their effective regulation (including through Environmental Impacts Assessments [EIAs] or licensing procedures) are crucial and offer considerable leeway for more integrated management.

> *On the 'theoretical side' of ICM, the systemic approach at the origin of the integration concept is essential to understand articulations, synergies, side effects, etc. On the 'action side' of ICM, a more ICM stems first and foremost – and maybe ironically – from strategic interventions on sectors and sectoral policies guiding them. The unquestionable need for more integration should not remove the focus from the main threats to the coastal zone – usually the lack of or appropriate implementation of sectoral policies. Think integrated but act strategic. (Billé and Marre 2015)*

Rather than being an all-encompassing solution for managing coasts and oceans, integrated ecosystem-based coastal management and more particularly national ICM strategies should be seen as facilitating and strengthening the implementation of diverse frameworks and processes given the commonality of their operational goals, approaches, and areas concerned. Actually, responsible management of any marine area should integrate the integrated coastal zone management (ICZM) or ICM strategy and the Ecosystem Approach of the Convention of Biodiversity as well as applying the same principles from the coast to the offshore waters. Then, instead of using different acronyms underpinned by the same principles, it would be most advisable to use the 'labelling' of 'integrated ecosystem-based coast and ocean management' as was recommended in 2009 throughout a national consultation ('Grenelle of the sea') in France. It is not about inventing a new concept but rather about reconciliation of biodiversity conservation and maritime knowledge and activities development towards the same final goal of coasts and seas sustainable development.[1]

21.4 Key principles and responses for building sustainable, hazard-resilient communities

The final section will provide concrete recommendations for guiding efforts and translating key principles into practice, very much similar to any ICM approach. The two first principles are particularly related to the two first disaster terms of 'reduction' and 'readiness' whilst the two others are more directly linked to the 'response' and 'recovery' phases[2]:

1. Put people first, especially the most vulnerable people, through implementing diverse forms of coordination, and institutional and non-institutional agreements. These should be characterized by transversal, non-hierarchical relationships and collective learning. Deliberative and participative approach, renewed principle of authority (negotiation, charter, partnership, contract, international convention, instruments for mediation and subsidiarity) in identifying and analysing long-term risks to human life and property from natural or non-natural hazards.
2. Put places and issues into context through nested systems of governance reflecting the various scales, entanglement, and interconnection of environmental and societal issues. The complex, multidimensional character of natural hazards necessitates the linking of natural, social, and technological considerations across scales. More generally speaking, all planning and decision making must recognize and analyse conditions, issues, and goals at least at the next higher or lower level in the governance system. This applies as well to developing operational systems and capabilities, including self-help and response programmes for the general public as well as specific programmes for emergency services.
3. Secure local awareness and implementation capacity through the mobilization of social networks and local authorities, which often lack the capabilities and resources to address adequate preparation for disasters and actually deal with the short- and long-term implications of a disaster. In addition to their own awareness and training, communities and local authorities will require the appropriate institutional arrangements to get the needed support from the next higher administrative scale (provincial, national, regional).
4. Monitor, review and adapt planning and decision-making processes, informed by initiatives that promote social learning through clear lines of responsibility and accountability of concerned organizations. In collaboration with local authorities and specialized agencies, coastal communities need to be empowered to work collaboratively to devise responses and recovering activities that translate these general principles and operational imperatives into locally relevant and culturally appropriate policies and practices.

21.4.1 Community's sense of ownership, at the core of planning and efficient implementation

Most of the time, hazard events like coastal storms (often doubled with flooding[3]) will be dealt with and talked of in a 'responsive-driven manner with a compelling focus on saving lives, providing emergency relief and marshalling resources for restoration and reconstruction', but such vital measures need to be put into a more holistic and proactive approach touching first of all upon the 'reduction' and 'readiness' phases. Following the two first principles mentioned earlier (putting people first and putting places and issues into context), the key is about building up more resilient communities through appropriate land-use planning securing 'meaningful participation'. Most of the countries, and more particularly EU member states, have developed land-use and local risks management plans, but the question of local stakeholders' sense of ownership of such plans remains largely open.

Glavovic (2000), drawing the lessons out of national and local participation process examples, identified some key enabling conditions to make it happen hence allowing a sense of ownership to develop amongst local stakeholders and communities:

- *The process should be designed in an inclusive, voluntary, and culturally sensitive manner*. Particular attention needs to be given to designing culturally sensitive and appropriate methodologies to engage diverse participants effectively in the participatory process. Different kinds of opportunities, forums, and participation methodologies need to be developed, tested and applied, depending on stakeholder needs. It should be an iterative process in which capacity and trust are progressively built over time, contributing to deeper insights and to enhanced stakeholder relationships. Locally networked and informed regional managers may play a key role in this regard.
- *The process should be aimed at empowering historically disadvantaged individuals, groups, and communities*. Socially and geographically distinct patterns of poverty and inequality will be perpetuated unless there is a commitment to empowering those who are marginalized. Creating opportunities for meaningful public participation can be a powerful means of mobilizing historically disadvantaged people.
- *The process should be conceptualized as a partnership-building endeavour*. A broadly owned policy outcome is based on a shared commitment to its implementation. Such partnership-like relationships provide the basis upon which stakeholders can learn about and appreciate the interests of others. Conceptualizing the process as a partnership-building endeavour helps to foster a common understanding of the issues and builds a share set of values that can be then translated into practical measures for cooperation.
- *The process should be designed and managed to deepen and extend public deliberation*. Promoting public participation presumes that participants are well informed about the issues at hand and are able to engage in group discussions that get to grips with the substantive nuances of the issues. It also presumes that participants are able to work through their differences of opinion and develop a common understanding of the issues. Usually, public meetings provide limited opportunity for in-depth discussion. Alternative forums and participatory methodologies are required to extend and deepen discussion, including small group discussion that facilitate

increased interaction between specialists and stakeholders, as well as deeper levels of interaction between stakeholders.

- *The process should be managed in an innovative, reflective, and deliberative manner that is responsive to changing circumstances and stakeholder interests*. From an operative point of view: (i) keeping the momentum requires *independent facilitators* who, depending on circumstances, may need to play different roles, ranging from mediator to negotiator, educator, advocate, and so forth; (ii) building stakeholders' interest, understanding, and trust necessitates timely, accurate and *regular feedback* that reflects the nature of their contributions and the manner in which they have been integrated into the products of the process; (iii) the process should be designed and managed to be *responsive* to the needs and interests of stakeholders and to the new insights that emerge in the course of the process, (iv) careful attention needs to be given to using the most *appropriate media* and means to make the outputs of the process widely accessible and reach particular target audiences, such as key decision-makers or the youth; (v) conducting such an extensive participatory process requires securing sufficient *financial resources* as well as a *reasonable timeframe* to engage stakeholders in formulating the coastal policy and plan.

Practically, such a planning process influences and is dependent at the same time on the governance arrangements at the different decision-making levels. A broad array of hybrid governance arrangements have being practised, sometimes since ancient times (Ostrom 1990), between the state, market actors, and communities, with a greater role for non-state actors such as resource users, private sector organizations, and non-government organizations (NGOs) (Figure 21.1).

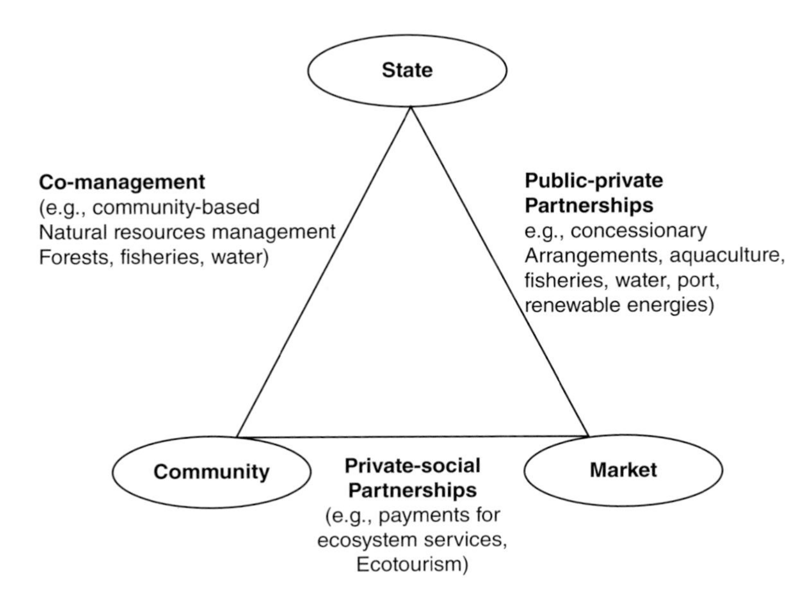

Figure 21.1 Hybrid governance arrangements. *Source*: Adapted from Lemos and Agrawal (2006).

21.4.2 Governance trajectory

As guidelines for community resilience building, Berkes and Ross (2013) identify nine important characteristics, leading to agency and self-organization in communities of place: people-place connections, values and beliefs, knowledge and learning, social networks, collaborative governance, economic diversification, infrastructure, leadership, and positive outlook. Adaptive capacity and agency can be facilitated by community members themselves through social learning, or by external change agents (NGOs) using well-known approaches in community development for building community strengths and resilience. Moreover, through history, each community-place has been going through varied forms of governance in response to political, economic, social, and environmental changes, highlighting the fundamental interconnection between political, social, economic, and cultural systems, and the way in which natural systems are coupled with and embed human systems. It is then important to know how the governance system in a specific place has responded through time (Olsen, Page, and Ochoa 2009). A careful documentation and analysis, carried out as a learning process with all stakeholders concerned, provide important insights and how best to design a forward-looking management and governance initiative. Such baselining methods are extremely useful to build a shared understanding of how current issues have evolved in a specific place. They have been successfully applied in a wide variety of social, political, and environmental contexts. One example is in the Caribbean island of Martinique (France) where a community analysis of governance trajectories has been recently carried out (de Cacqueray et al. 2015). The shared diagnostic led to the establishment of a 'timeline' covering the natural environmental changes, the policies, institutions, laws, and management tools related to the coastal area, socio-economic changes, and the overall historical and political events and context. The process led then to the common identification of two core capacity issues: policies and management practices related, (i) to climate change and coastal risks (erosion and storm surge) and, (ii) to conservation and creation of a network of marine protected areas. As said before, generating a community's sense of ownership looking in a learning process to the past, present, and future, is a key element in the 'reduction' and 'readiness' phase as well as the consecutive 'response' and 'recovery' phases. Put in an economic perspective, it refers to the strengthening of what may be called 'social capital' or the 'features of social organization, such as trust, norms, and networks that can improve the efficiency of society by facilitating coordinated actions' (Putnam 1993). The capacity to adapt to rapid and dramatic changes of governance largely depends on the balance between the stability provided by strong institutions on the one hand, and the capacity to experiment, innovate, and learn from changing circumstances through well-structured social networks (Duit and Galaz 2008). The latter is considered an inherent part of a 'robust' governance of complex social-ecological systems.

21.4.3 Communities and partnerships

'Ecosystem-based management is an integrated approach to management that considers the entire ecosystem including humans. The goal is to maintain ecosystems in a healthy, clean, productive, and resilient condition, so that they can provide humans with the services and benefits upon which we depend. It is a 1) spatial approach that builds around 2) acknowledging connections, 3) cumulative impacts and 4) multiple objectives. In this way, it differs from traditional approaches that address single concerns e.g., species, sectors or activities' (EEA 2014). In Europe, the frame is the Marine Strategy Framework Directive and the goal is the Good Environmental Status of European seas, the environmental pillar of the EU Integrated Maritime Policy.

Towards such an ambitious goal, partnerships between the three governance components (government, market, civil society) are a prerequisite to build up social-ecological resilience from local to national and regional seas levels. The varied governance arrangements as mentioned in Figure 21.1 are just some examples of these partnerships. Interestingly in the UK, over 60 voluntary Coastal Partnerships (Table 21.3) have developed since the 1990s (Stojanovic and Barker 2008). They commonly employ a coordinating officer and in some cases a small team (of up to five staff members) delivering core services plus projects depending upon funding availability. Other defining features of the Coastal Partnerships tend to be: a regular forum or conference bringing together decision-makers with sectoral interest groups to debate current issues; the use of topic/focus groups to carry out specific tasks such as problem solving, report writing, or policy development; and development of communication mechanisms such as workshops, websites, newsletters, and consultations to generate wide involvement from government, private, and voluntary sectors. These structures have provided the momentum to formulate and implement voluntary coastal and estuary management plans and

Table 21.3 Definition of coastal partnerships.

Coastal initiatives	Categories defined by the English Coastal Partnerships Working Group in 2007
CP	Coastal (including estuary) Partnership or forum bringing together all sectors to advocate sustainable management of a coastal area based on ICM principles. www.coastalpartnersips.org.uk
AONB/HC	Area of Outstanding Natural Beauty: a partnership or initiative set up to manage a designated landscape in the coastal zone. www.countryside.gov.uk/LAR/Landscape/DL/aonbs/index.asp or non-statutory, Heritage Coasts www.countryside.gov.uk/LAR/Landscape/DL/heritage_coasts
EMS	European Marine Site initiative set up to prepare and implement an EMS Management Scheme for a designated Special Protection Area/Special Area of Conservation. www.ukmarinesac.org.uk/uk-sites.htm
MNP/VMNR	Marine National Park or Voluntary Marine Nature Reserve set up to manage/protect an offshore park, reserve or protected area
CG	Coastal Group assisting production of shoreline management plans (SMPs) for flood and coastal erosion risk management. www.defra.gov.uk/environ/fed/policy/CoastalGroups.htm
Other	A variety of other initiatives which include local authority strategies for the coast, ad hoc partnerships based on topics such as beach care, litter, or marine wildlife

Source: From: Stojanovic and Barker (2008).

Not all of these programmes are operating across all of the constituent nations of the UK. Furthermore, the devolved administrations have their own approaches to engaging with Coastal Partnerships through the Scottish Coastal Forum, Wales Coastal and Maritime Partnership, and Northern Ireland Coastal Forum.

strategies. They are brought together into a national network through the Coastal Partnership Working Group (2006), which reports to the Annual Coastal Partnerships Forum.

At regional sea level, another worthwhile example is the 'Celtic Seas Partnership',[4] an EU-funded project aiming at drawing communities together from across the Celtic Seas to set up collaborative and innovative approaches to managing their coastal and marine social-ecological systems. It appears that the most commonly raised issue is, at the various scales, a 'crucial need to develop better relationships and trust' through, amongst others, the use of experienced unbiased facilitators, drawing together scientific evidence that stakeholders can collectively support, and holding meetings between officials from administrations at national and regional level (e.g. the regional convention OSPAR) to find areas of common ground (CSP 2015).

What is nascent in Europe already has a long history in Australia thanks to the Great Barrier Reef Marine Park Authority (GBRMPA) and its 'Extreme Weather Response Program'[5] to understand better the impacts of extreme weather and help coastal and marine industries and communities prepare for future extreme weather events. Building 'partnerships in resilience' with all stakeholders concerned (reef industries, researchers, communities, and Traditional Owners) is the policy developed by GBRMPA to proactively take multi-objective steps to support the resilience of the Reef in the wake of extreme weather events.

21.4.4 Measuring recovery from disasters

The recovery phase shares the three previous principles (put people first; put places and issues into context; secure local awareness and implementation capacity) with the reduction, readiness, and response phases, backed by an approved government policy (national/local), an enabling national and local system, the appropriate tools and the advocacy amongst all the actors including civil society as developed earlier in the chapter. The United Nations Development Programme (UNDP), 'Post-Disaster Recovery Guidelines' document (2012) states that 'the aftermath of a major disaster is frequently characterized by a multiplicity of actors promoting and initiating recovery activities. Coordination and information sharing thus become even more essential' to constitute a permanent dialogue and consensus-building mechanism between actors and 'a platform' for the strong, inter-sectoral coordination required to generate synergies between a large number of initiatives at the local, regional, and national scales.

But often, a lack of common measurements limits the ability to share information, to compare outcomes between disasters as well as to monitor changes over time. As with any management plan (including its pre-event baselines), the development of a robust and user-friendly set of recovery indicators can address such shortcomings and build the capacity of local practitioners by providing data for decision making. Actually, the task is complex and maybe long term when considering the overall 'recovery focus areas' (Dwyer and Horney 2014) in regard to the different kinds of 'capital' at stake as shown on Table 21.4.

These four 'capitals' and their recovery focus areas acknowledge that recovery extends well beyond restoring physical buildings and infrastructures or providing welfare services. As said before and to get on a sustainable path, recovery is best

Table 21.4 Some examples of recovery focus areas.

Social capital	Economic capital	Natural capital	Built capital
Increased participation in disaster planning	Business recovery	Restoration of natural resources	Repair or replacement of buildings and infrastructure
Healthy communities	Mobilization of recovery funding	Restoration of polluted areas	Communication of disaster-related information
Restoration of cultural sites/ resources	Economic stabilization	Restoration of amenity values	Implementation of hazard mitigation techniques
Population stabilization	Public sector recovery	Restoration of ecosystem functions and biodiversity	Use of recovery plan/ planning measures
Household Recovery	Social services recovery		

Source: Adapted from Dwyer and Horney (2014).

achieved when the affected communities exercise a high degree of self-organization, hence self-determination.

But, the reality in implementing plans of all kinds (recovery and others) is that one often sees only fragments of unconnected cycles with a major gap between repeated efforts at issue and stakeholder analysis and actual implementation of the plan or programme of action. Experience demonstrates repeatedly that a sound process, with appropriate participation, a technically competent programme or plan staff and sustained governmental support, may not deliver the desired outcomes as it has been often the case in post-disaster areas. Why? Because the only desired and well-identified outcomes are the final ones (of a project or programme) or 'end outcomes' (Figure 21.2) mainly touching upon the three capitals (Table 21.4), 'natural', 'built', and 'economic' capitals with very different timings, whilst the dynamic (Figure 21.2; 'intermediate outcomes') of the 'social' capital is rather ignored. It is to overcome such a limited vision of any integrated management plan implementation and monitoring that the 'orders of outcomes' approach was proposed.

The 'Orders of Outcomes' proposed framework (Olsen 2003; UNEP/GPA 2006) disaggregate the distant goal of sustainable forms of recovery and development into a sequence of more tangible outcomes (Figure 21.2) touching upon the dynamics of the four capitals including the 'social' one (Table 21.4).

The *First Order* examines whether a sufficient level of achievement – *an enabling framework* – has been attained that creates the preconditions required to successfully implement the plan of action of any policy or programme, e.g. whether or not the policy or programme has established unambiguous goals and whether user groups affected by the programme's actions understand and support its goals, management measures, and targets. Actually, the phases of 'reduction' and 'readiness' are supposed to contribute directly to the enabling framework.

Experience tends to show that an effective programme will be strengthening its First Order four categories of preconditions[6] as it generates some Second and Third Order

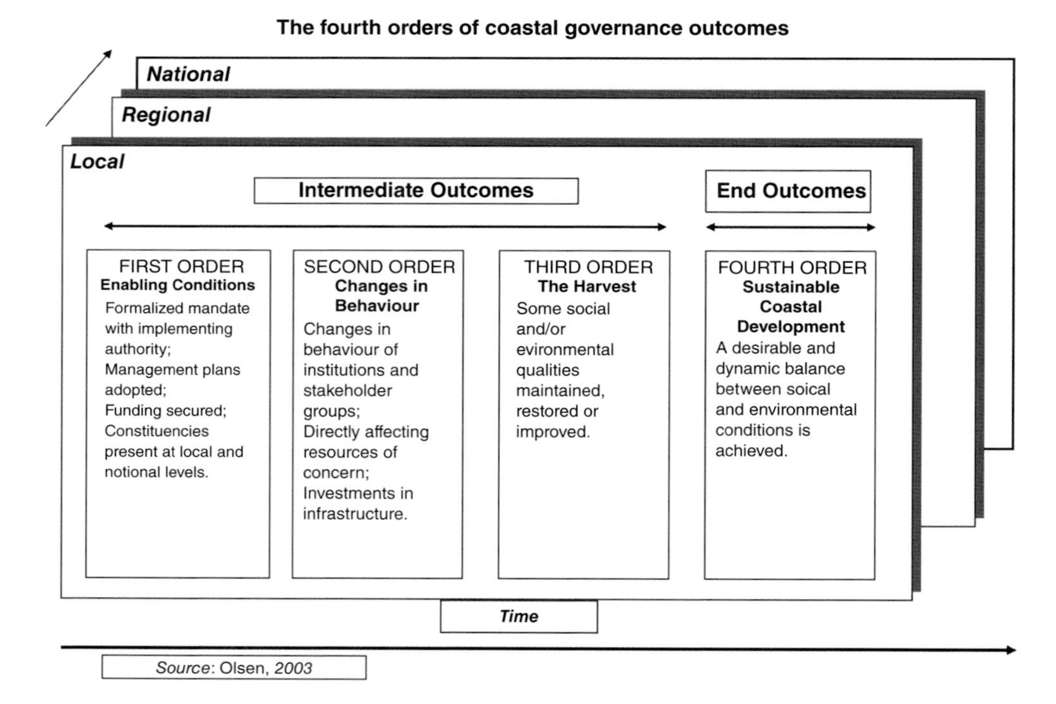

Figure 21.2 The four orders of coastal governance outcomes.

outcomes by addressing the most tangible issues. The key, therefore, is to build many bridges between the First and the Second Orders and not to structure a programme too rigidly into planning and implementation phases. Nonetheless, a well-informed understanding of the existing governance system and careful consideration of the indicators for the First Order preconditions will support sound judgments about readiness for implementation (Olsen et al. 2009).

The *Second Order* analyses *changes in behaviour* that occur during implementation: changes in the behaviour of target user groups, changes in the behaviour of key institutions, and changes in how and where financial investments are made.

The selection of *boundary partners* (see Strategy section: Identifying the boundary partners) enables a programme to specify what Second Order changes in behaviour is anticipated to generate progress towards the Third Order goals. The International Development Research Centre (IDRC) methods (Earl, Carden, and Smutylo 2001) suggest organizing such monitoring by identifying an *outcome challenge* for each category of boundary partners and then selecting graduated variables for assessing the degree to which those changes in behaviour are achieved. In that sense, the Second Order outcomes may be looked at as an expression of a *learning by doing* approach.

The *Third Order* measures *practical results and benefits*. These Third Order Outcomes, e.g. improved water quality, justify financial investments and motivates stakeholders and institutions to make changes in their behaviour that sustained success requires.

What will be monitored and how the monitoring will be done is logically to be determined by the specific Third Order targets or objectives supposed to be achieved. It is, therefore, important to describe and as much as possible quantify the environmental and social respective baseline conditions. From past experiences, it is strongly advisable to avoid getting lost in tracking changes on too many items (abundance of fish, water quality, income of target social groups, etc.) but instead to carefully select a very few indicators that will provide future comparison to the baseline conditions. As mentioned earlier, the relative simplicity of the monitoring system of the Mediterranean Strategy on Sustainable Development should be taken as an example.

The *Fourth Order* looks at the appropriate balance between environment and human society – *sustainable development*, i.e. the very-long-term utopian goal.

All in all, the purpose is to 'get people thinking' about specific community elements that need to be addressed to both prepare for a potential disaster as well as foster a successful recovery following a disaster event (Dwyer and Horney 2014). Generally speaking, it allows the shifting away from assessing a programme impact towards changes in behaviours and activities of people, groups, and organizations concerned. Here, the changes in behaviour are correlated to the changes in state which, of course, have still to be measured. The focus of measurements is therefore clearly outcomes not outputs (e.g. Euros spent on reconstruction).

In this very spirit, a funding agency, the Canadian IDRC, has developed a methodology called 'Outcome Mapping' (Earl et al. 2001), organized in three stages: (i) Intentional Design, asking the group or community concerned what is the vision to which the programme or project wants to contribute?, who are the programme/project's boundary partners?, what are the changes that are being sought?, and how will the programme/project will contribute to the change process?; (ii) Outcome and Performance Monitoring, provides a framework for monitoring actions/activities and boundary partners' progress towards the achievement of outcomes; and, (iii) Evaluation Planning to define the priorities for outcomes measurement (quantitative and qualitative) depending on the third order objectives and the choices made by the people/community to go back to what it was before the disaster or build up a 'new normal' to prevent such a disaster impact to happen again.

21.5 Conclusion: 'Hazard-resilient' communities vs. 'waves of adversity'

If the fact of increasingly frequent and potent hurricanes, typhoons, or coastal storms cannot be changed, the extent of their overall impact will most depend on the previous planning choices, institutional capacities (from local to national levels), and social vulnerability of people and communities. Most of the time and more particularly in Europe, technically sound risk and recovery management plans are produced, driven by a range of agencies and organizations from the national to local level. They usually focus on urban design issues and land-use options that would reduce specific hazard risks like flooding or coastal storms (or the two combined as mentioned in Section 21.4.1), prioritize redevelopment resources, and maintain services for the surviving population. But, like for any management plan, often the process does not create enough opportunities for meaningful public participation to create ownership of the plan and community engagement at every stage – planning, implementation, monitoring, and review – and at levels beyond mere consultation. Analysing the aftermath of the 'Katrina' hurricane in

New Orleans, Olhansky et al. (2008) reiterate the importance of pre-event planning, public participation, information infrastructure, and adequate external resources to recovery efforts. They highlight the difficulties experienced in rebuilding New Orleans due to the slow flow of Federal funding and inadequate local leadership in initial planning efforts and information management. The problem is that often (like in New Orleans), and in spite of the suffering, people desperately want to 'return to normal' even knowing that another hurricane or storm and flooding are inevitable. This added to persistent social vulnerability, could make future hazard impacts even more devastating.

It is, therefore, of the utmost importance to shift from response-driven policy to a more proactive approach that makes institutions and communities more 'hazard-resilient' and able to cope with and learn from hazard events. Land-use planning remains indeed essential but the challenge is to do it through meaningful public participation, i.e. how far citizens have a 'real say' in the planning outcome provided they are aware they are not separated from nature but in a totally interconnected 'social-ecological' system that requires appropriate governance institutions and processes (including property rights, access to education and health care, social exclusion, vulnerability, political marginalization, and power imbalances) in mediating access to its resources from the watershed (flooding) to coastal areas (storm surge and/or flooding). Finally, governance institutions and social processes play a fundamental role in determining the sustainability and resilience of coastal communities, more than any biophysical factor or hazard threat.

Rachel Brookie (2012), studying the governance aspects of the Canterbury earthquake recovery phase, has perfectly synthetized this approach with the following words: 'Though we cannot prevent most natural disasters, we can try minimising their impact and create governance arrangements to maximise recovery. The process for learning from disaster and recovery should consider not only institutional understandings, but also citizen expectations of what should be done better next time. Crucially, we must find ways to imbed learning from disasters into institutional memory'.

Notes

1. Yves Henocque, Integrated vs. sectoral: an assessment of coastal zone management integration, pressures, and new activities inputs. In preparation. Ed. Springer.
2. Reduction: identifying and analysing long-term risks to human life and property from natural or non-natural hazards, and taking steps to eliminate these risks or reducing their impact – Readiness: developing operational systems and capabilities before a civil defence emergency happens – Response: actions taken immediately before, during, or directly after a state of civil defence emergency – Recovery: the coordinated efforts and processes used to bring about the immediate, medium-term, and long-term holistic regeneration of a community.
3. Often, the major risk is not directly from sea-level rise, but from the intensification of storm surges and increased precipitation as secondary effects of climate change. The combination of heavier rainfalls with storm surges could make coastal flooding much worse in the future. This again shows how crucial it is to articulate the two management systems which apply to watersheds and coastal areas.
4. www.celticseaspartnership.eu
5. www.gbrmpa.gov.au/managing-the-reef/threats-to-the-reef/extreme-weather

6. (i) Specific goals for target environmental and societal outcomes; (ii) Supportive and informed constituencies; (iii) Required implementation capacity; (iv) Commitments for necessary authorities and resources for implementation.

References

Adger, W.N., Hughes, T.P., Folke, C. et al. (2005). Social-ecological resilience to coastal disasters. *Science* 309: 1036–1039. https://doi.org/10.1126/science.1112122.

Berkes, F. and Ross, H. (2013). Community resilience: toward an integrated approach. *Society and Natural Resources* 26: 5–20.

Billé, R.and Marre, J.B. (eds.) (2015). The RESCCUE approach. RESCCUE working paper. New Caledonia.: SPC Noumea.

Brookie, R. (2012). Governing the recovery from the Canterbury earthquakes 2010-11: the debate over institutional design. Institute for Governance and Policy Studies, Working Paper 12/01. Wellington, New Zealand: Victoria University of Wellington. https://core.ac.uk/download/pdf/30677697.pdf (accessed 27 February 2019).

de Cacqueray, M., Rocle, N., Meur-Ferec, C. et al. (2015). ATOUMO: vers une gestion intégrée de l'île de la Martinique et de son espace maritime. Trajectoires de gouvernance et adaptation aux changements passés, actuels et futurs. 80p. http://www-iuem.univ-brest.fr/pops/projects/atoumo (accessed 27 February 2019).

Celtic Seas Partnership. (2015). Working in partnership across the Celtic Seas. Workshop Report, 11-12 May 2015. Paris: Celtic Seas Partnership.

Chua, T.-E. (2006). *The Dynamics of Integrated Coastal Management. Practical Applications in the Sustainable Coastal Development in East Asia*, 431 p. Quezon City, Philippines: PEMSEA/GEF/UNDP/IMO.

Commission of the European Communities (2007). Report to the European Parliament and the Council: an evaluation of Integrated Coastal Zone Management (ICZM) in Europe Brussels: European Commission. COM(2007) 308 Final.

Defra (2005). *Charting Progress. An Integrated Assessment of the State of the UK Seas*. London: Department for Environment, Food and Rural Affairs.

Duit, A. and Galaz, V. (2008). Governance and complexity – emerging issues for governance theory. *Governance: An International Journal of Policy, Administration, and Institutions* 21 (3): 311–335.

Dwyer, C. and Horney, J. (2014). Validating indicators of disaster recovery with qualitative research. *PloS Currents* https://doi.org/10.1371/currents.dis.ec60859ff436919e096d51ef7d50736f.

Earl, S., Carden, F., and Smutylo, T. (2001). *Outcome Mapping – Building Learning and Reflection into Development Programs*, 139 p. Ottawa, Canada: International Development Research Center Ed.

EEA (European Environment Agency) (2014). *Marine Messages. Our Seas, our Future – Moving Towards a New Understanding*. Copenhagen: European Environemnt Agency.

Glavovic, B.C. (2000). *Building Partnerships for Sustainable Coastal Development. The South African Coastal Policy Formulation Experience: The Process, Perceptions and Lessons Learned*. South Africa: Common Ground Consulting, Department of Environmental Affairs and Tourism.

Glavovic, B.C. (2008). Sustainable communities in the age of coastal storms: reconceptualising coastal planning as a 'new' naval architecture. *Journal of Coastal Conservation* 12 (3): 125–134.

Henocque, Y. and Kalaora, B. (2015). Integrated management of seas and coastal areas in the age of globalization. In: *Governance of Seas and Oceans* (ed. A. Monaco and P. Prouzet), 235–279. London and Hoboken: ISTE Ltd and Wiley.

Holling, C.S. (1986). The resilience of terrestrial ecosystems: local surprise and global change. In: *Sustainable Development in the Biosphere* (ed. W.C. Clark and R.E. Munn), 292–317. Cambridge: Cambridge University Press.

Lemos, M.C. and Agrawal, A. (2006). Environmental governance. *Annual Review of Environment and Resources* 31 (1): 297–325.

Olhansky, R.B., Johnson, L.A., Horne, J. et al. (2008). Planning for the rebuilding of New Orleans. *Journal of the American Planning Association* 74 (3): 273–287. https://doi.org/10.1080/01944360802140835.

Olsen, S.B. (2003). Frameworks and indicators for assessing progress in integrated coastal management initiatives. *Ocean & Coastal Management* 46 (3–4): 347–361.

Olsen S. (2009). Building capacity for the adaptive governance of coastal ecosystems. A priority for the 21st century. Background paper for a LOICZ Dahlem-type workshop.

Olsen, S.B., Page, G.G., and Ochoa, E. (2009). The analysis of governance responses to ecosystem change: A handbook for assembling a baseline. LOICZ Reports & Studies No. 34. Geesthacht: GKSS Research Center, 87 pp.

Ostrom, E. (1990). *Governing the Commons. The Evolution of Institutions for Collective Action*. Cambridge, UK: Cambridge University Press.

Putnam, R.D. (1993). *Making Democracy Work. Civic Traditions in Modern Italy*, 258 p. Princeton, NJ: Princeton University Press.

Stojanovic, T. and Barker, N. (2008). Improving governance through local coastal partnerships in the UK. *The Geographical Journal* 174 (4): 344–360.

UNDP (2012). Post-disaster recovery guidelines – Crisis prevention and recovery New York: United Nations Development Programme.

UNEP/GPA (2006). *Ecosystem-Based Management: Markers for Assessing Progress*. The Hague: UNEP/GPA.

United Nations (1992). Convention on Biological Diversity. https://www.cbd.int/ecosystem/principles.shtml (accessed 26 February 2018).

USAID (2009). *Adapting to Coastal Climate Change: A Guidebook for Development Planners*. Washington, DC: United States Agency for International Development.

Wallace, J.M. and Gutzler, D.S. (1981). Teleconnections in the geopotential height field during the northern hemishpere winter. *Monthly Weather Review* 109: 784–812.

IV.2

Strategies, Instruments, and Resources Used to Face Coastal Risks Prevention

22

European Challenges to Coastal Management from Storm Surges: Problem-Structuring Framework and Actors Implicated in Responses

Suzanne Boyes and Michael Elliott
Institute of Estuarine and Coastal Studies, University of Hull, United Kingdom

22.1 Storm surge threats in European coasts

22.1.1 European regions most impacted

> *Projected sea level rise, possible changes in the frequency and intensity of storm surges, and the resulting coastal erosion are expected to cause significant ecological damage, economic loss and other societal problems along low-lying coastal areas across Europe unless additional adaptation measures are implemented. (EEA 2017)*

All areas face hazards, both natural and human-induced, and often these hazards are exacerbated by human actions (Elliott, Cutts, and Trono 2014). If those hazards affect assets which we value for human health and safety, then we regard these as risks. Due to global climate change, European seas are facing rising sea levels and more extreme and intense weather events such as heatwaves, flooding, droughts, and storms (EEA 2017). Europe is warming faster than many other parts of the world, with temperature on land over the past decade being on average 1.3 °C higher than in the pre-industrial era, compared with a global average rise of 0.8 °C (EC 2013a). Projections for northern and north-eastern Europe include greater precipitation and inland flooding, with a

Facing Hydrometeorological Extreme Events: A Governance Issue, First Edition.
Edited by Isabelle La Jeunesse and Corinne Larrue.

heightened risk of storminess, coastal flooding, and erosion. An increase in such events is likely to enlarge the magnitude of disasters, leading to significant economic losses, public health problems, and deaths.

Even though faced with these threats, coastal areas have continued to experience a rapid rise in population and associated developments within recent decades (Lutz and Samir 2010; Wang et al. 2014), with the current European coastal population exceeding 200 million, stretching from the North-East Atlantic and the Baltic to the Mediterranean and Black Seas (EC 2016a). In particular, low-lying coasts and coasts with impoverished populations are especially at risk (Barbier 2015). Therefore, any changes in water level caused by a rise in sea level and increased storminess poses a threat to human lives and infrastructure.

Of all the natural hazards occurring along European coastlines, storm surges can be one of the most devastating if precautions are not taken. Storm surges are at their most destructive when spring or equinoctial tides, sufficient fetch and wind direction, and low air pressure (producing higher water levels) all combine to produce hazardous and potentially catastrophic conditions (Baxter 2005). Storm surges present a natural and increasing hazard throughout European coastal areas with an effect which may be exacerbated because of land subsidence due to isostatic rebound, a feature particularly notable around the southern North Sea (Wolanski and Elliott 2015). Hence because the land level is changing at the same time as the sea level, in these areas the term Relative Sea-Level Rise is used. The rate of global warming has increased leading to great concern over the associated relative sea level rise and the expected increase in the frequency of storms and tidal surges (CPSL 2001). Coastal flooding and extreme erosion events caused by increasingly variable and stormy weather and surge events are generally anticipated to increase as a result of climate change.

On a European scale, the regional seas all experience storm surges to different degrees. Research into climatic hazards across Europe has indicated that over the next century, environmental hazards are likely to increase, particularly along coastlines and on floodplains (Forzieri et al. 2016). A number of studies have investigated the impacts of storm surges at the local or regional scale, with a recent study by Vousdoukas et al. (2016) giving projections for Europe for 2010–2040 and 2070–2100. This study shows that the North and Baltic Sea coastlines show the largest increase in storm surges, where in contrast the southern European Mediterranean coastal areas can expect minimal changes. Projections indicate that storm surge levels in Europe will increase on average by 15% by 2100 under a high-emissions scenario producing higher seas by sea level rise and increased storminess.

Elliott et al. (2014) explain the concept of risk and present a typology for risk and hazards. Natural hazards which can be exacerbated by human activities include 'climatological hazards' which can be (i) acute/short term in nature but nevertheless devastating if unpredicted and unmanaged (e.g. storm surges, hurricanes), or (ii) chronic/long term in nature and again exacerbated by human activities (e.g. sea level rise, storminess). The severity of the risk is then proportional to the number of people or the value of the assets affected.

Modelling of hazards, taking into account frequency of climatic extremes and regional variation, was carried out by Forzieri et al. (2016) for three periods to the end of the present century (2011–2040, 2041–2070, and 2071–2100). The study predicted the threats posed by heat- and cold-waves, river and coastal floods, droughts, wildfires, and windstorms. Climate simulations were based on 'business-as-usual' greenhouse-gas emissions under the A1B emissions scenario of the Intergovernmental Panel on Climate Change (IPCC). Forzieri et al. (2016) found that coastlines and floodplains will be susceptible to experiencing multiple hazards where flooding combines with temperature-related climate changes. Vulnerable areas include the British Isles, the North Sea margin area, north-western parts of the Iberian Peninsula, as well as parts of France, northern Italy and the Balkan countries along the Danube River. There will also be increased hazards along European coastlines from coastal floods, mainly due to sea level rise for example the Danube Delta in eastern Europe.

When low atmospheric pressure (which causes the sea level to rise) combines with a high spring tide, which also coincides with strong winds which maximize wave height, the resulting storm surge has the potential to damage coastal landforms, ecosystems, and infrastructure, accompanied by extensive sea flooding (Spencer et al. 2015). Storm surges have occurred throughout history with one of the most catastrophic being the so-called St. Lucia's flood (Sint-Luciavloed) which arguably formed the character and response in the Netherlands and the German North Sea coast to such events. On 14 December 1287, the day after St. Lucia's Day, up to 80000 people were killed during a time when the coastal population of Europe was much less than today.

22.1.2 Management responses over the last century

Such storm surges then require management measures, for example Table 22.1 shows the larger storm surges to hit the North Sea over the last two centuries and the subsequent management measures undertaken to minimize loss of life.

22.1.2.1 Baltic Sea The effects of storm surges on countries around the Baltic Sea have also been extensively studied (Markus Meier 2006; Gräwe and Burchard 2012; Wolski et al. 2014; Snoeijs-Leijonmalm, Schubert, and Radziejewska 2017). In the last 50 years, the number of storm surges along various Baltic coasts has been increasing steadily, with the probability of storm surge flooding doubling towards the end of the twentieth century (Sztobryn et al. 2005). Due to the bathymetry of the Baltic Sea, Wolski et al. (2014) showed that the south-western coasts (the Bay of Mecklenburg) and the eastern coasts (the Gulf of Riga with Pärnu Bay, the Gulf of Finland, the northern part of the Gulf of Bothnia) were particularly exposed to extreme sea levels. With the exchange of water from the North Sea, the Danish Straits as well as the bays of the Baltic Sea are areas with a high number of storm surge events (200–300 surges). Significant increases in storm surge height during the twentieth century were reported along the Estonian coast of the Baltic Sea (Suursaar, Kullas, and Szava-Kovats 2009). In contrast, the Swedish coastlines in the central Baltic (Northern and Southern Baltic Proper, Western Gotland Basin) are the coastlines least exposed to extreme sea levels (Wolski et al. 2014). Gräwe and Burchard (2012) state that a rise in sea level has greater potential to increase surge levels in the Baltic Sea than increased wind speed. Such storm surges in the Baltic have resulted in the Inner Neva Estuary in the eastern Gulf of Finland having a 16 km storm surge barrier constructed from 1978 to 2011 (Snoeijs-Leijonmalm et al. 2017).

Table 22.1 Significant European storm surges over the last century and management responses.

Year	General description and Impacts	Management improvements (post flooding event)
1953	High tide combined with high winds and heavy rainfall in catchments resulted in seawater overflowing defences, wave overtopping, breaching, and failure of defences North Sea wide. Dikes and seawalls collapsed with over 1800 deaths in the Netherlands and over 300 deaths in England.	• Advances in coastal engineering. Defences were raised and strengthened along the North Sea coastlines. • Delta Project formalized by an Act of Dutch Parliament in 1957 to reduce the risk of continuous flooding – closure of the main tidal estuaries and inlets in the SW Netherlands (Nienhuis and Smaal 1994)

1962	Flooding of Hamburg, Germany. 60 000 homes were destroyed with a death toll of 315 in Hamburg.	• Building of Hamburg flood barrier, Germany (operational since 1968)
		• Rhine-Meuse Delta works were completed in 1997 which consisted of 13 dams, including barriers, sluices, locks, dikes, and levees. This to reduce the Dutch coastline's size and protect the areas within and around the Rhine-Meuse-Scheldt delta from North Sea floods
		• Building of major coastal barriers and barrages:
		  ○ Hull Tidal Barrier, UK (operational since 1980) which provides a 1 in 200 year standard of protection
		  ○ Thames Barrier, UK (operational since 1982)
		  ○ Eastern Scheldt storm surge barrier, NL (operational since 1986)
		• Major advances in storm surge forecasting, flood warning systems, and emergency planning
		• Investment in more natural flood alleviation schemes around the North Sea (e.g. managed realignment within estuaries and coastal areas)
		• 1994 – UK Shoreline management plans (SMPs)
		• Flood and coastal erosion risk management and investment

2012/2013	December and January storms resulted in around seven fatalities and 1700 properties flooded across England. The surge resulted in localized breaching, overtopping, and back-barrier flooding associated with gravel ridges and relatively low earthen banks as well as failure in more highly-engineered coastal defences.	• Continued improvements to combat coastal flooding and sea level rise through climate change • Continued large scale engineering projects - coastal barriers and barrages: ○ Boston Barrier, UK (due to be completed in 2019) • Continued investment in flood and coastal erosion measures

22.1.2.2 Mediterranean coast Storm surges in the Mediterranean Sea have caused flooding events in many areas including the Greek Islands of Crete and Cyprus, the Venice Lagoon and deltaic areas of Egypt, France, Italy, and Spain (Nicholls and Hoozemans 1996; Conte and Lionello 2014). These low-lying areas are at high risk in cases of extreme storm surge events producing the environmental impacts of coastal flooding, erosion, and destruction of coastal vegetation, and the economic impacts of saltwater intrusion into freshwater supplies and damage to infrastructure (White 1974). Due to the varying topography of the Mediterranean basin, there are various ways in which tidal surge events impact on the coastline from the west to the east. For example, Lionello et al. (2012) reported that the shape of the North Adriatic basin together with its physical characteristics of unique sea surface dynamics, astronomical tides, and the long fetch of strong south-easterly Sirocco winds may favour intense storm surge conditions. This regional sea has been extensively studied in terms of projected storm surge dynamics, with numerical modelling predicting a general decrease in the frequency and intensity of extreme events and storminess (Marcos et al. 2011; Conte and Lionello 2014; Androulidakis et al. 2015). However, rising sea levels and isostatic changes combined with low pressure systems will still produce tidal surge events, likely to have the greatest impact on the Aegean and Adriatic coastlines (Conte and Lionello 2014). Therefore, coastal management in this regional sea is addressing climate change and rising sea levels.

22.1.2.3 Atlantic coast and North Sea Vousdoukas et al. (2016) modelled extreme storm surge levels within Europe and concluded that the highest levels in Europe occur along the North Sea coast. The geomorphology, hydrology, and climatic conditions of the North Sea make it susceptible to storm surges, with its shallow water, and prevailing wind blowing from north to south. The funnelling effect of the North Sea, which both narrows and becomes shallower with a progression southwards and which under normal circumstances has an anti-clockwise gyre, pushes the surge tide along the western edge of the North Sea, affecting the east coast of the United Kingdom, and then up the eastern continental coasts of Belgium, the Netherlands, Germany, and Denmark. The most notable surge in recent times is the 1953 North Sea storm surge where over 2000 people lost their lives in eastern England, Belgium, and the Netherlands (Koppel 2007) and this prompted the extensive protection works known as the Dutch Delta Works (Nienhuis and Smaal 1994). Significant overtopping or breaching of flood and coastal defences occurred on low-lying coastal areas. The latest significant storm surge to affect North Sea countries occurred in December 2013, when strong winds and elevated water levels in the North Sea caused by low pressure, coincided with a high spring tide, which all combined to create the largest tidal surge to impact along England's East Coast between the Humber and the Wash since 1953. Elsewhere in northern Europe, the hurricane-force winds and same tidal surge killed at least five people and flooded parts of Hamburg in Germany.

In 2016, Vousdoukas et al. projected high storm surge level values along the west-facing coastline of the Irish Sea. Due to the topography of the coastline, the prevailing westerly and north westerly winds and shallow water depths in the eastern Irish Sea (average 40 m), the surrounding coastlines are prone to large surges (Brown and Wolf 2009). The influence of North Atlantic storm tracks and depression characteristics on propagation of surge waves around the Irish coast have also been investigated by Olbert and Hartnett (2010), who reported that the semi-enclosed nature of the Irish Sea propagates surge tides from both the north and south. Jones and Davies (1998) reported that the generation of surges in the Southern Irish Sea, Celtic Sea, and SW Approaches also play an important part in the consequences of surges on coastlines within the Irish Sea.

In order to understand how hazards such as storm surges can be most effectively addressed using governance and other management approaches, the following sections look at the legislative responses (as measures to control the situation) used in Europe along with the actors or stakeholders who implement these. Some of the strategies employed to adapt to the coastal surge incidents are also assessed.

22.2 European governance

22.2.1 An adapted problem-structuring framework to European coastal storms

The European Environment Agency (EEA) is a European body that provides independent information on the environment for those involved in developing, adopting, implementing, and evaluating environmental policy, and also the general public. One of these areas is the effects of climate change on coastlines and flood risk. As an executive agency of the European Commission, the EEA, in close collaboration with the European Environmental Information and Observation Network (EEION), gathers data and produces assessments on a wide range of topics related to the environment, current trends and pressures, economic and social driving forces, policy effectiveness, and identification of future trends and problems using scenarios and techniques.

In order to analyse and address the complexity of the marine environment, scientists and policy makers have used problem-structuring frameworks to assess the causes, consequences, and responses to change in a holistic way (EC 1999; Atkins et al. 2011; Gregory et al. 2013; Pinto et al. 2013). The original and well-used DPSIR (Driver-Pressure-State-Impact-Response) framework (Figure 22.1) (Patrício et al. 2016) has been adapted over the years to the more comprehensive DAPSI(W)R(M) framework (pronounced *dap-see-worm*), in which *D*rivers of basic human needs require *A*ctivities which lead to *P*ressures (Elliott et al. 2017). The Pressures are the mechanisms of *S*tate change on the natural system which then leads to *I*mpacts (on human *W*elfare). Those then require *R*esponses (as *M*easures) (Figure 22.2).

Using the new DAPSI(W)R(M) framework and applying it to the hydro-meteorological event of coastal surges, we consider that the main societal *D*rivers relate to the basic human need of security (e.g. protection from elements and hazards) and space (e.g. places to live and work). Each of these can then be achieved through human *A*ctivities such as building coastal infrastructure including artificial reefs and barrages, beach replenishment, groynes, land claim, constructing sea walls/breakwaters, urban dwellings and other buildings. These activities either individually or cumulatively cause *P*ressures which are reflected in the mechanisms of change and can result in changes to the natural system (*S*tate changes) and subsequently the societal system (*I*mpacts on human *W*elfare) (Elliott et al. 2017). Elliott (2011) defines *P*ressures which are created

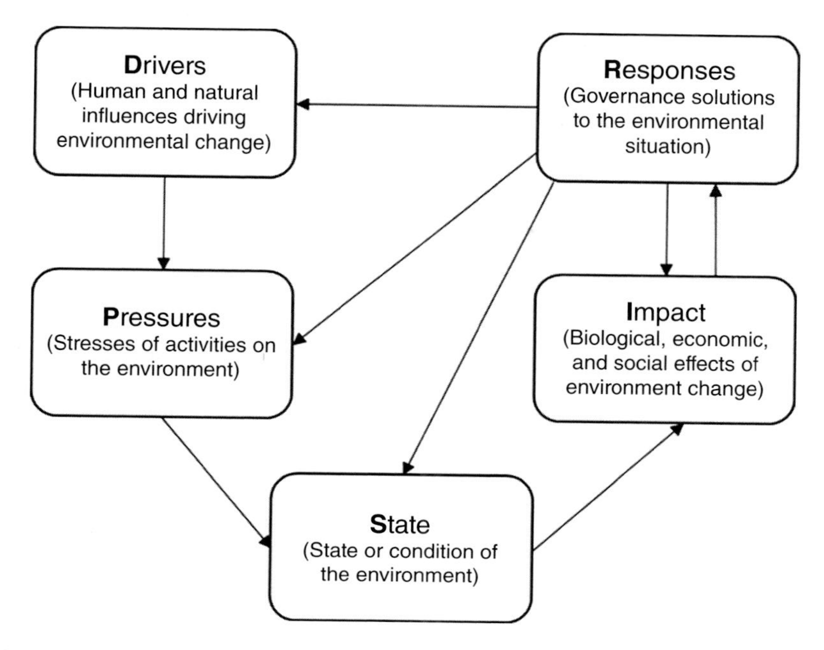

Figure 22.1 DPSIR framework. *Source*: Adapted from Gari, Newton, and Icely (2015).

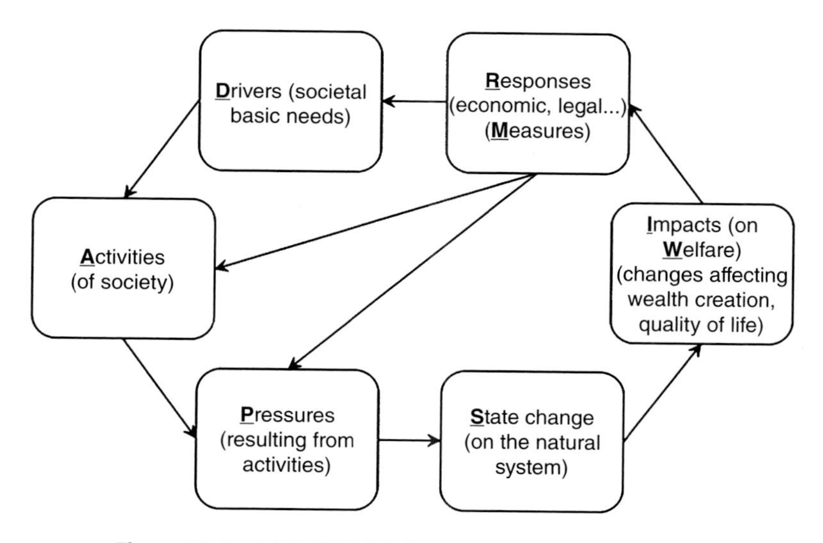

Figure 22.2 DAPSI(W)R(M) framework Elliott et al. (2017).

as either *exogenic unmanaged pressures* (i.e. those caused and originating from outside the immediate area, but whose causes can only be managed at a larger scale; however the consequences of which still have to be managed, e.g. climate change) and *endogenic managed pressures* (i.e. those occurring within the area of concern as a result of a licensed activity and where the causes and consequences need to be managed, e.g. building sea defences). Examples of pressures resulting from building coastal infrastructure can include seabed smothering, changes to the natural sea level and water flow rates, changes in wave exposure which affects the size, number, and distribution of the waves (Elliott et al. 2017). *S*tate changes relate to changes in the natural environmental system as a result of a single or multiple pressures. Once those pressures, state changes and *I*mpacts on human *W*elfare occur then they are addressed using *R*esponses from *M*easures involving governance (policies, politics, administrations, and legislation), economics, technologies, and other means.

22.2.2 European legislative measures of response to climate change threats and coastal surges

22.2.2.1 International and European level At an international level, to avoid the most serious risks and consequences of climate change, the 2016 Paris Agreement was reached by world leaders to keep global warming below 2 °C compared to pre-industrial temperatures. This builds on the United Nations Framework Convention on Climate Change (UNFCCC) and requests all nations to undertake an ambitious effort to combat climate change and adapt to its effects (UNFCCC 2017). To meet this target, international action to reduce greenhouse gas emissions will be needed for many decades. The Paris Agreement takes an integrative approach by complementing the activities of signatories and promotes greater coordination, coherence, and information-sharing.

In April 2013, a white paper introducing 'An EU Strategy on adaptation to climate change', which supports an integrated and coordinated approach at the Community level, was adopted by the European Commission (EC 2013a). This aims to promote greater coordination and information sharing between Member States on flood risks and includes guidance outlining actions to strengthen the European ability to be more climate-resilient. In the case of Europe, it ensures that climate adaptation considerations are addressed in all relevant EU policies including marine and inland water issues, forestry, agriculture, biodiversity infrastructure as well as buildings, migration, and social issues. The Strategy encourages Member States to adopt comprehensive adaptation strategies, where funding will support their implementation. Member States should undertake 'climate proof' action by further promoting adaptation strategies in key vulnerable sectors such as fisheries and agriculture. Adaptation should ensure that European infrastructure is made more resilient and insured against natural and man-made disasters. An accompanying document looks at climate change, environmental degradation, and migration (EC 2013b). Initiatives such as Climate-ADAPT, an EU climate adaptation platform, should be used to make decision makers better informed and address gaps in knowledge (EC 2017).

At a European level, the EU Directive 2007/60/EC on the assessment and management of flood risks ('the Floods Directive') entered into force in November 2007. It aims to reduce and manage the risks that floods pose to human health, the environment, cultural heritage, and economic activity by ensuring that flood risk from all sources is assessed and managed in a consistent way. It has given a global approach on how to prevent and manage the impacts of climate change on the occurrence of floods and to enhance the public awareness of flood risk. This Directive requires Member States to assess all water courses and coastlines for their potential risk from flooding, to map the potential flood extent, identify the assets and humans at risk in these areas and take adequate and coordinated measures to reduce this flood risk. The Directive applies to inland waters as well as all coastal waters across the whole territory of the EU (EC 2016b). The Directive needs to be coordinated with the environmental protection-based Water Framework Directive (WFD), primarily by aligning flood-risk management plans (FRMP) with River-Basin Management Plans (RBMPs), and by consulting with the public on the content of FRMPs. The FRMPs should also be aligned with the conservation objectives of the area in that Article 6(3) of the Habitats Directive (92/43/EEC) defines the requirements for assessment of plans and projects potentially affecting European sites under a 'Habitats Regulations Assessment'.

The WFD (2000/60/EC) established a framework for community action in the field of water policy. The WFD aims to establish a framework for the protection of inland surface waters (rivers and lakes), transitional waters (estuaries), coastal waters, and groundwater, so that Member States should aim to achieve 'Good Chemical Status' and 'Good Ecological Status' and to prevent deterioration in the status of those water bodies by 2015, or 2021 under the second management cycle. Successful implementation of the WFD will help to protect all elements of the water cycle and enhance the quality of all water bodies including estuaries and coastal waters (Borja and Heinrich 2005; Hering et al. 2010; Boyes and Elliott 2014). Member States are asked to identify individual river basins and assign them to RBMPs and identify 'competent authorities' responsible for that management; coastal waters are identified and assigned to the nearest or most appropriate river basin district. Environmental objectives are set for each classification of water body and then monitored according to the health of various ecological or chemical components, and including hydrology as a supporting element. With regard to coastal waters (designated out to one nautical mile and thus overlapping with the EU Marine Strategy Framework Directive), Member States should implement necessary measures to prevent their deterioration and to protect, enhance, and restore them, with the aim of achieving 'Good Ecological and Chemical Status' by 2015, or 2021 at the latest. The WFD takes note of Heavily Modified Water Bodies, i.e. those which have been altered by engineering and so may not be able to meet Good Status but rather they should have Good Ecological Potential (Borja and Elliott 2007). Hence, engineering works for flood protection and safety can be regarded as taking precedence over measures to achieve Good Ecological Status.

The Environmental Impact Assessment (EIA) Directive was first introduced in European law in 1985 (85/337/EEC), with the most recent revision (2014/52/EU) adopted May 2014 to simplify the rules for assessing the potential effects of new projects on the environment (Lonsdale et al. 2017). The revised Directive pays greater attention to threats and challenges that have emerged since the original version entered into force some 25 years ago with the amendments aiming to improve consideration of resource efficiency, climate change, biodiversity, and disaster prevention in the assessment process. Specific consideration should now be given to impacts of a project on, and its resilience to, climate change, biodiversity, and disaster prevention (such as flooding, sea level rise, storm surges, earthquakes) in the assessment process, but limiting this to significant effects only. This places a much greater emphasis on developers to incorporate external hazards into their long-term planning for any development (Lonsdale et al. 2017). Member States should decide either on a case-by-case basis or in terms of thresholds and criteria whether coastal works require an EIA to be conducted as listed under Annex II of the Directive.

22.2.3 Actors involved in both pressures and responses

Coastal risk prevention and management centres on the actors (stakeholders) involved in all aspects of the risk assessment and its management. Involving all stakeholders in management solutions is crucial to ensure that those actions are implemented, accepted, and successful. The defined set (typology) of stakeholders in any environmental challenge includes several classes (Elliott 2014; Newton and Elliott 2016). First, those stakeholders using (and abusing) the environment; second, those regulating the users and uses; third, those benefitting from the uses and users; fourth, those affected by the uses and users, and fifth, those influencing the management, use, and perception of the environment. Table 22.2 provides a description of these typologies and examples.

In the case of flood, erosion, and storm-surge protection, several organizations within Europe provide support, advice, and policy for coastal risk prevention and management. At a national level, Member States have their own legislation and policies to

Table 22.2 Actors at the national, regional, and global levels for coastal risk prevention and management.

Stakeholder typology	Description	Actors
Users	Those using space or taking resources from the marine system who cause / exacerbate the problem	Land occupiers (industry, urban, infrastructure) who expect flood protection Removing aggregates and natural materials which would help maintain coastline equilibrium
Abusers	Those removing natural habitat from the system which cause / exacerbate the problem	Removers of wetland for agriculture or land claim
Regulators	Those giving permission to occupy space or extract/input materials. Those who control the funding and use science to make decisions on flood defence and coastal protection	Government Administrations, legislative bodies, international policy makers, national and European legislators, statutory bodies (environmental protection agencies, statutory nature conservation bodies, harbour authorities, planning authorities)
Beneficiaries	Those positively affected from the ecosystem services and goods created by the system and delivered by the users	Society who are protected from future flooding events, insurance companies, asset holders, land owners, infrastructure users (e.g. transport)
Affectees	Those affected by the uses and users, affected by the policy decisions, impacted by the decisions whether positive or negative	Society who lose their land to erosion, incur costs of flooding or loss of life. Society (industry and land owners) who benefit from flood defence policy decisions, to protect their property
Influencers	Those who raise awareness and perception of the problems and the consequences of flooding (e.g. climate change, sea level rise, flood risk)	Politicians, non-governmental organizations, scientists, researchers, pressure groups, society, media, educators, insurance companies

Source: Adapted from Newton and Elliott (2016).

address the impacts of flooding and to implement the EU legislation. If protection works are proposed as a management option, then Member States should decide whether an EIA is required. However, in some cases, such as emergency works required because of coastal failure, flooding, or storm damage in which public safety or human assets are at risk, then the EIA process may be curtailed or even not required. In this case, the causes of damage may be regarded as *force majeure*, external factors outside the control of the regulators, and so be treated separately from planned intervention (Elliott et al. 2015; Saul et al. 2016).

22.2.4 Management measures

The coast is a dynamic environment and people living and working in coastal areas have always had to adapt to change. Climate change is regarded as an 'exogenic unmanaged pressure' in that we are not managing the causes of climate change, but managing the consequences of it and our responses to it (Atkins et al. 2011; Elliott 2011). Although unique conditions have to occur for a storm surge to create maximum impact, European policy makers need to be able to predict and plan appropriate action for these events and the effects of climate change. Increasing societal demand for living at the coast requires reducing risks from hazards such as storm surges, either by accommodating or minimizing risks through engineering solutions or adapting to changes. However, for these solutions to be successful, governance of coastal hazards and associated management decisions need to follow a set of rules that ensure the protection of the natural system and at the same time its exploitation by society.

The available and possible responses that aim to address the actual and potential problems on the coast and their causes related to flooding and erosion, cover all aspects of society, the so-called 10-tenets (Mee et al. 2008; Gray and Elliott 2009; Elliott 2013; Barnard and Elliott 2015). Many of these relate to the governance of the system, which covers the policies, politics, administration, and legislation; it covers agreements, guidelines, and laws and operates at local, national, regional, and international levels. Most importantly that governance is to protect public safety, the economy and assets, the ecology and the natural system (Boyes and Elliott 2014, 2015). It is emphasized here that although the following focuses on the European system, many of the main messages also pertain to other developed areas.

For managers to successfully respond to and manage extreme marine events such as storm surges, it is suggested that measures need to follow the 10 tenets, a set of requirements which cut across all aspects of the natural and societal aspects of the environment (Table 22.3). The defence option, in which prevention is paramount, needs to be sustainable but often the economic tenet is the most important, hence the need to quantify the costs and benefits of the systems created (Elliott et al. 2016). Cost–benefit analyses and cost-effectiveness analyses are undertaken for each protection scheme and options.

22.2.4.1 Reactive or defensive measures Hard defence (also termed hard engineering) options including concrete seawalls, revetments, and gabions have been the preferred option in the past to provide increased protection against more extreme events. Investment has also been made by some European governments into storm surge gates

Table 22.3 The 10-tenets framework with examples of management responses related to coastal storm surges.

Successful management Responses (as Measures) to changes should be:	Examples of measures related to storm surges
Ecologically sustainable	Reduce negative impacts to the ecology of the area (e.g. through minimizing impacts on water quality (hydrography and chemical), physical nature of the system (morphology, biology)
Technologically feasible	Ensuring major infrastructure projects (e.g. tidal barrages) are capable of being accomplished with the technology available
Economically viable	Sufficient funding available for the measure, acceptable cost–benefit analysis/ratio
Socially desirable/tolerable	Societal requirement for protection from future flood events
Legally permissible	Compliance with laws and regulations; marine licence, Environmental Impact Assessment (EIA), Strategic Environmental Assessment (SEA), Habitats Regulations Assessment
Administratively achievable	Agreement from administrative and statutory bodies such as an environmental protection agency
Politically expedient	Agreement with the manifesto commitments of the ruling party
Ethically defensible (morally correct)	Funding mechanisms are not a burden on future generations, discounting mechanisms are not a liability
Culturally inclusive	Protection of culturally and aesthetically important areas, no interference of indigenous human population areas
Effectively communicable	Agreement by stakeholder consultation, advertised decision-making

Source: Expanded from Elliott (2013), Barnard and Elliott (2015).

and flood barriers across estuary and river mouths which now give protection from flooding during extreme events. Examples include Venice; the Neva Estuary in the Gulf of Finland; the Eastern Scheldt in the Netherlands; Hamburg in Germany, and London and Hull in the UK. Major infrastructure systems such as barriers will be subject to a Strategic Environmental Assessment (SEA) under Directive 2001/42/EC.

22.2.4.2 Adaptive measures Together with the more extensive hard-engineering options come other alternatives. Whilst we have the capacity to engineer the coastline to protect us from natural hazards such as storm surges, this may result in an artificial system which is difficult and costly to maintain in the long run. It may also be working against, rather than with nature. Protecting the human assets at one part of the coastline could have repercussions downstream if the nature of the defence undermines the coast elsewhere. For example, building groynes to trap sediments and create a defensive beach in one place could starve another area of sediment and exacerbate erosion. Hard engineering options may involve greater costs than the value of assets being protected. Conventional hard defences costing several million Euros per 100 m are not economically viable when the buildings or farmland behind them are valued less. The question of adaptation of coastlines becomes more prevalent and in particular the policy of working with nature or the so-called 'giving space for water'. When there is no justifiable argument to defend a length of coastline for technical, economic, social, or environmental reasons, communities must adapt and learn to live with the risk. The Stern Review on the Economics of Climate Change said 'Adaptation will be crucial in reducing vulnerability to climate change and is the only way to cope with the impacts that are inevitable over the next few decades ... without early and strong mitigation, the costs of adaptation rise sharply ... governments have a role to play in making adaptation happen, starting now, providing both policy guidelines and economic and institutional support' (Stern 2007). This may involve communities being encouraged to 'roll-back' from the coast and planning restrictions being placed on proposed new developments.

In order to respond to climate change and its impacts across Europe, the EEA (2017) advises that better and more flexible adaptation strategies, policies, and measures will be crucial to lessen these impacts. Similarly, if options can deliver benefits for human safety, economy, and ecology, the so-called 'triple wins', then it has a greater chance of being implemented. Countries in Europe are now aware that they need to look at alternative solutions to addressing coastal erosion and flooding. Mitigating storm surge risks by applying ecoengineering solutions have been discussed by Van Staveren et al. (2014), Wolanski and Elliott (2015), and Elliott et al. (2016). More natural defence options are now being used which allow coastal zones to adapt to natural processes and reduce the need for long-term intervention. These options may include the management of saltmarshes, sand dunes, shingle banks, or sand beaches. Maintaining these natural features can negate the need for more costly options such as building hard defences. Management methods may include beach nourishment or recharging beaches where they are naturally eroding and undertaking dune management such as grass planting and fencing.

The managed realignment of defences along coastlines and within estuaries has become a common management technique to counteract the impact of sea level rise and local increase in storminess related to global warming (Wolanski and Elliott 2015). This allows the shoreline to evolve in a more flexible natural manner with the objective of promoting more sustainable flood and erosion risk management (Esteves 2014). The existing line of defence is allowed to degrade or is intentionally breached, allowing water on high water events (high tides, storm surges) to occupy the area of land behind, increasing the capacity of the estuary. Saltmarshes that develop as part of the realignment, help to absorb wave energy, and offer soft coastal protection in themselves. This means that inland defences do not need to be as high as if new seawalls were built further seaward. The importance of intertidal mudflats and saltmarsh in coastal defence is now widely accepted as a means of absorbing energy of incoming waters and, in many areas, natural sea defences are considered to be a more effective, ecologically sustainable and economically viable method of coastal protection (CPSL 2001; Möller and Spencer 2002; Wolanski and Elliott 2015).

22.2.4.3 Preventative measures Greater emphasis is now being placed upon adaptation strategies, such as 'roll-back' options, which offers a sustainable alternative to engineered coastal defences where the defences are not considered to be viable and have little or no realistic chance of being defended. Roll-back entails the political and economic encouragement for the physical movement of assets further inland, away from the threat posed by coastal change. This preventative measure is now being used along English and Welsh coastlines. For example, the East Riding of Yorkshire Council, UK was one of the first local authorities to adopt coastal roll-back policies which, since

2003 and 2005, have supported the owners of caravan parks and residential properties respectively. The policies offer a proactive and consistent approach within the planning system to enable the roll-back of permanent residential and agricultural dwellings/farmsteads and business facilities (including caravan parks) at risk from coastal erosion (Boyes, Barnard, and Elliott 2016). This is subject to individuals having suitable land available for roll-back, and securing the necessary planning permission for its development. Roll-back allows infrastructure to be relocated in a managed way, with the potential for a number of benefits relating to community cohesion and sustainability, regeneration and improving the quality of the local environment, where the provision of coastal defences is not considered to be economically viable or environmentally sustainable. The policy is accompanied by a moratorium on building within the areas at risk, often within a 50 or 100 year time horizon.

22.2.4.4 Predictive measures: Monitoring, research, and forecasting Once protection and prevention measures have been agreed then monitoring is required to determine whether those measures are successful and also assess for future issues. Coastal monitoring around most European coastlines occurs using Global Positioning System (GPS), aerial photography (and increasingly drones) and Light Detection and Ranging (LiDAR) technologies to assess changes in cliff erosion rates, beach and intertidal profiles. The results of these surveys help to make decisions about managing the impacts of coastal change and provide future protection to extreme events such as storm surges and increased storminess. Coastal structures designed to protect larger communities and infrastructure are regularly inspected and maintained as required.

The communication of flooding events caused by storm surges and other climatic events has now improved with flood warning systems developed around Europe to give the general public warning of potential flood events caused by coastal and inland flooding. For example, the detection of a developing storm surge at the north of Scotland allows the southern areas in eastern England, Belgium, the Netherlands and Germany to prepare. Governments have invested time and money to anticipate future storm events in order to protect lives, property, industry, and infrastructure from the devastating events of coastal flooding events. Advances in forecasting systems using numerical modelling allow a greater understanding of when these events are likely to occur. This underlines the urgency to enhance current understanding and forecasting capacity of coastal hazards and their impacts such that European policymakers and relevant authorities can plan appropriate and timely adaptation and protection measures.

22.3 Discussion and conclusions

22.3.1 Challenges to coastal management from storm surges in Europe

Coastal areas are always going to be at risk of flooding. The consequences of more people moving to the coast and having larger cities, ports, and industries on the low-lying estuaries and on soft coasts means more assets are at risk and therefore there is a greater requirement for protection. These risks are exacerbated by a feedback loop in which the nature of the coast and estuaries encourages settlement, industry and port development and yet these may then exacerbate the problems, by preventing expansion space for surge water as well as requiring even greater defences because more assets are at risk.

22.3.2 Management and governance aspects

Management and governance are then required to tackle the causes and consequences of these pressures with various tools at its disposal. Expertise and technological aspects can address the actual or potential causes, for example by using hard or soft engineering. Although more expensive, hard engineering (using concrete sea walls, groynes, and breakwaters) is more permanent and may involve less maintenance than soft engineering. However, it can create coastal squeeze by fixing the upper shore level and only allowing water levels to rise, especially with relative sea level rise. This is an increasing problem which removes intertidal habitat even if it increases the amount of subtidal habitat. It may also adversely affect land and assets downstream, such as in the direction of longshore drift on coastlines, thus being open to legal challenges. In contrast, soft engineering which involves 'working with nature', and enhancing the adaptability in the system, perhaps creates resistance to change and resilience from change (Elliott et al. 2016) but it may involve sacrificing agricultural land for wetlands (e.g. by using beach nourishment or managed realignment). However, as described above, the success of the governance might be dictated by the fact that an area is designated for specific habitats which have to be protected (e.g. under nature protection legislation) or has a set of users and uses which should not be impaired.

Much of the legislation and governance described here requires the natural system to be protected whilst also allowing an area to deliver the ecosystem services and societal goods and benefits (Elliott et al. 2017). Hence the governance promotes the maintenance of use, public safety, economic benefits, and protection of the natural environment. This increasingly involves ecoengineering approaches to aid the restoration of estuarine/coastal systems which in turn involves both adding but also removing structures and impediments to natural ecosystem processes that are most likely to promote successful and sustainable ecology (Wolanski and Elliott 2015; Elliott et al. 2016). Both soft and hard engineering are valuable in different ecoengineering schemes depending on the legislative requirements, the circumstances, and available information and allows flexibility in management to address the situation. The overall aim is to achieve the 'triple wins' for human safety, economy, and the ecology and a soft engineering approach is preferable as a first option. It will often be cheaper than a 'hard' approach although it may require more maintenance for a longer period.

Resilience and adaptation of coastal communities to the pressures of flood and coastal erosion will be the key driver of the design of sustainable coastal management solutions. With European coastlines under great pressure from human uses, intensified by the effects of a changing climate with rising sea levels and increased storminess, the need to manage coastal activities in a sustainable way is becoming critical. There is increasingly a proactive approach to the control of erosion and flooding given societal acceptance that it occurs, and indeed could become more commonplace with climate change.

22.3.3 Economic and political aspects

The governance will demand that finance is made available for coastal protection works, but European and national policies will determine the priorities for protecting urban areas for public safety, industry, and amenities (e.g. lighthouses) which are in the national interest. Hence, low-grade farmland may not be protected following a cost–benefit analysis, although farmland in one area may be protected if the aim is to stop flooding elsewhere; for example, farmland in the outer Thames estuary east of London has to be protected to stop flooding bypassing the defences and barrage further upstream. Whilst the governance (the laws, administrations, policies, and politics) requires areas to be protected and gives the statutory mechanisms for protection, there are economic and political consequences of protecting all areas as the cost of defences may be more than the cost of assets being protected. However, paradoxically once a set of assets is protected, its value increases and so there is an even greater benefit–cost ratio requiring more protection in the future. In often-flooded areas, and where householders are not given compensation from the municipality or government, then it may fall to the insurance industry to protect them. Therefore, there is the possibility that it may be the insurance companies that influence building (or discourage building) on flood-prone land rather than environmental agencies or municipalities.

As a further repercussion of the management and governance, or perhaps as the driver of this action, society, the general public, and industry require protection against flooding and do not tolerate reduced levels of protection. The laws require an area to be protected but at the same time, because of legal repercussions, authorities are wary of carrying out measures which may solve the problem in one area but make it worse elsewhere. Under national law an owner is responsible for his/her own property and land and does not get compensation for damage unless they can prove someone else is responsible. If damage or risk is due to a natural phenomenon or a factor outside the owner's control (the claim of *force majeure* or an Act of God) (Saul et al. 2016) then compensation is at the discretion of the government or an insurance claim. However, if third-party actions are deemed to be the cause of the risk or damage then there can be a claim for compensation against that other party whose actions made things worse. Thus the legal complexity, and management responsible for an area, depends on who controls building on land liable to flooding compared to who is responsible for flood control – a municipality may want more building and development and so give permission but an environment protection agency or flood protection authority may then be responsible for protection.

22.3.4 Societal aspects of flood protection

The demand for flood and erosion prevention also has societal and perception repercussions and if society believes there to be a problem then by definition there is one and this requires some response, if only a degree of reassurance by the authorities. As indicated earlier, wealthier areas would have a higher value in a cost–benefit analysis of flood protection costs versus assets protected compared to poorer areas. Similarly, more affluent areas may have a more articulate population, more confident of pressuring authorities and so demanding greater support and action. Hence the higher social classes (and the more expensive parts of a country) both demand and may be given greater protection.

A second feature is 'societal memory', in which if the local population can remember or has experienced the consequences of flooding and storm surges, then they will create greater pressure to have protection. In contrast, if no one can remember the consequences of flooding, then they may be more reluctant to pay higher taxes to protect areas. For example, in the Netherlands there is still an ageing population that can remember the devastating 1953 storm surge and so would be willing to lobby for funding to maintain high dikes. Once that societal memory decreases then so will the pressure for maintaining defences. Previous generations would have accepted that they cannot beat nature and so tolerated flooding (building churches and other important buildings on higher land) or moved to other areas. In contrast, society today demands insurance or protection against flooding and so relies on authorities. Increased communication and forecasting of floods has ensured that local people have the correct information and in the right format to make informed judgements.

22.4 Conclusions

The analysis here has first shown that there is a need for a rigorous problem-structuring framework to storm surge protection in the form of flood and coastal defences, centred around a risk assessment and risk management approach. Such an approach requires an input by all types of stakeholders, although different groups of stakeholders have both different priorities and also differing abilities to achieve coastal protection.

Second, protecting European coastlines has an input from many authorities and other aspects of governance, including legislative instruments at the national, European, and international levels. However, to be effective these need to be coordinated and able to provide protection and wide-ranging spatial and temporal scales.

Third, there is a well-developed approach to protection, using both hard and soft engineering techniques. However, these require both a rigorous cost–benefit analysis and the acknowledgement that hard defences can have unwanted repercussions on adjacent areas. This could include legal challenges.

Fourth, societal memory plays a role in protection but this may also create resistance to moving from at-risk areas. It is widely acknowledged that the risks from storm surges due to climate change will increase, especially with greater migrations to vulnerable areas. Hence, governments may continue to give priority to urban areas and industrial areas in the national interest and consequently less protection to low-vale agricultural areas. Adaptation to climate change encourages the development of public policies that are more engaged with the prevention and management of flooding risks. Chaumillon et al. (2017) suggest adaptation measures (adapted buildings, installation of easily accessible refuges, relocation/retreatment) should be developed. This in turn will create more resilient territories but will require a strong institutional basis. Hence, there is an increasing awareness of the effect of climate change on assets from increased flooding and storm surges (Stern 2007).

Lastly, it is of importance that recent EU EIA legislation (2014/52/EU) requires developers to consider the effects of their development on mitigation for climate change and also the effect of climate change on their development. This therefore requires, inter alia, the effects of building and infrastructure developments on flood and erosion protection schemes and whether, for example, building a structure will make flooding worse elsewhere. This is a case of responding to the consequences of climate change, as an exogenic unmanaged pressure, rather than addressing the causes which has to be left to global action.

References

Androulidakis, Y.S., Kombiadou, K.D., Makris, K.V. et al. (2015). Storm surges in the Mediterranean Sea: variability and trends under future climatic conditions. *Dynamics of Atmospheres and Oceans* 71: 56–82.

Atkins, J.P., Burdon, D., Elliott, M. et al. (2011). Management of the marine environment: integrating ecosystem services and societal benefits with the DPSIR framework in a systems approach. *Marine Pollution Bulletin* 62: 215–226.

Barbier, E. (2015). Climate change impacts on rural poverty in low-elevation coastal zones. *Estuarine, Coastal and Shelf Science* 165: A1–A13.

Barnard, S. and Elliott, M. (2015). The 10-tenets of adaptive management and sustainability: an holistic framework for understanding and managing the socio-ecological system. *Environmental Science and Policy* 51: 181–191.

Baxter, P.J. (2005). The East Coast Big Flood, 31 January–1 February 1953: a summary of the human disaster. *Philosophical Transactions of the Royal Society of London* 363A: 1293–1312.

Borja, A. and Elliott, M. (2007). What does 'good ecological potential' mean, within the European Water Framework Directive? *Marine Pollution Bulletin* 54: 1559–1564.

Borja, A. and Heinrich, F. (2005). Implementing the European Water Framework Directive: the debate continues. *Marine Pollution Bulletin* 50: 486–488.

Boyes, S.J. and Elliott, M. (2014). Marine legislation – the ultimate 'horrendogram': international law, European directives and national implementation. *Marine Pollution Bulletin* 86: 39–47.

Boyes, S.J. and Elliott, M. (2015). The excessive complexity of national marine governance systems – has this decreased in England since the introduction of the Marine and Coastal Access Act 2009? *Marine Policy* 51: 57–65.

Boyes, S.J., Barnard, S., and Elliott, M. (2016). *The East Riding Coastline: Past, Present and Future.* Prepared for East Riding of Yorkshire Council (ERYC) by the Institute of Estuarine and Coastal Studies (IECS), University of Hull. Funded through the Defra Coastal Change Pathfinder project and the East Riding Coastal Change Pathfinder (ERCCP). Institute of Estuarine and Coastal Studies, University of Hull, Hull, UK.

Brown, J.M. and Wolf, J. (2009). Coupled wave and surge modelling for the eastern Irish Sea and implications for model wind-stress. *Continental Shelf Research* 29: 1329–1342.

Chaumillon, E., Bertin, X., Fortunato, A.B. et al. (2017). Invited review: storm-induced marine flooding: lessons from a multidisciplinary approach. *Earth-Science Reviews* 165: 151–184.

Conte, D. and Lionello, P. (2014). Storm surge distribution along the Mediterranean coast: characteristics and evolution. *Procedia – Social and Behavioural Sciences* 120: 110–115.

CPSL (2001). Final report of the Trilateral Working Group on Coastal Protection and Sea Level Rise. Wadden Sea Ecosystem No.13. Wilhemshaven, Germany: Common Wadden Sea Secretariat, 63pp.

EC (1999). *Towards Environmental Pressure Indicators for the EU*, 1e. Luxembourg: Office for Official Publications of the European Communities.

EC (2013a). EU strategy on adaptation to climate change, 2013, 16/04/2013. SWD(2013) 133 final. Brussels: European Commission.

EC (2013b). Climate change, environmental degradation, and migration, Accompanying the document: An EU Strategy on adaptation to climate change. 16.4.2013 SWD(2013) 138 final. Brussels: European Commission.

EC (2016a). Integrated coastal management. environment. Brussels: European Commission. http://ec.europa.eu/environment/iczm/index_en.htm (accessed 28 February 2019).

EC (2016b). The EU Floods Directive. Brussels: European Commission http://ec.europa.eu/environment/water/flood_risk/index.htm (accessed 28 February 2019).

EC (2017). Climate-ADAPT – European Climate Adaptation Platform. Brussels: European Commission http://climate-adapt.eea.europa.eu (accessed 28 February 2019).

EEA (2017). Climate change, impacts and vulnerability in Europe 2016. An indicator-based report. Copenhagen: European Environment Agency.

Elliott, M. (2011). Marine science and management means tackling exogenic unmanaged pressures and endogenic managed pressures – a numbered guide. *Marine Pollution Bulletin* 62: 651–655.

Elliott, M. (2013). Editorial: the *10-tenets* for integrated, successful and sustainable marine management. *Marine Pollution Bulletin* 74: 1–5.

Elliott, M. (2014). Integrated marine science and management: wading through the morass. *Marine Pollution Bulletin* 86: 1–4. https://doi.org/10.1016/j.marpolbul.2014.07.026.

Elliott, M., Cutts, N.D., and Trono, A. (2014). Review: a typology of marine and estuarine hazards and risks as vectors of change: a review for vulnerable coasts and their management. *Ocean and Coastal Management* 93: 88–99.

Elliott, M., Borja, Á., McQuatters-Gollop, A. et al. (2015). *Force majeure*: will climate change affect our ability to attain Good Environmental Status for marine biodiversity? *Marine Pollution Bulletin* 95: 7–27. https://doi.org/10.1016/j.marpolbul.2015.03.015.

Elliott, M., Mander, L., Mazik, K. et al. (2016). Ecoengineering with ecohydrology: successes and failures in estuarine restoration. *Estuarine, Coastal and Shelf Science* 176: 12–35. https://doi.org/10.1016/j.ecss.2016.04.003.

Elliott, M., Burdon, D., Atkins, J.P. et al. (2017). Viewpoint: "*And DPSIR begat DAPSI(W)R(M)!*" – a unifying framework for marine environmental management. *Marine Pollution Bulletin* 118 (1–2): 27–40.

Esteves, L.S. (2014). *Managed Realignment: A Viable Long-Term Coastal Management Strategy?* Springer Briefs in Environmental Science. Dordrecht: Springer Science+Business Media https://doi.org/10.1007/978-94-017-9029-1_2.

Forzieri, G., Feyen, L., Russo, S. et al. (2016). Multi-hazard assessment in Europe under climate change. *Climate Change* 137 (1): 105–119.

Gari, S.R., Newton, A., and Icely, J.D. (2015). A review of the application and evolution of the DPSIR framework with an emphasis on coastal social-ecological systems. *Ocean and Coastal Management* 103: 63–77. https://doi.org/10.1016/j.ocecoaman.2014.11.013.

Gräwe, U. and Burchard, H. (2012). Storm surges in the Western Baltic Sea: the present and a possible future. *Climate Dynamics* 39: 165–183.

Gregory, A.J., Atkins, J.P., Burdon, D. et al. (2013). A problem structuring method for ecosystem based management: the DPSIR framework. *European Journal of Operational Research* 227: 558–569.

Gray, J.S. and Elliott, M. (2009). *Ecology of Marine Sediments: Science to Management*, 260 p. Oxford: Oxford University Press.

Hering, D., Borja, A., Carstensen, J. et al. (2010). The European Water Framework Directive at the age of 10: a critical review of the achievements with recommendations for the future. *Science of the Total Environment* 408 (19): 4007–4019.

Jones, J.E. and Davies, A.M. (1998). Storm surge computations for the Irish Sea using a three-dimensional numerical model including wave-current interaction. *Continental Shelf Research* 18 (2): 201–251.

Koppel, T. (2007). *Ebb and Flow. Tides and Life on Our Once and Future Planet*. Toronto: Dundurn Press ISBN: 978-1-55488-148-2.

Lionello, P., Cavaleri, L., Nissen, K.M. et al. (2012). Severe marine storms in the Northern Adriatic: characteristics and trends. *Physics and Chemistry of the Earth Parts A/B/C* 40: 93–105.

Lonsdale, J., Weston, K., Elliott, M. et al. (2017). The Amended European Environmental Impact Assessment Directive: UK marine experience and recommendations. *Ocean and Coastal Management* 148: 131–142.

Lutz, W. and Samir, K.C. (2010). Dimensions of global population projections: what do we know about future population trends and structures? *Philosophical Transactions of the Royal Society B* 365: 2779–2791.

Marcos, M., Jordà, G., Gomis, D. et al. (2011). Changes in storm surges in southern Europe from a regional model under climate change scenarios. *Global and Planetary Change* 77: 116–128.

Markus Meier, H.K. (2006). Baltic Sea climate in the late twenty-first century: a dynamical downscaling approach using two global models and two emission scenarios. *Climate Dynamics* 27 (1): 39–68.

Mee, L.D., Jefferson, R.L., Laffoley, D.d.'A. et al. (2008). How good is good? Human values and Europe's proposed marine strategy directive. *Marine Pollution Bulletin* 56: 187–204.

Möller, I. and Spencer, T. (2002). Wave dissipation over macro-tidal saltmarshes: effects of marsh edge typology and vegetation change. *Journal of Coastal Research* 36: 506–521.

Newton, A. and Elliott, M. (2016). A typology of stakeholders and guidelines for engagement in transdisciplinary, participatory processes. *Frontiers in Marine Science* 3: 13. https://doi.org/10.3389/fmars.2016.00230.

Nienhuis, P.H. and Smaal, A.C. (1994). *The Oosterschelde Estuary: A Case-Study of a Changing Ecosystem*. Dordrecht, NL: Kluwer Academic Publishers.

Nicholls, R.J. and Hoozemans, F.M.J. (1996). The Mediterranean: vulnerability to coastal implications of climate change. *Ocean and Coastal Management* 31 (2–3): 105–132.

Olbert, A.I. and Hartnett, M. (2010). Storms and surges in Irish coastal waters. *Ocean Modelling* 34 (1–2): 50–62.

Patrício, J., Elliott, M., Mazik, K. et al. (2016). DPSIR – two decades of trying to develop a unifying framework for marine environmental management? *Frontiers in Marine Science* 3: 177. https://doi.org/10.3389/fmars.2016.00177.

Pinto, R., de Jonge, V.N., Marques, J.C. et al. (2013). Temporal stability in estuarine systems: implications for ecosystem services provision. *Ecological Indicators* 24: 246–253.

Saul, R., Barnes, R., and Elliott, M. (2016). Is climate change an unforeseen, irresistible and external factor – a *force majeure* in marine environmental law? *Marine Pollution Bulletin* 113 (1–2): 25–35.

Snoeijs-Leijonmalm, P., Schubert, H., and Radziejewska, T. (eds.) (2017). *Biological Oceanography of the Baltic Sea*. Dordrecht, NL: Springer.

Spencer, T., Brooks, S.M., Evans, B.R. et al. (2015). Southern North Sea storm surge event of 5 December 2013: water levels, waves and coastal impacts. *Earth-Science Reviews* 146: 120–145.

Stern, N.H. (2007). *The Economics of Climate Change: The Stern Review*. Cambridge, UK: Cambridge University Press.

Suursaar, Ü., Kullas, T., and Szava-Kovats, R. (2009). Wind and wave storms, storm surges and sea level rise along the Estonian coast of the Baltic Sea. *WIT Transactions on Ecology and the Environment* 127: 149–160.

Sztobryn, M., Stigge, H.-J., Wielbińska, D. et al. (2005). *Storm surges in the Southern Baltic Sea (Western and Central Parts), Berichte des Bundesamtes für Seeschifffahrt und Hydrographie*, Nr. 39. ISSN 0946–6010. Hamburg: BSH.

UNFCCC (2017). Paris Agreement: Status of Ratification. Bonn: United Nations Framework Convention on Climate Change. http://unfccc.int/2860.php (accessed 12 June 2017).

Van Staveren, M.F., Warner, J.F., van Tatenhove, J.P.M. et al. (2014). Let's bring in the floods: de-poldering in the Netherlands as a strategy for long-term delta survival. *Water International* 39 (5): 686–700.

Vousdoukas, M.I., Voukouvalas, E., Annunziato, A. et al. (2016). Projections of extreme storm surge levels along Europe. *Climatic Dynamics* 47: 3171–3190.

Wang, G., Liu, Y., Wang, H. et al. (2014). A comprehensive risk analysis of coastal zones in China. *Estuarine, Coastal and Shelf Science* 62: 131–140.

White, G.F. (1974). *Natural Hazards: Local, National, Global*. New York: Oxford University Press.

Wolanski, E. and Elliott, M. (2015). *Estuarine Ecohydrology: An Introduction*. Amsterdam: Elsevier. ISBN: 978-0-444-63398-9.

Wolski, T., Wiśniewski, B., Giza, A. et al. (2014). Extreme sea levels at selected stations on the Baltic Sea coast. *Oceanologia* 56 (2): 259–290.

23

Perceptions of Extreme Coastal Events: The Case of the French Atlantic and Mediterranean Coasts

Lydie Goeldner-Gianella[1] and Esmeralda Longépée[2]

[1] *Laboratory of physical geography CNRS 8591 LGP, Université Paris 1 Panthéon-Sorbonne, Paris, France*

[2] *Laboratory 228 Espace-DEV, University of Mayotte, Mayotte, France*

This chapter looks at perceptions and social representations of extreme coastal events, in this instance storms and their direct impacts, i.e. erosion and coastal flooding. The perception of storms and their consequences has changed considerably over the last few centuries. In ancient times, the sea and its movements were frequently personified as gods or monsters and aroused great fears (Corbin and Richard 2004). During the Renaissance, the perils Christians encountered during their lifetime were compared to the dangers met at sea or on the coast, especially shipwrecks and storms (Corbin 2014). These events were increasingly represented in sixteenth- and seventeenth-century paintings and inspired countless votives offered in churches. However, from the eighteenth century onwards, perceptions changed somewhat: the sea became more than just a source of anguish and was increasingly seen as a spectacle to admire, stirring emotions: a 'social spectacle' in the words of Bousquet and Miossec (1991). The sea was glorified, awe-inspiring in its breadth and power: 'it is a landscape that can be neither developed nor tamed' (Corbin 2014). Painters and writers, especially the Romantics, were clearly aware of this dual perception of the sea, combining admiration and fear. This was in fact the subject of an exhibition held by the Bibliothèque Nationale de France in 2004: *La mer, terreur et fascination* (Corbin and Richard 2004). There was also a lot of fear surrounding the depths of the oceans, of which people were far less aware before the late nineteenth century. Popular imagination filled the deep seas with monsters and evil creatures, as can be seen in the legends featuring the French coast listed by historian P. Sébillot in the early twentieth century (Sébillot 2015).

Facing Hydrometeorological Extreme Events: A Governance Issue, First Edition.
Edited by Isabelle La Jeunesse and Corinne Larrue.

Our present interest in the social representations of extreme coastal events lies in the fact that these events, which are recurrent in mainland France, are a priori easily visible to people that live on or visit the coast. For example, 250 m from the shore, a quarter of urbanized coastal areas in France are areas where erosion is clearly visible (Colas 2007). In addition, 1.4 million full-time residents are currently concerned by the risk of coastal flooding and a fifth of buildings are single-storey dwellings in areas deemed to be exposed to the flood risk (MEDDE 2011). As well as experiencing extreme events in this way, populations are also better informed about marine risks by the authorities and/or media. Public policies established after the end of the first decade of the millennium encourage increasing numbers of coastal stakeholders – including coastal scientists – to focus on the public and inform them of coastal hazards, resulting in a number of strategies and schemes: the *Stratégie nationale d'adaptation au changement climatique* (2007), *Rapport sur les conséquences de la tempête Xynthia* (Anziani 2010), *Plan national d'adaptation au changement climatique* (2011–2015), *Stratégie nationale de gestion intégrée du trait de côte* (DGALN 2012), *40 mesures d'adaptation au changement climatique et à la gestion intégrée du trait de côte* (Comité national de suivi de la Stratégie nationale de gestion intégrée du trait de côte, 2015), and the *Stratégie nationale pour la mer et le littoral*, adopted by decree in February 2017.

Against this background, numerous scientific works have studied the perception and representations of coastal risks, mainly in mainland France, over the past 15 years. In this chapter, by 'perception' of coastal risks, we mean a sensory – affecting the senses, and sight in particular – and cognitive act through which the individual develops their inner vision or their mental image of an object or an event. 'Social representations' of coastal hazards mean 'a form of knowledge that is socially generated and shared, [...] helping to shape a reality common to a social group' (Jodelet 1989). We can also consider that perception of coastal hazards helps form social representations of those risks, but that the latter also influence perception, a partially cognitive act. Studying these two forms of relationship with risks also implies questioning other mechanisms and factors related to perceptions and representations, namely sensitivity to coastal risks, knowledge of coastal hazards, the memory of the risk and practices, and so on. Perceptions and representations are often studied using questionnaires or interviews, depending on whether the information sought is quantitative or qualitative. In the case of extreme coastal risks, surveys have been conducted in France since 2008 by researchers from various teams on a dozen or so research programmes covering the three coastlines of mainland France and almost 20 municipalities (Figure 23.1), all affected by at least one stand-out event, such as storm Xynthia in 2010. Depending on site, these surveys used samples of 200–900 people, made up of full-time and part-time residents and/or tourists and visitors. Whilst the analysis of perception and representations of extreme events will, as a priority, draw on these scientific results, in this chapter special focus will be put on the Mediterranean town of Leucate where, in 2015, our team surveyed 493 people on the risk of storms and coastal flooding as part of the CRISSIS programme.[1] The town is a large seaside resort on the Languedoc coast (4000 full-time residents and 80 000 tourists a day in summer) and is particularly sensitive to the coastal flooding risk due to its shoreline's exposure to the Mediterranean storms and the location of a significant proportion its dwellings and infrastructures in low-lying coastal areas (Figure 23.2).

After demonstrating that awareness of storms and their direct consequences (flooding and erosion) is currently inadequate in mainland France, we will look at the factors that may explain the gradual decline in the notion of coastal hazard. We will then put

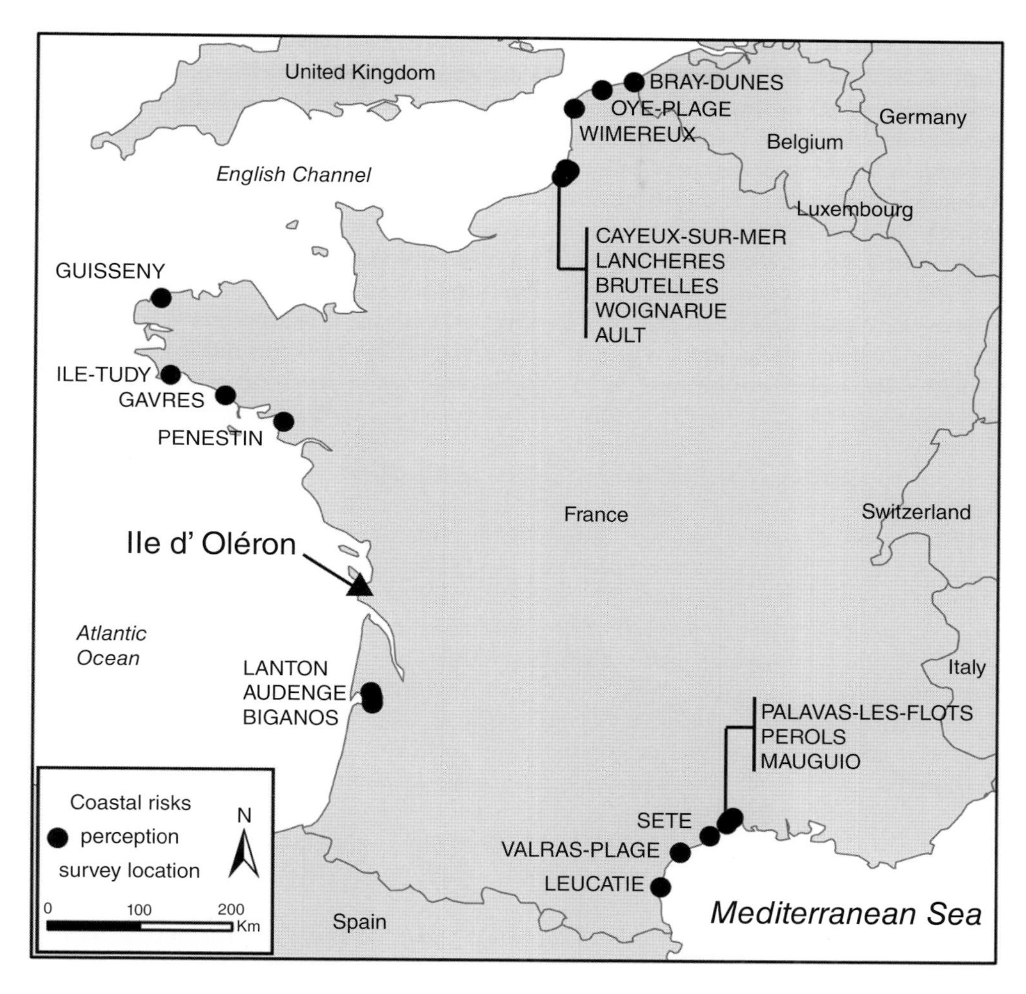

Figure 23.1 Location of scientific studies conducted in mainland France into the perception of coastal risks between 2008 and 2015.

forward some recommendations based on these studies into perceptions and social representations to guide public policies on extreme coastal hazards.

23.1 Contemporary society is increasingly unaware of risks related to the sea

23.1.1 Residents and visiting populations relatively unaware of risks related to the sea

Most of the surveys carried out in mainland France after 2008 (Table 23.1) show that there is genuine sensitivity to coastal hazards – for example, erosion in Nord-Pas-de-Calais (Meur-Férec et al. 2010) – or a certain recollection of the risk. Amongst those

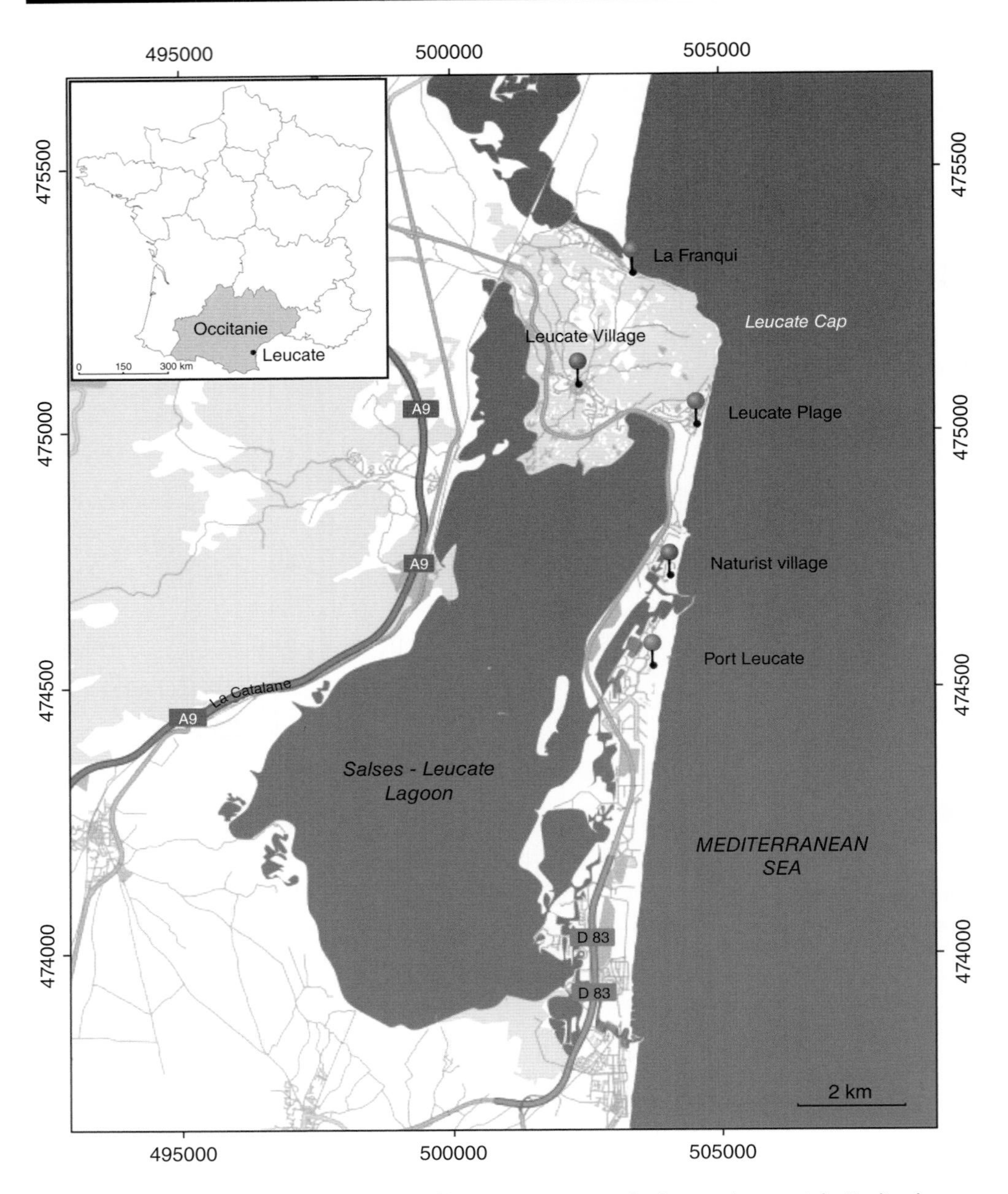

Figure 23.2 The town of Leucate and its resort areas on the Languedoc coast in Occitanie.

surveyed, some people have relatively good knowledge of the coastline honed through regular observation of sand movements, sea swell, and the wind, or through experience (Tricot 2013). Nonetheless sensitivity to the risk and a certain degree of knowledge do not necessarily go hand-in-hand with a fear of the sea. The surveys most often point to an absence of fear, evoking the phenomenon that sociologists call an 'optimism bias' – a bias broadly confirmed by the fact that buildings are poorly adapted to the flood risk,

Table 23.1 Resident and visiting populations relatively unaware of risks related to the sea (results of surveys conducted in mainland France since 2008):

Bibliographic reference	Risk and region studied (see Figure 23.1)	'Residents' capacity', observation of the sea	Sensitivity to risks or fear of the sea	Memory of risks	Knowledge of the hazards	Knowledge of management methods (hard or soft defence systems)	Preference for...	
							...hard defence	...soft defence
Anselme et al. (2008)	coastal flooding: Leucate	– (not asked question)	–	average	low	low	–	–
Roussel et al, (2009)	erosion: Palavas-les-Flots, Sète, Valras-Plage	–	–	–	low	low	–	–
Meur-Férec et al. (2010)	erosion + coastal flooding: Bray-Dunes, Oye-Plage, Wimereux	–	yes	–	average	–	–	–
Goeldner-Gianella (2011)	coastal flooding: from Cayeux-sur-mer to Ault	–	no	high	–	–	high	low
Tricot (2013)	erosion + coastal flooding: Gâvres, Guissény	yes	–	–	–	–	in decline	–
Rey-Valette et al. (2012); Hellequin et al. (2013); Rulleau et al. (2015)	coastal flooding: Palavas-les-Flots, Mauguio, Pérols	low for part-time residents	no	–	–	–	Fairly balanced preferences (but soft defence in 2nd position)	
	erosion: Palavas, Carnon, Petit travers, Boucanet	–	–	–	low	–	high	low, but coastal retreat is growing
Rey-Valette et al. (2013)	erosion: Palavas, Carnon, Petit travers, Boucanet	–	–	–	low	–	high	low, but coastal retreat is growing

(*Continued*)

Table 23.1 (*Continued*)

Bibliographic reference	Risk and region studied (see Figure 23.1)	'Residents' capacity', observation of the sea	Sensitivity to risks or fear of the sea	Memory of risks	Knowledge of the hazards	Knowledge of management methods (hard or soft defence systems)	Preference for...	
							...hard defence	...soft defence
Goeldner-Gianella et al. (2013, 2015)	coastal flooding: Lanton, Audenge, Biganos	–	no	average	–	–	high	low
Krien (2014)	erosion + coastal flooding: Pénestin, Tudy island	yes	no	–	–	–	–	–
Chionne (2015)	erosion + coastal flooding: Oléron island	–	no	yes	link with memory	–	high	high too but to maintain the coast
Longépée et al. (2015)	coastal flooding: Leucate	yes	sensitivity to the risk but no fear of the sea	average	–	low to average	–	–

as proven by storm Xynthia in France in 2010. Hellequin et al. (2013) showed that most people surveyed on the Languedoc coast fear individual (illness, road accidents, and so on) rather than environmental risks; the flood risk is only ranked fourth! Concerning environmental risks specifically, pollution (sea, air, and water) is a concern for half of those surveyed, ahead of gradual risks (such as rising sea levels) or more occasional risks (coastal flooding [43%], storms [27%], continental flooding and tsunami [12%]). In the Arcachon basin, the sea and coastal flood risks are also weakly considered by the resident population and tourists (Goeldner-Gianella et al. 2015).

In addition, retaining a memory of a risk – remembering a particular storm, flood, or erosion event – is not necessarily synonymous with sound knowledge of the hazard or awareness of its aggravating effects and management methods (the role of dikes, depolderizing, sand fences, beach replenishment, and so on). This was highlighted in several of the areas studied, for both beach erosion and flooding. We also see that tourists, visitors, and full-time/part-time residents on the mainland French coast do not demonstrate any fear of the coastal risk and do not have adequate knowledge about storm, erosion, and flooding hazards or of management methods; this is surprising amongst residents that have much lengthier contact with the sea and coastal events. This finding is broken down in Table 23.1, survey by survey.

The specific case of Leucate was chosen to illustrate these general ideas. The town comprises five territorial entities. 'Port Leucate' and the 'naturist village' are located entirely on the sand barrier – locally known as the *lido* – separating the Mediterranean sea from the lagoon of Salses-Leucate, whilst 'Leucate Plage' sits below a former active cliff (Figure 23.2). These three entities, affected differently by storms and coastal flooding, were selected for a survey into perception and representations amongst their occupants. This survey served as the basis for an analysis of differences in knowledge, behaviour, and perception depending on living area and the 'resident profile' of the people surveyed: full-time or part-time residents or tourists.[2] In terms of the distribution, there was a relatively good balance between categories of interviewed people and types of location (Table 23.2). Knowledge and perceptions of climate and marine risks amongst local stakeholders were also assessed via semi-structured interviews.

The Leucate survey shows that all the people interviewed live in potentially floodable zones and that 68% of them say that they have observed changes in the state of the beaches caused by storms. However, nearly half of residents that answered the survey are not aware that they 'live in an area at risk'. The responses from full-time and part-time residents are very similar, whilst those from tourists are significantly different.[3]

Table 23.2 Distribution of the surveys in Leucate according to living area and residential or user profile of the person interviewed.

Investigation sector	Full-time residents (%)	Part-time residents (%)	Tourists (%)	Total (%)
Leucate Plage	2 (n = 12)	4 (n = 18)	9 (n = 42)	15 (n = 72)
Naturist Village	5 (n = 23)	4 (n = 20)	1 (n = 3)	9 (n = 46)
Port Leucate	32 (n = 158)	35 (n = 174)	9 (n = 43)	76 (n = 375)
Total[a]	39 (n = 193)	43 (n = 212)	18 (n = 88)	100 (n = 493)

[a] Totals are not exact due to rounding.

For example, only 23% of tourists know that Leucate is affected by coastal flooding whilst half of the full-time and part-time residents questioned are aware (Figure 23.3a). If these results associate full-time and part-time residents, their awareness of the coastal flooding risk remains rather moderate.

Bearing in mind that, over the last 10 years, Leucate has been affected by storm floods (in 2010, 2011, 2013, and 2014), it may appear surprising that half of the residents interviewed responded negatively. These answers can be explained by the representation that

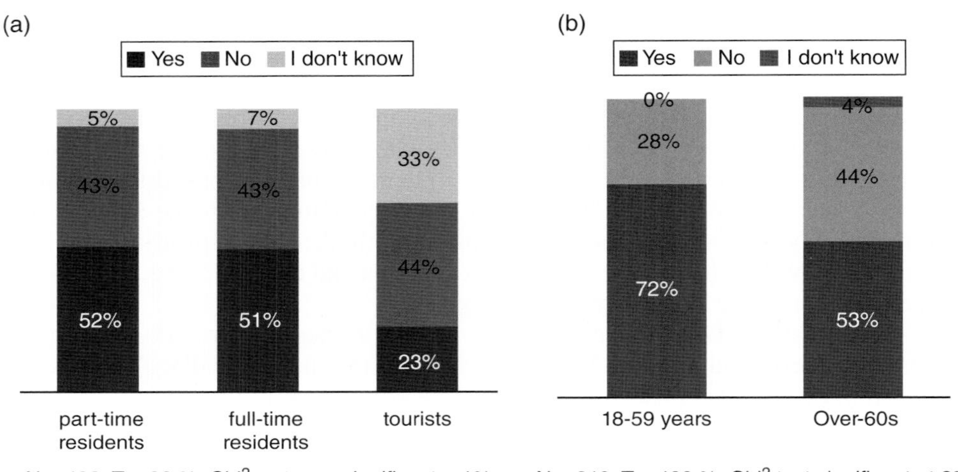

N = 493, T = 99 %, Chi^2 test very significant < 1%

N = 212, T = 100 %, Chi^2 test significant at 3%

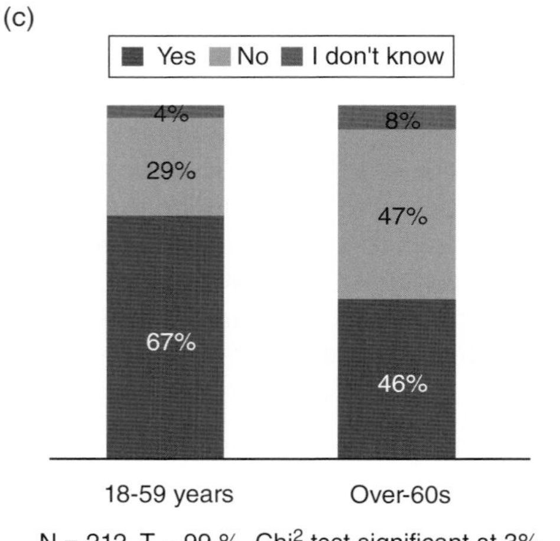

N = 212, T = 99 %, Chi^2 test significant at 3%

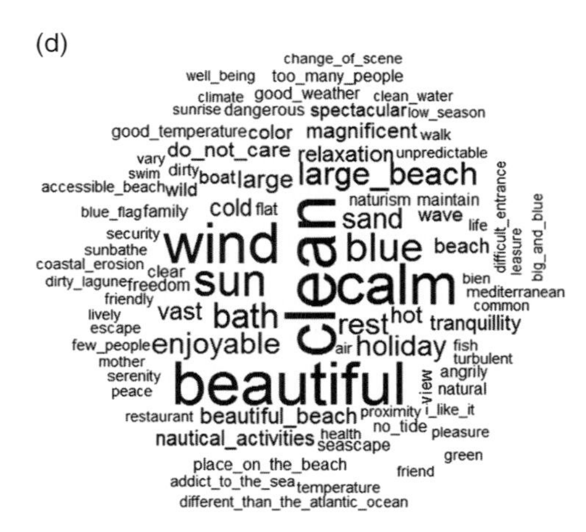

N = 493, T = 100 %, words presents in the word cloud were mentionned at least three times

Figure 23.3 Perceptions of the sea and the risk of coastal flooding in Leucate amongst the people surveyed in the town. (a) Response according to the personal profile of the people asked: 'In your view, is Leucate affected by coastal flooding?' (b) Awareness amongst part-time residents that they live in an area at risk from coast flooding, according to age. (c) Response from part-time residents according to age to the question: 'In your view, is Leucate affected by coastal flooding?' (d) Words freely associated with the expression 'the sea at Leucate'.

people have of coastal flooding and of their place of residence. From the interviewees' comments, it appears that some of them do not class low levels of seawater intrusion in the town's streets as flooding. The area lived in is also decisive: 73% of the residents interviewed at Leucate Plage and 70% at the naturist village declared that Leucate is affected by flooding, versus just 47% of those from Port Leucate. In fact, the severest floods, causing the most damage, have occurred at Leucate Plage and, to a lesser degree, at the naturist village where mainly seafront houses are exposed to damage. On the other hand, at Port Leucate, coastal flooding has not yet reached the residential areas. Therefore, Port Leucate residents, who dominated our sample, appear to know little about the floods affecting the other two sectors or tend to forget that Port Leucate does not represent the entire town. Finally, there is often very little knowledge of the hazard itself and the physical processes that characterize it.

In their study, Rulleau et al. (2015) talk about the specific perceptions of part-time residents, who are mainly retired people[4] and who have quite a fatalist outlook, feeling little concerned by coastal flooding, a risk that is likely to increase over the long term. Our sample at Leucate is mainly made up of retired people, with 71% aged over 60: 74% of the full-time residents and 77% of the part-time residents. Our results match those obtained in other municipalities in Languedoc proving that, amongst part-time residents, ageing does not lead to greater awareness of the risk of coastal flooding: the 18–59 age group is very different from the over-60s group in that it is more aware of living in an area at risk of coastal flooding (Figure 23.3b) and more readily acknowledges that Leucate is affected by coastal flooding than the over-60s do (Figure 23.3c). Does the part-time residents' level of education, which varies with age, have an influence on their replies? 60% of 18–59-year-olds followed two years of higher education compared to just 35% of over-60s.

23.1.2 Stakeholders from the political, professional, and associative spheres slightly more aware of coastal risks

The perceptions and representations amongst people other than the general public, i.e. stakeholders in the political, professional, and associative spheres concerned with the environment, development, or risks, have also been studied over recent years in some of the surveys previously mentioned (Table 23.1). The results show that these stakeholders are usually more aware of coastal events, better remember past events, and have better knowledge of the risks and how to manage them (Roussel et al. 2009; Goeldner-Gianella et al. 2015) than other residents and tourists. However, there are distinctions depending on the type of stakeholder.

A study into social representations of the marine environment held by elected representatives in French coastal communities showed that people in towns along the Mediterranean or Channel had fairly balanced representations of the coast. More of them are interested in (ecologically) protecting the marine areas of their community rather than developing it or focusing on the risks. It is only along the Atlantic coast, hit by storm Xynthia in 2010, that the focus on risks is much greater than the interest in development, but it is still surpassed by the issue of general protection (Le Moel, Moliner, and Ramadier 2015). However, the marine part of the coastal strip is generally barely assimilated in the iconographic representations of their territory that the elected representatives draw, as if these marine areas were of little importance. Only the Brittany and Provence-Alpes-Côte d'Azur (PACA) regions appear to take a greater interest (Le Moel 2016, pers, comm.). The group of elected representatives surveyed tends to suggest that the coastal risk issue is not a priority for this type of stakeholder.

Roussel et al. (2009) interviewed a varied panel of stakeholders on the issue of beach erosion and soft or hard measures to overcome it. They saw that the erosion hazard (as a physical process) and its management methods are familiar enough to agents appointed to manage coastal erosion, environmental associations, farmers, and fishers, but less well known amongst other types of management agent and association, and other economic stakeholders on the coast; the latter focus more on impacts and tourist issues related to erosion. However, the causes of erosion, and especially aggravating human practices, are well-known to all stakeholders (more than amongst beach users), who mainly mention the alteration of the dunes, seafront buildings, excessive visitor numbers, and the presence of a hard defence system (dikes and breakwaters). Only the economic stakeholders appear to be less well informed on the subject. Thus, the knowledge deficit surrounding this risk and its causes concerns certain stakeholder groups, and especially economic stakeholders on the coast.

Chevillot-Miot, Chadenas, and Mercier (2016) interviewed professional and institutional stakeholders and other parties in the field in Charente Maritime to gather their perception of the coastal flooding risk on that part of the coast after storm Xynthia. These people were asked to take a map of the Charente coast and draw the areas they believed to be exposed to a flooding risk. The study showed certain disparities in knowledge of the risk, linked to the category of stakeholder, the significance of past events and the level of media coverage of the places, but also to knowledge of the regulatory documents indicating flood zones. Stakeholders close to the ground (farmers, the fire brigade, and associations) are well familiar with the risk, but mainly in their own areas of work or in the areas where they experienced the 2010 flooding. Institutional stakeholders have very varied levels of knowledge. They have also better knowledge of the regulatory zoning – often having contributed to it themselves (Territoire à risque important d'inondation [TRI], Plan de prévention des risques [PPR], and the Programme d'actions de prévention des inondations [PAPI])– than of the Xynthia event, which they did not necessarily live through on the ground. This study shows that, broadly speaking, whilst a certain memory of the risk is shared by stakeholders on the coast, knowledge about the full extent of the risk remains disparate, varying from one group to another. Awareness-raising work on the marine risks appears necessary for these stakeholders in coastal regions: although they are more familiar with the issues than residents or tourists, they are not all fully and equally informed.

23.1.3 Past perceptions and representations were a closer match to the reality of marine risks

The low awareness of marine risks can also be seen in changes to building design and location. The multiplication of beachfront constructions, the lower design of the buildings and/or the abandoning of preventive actions in the event of storms all point to the dismissal of ancestral fears. These days, buildings are often poorly adapted to extreme events. It is estimated that in coastal communities currently at risk from coastal flooding, 20% of buildings are single-storey dwellings, offering nowhere to retreat to in the event of rising water levels (Ministère de l'Écologie du Développement Durable et de L'Énergie 2011). Recent inventories of buildings in two municipalities in the Arcachon basin give even higher figures than this estimate. In 2011, to the west of the D3 road (next to the coast) in the town of Audenge, 77% of buildings were single-storey dwellings; this figure stood at 63% in Lanton in 2012 – where 80% of houses also lacked on opening from the roof (Goeldner-Gianella et al. 2013). Another factor that changed considerably over the twentieth century is the location of dwellings in coastal communities and, more specifically, their average distance from the sea. A. Creach (2015) very clearly maps the evolution of building vulnerability in seven municipalities on the French Atlantic coast from 1950 onwards, with an increase in built-up areas in zones subject to a potentially life-threatening risk. This risk evidently existed in the past: in 1705, for example, the coastal communities of Barbâtre, La Guérinière, and L'Épine already faced a high risk of coastal flooding (Creach 2015). However, the risk was less widespread than it is today. Lastly, commonsense preventive measures, well-known to farming and fishing families in Charentes and Vendée, included owning small boats, maintaining sea dikes, ditches, and channels to enable water to be removed quickly in the event of flooding, and planting and maintaining trees to consolidate the dikes or anticipate future repair work, maintaining natural or manmade defences (afforestation of the dunes and wetlands, fisheries and fish sluices), building further back from the coast, and using a network of bells to raise the alert (Garnier and Surville 2010; Duvat and Magnan 2014; Péret and Sauzeau 2014; Chionne 2015). The breakdown in the ancestral relationships between the 'historically resilient' (Chionne 2015) coastal communities and their environment coincides with the emergence of a more peaceable perception of the sea, but one that has potentially more lethal consequences.

In Leucate, exposure to marine hazards is just as recent. Until the early twentieth century, the *lido* was only occupied by vines and fishing huts because residents of Leucate Village were aware of the risk of flooding: 'In living memory, the largest flooding event to hit Leucate was in 1929. At that time, the coast was wild and uninhabited. There were just a few fishing huts built on the beachfront and people only lived in them in summer – fortunately, because one autumn morning in 1929, the people of Leucate saw that the sea had risen to within a few hundred metres of the entrance of the village' (Hiron 1998, p. 257). Seaside tourism first came to Leucate 'Plage' – named accordingly – in the early twentieth century, and building continued in the lower part of the village from the 1950s to the 1980s. However, the interministerial mission for tourist development on the Languedoc-Roussillon coast (known as the Racine mission), initiated in the 1960s, gave Leucate its true seaside resort status with the development of Port Leucate, continuing from the 1950s to the 2000s with a peak from the mid-1970s to the mid-1980, and construction of the naturist village from the 1970s to the 2000s (Pont 2015). The Racine mission connected these strips of sand to the 'continent' via partially floodable roads to turn them into residential areas, totally ignoring the fact that these beach barriers are highly exposed to climate and marine hazards.

23.2 Multiple factors behind the gradual dwindling of the 'culture of coastal risks'

Several factors may explain the change in social representations of the sea and coastal risks over the twentieth century, which some have referred to as the decline of the culture of risk (Anziani 2010). We will first look at the explanations concerning a change in coastal practices and the gradual disempowerment of coastal societies, before examining the occurrence of storms and the advance in scientific knowledge about the sea.

23.2.1 Practices and social representations of the sea increasingly detached from the notion of danger

The acknowledged decline in a certain risk culture is related to the change in professional and leisure practices involving the sea and to the change in social representations and images of the sea. The decline in primary activities linked to the sea is most apparent in the agricultural sector and has led to the abandonment of the traditional practices of monitoring and maintaining dikes and drainage systems mentioned earlier. On the French coast in recent decades, we have seen a considerable reduction in the land devoted to agriculture and in the number of holdings, due to difficulties in finding people to take over farms. There has also been a marked fall in aquaculture activities and salt production in the past few years. Undeniably, the coastal economy is increasingly focused on services for full-time and part-time residents and tourists – what we call the residential economy. In 2005, the primary sector only accounted for 4% of jobs in coastal 'departments', compared to 21.7 and 74.3% respectively for the secondary and tertiary sectors. In addition to this change in business activity, a strong social and demographic trend has been observed in all mainland French coastal areas since the 1970s (Observatoire du littoral): coastal communities - i.e. costal 'cantons' – gained an additional one million full-time residents between 1975 and 2005, plus an increasing number of tourists and part-time residents. 75% of the growth in the coastal population can be explained by the migratory flows observed between 1999 and 2005. This went hand-in-hand with a rise in the number of over-60s on the coast over the same period. Today's coastal community therefore only has a weak link with the sea in terms of business activity and people do not have a long experience of living in such an environment.

Nonetheless, the 'lure of the sea', which was first felt in the eighteenth and nineteenth centuries in Western Europe (Corbin 1990), has gradually seen the coastal population move closer to the sea and practice an increasing number of leisure activities there. These activities have changed over time, ranging from contemplative walks to bathing and water sports – three activities that are still enjoyed today. In fact, the first seaside activity that the French people interviewed mention is walking and watching the sea (85%), followed by bathing and related activities (72%), seafood consumption (55%) and water sports (30%) (IFOP 2014). At Leucate, the people interviewed said that they bathe (54%), walk (51%), relax, read, or sunbathe on the beach (39%), or practise sports such as cycling or jogging (19%). What is more, when we ask people in mainland France where they would rather live, since 2009 the first response has been the seaside, ahead of the countryside, town, and the mountains (IFOP 2014). The sea view has thus become a major component of the coastal economy (Robert 2009) and a factor in the residential choices of certain population categories. Moreover, the coastal population shows a growing interest in different kinds of connection with the sea, expressed in cultural events centred on maritime identity.[5] Over recent decades, there has been an increase in the number of maritime ecomuseums focused on activities practiced on land and at sea (*Musée des Terres Neuvas* in Fécamp, *Musée des Cap-Horniers* in Saint-Servan, the *Maisons de corsaires* in Saint-Malo or the many salt museums), and restorations of old sailing ships that can be visited in the harbour or boarded for trips out to sea. The rebuilding of the Hermione,[6] an eighteenth-century French warship is the perfect example. However, this marked interest in the sea does not usually result in the various population groups grasping the most dangerous aspects of the sea, which are usually experienced in winter.

Representations of the sea and marine hazards also changed significantly during the twentieth century. The historian Alain Corbin explains (2014), however, that what people commonly call 'the sea' is not what can be perceived whilst out sailing, but what they see from a fixed point on the shore, the sea that you can observe whilst on land, 'in its union with the shore'. Nonetheless, the high seas have their own representations – often connected to the idea of conquest by and freedom for humankind (Roux 1997; Parrain 2010). On land, the sea is now associated with values with positive connotations (beautiful, clean, calm, enjoyable), leisure (large beach, bathe, holiday, rest) and the natural elements used to describe it (wind – highly present on this part of the Mediterranean coast –, sun, blue, sand, vast), as shown in the word cloud obtained using the words spontaneously cited by the people questioned in Leucate (Figure 23.3d). Negative elements or those concerning risks are barely mentioned spontaneously. It is difficult to distinguish the more negative-sounding words or those linked to marine hazards (cold, dangerous, dirty, anger, coastal erosion) in this word cloud. The words storm and flooding were not mentioned at all by the almost 500 people surveyed! In the extensive survey on the relationships between French people and the sea conducted by the Agency for Protected Marine Areas (AAMP) and the Institut français d'opinion publique (IFOP) in 2014, there is a similar impression of indifference concerning marine hazards: 84% of French people believe that the world's seas and oceans are in poor condition today (when asked a direct question on the matter) and associate the dangers that threaten it with 'pollutions' from activities on land and boats, 'oil slicks', 'intensive fishing', and 'climate change'. As such, 'to protect the sea', 68% of French people consider that 'there is need to develop human activities that are more respectful of the marine environment', though they do not go as far as suggesting 'significantly reducing human activities at sea' (11%). It turns out that the sea is no longer seen as a danger to humankind (although the survey did not cover this aspect), but is instead perceived to be threatened by humankind – an idea that was also indirectly reflected in the Leucate survey, in the lack of terms evoking the danger for humans.

These social representations are now associated with a high number of images of the sea, available as postcards, photographs, or films, all conveying the same impression that the sea is tamed and/or aesthetically pleasing. The photographer Philippe Plisson manages an online image database containing 35 000 pictures of storms, lighthouses, racing boats and merchant vessels (and their crews), well-known coastal regions (Brittany, Venice, and the French coasts in general) and the like. His images of storms do not arouse fear: they are usually highly aestheticized and, as such, inspire admiration or fascination before the Sublime, as in the eighteenth century. Recent French films featuring the sea are either documentaries demonstrating its beauty or the threats that surround it – *Océans* (2007), *Planète Océan* (2012), *Méditerranée, notre mère à tous* (2013) – or films showing life on the high seas and all that it entails – *En solitaire* (2013) or *Fidélio, l'odyssée d'Alice* (2014). Only rarely do these films evoke the threats from the sea to people on the coast.

23.2.2 The gradual disempowerment of coastal societies

Another factor that may contribute to this indifference to the risks is a gradual disempowerment of the coastal society. From the late eighteenth century onwards, civilian security came to be seen as a right, included in the Declaration of the Rights of Man and of the Citizen. This right was confirmed in the 1793 French constitution: its Article 8 stated that 'Security consists in the protection afforded by society to each of its members for the preservation of his person, his rights, and his property'. The following century, a right to solidarity within society was introduced. This explains why the public authorities embarked upon an active policy to defend civilians against natural hazards on the coast, resulting in the construction of hard sea defence systems from the nineteenth century to the early 1990s. Whilst the law dated 16/09/1807 made land owners on the coast responsible for their own protection (Article 33), it nonetheless stated that 'the need' to build sea defence dikes 'will be taken on-board by the government' and that 'aid from public funds' could be used for this purpose where 'the government deems it fair and useful'. In the nineteenth century therefore, the public civil engineering department employed dike attendants to monitor structures and carry out minor repairs (Boucard, in Péret and Sauzeau 2014). Ile de Ré provides a good example of this kind of change in responsibility: sea dikes were maintained by islanders from the Middle Ages up until the seventeenth century, when the province (or généralité) intendant took responsibility for decisions on the work to be done, following reports and opinions from the king's engineers. Public monitoring and maintenance remained the norm on the island until the 1930s (Boucard, in Péret and Sauzeau 2014).

This policy of protection evolved in the late twentieth century, becoming part of a wider policy of prevention and, in recent years, encompassing new thinking on the need to adapt to climate change and its impacts – two trends leading to a lower level of State involvement. However, the public authorities continue to exercise a number of responsibilities, which are well-known to the public. For example, the mayors of coastal municipalities have to 'ensure public safety from day to day, through information, preparation and a warning system' (Meur-Férec and Rabuteau 2014). As the general administrative police authority responsible for public order, they are required to install signage indicating risks on the coast (e.g. cliff landslides or dikes in a state of disrepair), to listen to weather warnings from Météo-France – especially the new 'waves/flooding' warning introduced after storm Xynthia – to follow the prefect's instructions in the event of a red warning, to take charge of ownerless dikes – which account for nearly a quarter of all dikes along the mainland French coast – to oversee application of the plan for the prevention of coastal risks (PPRL or PPR) within the framework of local urban planning (PLU), to produce a communal information file on major risks (document d'information communal sur les risques majeurs DICRIM) and, once the PPR has been approved, a communal response plan (PCS) to facilitate crisis management. The State's decentralized services also play a role in this respect with regard to coastal hazards: the DDTMs (departmental territories and sea directorates) rank protection systems such as dikes whilst the DREAL (regional directorates for the environment, town planning and housing) inspect the systems. Given this array of information and prevention initiatives, it is understandable that residents and tourists have the impression that they are adequately protected from the sea, regardless of their level of trust in the authorities or their level of knowledge of these various documents and schemes. In Leucate, 63% of the residents surveyed think that the municipality handles the risk of coastal flooding. When they are asked what measures would require them to evacuate their home in the event of a flood risk, they mainly refer to the town council, with a majority mentioning a preference for telephone warning (45%) or sirens (24%).

23.2.3 The focus on coastal risks is proportional to the events experienced: The case of the French and North Sea coasts

The fact that extreme coastal events were rare on the French coast in the twentieth century also explains why people in coastal communities know little about them. For example, prior to Xynthia, there had only been two storm floods on Ile de Ré (in 1941 and 1957); if the 1941 storm was a significant event, the 1957 episode was much less severe. These events were largely forgotten about until 1999 and then 2010, after 70 years of calm weather! 'Other than the collective memory and a little historical research, there was nothing to remind newcomers that the sea could be wild in winter and flood part of the island', (Boucard, in Péret and Sauzeau 2014). It is the same storey on Noirmoutier Island, where there had been no serious flooding since 1941, apart from the Sébastopol agricultural polder in 1971. The built-up area has spread to the low-lying salt marshes, dangerously close to the dikes (Creach 2015). The people interviewed at Leucate also have very little personal experience of significant events in the town. Yet there have been 14 orders formally recognizing natural disasters following storm events since 1980. The event most likely to have remained in people's memories is the beaching of three deep-draught vessels to the north of the town during the November 1999 storm, showing just how high the water rose during that event. It was mainly during the second half of the twentieth century that storm floods were rare in France; this kind of extreme event was more frequent in the earlier half of the last century, with 63 occurrences between 1500 and 2010 (Garnier and Surville 2010). The Saintongeais and Vendeen coasts experienced six storm floods in the first half of the twentieth century, but just two between 1950 and 2010; the figures are four versus one for the Mediterranean Coast, and six versus one on the Lower Normandy coast (Garnier and Surville 2010). There are also spatial polarizations on the coastal strip, which explains why storm events are better known in some places than others. The Cotentin peninsula and its surrounding areas have been the most affected by storm floods since the eighteenth century, as well as the northern part of the Gulf of Lion between Agde and Villefranche since the seventeenth century, and especially during the nineteenth century, and then the areas around the island of Noirmoutier, Les Sables-d'Olonne, Ile de Ré and the mouth of the Gironde since the seventeenth century, with an increasing number of events in the twentieth century (Garnier and Surville 2010). The uneven distribution of the risk over time and space is thus another factor contributing to the lack or gradual loss of the risk culture.

Some of France's European neighbours around the North Sea appear better prepared for extreme coastal events, no doubt due to the recurrence of violent storms, erosion processes and flooding that their coastal areas experience. These countries have suffered several disasters similar to – although even more dramatic than – storm Xynthia in France. The most devastating events in the North Sea in recent decades were the storm that raged from 31 January to 2 February 1953 in the UK and the Netherlands, and the February 1962 storm in Germany with several hundred dikes breached and failing to protect the coasts and estuaries from the sea. Apart from the number of breaches, the number of victims and the extent of the flooded areas reflect the scale of the disasters: there were hundreds of victims in the UK and Germany (over 300 in each country) and thousands in the Netherlands (1850 victims, mainly in Zeeland). According to reports, the flooded areas stretched some 38 000 ha in Germany, 60 000–100 000 ha in southeast England and 200 000 ha in south-eastern parts of the Netherlands where entire islands were submerged along the Eastern Scheldt, Grevelingen, and Haringvliet estuaries. An estimated 45 000 farm animals were lost in the UK and 150–200 000 in Zeeland. 47 000 buildings were damaged in the latter region, 24 000 in the UK and several thousand in Germany. A comparison of the three countries' situations in 1953 and 1962 (Goeldner-Gianella 2013) shows that the Netherlands were worst affected, with Zeeland suffering a 'national disaster'. The place that the 1953 storm still holds in Dutch memories can be explained by its strong geographical concentration in virtually one region (Zeeland and the southern part of Zuid Holland), compared to the east coast of England where the flooded areas were more widely spread. In addition, the sea also did not rise as far inland in the UK as it did along the Zeeland estuaries, still open to the sea at that time. Since these extreme events, considerable protective measures have been implemented in these regions, for example the construction of immense estuary dams as part of the Delta Works in the Netherlands, the building of new dikes in the German coastal Länders and the Thames barrier that has protected London since 1984. These policies are continually reviewed and updated to take on-board the new impacts of climate change.

Despite their disastrous consequences, the 1953 and 1962 storms had contrasting psychological impacts on the countries' populations. The British soon appeared to forget the severity of the 1953 storm (Rupp-Armstrong and Nicholls 2007). Several surveys revealed a lack of fear about future flooding in the UK (Myatt, Scrimshaw, and Lester 2003). At the same time, researchers pointed to a high level of trust in the sea defences and in the warning systems in place: these days, few inhabitants think that the British coast poses a life-threatening risk. This is not true of people on the German and Dutch coasts where the storms left a profound mark on collective memories. It is true that similarly extraordinary disasters had already occurred in Europe in previous centuries. In Germany, there is often mention of the mediaeval village of Rungholt, which lay on an island off Northern Friesland. It was wiped off the map during a storm in 1362. Going back even earlier, there were 10 000 storm victims in Eiderstedt and Dithmarschen in 1216. Later on, in 1634, a large part of Northern Friesland was again flooded and this time split into three islands during a storm, causing 6000 deaths and the loss of 50 000 animals. German culture has been strongly marked by these events, as can be seen in the tales, legends, and proverbs linked to the sea. Human and animal sacrifices actually took place in the German coastal regions in an attempt to appease the spirits during storms and to ask them to keep the dikes strong; these went on until the sixteenth and seventeenth centuries respectively. In 1421, the Dutch coasts suffered a storm and flooding, submerging 70 villages and 42 000 ha, and causing 100 000 fatalities (Wagret 1959). Geographer A. Miossec (1993) describes the Netherlands as a country 'at the peril of the sea', where 'the cruel memories of 1953 will long weigh on people's behaviour. Flood control is the cardinal rule for good coastal management' [...]. The Zeelanders who were personally affected by the events of 1953 or concerned by the post-storm rebuilding efforts felt like they were living through a 'war', their region as devastated as a battlefield. Understandably, they cannot envisage giving in to the sea today, for example through the depolderization programmes (Goeldner-Gianella 2013).

In addition, considerable and varied efforts are made to sustain a memory of risk amongst younger generations, more than 60 years after the 1953 storm. For example, signs have been installed in the western part of the Netherlands reminding people of where the natural shoreline would lie if the 7000 km^2 of polders had not been established to gradually reclaim one third of the country's surface area from the sea. On the Zeeland island of Schouwen-Duiveland, which was almost entirely submerged in 1953, there is a museum commemorating the disaster (Museum Watersnood 1953). In the Netherlands (Figure 23.4) and Germany, there are also several monuments along the coast that

Figure 23.4 Monument commemorating the February 1953 storm in Zeeland, clearly illustrating humankind's vulnerability before the sea. *Source*: Photo: L. Goeldner-Gianella.

serve as a constant reminder of humankind's intrinsic vulnerability before the sea. There is still a strong focus on extreme coastal events in these two countries, amongst public and other coastal stakeholders and the general public. It is true, however, that these areas have experienced some particularly dramatic events over the centuries.

23.2.4 Inadequate information on coastal hazards and only recent scientific understanding of these phenomena

Other than the social, economic, political, and situational factors mentioned above, the fact that scientific and technological advances have been made only recently may also explain the continued lack of understanding of marine hazards amongst the public and the key stakeholders on the coast.

Putting aside events that may have affected or concerned them directly, coastal populations have access to various types of information there to remind them of how dangerous the sea can be. Since Xynthia, some municipalities have taken measures to increase awareness of the risk. For example, in La Rochelle the lower parts of the trees in one park near the harbour have been painted blue to show the level the water reached in the town during the storm (Figure 23.5). But do today's residents and tourists really know

Figure 23.5 Trees pained blue in La Rochelle to demonstrate the water level reached during storm Xynthia (lower band) and the level that coastal flooding could reach in the future with a 1-m rise in sea level (upper band). *Source*: Photo: L. Goeldner-Gianella.

what the blue lines mean, since there are no explanatory information boards? Elsewhere on the coast, flood markers remain quite rare. A. Creach observed that out of the 2000 flood markers distributed to municipalities after Xynthia, only one sixth had been installed by 2015! This is especially worrying given that back in 2003, law 699, known as the 'law on hazards' required mayors and the competent State departments to carry out an inventory of flood markers in their municipalities and to set up markers to indicate coastal flooding. Another example: in Leucate, a text analysis using a digital version of *Cap Leucate* magazine (46 issues published between January 2010 and June 2015) reveals a similar lack of information on extreme coastal events. The words 'wind', 'sea' and 'protection' are frequently mentioned in the magazine (respectively 326, 212, and 100 times) but the same cannot be said of the words 'storm', 'flood', 'submersion', and 'erosion', which only occur 32, 16, 9, and 7 times respectively and are often only mentioned after really significant events such as the 2013 storm (Pont 2015). In addition, the magazine is not necessarily read by everyone: as shown by the Leucate survey, few part-time residents do read it or automatically receive it, since the council only delivers the magazine to part-time residents who specifically request it.

Stakeholders on the coast may have demonstrated difficulties in understanding the marine processes and in adopting effective protective measures because knowledge of the processes is actually quite recent, as is its dissemination outside of scientific circles. Admittedly, thanks to André Guilcher's work, the overall functioning of the coastline, erosion, and dune nourishment were partially explained in the 1950s and, early on, Guilcher understood the importance of studying the more immediate underwater environment (Pinot 2002). Physical models were produced in the 1960s (with the wave flume, for example) and the notion of 'sediment cell', crucial to coastal management and development,[7] emerged in the 1970s in the English-speaking sphere (Cohen et al. 2002). However, it was only in the 1980s and 1990s that a more 'holistic' approach to the coast developed there and in France, namely the 'morphodynamic' approach applied by coastal geomorphologists (Cohen et al. 2002) enabling better quantification of sediment flows and deeper understanding of the factors at work. The advances made in topography measurement, marine weather conditions, and granulometry since that time, along with the modelling of the processes at work, have all helped take things forward. Modelling is used to test hypotheses that come from the field, and digital modelling is increasingly preferred over physical and semi-quantitative modelling. Nonetheless, field work is still essential when it comes to calibrating these models which do not yet take the high degree of spatial and temporal variation into account (Gueben-Venière 2014). Progress also needs to be made with predictive studies for better integration of the physical elements that are now vital, such as the rise in sea levels or changes in storms likely to occur in the future. The results of these digital models are not necessarily intelligible to most political, economic, or environmental stakeholders on the coast and these parties may therefore doubt them or ignore them. We should reiterate the fact that these stakeholders are not yet all aware of how the hazard occurs or of methods used to tackle coastal retreat, even though improvements in scientific knowledge in the field date back some 20–30 years.

All in all, there is still room for progress by scientists[8] and management stakeholders on the French coast if they are to increase awareness of the risk and help revive a real culture of risk.

23.3 What recommendations for public policy emerge from this research into the perceptions and representations of risks?

Having understood the importance of changes in coastal practices and representations of the sea, the disempowerment of the population with regard to the sea, the recurrence of extreme coastal events, and the advance of science and technologies, we can now establish some recommendations for public policy based on improved knowledge of the perceptions and representations of the risk. First, this knowledge provides a better understanding of the coastal populations' and stakeholders' preferences regarding 'hard' or 'soft' defence measures and risk management, and it may help in influencing them. Next, these surveys have had beneficial effects, not only on the people surveyed, but on the public stakeholders who were sent the results.

23.3.1 Are preferences with regard to sea defence systems linked to the low risk culture?

We can justifiably question whether the lack of attention the French pay to storms and their impacts has an influence on the type of defence system they prefer. Based on the surveys we referred to earlier (Figure 23.1, Table 23.1), we can see that although closed, well-maintained dikes are given preference over open dike systems (in depolderizing projects), these preferences are paradoxically not usually linked to the dikes' role in sea defence. Some sites are no doubt exceptions, such as the Bas-Champs areas of Picardy between Cayeux and Ault (Figure 23.1) that now link up to the shingle bank (perceived as a dike) after significant flooding in 1990.[9] However, broadly speaking, as discussed earlier, the sea does not spark fear these days. The dikes are mainly appreciated for the sporting or leisure possibilities they offer (jogging, walking, or cycling) or for hunting (Goeldner-Gianella et al. 2015), for instance in the marshlands that are home to numerous bird and duck species in the polders sheltered by the dikes, or on the salt marshes immediately next to the dikes on the Picardy coast or in the Arcachon basin. Conserving the dikes is a way of ensuring a future for these activities, which have often been popular for several decades. It is perhaps the absence of fear of the sea that explains, at least in part, the acceptance of softer defence systems on certain coasts, such as beach replenishment on the sandy coastline of Languedoc (Table 23.1). Coastal risk managers favour these soft management modes which fit well with their wish to implement less costly and more sustainable coastal defence systems, but it is worth questioning whether they are merely accepted by the public because of an 'indifference to the risk' rather than a marked 'preference for more integrated management of the coastal zone'? As it is, few residents and tourists subscribe to adaptation policies, whether they include the relocation of properties and activities, or a laissez-faire approach (Rey-Valette and Rulleau 2016) – which again results in a lack of knowledge about marine risks and their potential worsening in the future.

At Leucate, we cannot discuss any preference for risk management methods because the people interviewed were barely aware of the presence of sea defence structures. In fact, 38% of them thought that the town did not have any such structures whilst 15% said they were unable to respond. Those who claimed to be aware of these structures

(47%) are the same people that declared that Leucate is affected by coastal flooding (63% of them). They are mainly from Leucate Plage (73%) and the naturist village (70%), the two parts of the town the worst affected by coastal flooding. In the naturist village, the inhabitants of three residential complexes came together to fund the riprap above the beach. At Leucate, there is therefore a clear link between the focus put on coastal risk management methods and the observation or experience of coastal flooding: only the people most concerned by flooding are aware of the defence structures.

Most stakeholders on the coast have a position that contrasts with that of the general public: they have a preference for soft coastal defence techniques (depolderizing, beach replenishment, and so on). Yet there are distinctions between stakeholders, depending on their level of attachment to or interest in the coastal area in question. For example, when it comes to the management methods recommended for sandy beaches threatened by erosion, the stakeholders that are the least involved in the coast or the least committed to local policies are the most in favour of a strategic retreat, contrary to parties with a high stake in these areas who are more favourable to a stabilization of the coastline (Roussel et al. 2009). There is a similar split when it comes to coastal flooding and its management methods: the stakeholders with a lower involvement in the territory and considering that the Arcachon basin polders are natural spaces to be integrated in an overall environmental management plan are therefore more in favour of depolderizing than people with firmer roots in the territory and a greater attachment to the polders and their history. The latter are more likely to opt for traditional defence strategies, namely maintenance of the sea dikes (Goeldner-Gianella and Bertrand 2013). These two examples show that sensitivity to marine risks is often more marked amongst stakeholders with a direct interest in the territories concerned, who therefore usually opt for traditional defence techniques.

23.3.2 Sound knowledge of perceptions is vital in establishing relevant local policies

Finally, we will look at the overall lessons that can be learned from the surveys into the perception and representations of coastal risks. There are several lessons and they appear to have a positive impact on the populations surveyed and on the public authorities.

First of all, through the questions that the surveys raise and the themes that they deal with, the surveys no doubt raise awareness about a potential risk amongst the residents and tourists questioned, a risk that they had not previously considered. In addition, when they ask questions about the sea defence methods (known or preferred methods, and so on) or initiate a dialogue between the interviewer and interviewee after the survey itself, they disseminate a form of knowledge on existing or possible defence and adaptation systems. Finally, the questions asked during the survey may trigger subsequent reflection or incite interviewees to seek out further information or take preventive or adaptive action themselves. In Leucate, the people surveyed often asked for more details which were discussed once the questionnaire had been completed. Our survey served to remind Leucate residents of the importance of providing the town council with their contact details to receive warnings, to stay informed on the role of the sirens and to find out where the nearest refuge point is. What is more, information was shared during discussions between neighbours after our visit. In the end, 35% of the people interviewed thought they would find out more about the risk of coast flooding in Leucate after our survey and some of them asked where they could find further information. Some interviewees in Leucate looked out and commented on photographs at the end of the interview (Figure 23.6).

These surveys are also of benefit to the public authorities as they may form the first step in a public information policy. Leucate town council took advantage of the announcement about the CRISSIS survey in the *Cap Leucate* municipal magazine in March 2015 (www.leucate.fr) to provide its population with information on the town's response plan. On Ile d'Oléron, the town council also shared the preliminary results of a survey into perceptions of marine hazards (Chionne 2015) alongside information on the flood prevention action plan and the natural risk prevention plan for the island in the *Vent Portant* monthly magazine published by the communities' association (January 2016). Surveys into the perception and representations of risk also help to identify more accurately the type of people that the authorities need to inform or alert as a priority. In Leucate, the three resort sectors are inhabited by many full-time and part-time residents aged over 60. These people do not feel particularly concerned by the risk of coast flooding, which they believed to be linked to rising sea levels and – in their eyes – a problem for future generations. Storm Xynthia did not give them much cause for concern as many of them believe the Mediterranean to be a calm – or indeed 'dead' – sea (because there are no tides), so storms like those experienced on the Atlantic coasts are unlikely to occur there (a feeling that the younger generations

Figure 23.6 During a storm in the 2010s, the seafront houses in the Eden residential complex at the naturist village suffered significant damage by the sea. *Source*: Photo: L. Guyot, Leucate resident.

share). However, the older population – as seen during storm Xynthia – is one of the most vulnerable to this type of risk. The town council could make use of these perception biases in its information campaigns. They could also take into account the communication channels preferred by older people: amongst the over-60s surveyed, the preferred communication channels are letters or brochures in the letterbox (42%) or, more rarely, door-to-door campaigns (12%), conferences or informative meetings (11%) and a website (10%). Another population group that needs to be alerted as a priority or differently is the group of part-time residents. The CRISSIS survey shows that they are not very familiar with the warning systems recently introduced in Leucate. Of the part-time residents interviewed, 74% were not affected by the information campaign explaining the town response plan, which includes a telephone warning system, and only 7% of them have provided their phone number so they can be contacted by phone in the event of a flood risk, compared to 32 and 37% respectively of full-time residents. Likewise, fewer part-time residents (16%) than full-time residents (46%) know that sirens have been installed, although many people are unaware that their role is to raise the alert in the event of flooding. Only 5% of them know what to do if the siren sounds: i.e. go to one of the refuge points designated by the town council. These results strongly suggest that Leucate town council needs to introduce a communication campaign targeting part-time residents, many (71% of those surveyed) of whom are present from October to March when the risk of flooding is highest. The council also needs to continue to inform full-time residents about the systems in place and especially what to do in the event of an alert. All in all, the majority of people interviewed and aware of the warning systems are not particularly interested in the procedure to follow in the event of a crisis, perhaps due to excessive confidence in the response systems or the fact that they are not afraid of the coastal flooding risk.

Finally, surveys into the perception and representations of risks help to locate the most 'socially vulnerable' communities, i.e. those where a culture of risk is most lacking, on condition that the methodology applied – i.e. mental maps – is used to gather spatialized information. This information can be marked by the interviewees on the maps of the town or village provided by the interviewer and should cover zones previously affected by flooding or erosion, or zones potentially concerned by these risks. In this way, the people surveyed help build a running memory or on the contrary, help shape forecasts, something that will inevitably raise their awareness of the issue. Once all this data is put together on a single map, the authorities can compare the perceived flooding zones with actual events and thus get a better picture of the districts where residents need to be better informed of the risks. For example, in the resort of Leucate Plage (Figure 23.7), a comparison of the flooding risk perceived by residents and the real risk shows that the perception is somewhat exaggerated to the north of the resort, but underestimated in the south (an area that is actually uninhabited) and the inner parts of the resort near the lagoon or cliff, where fewer people apprehend a flood than in the seafront areas. Other geography research teams have drawn up mental maps on the perception of marine hazards. Chevillot-Miot et al. (2017) questioned institutional and professional stakeholders and people in the field about the location of areas exposed to coastal flooding risks on the Charente Maritime coast. The mind maps produced by the interviewees during the survey show real knowledge of the risk in certain sectors but a lack of knowledge in other, less urban zones or zones that were less affected by storm Xynthia or subject to less media coverage during the storm. In this case, the survey helped to spatialize risk culture at a more regional scale amongst stakeholders who are

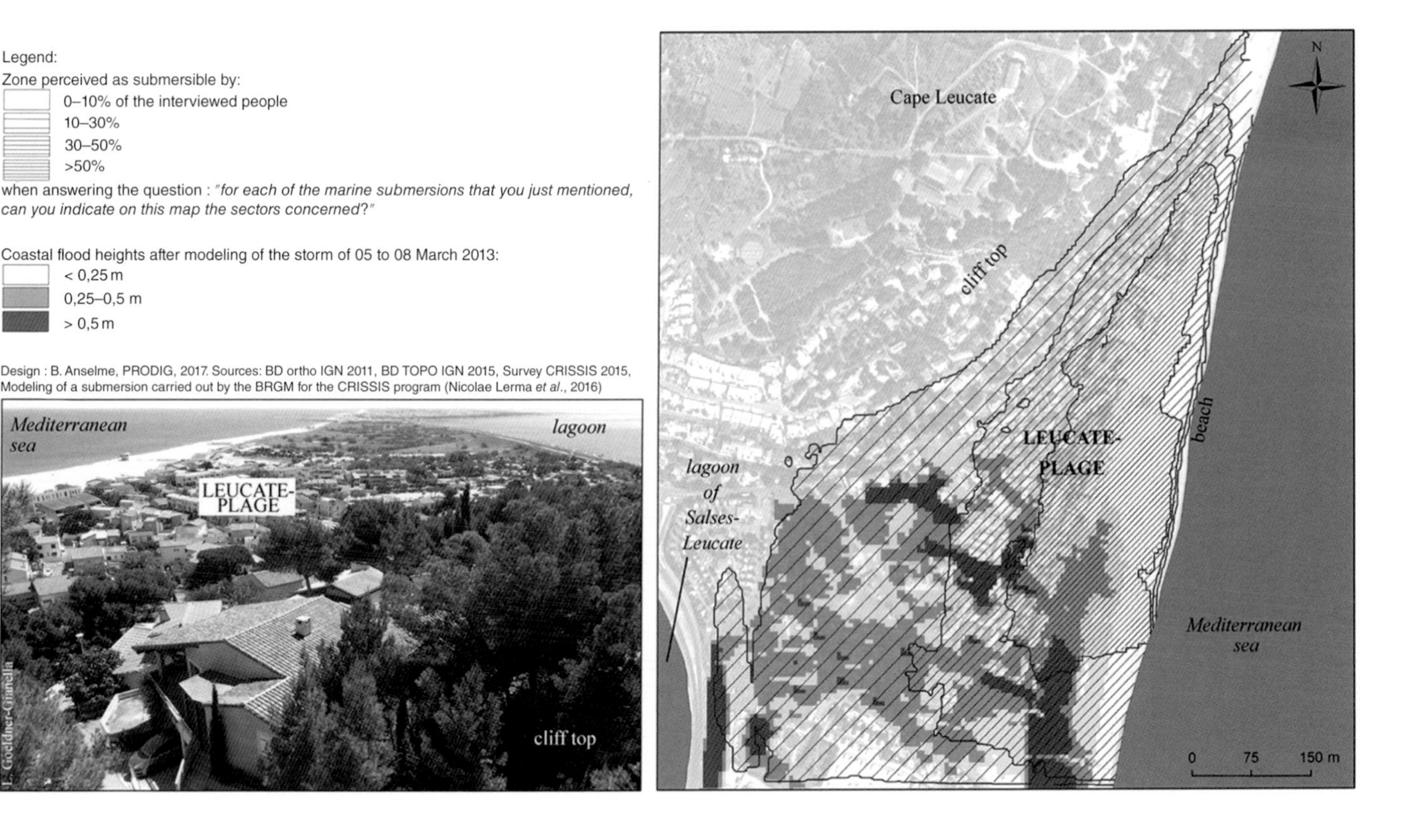

Figure 23.7 Comparison between perception of the flood risk at Leucate-Plage and the real risk, modelled by the Bureau de Recherches Géologiques et Minières (BRGM) for the storm of 5–8 March 2013.

a priori better informed than residents and tourists. Regardless of the scale, however, spatialization of the perception and representations of the risk can be used to adjust local information policies to make them more useful and more cost-effective.

23.4 Conclusion

We do not believe it is necessary to increase the number of surveys as many have now been conducted on certain sections of the French coast, especially in Languedoc and Brittany. However, we think it is now important to convince the public authorities of the usefulness of the surveys when it comes to making local information policies on extreme coastal risks more effective and better adapted in spatial terms. This is all the more important since some municipalities are already starting to consider the retreat of their activities and dwellings, in the long term at least. In French overseas regions however, the widespread use of surveys appears still important in helping the territories most concerned by extreme coastal risks and climate change to adapt, e.g. the Polynesian atolls.

Overall, all knowledge about the sea and its hazards needs to be increased, as indicated in the new national strategy for the sea and the coast (SNML) adopted in France in February 2017 after several years' delay. Its Focus I.E. entitled 'Building a society of marine and maritime knowledge' recalls the importance of 'improving the discovery, knowledge and understanding of these specific areas', whilst its Focus II.E. on the adaptation of coastal development to climate – which reiterates the content of the national strategy for integrated management of the coastline – expresses the need to 'make data on coastal hazards and ecosystems available to all stakeholders concerned'. We can probably consider that coastal populations are now counted amongst these stakeholders in the authorities' view. The SNML concludes by emphasizing that efforts to adapt to coastal risks 'imply the acquisition of more in-depth knowledge'. As we have pointed out here, knowledge of the sea, past risks, and future changes, should be expanded across the board: in scientific work, amongst political and economic stakeholders on the coast, and amongst its residents and visitors. It would also be useful if schools included natural risks in the region on their curricula.

Acknowledgements

The CRISSIS programme, concerning Leucate, was backed by the *Conseil Supérieur de la Formation et de la Recherche Stratégiques*. The authors thank the reviewers of this chapter and all the residents and actors of Leucate who provided them with scientific comments or local support and information.

Notes

1. CRISSIS programme (*Caractérisation du RIsque Submersion marine sur des SItes sensible* – characterization of the coastal flooding risk at sensitive sites) backed by the *Conseil Supérieur de la Formation et de la Recherche Stratégiques*. URL: http://crissis2015.free.fr.
2. Fewer tourists were interviewed as they are less likely to be exposed to coastal flooding, which occurs from October to March, the low season for tourism.

3. All the results of the bivariate analysis presented for Leucate have been chi-squared tested. Only the significant results are presented in this chapter.
4. In this sample, part-time residents are older (aged 62 on average) than full-time residents (aged 51.5 on average).
5. Maritime identity is defined as the nature and evolution of the links that have historically existed between human societies and the maritime and coastal environment, in other words, the diversity of sensitivities to the coastal and marine environment (Péron and Rieucau 1996).
6. The *Hermione* is a French warship dating from the late eighteenth century, which took the Marquis de La Fayette to the United States in 1780, where he intended to join the American patriots to fight for their independence. The building of a replica of the vessel began in Rochefort in 1997. It set sail again in 2014, crossed the Atlantic and is open to visitors.
7. 'Section of the coast with an overall balanced sediment budget and along which sediments flow' (Cohen et al. 2002).
8. URL: http://crissis2015.free.fr/wp-content/uploads/colloques/poster-anrn.pdf.
9. Artelia, Atelier de l'île (2011) Feasibility study: possible partial depolderising of Bas-Champs du Vimeu (Picardy). The search for a future in a sustainable territory. Unpublished document.

References

Anselme, B., Durand, P., Defossez, S. et al. (2008). Risque de submersion marine: comment améliorer l'approche opérationnelle? Cybergeo, 12 p.

Anziani, A. (2010). *Rapport sur les conséquences de la tempête Xynthia*. Paris: Sénat, Les rapports du Sénat 554, 100 p.

Bousquet, B. and Miossec, A. (1991). La tempête, du moment de nature au spectacle de société. La place du phénomène dans le géosystème au XXe siècle. *Bulletin de l'Association des Géographes Français* 3: 241–251.

Chevillot-Miot, E., Chadenas, C., and Mercier, D. (2017). La résilience proactive par la carte mentale: exemple du territoire de la Charente-Maritime. *Les Cahiers de l'Association Tiers-Monde* 32: 153–162. https://www.researchgate.net/publication/304625551_LA_RESILIENCE_PROACTIVE_PAR_LA_CARTE_MENTALE_EXEMPLE_DU_TERRITOIRE_DE_LA_CHARENTE-MARITIME (accessed 28 February 2019.

Chionne, D. (2015). Perception et gestion du changement climatique: le cas de la société insulaire oléronaise (France). *Bulletin de l'Association des Géographes Français* 3: 287–302.

Cohen, O., Dolique, F., Anthony, E.J. et al. (2002). L'approche morphodynamique en géomorphologie littorale. In: *Le littoral, regards, pratiques et savoirs* (ed. N. Baron-Yelles, L. Goeldner-Gianella and S. Velut), 191–214. Paris: Éditions Rue d'Ulm.

Colas, S. (2007). Analyse statistique et cartographique de l'érosion marine. Orléans: IFEN. *Les dossiers*, 06, 36 p. http://www.onml.fr/uploads/media/dossier_erosion.pdf (accessed 28 February 2019).

Comité national de suivi de la Stratégie nationale de gestion intégrée du trait de côte (2015). 40 mesures d'adaptation au changement climatique et à la gestion intégrée du trait de côte. Rapport du Comité national de suivi de la Stratégie nationale de gestion intégrée du trait de côte, 30 p. Paris: Ministere de l'Écologie, du Développement Durable, des Transports et du Logement (MEDDTL).

Corbin, A. (1990). *Le territoire du vide. L'occident et le désir du rivage (1750–1840)*, 407 p. Paris: Flammarion.

Corbin, A. (2014). *Le ciel et la mer*. Paris: Flammarion.

Corbin, A. and Richard, H. (2004). *La mer, terreur et fascination*, 199 p. Paris: Seuil, Bibliothèque Nationale de France.

Creach, A. (2015). Cartographie et analyse économique de la vulnérabilité du littoral atlantique français face au risque de submersion marine (volume 1), Thèse de Doctorat de Géographie, Université de Nantes, 320 p. https://tel.archives-ouvertes.fr/tel-01275600/document (accessed 28 February 2019).

DGALN – Direction Générale de l'Aménagement, du Logement et de la Nature (2012). Stratégie nationale de gestion intégrée du trait de côte. Vers la relocalisation des activités et des biens. 20 p. http://www.developpement-durable.gouv.fr/IMG/pdf/12004_Strategie-gestion-trait-de-cote-2012_DEF_18-06-12_light.pdf (accessed on 3 April 2015).

Duvat, V. and Magnan, A. (2014). *Des catastrophes… 'naturelles'?*, 312 p. Paris: Éditions Le Pommier.

Garnier, E. and Surville, F. (eds.) (2010). *La tempête Xynthia face à l'histoire. Submersion et tsunamis sur les littoraux français du Moyen Âge à nos jours*, 174 p. Saintes: Le Croît Vif.

Goeldner-Gianella, L. (2011). *Approche de géographie sociale (perceptions et représentations)*, 24 p.,. In: *Etude de faisabilité, dépoldérisation partielle et éventuelle des Bas-Champs du Vimeu, La recherche d'un avenir sur un territoire pérenne*. Expertise pour le Syndicat mixte Baie de Somme/ Grand littoral picard. http://www.baiedesomme.org/milieuxnaturels/docs/51761tude-sociale.pdf (ed. A.d.l.'î. Sogreah, B. Consulting, L. Goeldner-Gianella, et al.). accessed 10 March 2019.

Goeldner-Gianella, L. (2013). *Dépoldériser en Europe occidentale. Pour une géographie et une gestion intégrées du littoral*. Paris: Publications de la Sorbonne 350 p.

Goeldner-Gianella, L. and Bertrand, F. (2013). La submersion marine et ses impacts environnementaux et sociaux dans le Bassin d'Arcachon (France): est-il possible, acceptable et avantageux de gérer ce risque par la dépoldérisation? Rapport de synthèse du programme de recherches BARCASUB (programme LITEAU), 17 p. http://isidoredd.documentation.developpement-durable.gouv.fr/documents/Temis/0079/Temis-0079878/21366_B.pdf (accessed 28 February 2019).

Goeldner-Gianella, L., Bertrand, F., Oiry, A. et al. (2015). Depolderisation policy against coastal flooding and social acceptability on the French Atlantic coast: the case of the Arcachon Bay. *Ocean & Coastal Management* 116: 98–107.

Goeldner-Gianella, L., Bertrand, F., Pratlong, F. et al. (2013). Submersion marine et dépoldérisation: le poids des représentations et des pratiques sociales dans la gestion du risque littoral. *Espace, Populations, Sociétés* 1–2: 193–209.

Gueben-Venière, S. (2014). Vers une gestion renouvelée du littoral européen: des ingénieurs néerlandais, anglais et français de plus en plus 'verts'?, Doctorat de Géographie, Université Paris 1, 458 p.

Hellequin, A.P., Flanquart, H., Meur-Férec, C. et al. (2013). Perceptions du risque de submersion marine par la population du littoral languedocien: contribution à l'analyse de la vulnérabilité côtière. *Natures Sciences Sociétés* 21: 385–399.

Hiron, J. (1998). *Il était une fois Leucate*, 209 p. Leucate: Editions du Cap Leucate.

IFOP (2014). Les Français et la mer. Agence des aires marines protégées, Le Marin, Ouest France, 58 p. http://www.aires-marines.fr/Actualites/Sondage-exclusif-les-francais-et-la-mer (accessed 27 September 2016).

Jodelet, D. (1989). Représentations sociales: un domaine en expansion. In: *Les représentations sociales* (ed. D. Jodelet), 31–61. Paris: Presses universitaires de France.

Krien, N. (2014) *Place des risques côtiers dans la représentation du cadre de vie d'individus possédant des enjeux sur des communes 'à risque'*. Thèse de psychologie sociale et environnementale, Université de Bretagne occidentale, 237p. https://tel.archives-ouvertes.fr/tel-01140785/document (accessed 10 March 2019).

Le Moel, B., Moliner, P., and Ramadier, T. (2015). Représentation sociale du milieu marin et iconographie du territoire chez des élus de communes littorales françaises. *VertigO – la revue électronique en sciences de l'environnement* 15 (1): https://doi.org/10.4000/vertigo.16014.

Longépée, E., Goeldner Gianella, L., Defossez, S. et al. (2015). Perception, représentation et gestion du risque du submersion marine sur la côte languedocienne française: le cas de la station balnéaire récente de Leucate, Symposium international, Vulnérabilité des littoraux méditerranéens face aux changements environnementaux contemporains 20–24 October 2015, Kerkennah, Tunisie, 7 p.

Meur-Férec, C., Deboudt, P., Morel, V. et al. (2010). La vulnérabilité des territoires côtiers à l'érosion. Vers une prise en compte des risques dans la GIZC. In: *Agir ensemble pour le littoral*, 113–129. Paris: MEEDDAT.

Meur-Férec, C. and Rabuteau, Y. (2014). Plovenez-Les-Flots: un territoire fictif pour souligner les dilemmes des élus locaux face à la gestion des risques côtiers. *L'Espace géographique* 43 (1): 18–34.

Ministère de l'Écologie, du Développement Durable, des Transports et du Logement (2011). *Plan national d'adaptation de la France aux effets du changement climatique (2011–15)*. Paris: MEDDTL 187 p.

Ministère de l'Écologie, du Développement Durable et de l'Énergie (2011). *Première évaluation nationale des risques d'inondation, Principaux résultats - EPRI 2011*. 16 p. https://www.ecologique-solidaire.gouv.fr/sites/default/files/EPRI-Principaux-resultats_120712.pdf (accessed 28 february 2019).

Miossec, A. (1993). La gestion de la nature littorale en France Atlantique. Etude comparative (Royaume-Uni, Pays-Bas, Espagne, Etats-Unis). Doctorat d'Etat, Université de Bretagne Occidentale, vol. 1, 469 p.

Myatt, L.B., Scrimshaw, M.D., and Lester, J.N. (2003). Public perceptions and attitudes towards a forthcoming managed realignment scheme: Freiston shore, Lincolnshire, UK. *Ocean & Coastal Management* 46: 565–582.

Parrain, C. (2010). *Territorialisation des espaces océaniques hauturiers*. Thèse de Géographie, Université de La Rochelle, 492 p.

Péret, J. and Sauzeau, T. (2014). *Xynthia, ou la mémoire réveillée. Des villages charentais et vendéens face à l'océan (XVIIe-XXIe siècle)*, 289 p. La Crèche: Geste éditions.

Péron, F. and Rieucau, J. (1996). *La Maritimité aujourd'hui*. Paris: L'Harmattan 336 p.

Pinot, J.P. (2002). Géographie des littoraux en France. Évolution d'une discipline. In: *Le littoral, regards, pratiques et savoirs* (ed. N. Baron-Yelles, L. Goeldner-Gianella and S. Velut), 27–58. Paris: Éditions Rue d'Ulm.

Pont, C. (2015). *Aborder et caractériser le risque de submersion marine sur les littoraux sableux méditerranéens: l'exemple de la commune de Leucate*. Master 1 Risques et Environnement, Université Paris 1, 150 p.

Rey-Valette, H., Fraysse, N., Richard, A. et al. (2013). Perception des plages et des politiques de rechargement: réflexions à partir du cas du golfe d'Aigues-Mortes (Hérault/Gard). *Espace, Populations, Sociétés* 1/2: 177–192.

Rey-Valette, H. and Rulleau, B. (2016). Gouvernance des politiques de relocalisation face au risque de montée du niveau de la mer. *Développement durable et territoires* 7/1 https://developpementdurable.revues.org/11282 (accessed 28 February 2019).

Rey-Valette, H., Rulleau, B., Meur-Férec, C. et al. (2012). Les plages du littoral languedocien face au risque de submersion: définir des politiques de gestion tenant compte de la perception des usagers. *Géographie, économie, société* 14: 369–391.

Robert, S. (2009). *La vue sur mer et l'urbanisation du littoral*. Thèse de Géographie, Université de Nice Sophia Antipolis, 456 p. https://tel.archives-ouvertes.fr/tel-00442279/file/These_SRobert.pdf (accessed 28 February 2019).

Roussel, S., Rey-Valette, H., Henichart, L.M. et al. (2009). Perception des risques côtiers et Gestion Intégrée des Zones Côtières (GIZC). *La houille blanche* 2: 67–74.

Roux, B. (1997). *L'imaginaire marin des Français*, 224 p. Paris: Editions L'Harmattan.

Rulleau, B., Rey-Valette, H., Flanquart, H. et al. (2015). Perception des risques de submersion marine et capacité d'adaptation des populations littorales. *VertigO* Hors-série 21. https://doi.org/10.4000/vertigo.15811. http://vertigo.revues.org/15811 (accessed 28 February 2019).

Rupp-Armstrong, S. and Nicholls, R.J. (2007). Coastal and estuarine retreat: a comparison of the application of managed realignment in England and Germany. *Journal of Coastal Research* 23 (6): 1418–1430.

Sebillot, P. (2015). *Le folklore de France*, 456 p. Paris: La mer et les eaux douces, Editions Ligaran.

Tricot, A. (2013). *ADAPTALITT, Rapport de synthèse*, 11 p. http://halshs.archives-ouvertes.fr/halshs-00803599 (accessed 28 February 2019).

Wagret, P. (1959). *Les polders*, 316 p. Paris: Dunod.

IV.3

Lessons from Cases of Coastal Risks Governance

24

After Xynthia on the Atlantic Coast of France: Preventive Adaptation Methods

Denis Mercier[1], Axel Creach[1], Elie Chevillot-Miot[2], and Sophie Pardo[3]
[1] *Laboratory CNRS 8185 ENeC, Paris Sorbonne University, Paris, France*
[2] *Laboratory CNRS 6554 LETG-Nantes, University of Nantes, Nantes, France*
[3] *Economics and management laboratory LEMNA, University of Nantes, Nantes, France*

24.1 Introduction

Xynthia (February 2010) was the most severe coastal surge in the recent history of the French Atlantic coast in terms of hazard, disasters, fatalities, and consequences for coastal management policy. Forty-one people died by drowning due to the flood event on the French Atlantic coast. It represents one of the deadliest events in France in recent decades (Vinet 2010; Magnan and Duvat 2014). Xynthia could be considered a tragedy because of its human disaster, but a predictable one (Mercier and Acerra 2011) because of the location of the areas that were flooded and in terms of the location of these deaths. In France, it is estimated that five million people are living in a coastal flood zone, following a rise of two metres (Kolen et al. 2010). The Xynthia event cost around 2.5 billion euros (Cour des Comptes 2012) and for individual houses, 75% of reconstruction costs were associated with interior elements, with damage to structural components remaining very localized and negligible (André et al. 2013).

This coastal flood case raises the question of the coastal development policy in France and, more broadly, of all submersible areas of the national territory. Since this tragedy, several geographical studies have been published and several tools proposed for a better knowledge of this natural hazard. The aim of these chapter is (i) to identify the natural causes of this disaster, (ii) to identify the causes of the fatalities, (iii) to analyse the new strategy for coastal management of France, and (iv) to propose life-saving maps as a geographical tools for a better coastal management that will be able to reduce the fatal toll in the case of future events, and to improve coastal preparedness.

Facing Hydrometeorological Extreme Events: A Governance Issue, First Edition.
Edited by Isabelle La Jeunesse and Corinne Larrue.

24.2 A normal storm in terms of natural hazard but a major coastal flood due to the concomitance of the meteorological and marine agents

Coastal flooding is observed on lower coasts when an atmospheric depression, generating strong wave agitation and a storm surge inversely proportional to the decrease in air pressure (a decrease of one hectopascal induces an elevation of one centimetre of water), combines in time with strong tides and leads to an over wash (Chaumillon et al. 2017).

24.2.1 Xynthia was a normal storm in terms of natural elements

The tidal coefficient is the theoretical measurement of the difference between high tides and low tides; it varies on a scale between 20 and 120. A coefficient of 70 is equivalent to an average tide and one of 95 is equivalent to a mean spring tide. During Xynthia, the tidal coefficient was 102 at 5.18 a.m. on 28 February 2010. The low pressure was at a minimum of 977 ha and the wind speed did not exceed 130 km/h on the Vendée coastline whilst reaching up to 158 km/h in bursts in the western part of Ré Island during the night (Chauveau et al. 2011; Bertin et al. 2012, 2014).

Although these levels were not exceptional, the concomitance of these natural elements was rare for the French coast. Feuillet, Chauveau, and Pourinet (2012) analysed the instantaneous maximum wind data (more than 89, 102, and 118 km/h) at nine stations along the French Atlantic coast to determine the frequency of the sea levels reached during the Xynthia disaster. They crossed these series with tide coefficients. They also selected specific days on which the maximum wind speed occurred at less than three hours from high tide. In this way, they estimated the frequency and return time of storm tides and situations breeding a high coastal flooding risk. Consequently, they showed that events comparable to Xynthia are rare (only 10 cases observed during 27 years), or even exceptional at some stations, like La Rochelle. The return period of storm tides varies according to the site, from 177 years at La Rochelle for a storm of 140 km/h knowing that high tide occurs, to only five years at Belle-Île. Breilh et al. (2013) analysed historical archives since 1500 CE to produce a database of storm-induced coastal flooding in the French central part of the Bay of Biscay. From this new database, 46 coastal floods have been reported (1 event every 11 years on average), which demonstrates the high vulnerability of this region to coastal flooding.

Storm Xynthia resulted in the flooding of low-lying areas, mainly polders in the Poitevin marshes, the Charente estuary, and Ré and Oléron Islands (Figure 24.1, Breilh et al. 2014). More than 50 000 ha were flooded due to Xynthia (Chauveau et al. 2011; Verger 2011; Magnan and Duvat 2014).

24.2.2 The site effect

The geographical approach highlighted by several analyses (Perherin 2007; Chevillot-Miot and Mercier 2014; Creach 2015) demonstrates the importance of site effects. For Xynthia, the bay bottoms were more subject to the risk of marine flooding. The estuaries and the bays have a low topography and sandy coasts, which can lead to a phenomenon

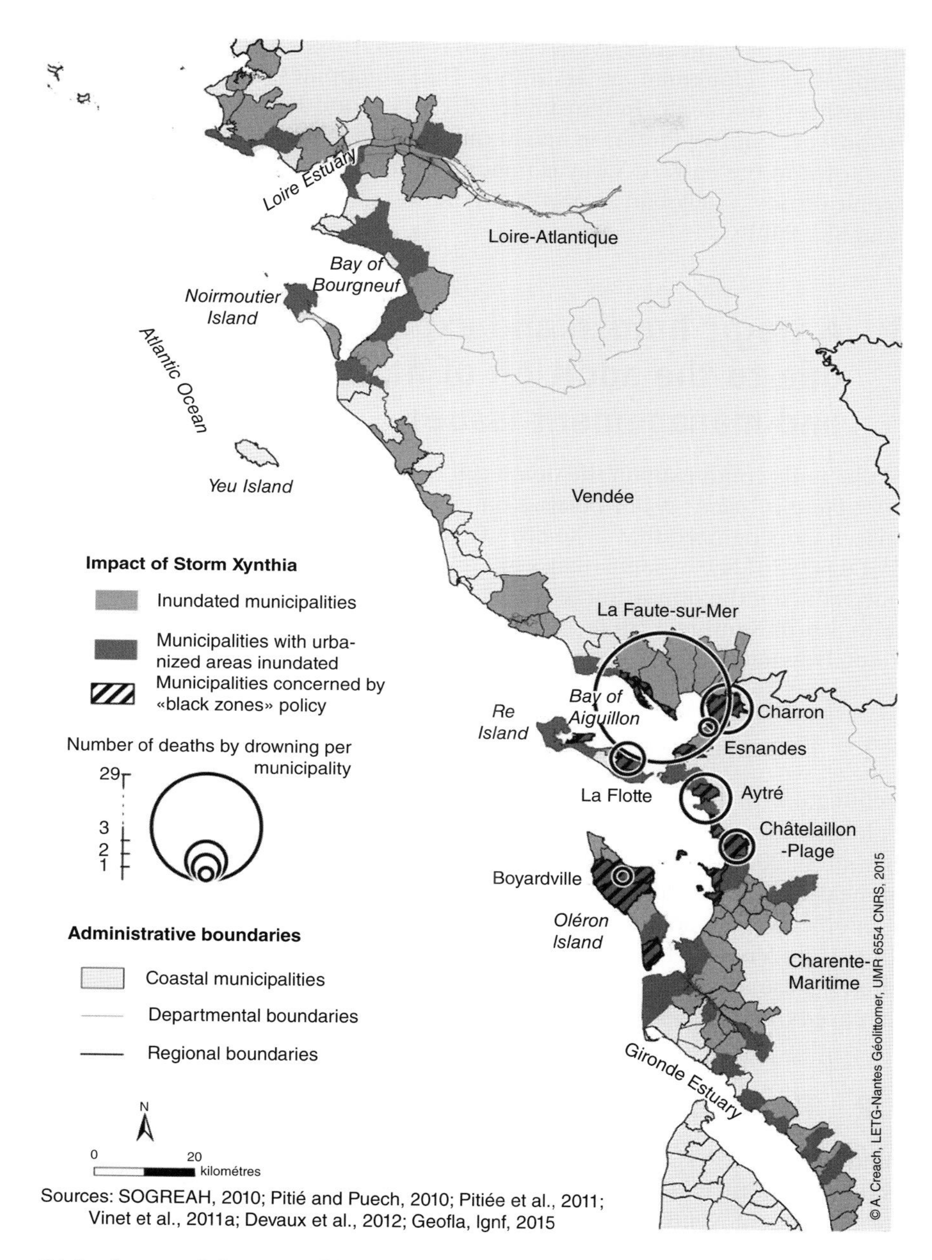

Figure 24.1 Impact of Storm Xynthia on the French Atlantic coast, between the Loire and Gironde Estuaries. *Source*: From Creach (2015), modified.

of wave resonance during a storm (Figure 24.1). These are trapped, one after the other, which results in an increase in the water level in a small bay (Perherin 2007). These marine flooding phenomena have also occurred in other parts of the world. New Orleans, submerged after the passage of Hurricane Katrina (Category 3), St-Louis Bay in the Gulf of Mexico in 2005, is one example. The city, in the shape of a basin, is situated at sea level, surrounded by protective dykes rising to four metres in height in the vicinity of the Mississippi River (Jonkman et al. 2009). The Philippines offers another example illustrating this spatial logic. In November 2013, Typhoon Haiyan (Category 5) caused submersion waves of five metres in height on Tacloban (Mori and Mase 2013), the most populous town in the eastern Visayas region of the country. This coastal community is topographically low (Tacloban airport is two metres a.s.l.) and located in the bay of San Pedro and San Pablo.

24.3 A tragic human and expensive material toll due to the addition of natural factors and management issues

Several elements could explain the death toll of Storm Xynthia.

The problem of water depth poses a danger to coastal populations. Beyond 1 m of water rise, it is recognized that a 'healthy, fit' adult experiences great difficulty in moving (MEDDTL 2011c). In some places in La Faute-sur-Mer, the sea level reached 2.5 m in height, reducing the opportunities for shelter. For example, all the deaths by drowning due to Storm Xynthia were recorded in houses in which the water depth was above 1 m (Vinet, Boissier, and Defossez 2011).

24.3.1 The problem of architecture

The architecture of the buildings can be an additional risk along rivers and coastal areas (Jousseaume, Landrein, and Mercier 2004; Jousseaume and Mercier 2009; Jonkman et al. 2009; Leone et al. 2011; Vinet et al. 2011, 2012a, 2012b). For Xynthia, 78% of the deaths occurred in single-storey constructions, the main architectural model, where people were trapped by the fast-rising water with no chance of escaping (Lumbroso and Vinet 2011; Vinet et al. 2012a). The absence of a roof window and the widespread presence of electric shutters that became stuck were aggravating factors that increased the risk.

24.3.2 The problem of dykes

The rapid rise of the water and the surprise effect this has on occupants are important factors that can cause casualties. The effect of dyke failure is paramount, as illustrated by the fact that 90% of Xynthia deaths occurred within 400 m of dykes (Vinet et al. 2012b).

Improving dykes is not the solution. Dykes have been built since the mediaeval period to protect productive areas of saliculture and agriculture from the whims of the sea. The dykes have often been destroyed as Sarrazin (2012) has shown and have not been built for the protection of residential houses, especially second homes, which were built recently on a desirable coastline with the expansion of tourism (Renard 2005). This was particularly true in La Faute-sur-Mer as shown by Perret and Sauzeau (2014). Would the same policy of 'colonization' of the littoral occur if the reconstruction of the dykes was not paid for by the community, but by the individuals who benefit from it?

24.3.3 The age of the population

The age of the population had a negative impact on their ability to escape and be rescued (Vinet et al. 2011). The median age of people in La Faute-sur-Mer, where 29 people died during the Xynthia event, is significantly higher (52 years) than in the rest of France (38 years) (Vinet et al. 2012b). Except for children, three-quarters of the deceased were over 60 years old and 20% over 80 years (median age: 74 years). The explanation may be that when people who own two houses retire, they settle down in their second home.

It can also be noted that Storm Xynthia occurred during the night and people were surprised during their sleep. No evacuation order was given by the authorities but people were told to close their shutters for protection against the wind (Vinet et al. 2011; Kolen, Slomp, and Jonkman 2013).

24.3.4 The problem of urbanization and local policy

The human and material toll of Storm Xynthia could be explained by the inadequacy between the architecture and the potential coastal flood areas as 100% of the deaths were located inside houses (Vinet et al. 2011). It was one of the most important lessons learned from the storm and one that symbolizes the difficulty public policies have in controlling urbanization in areas at risk (Vinet et al. 2012a, 2012b). The occupation of low-lying areas for the development of housing estates is a sign of urbanization unaware of coastal hazards. This development was made possible by the lack of clear regulations to control urbanization in exposed areas.

24.4 Post-Xynthia policy: A new strategy for coastal management in France

It raises the question of the potential consequences of future storms, in a context of sea-level rise. In particular, 590 000 acres of land are known to lie below the level of 1 m (CETMEF et al. 2009), 864 towns are particularly vulnerable to coastal flooding, whilst 850 000 people and 570 000 buildings are exposed to flood in France (ONML 2012). Moreover, 20% of the residential buildings exposed to flooding are single-storey houses (MEDDE 2012). In this context, the question of how to manage this type of area needs to be addressed.

Following Storm Xynthia, the French government first provided a list of houses that were considered too dangerous for people and that were destroyed, initiating a process of withdrawal. This planning was defined by the MEDDE on the basis of five criteria (Figure 24.2): (i) a building that received one metre or more of water during the storm; (ii) a building located between 90 and 110 m from a dike; (iii) the kinetics of the flood; (iv) the level of damage to the building and the architecture; and (v) the coherence of the delimited area to avoid the dispersion of constructions. These criteria were criticized because of a lack of transparency in the method (Verger 2010; Mercier and Chadenas 2012; Magnan and Duvat 2014). Finally, 1628 houses considered extremely dangerous were purchased by the French Government and destroyed for a total cost of more than 315.7 million Euros (Pitié and Puech 2010; Pitié et al. 2011; Cour des Comptes 2012).

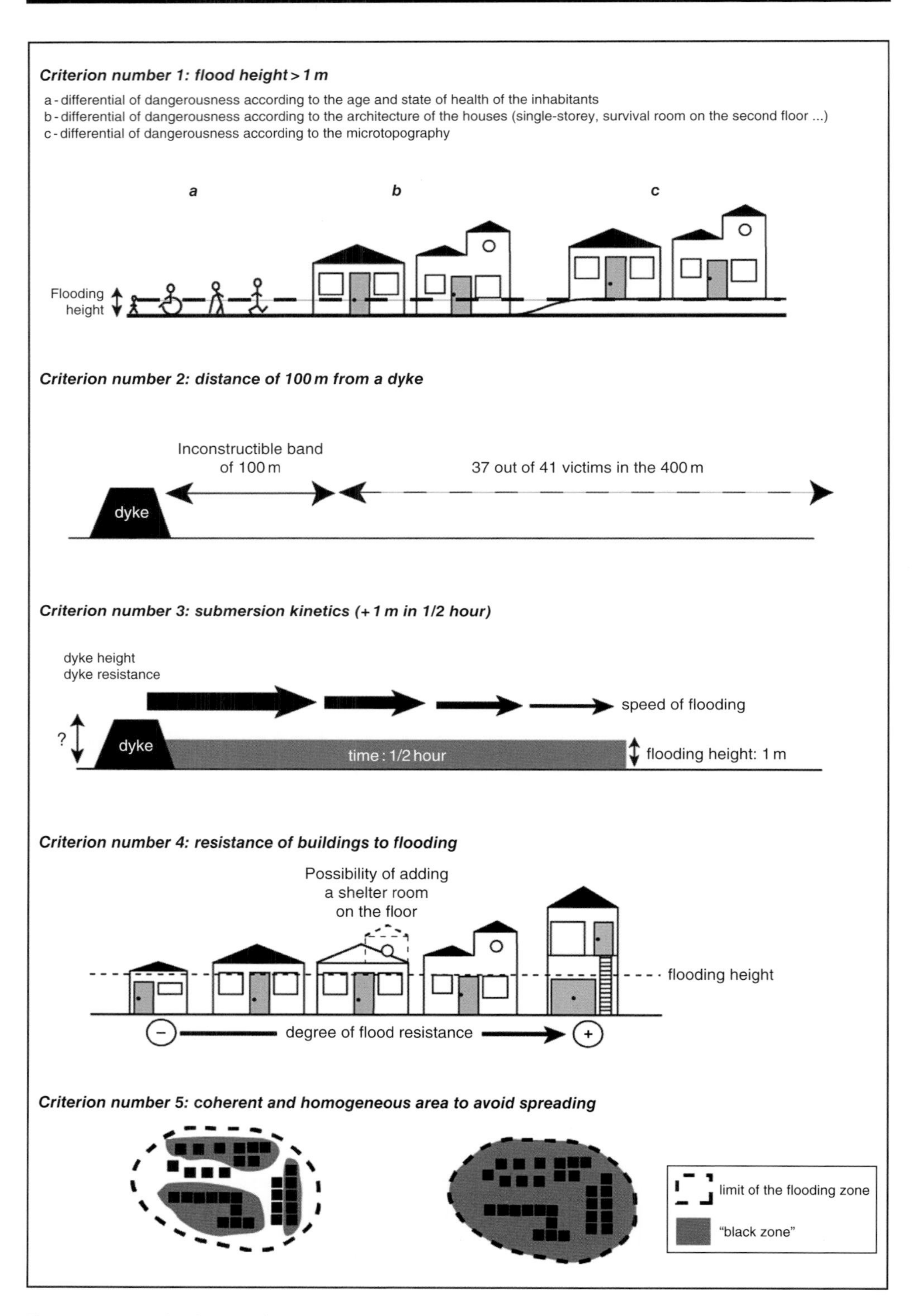

Figure 24.2 Criteria to define the 'black zone' where houses were destroyed after Storm Xynthia and its limits. *Source*: From Mercier and Chadenas (2012), modified.

However, this policy concerned only those municipalities that were impacted by the storm and was not applied on the same scale in other tragedies in France that occurred after Xynthia, except in the Var department in south-eastern France after the inundation of June 2010, in which few houses were destroyed (Vinet et al. 2011). Furthermore, our contemporary societies are not always ready to cede territories to nature. No alternative solutions were explored by the 'black zones' policy, such as architectural adaptation, for example.

Note that there were previous documents that available to manage urbanization in low-lying areas. The main one is the Risk Prevention Plan (RPP), which defines areas as unconstructible or constructible under certain conditions, depending on the hazard level. Due to the high level of marine submersion, the Xynthia level is now the reference value often used for the RPP, in which it is recommended to use the 1 in 100 years sea-water level or the historical one if it is higher (Chadenas, Creach, and Mercier 2014).

However, the RPPs are insufficient and sometimes poorly adapted to the stakes. As Figure 24.3 shows, the adoption of RPPs since the law of 1995 for coastal areas in France depends more on hazards, such as Storm Xynthia, than on political will. Furthermore, the communal safeguarding plan (PCS), imposed by the law of 13 August 2004, is far from being adopted by all communities (Creach and Pourinet 2015).

In addition, many new elements appeared after Storm Xynthia. Action plans for flash flood management were established (MEDDTL et al. 2011), new guidelines were given for carrying out RPPs (MEDDE 2013), a new strategy for the management of flood risk was produced (MEDDE 2014) and Action Plans for Flood Prevention were reinforced

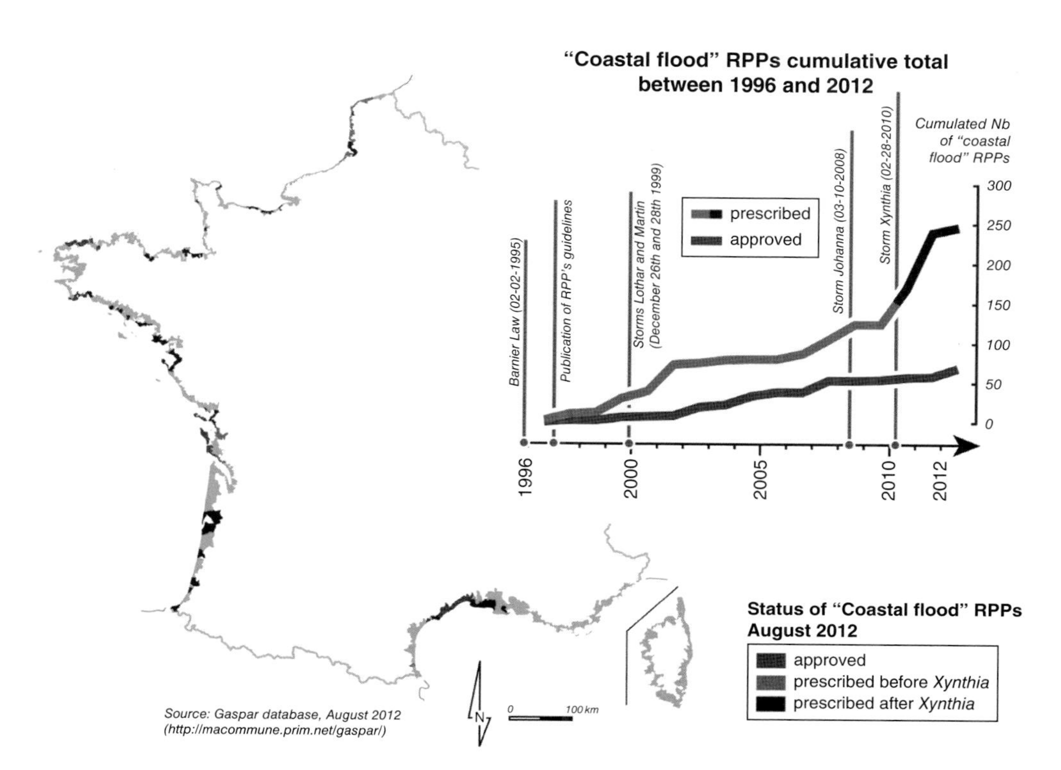

Figure 24.3 RPP coastal flooding adoptions from 1995 to 2012. *Source*: From Creach and Pourinet (2015), modified.

(MEDDTL 2011a). Moreover, a Business Continuity Plan (BCP) was introduced, which maintains stable activities for all professional and institutional organizations during a period of crisis (CEPRI 2010). At the same time, the French State implemented the European Directive for Flood Management (European Commission 2007). The major goal of all these policies is to protect human life.

However, all these elements look like a reactive rather than a preventive adaptation. A reactive adaptation could be defined as 'one shock, one reaction', whereas a preventive adaptation is more like 'building better for the future' (Nicholls 2011; MEDDTL 2011b). Reactive adaptation is reputed to be costly and sometimes only moderately efficient (Nicholls 2011; MEDDTL 2011b).

Today, it appears essential to plan preventive adaptations as we know that other territories are exposed to an increasing coastal flood risk due to sea level rise (Cazenave and Le Cozanet 2013; IPCC 2014).

This ability to plan preventive adaptations is partly linked to the question of the shared risk culture (state, elected representatives, populations). Should we rely on the Act of 13 July 1982, which is based on national solidarity for known risks, the exposure to which is avoidable, when the perverse and disempowering effects of this insurance system are known?

24.5 Life-saving maps: New geographical tools for a better coastal management

In order to plan preventive adaptation and protect human life, it is necessary to carry out a vulnerability assessment of residential buildings. Several geographical approaches could be useful to reduce the potential damage in case of future natural hazards. These approaches could be more or less complex.

24.5.1 Water depth assessment

Water depth was identified as a factor of vulnerability for people after Storm Xynthia, in which all deaths by drowning occurred in houses that received more than 1 m of water (Vinet et al. 2011) as in other coastal disasters (Jonkman and Penning-Rowsell 2008).

The first approach consists of inventorying and mapping the houses potentially affected by a coastal flood by using a 'static' method of superimposing high resolution topography data and a sea water reference level. Such an approach was proposed by Chevillot-Miot, Creach, and Mercier (2013) for Noirmoutier Island, on the French Atlantic coast. The altitude of buildings was measured for each house using LiDAR data from the Litto-3D programme, which provides a vertical accuracy of 20 cm (IGN & SHOM 2012). The altitude for each building corresponds to the elevation of the natural ground at the house location. This calculation gives the difference between the altitude of the ground of houses and the potential water level that could be reached during Storm Xynthia. Thus, it provides the potential depth that could be reached by water in every house for this type of event. This study reveals the hypothetically vulnerable neighbourhoods of the four towns of the island (Figures 24.4 and 24.5). The island has 10 716 buildings (48% of all constructions on the island) in a flood zone for a reference

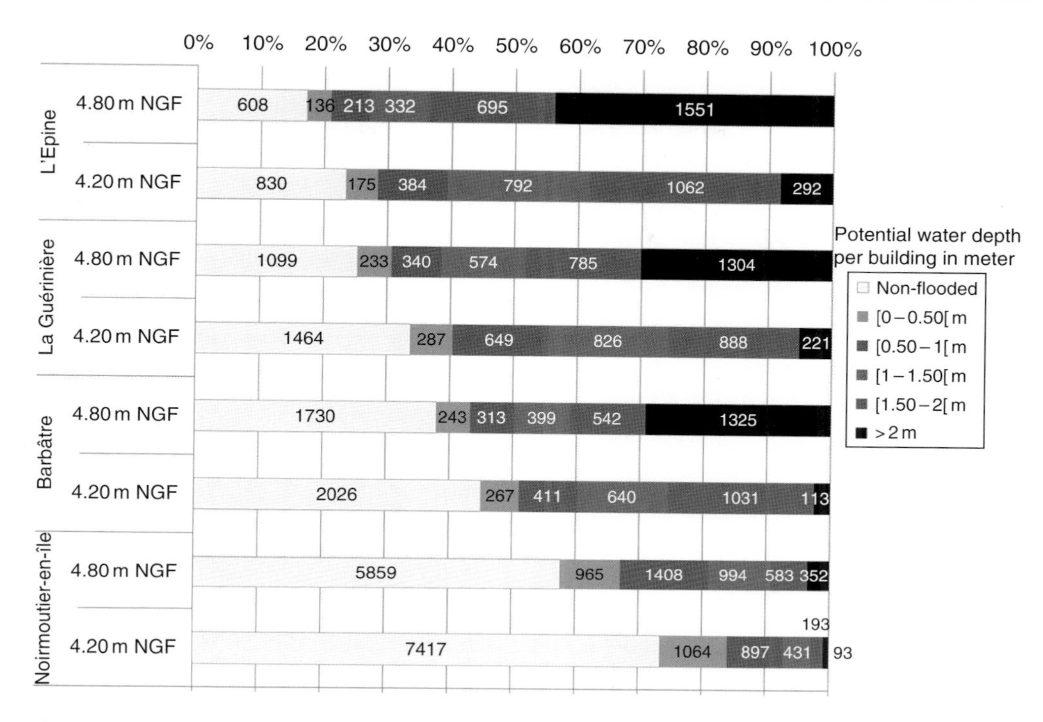

Figure 24.4 Number of buildings in Noirmoutier Island for the four municipalities with the potential depth that could be reached by water in every house for each type of event (4.20 m NGF for a Storm Xynthia event and 4.80 m NGF with the potential sea-level rise in 2100. *Source*: After Chevillot-Miot et al. (2013) modified.

of 4.20 m NGF (Nivellement Général de la France), which is the water level measured during Storm Xynthia on this island (Devaux et al. 2012), of which 6582 buildings (29%) are likely to be flooded by water heights greater than 1 m. For the reference of 4.80 m NGF, due to the sea level rising by 2100, this total reaches 13287 (58%), of which 9436 are affected by more than 1 m of flooding (42%).

24.5.2 Link between water depth and architectural typology

The second approach is to map the links between potential water depth in low-lying areas (Chevillot-Miot et al. 2013) and the architectural typology of constructions at the scale of existing buildings. The same field study on Noirmoutier Island was analysed by Creach et al. (2015a) with this new approach.

The architectural typology of residential buildings was another major factor that explained the toll of Storm Xynthia: 78% of the deaths by drowning in La Faute-sur-Mer occurred in single-storey buildings (Vinet et al. 2011; Vinet et al. 2012a, 2012b). Based on previous work (Jousseaume and Mercier 2009; Vinet et al. 2011, 2012a; Chevillot-Miot et al. 2013), residential buildings were divided by Creach et al. (2015a) into three categories, depending on the risk to the occupants: single-storey construction, construction with a rescue-level and construction with one floor or more.

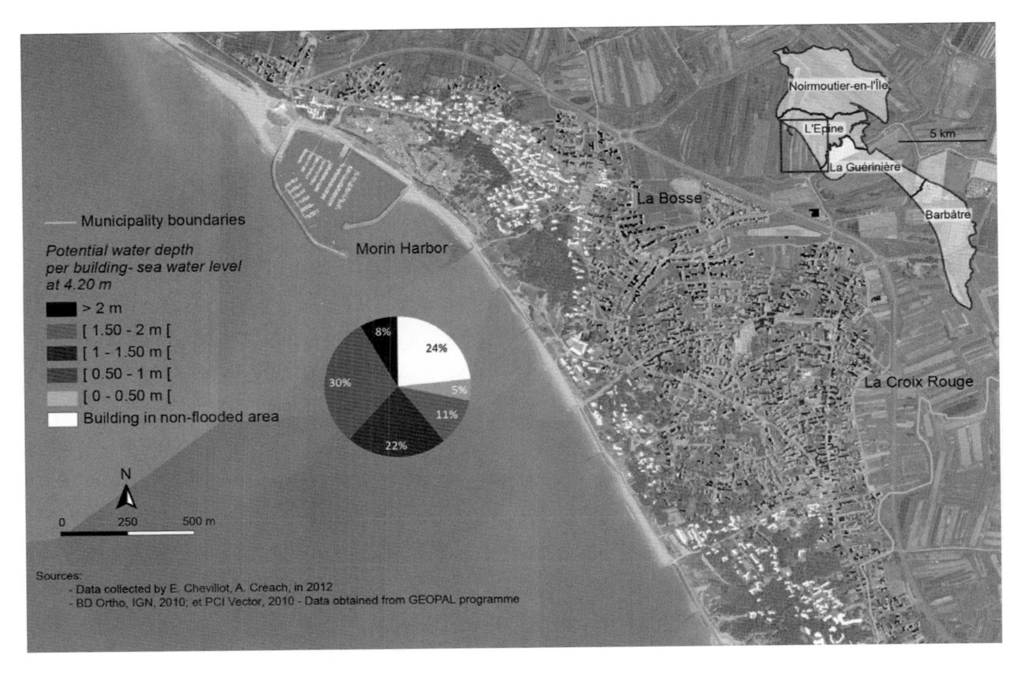

Figure 24.5 Number of buildings in the municipality of L'Epine on Noirmoutier Island with the potential depth that could be reached by water in every house for a Storm Xynthia type of event (4.20 m NGF). *Source*: After Chevillot-Miot et al. (2013), modified.

By combining the two types of information (potential water depth and architectural typology), Creach et al. (2015a) assumed that a single-storey construction in which the water depth can reach 2 m or more presents a potentially lethal risk to its dwellers. In contrast, a residential building with one floor or more in which the water depth cannot exceed 1 m is potentially a much safer place (Figure 24.6).

With this method, Creach et al. (2015a) revealed that 37% of residential houses on Noirmoutier Island are located in potential flood areas (in reality this number reaches 48% as for the above approach but, because some buildings are not visible from the street, there architectural type is unknown). 65% of them are single-storey constructions and 60% could be flooded by 1 m or more of water.

Two kinds of presentations of the results could be proposed. The first one is a graph of the quantitative results of vulnerable buildings per municipality and the second is a map at a fine scale (Figure 24.6).

24.5.3 Assessment of potentially lethal houses for inhabitants: The VIE index

The third approach is a new tool called the VIE index ('LIFE index' in English) proposed by Creach et al. (2015b). The aim of this tool is to identify dangerous areas for the population through characteristics of houses in order to prevent future deaths by drowning. The VIE index, for *Vulnérabilité Intrinsèque Extrême*, is based on four criteria

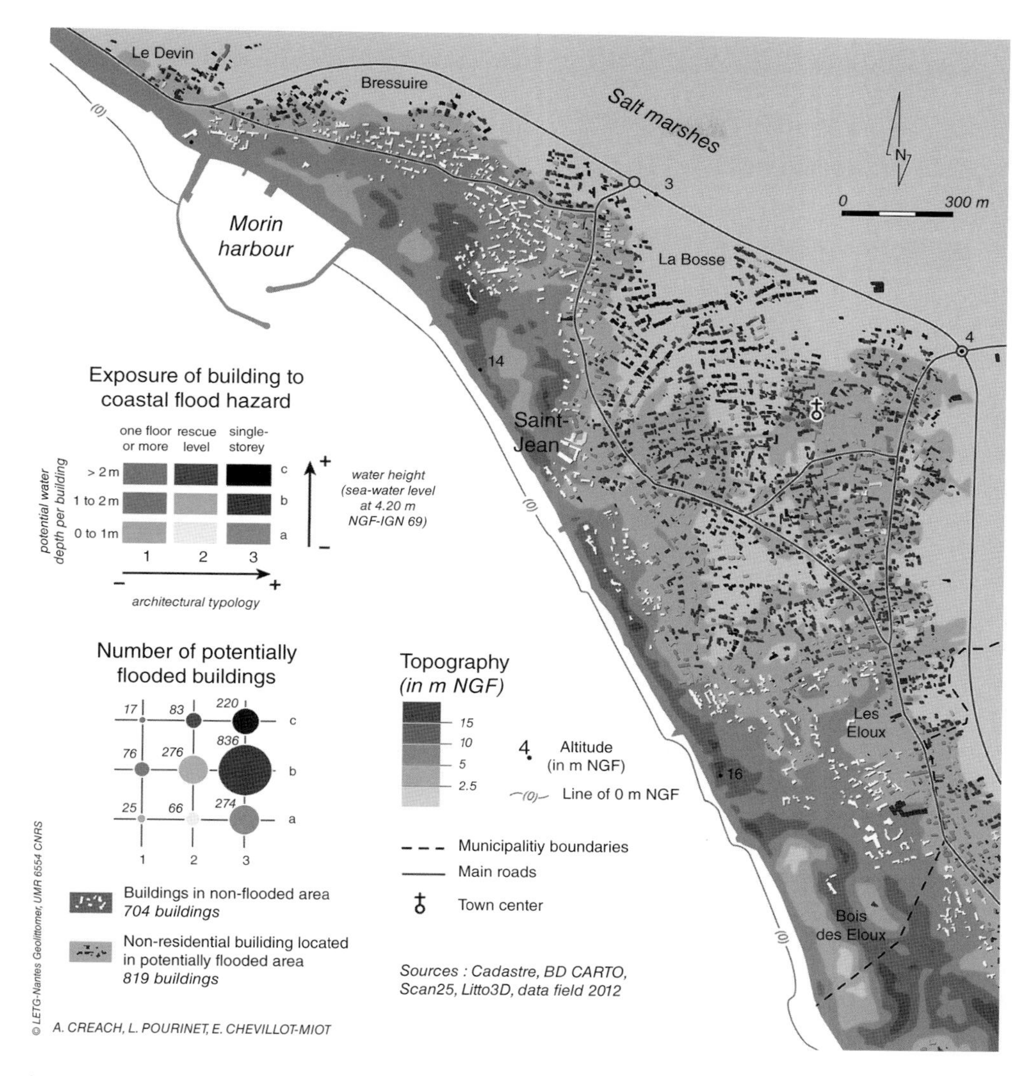

Figure 24.6 Map of vulnerable buildings in the municipality of L'Epine on Noirmoutier Island (Creach et al. 2015c).

(Figure 24.7): (i) potential water depth per house, (ii) distance between dykes and houses, (ii) architectural typology, and (iv) closeness to rescue point.

Based on these four criteria, the aim of the VIE index is to score every house with regard to its vulnerability to coastal flood hazard. As presented in Figure 24.7, the scores range from 0 to 4 for each criterion. To classify houses, the scores obtained for each criterion are added together to give a sum that defines the overall level of vulnerability. Criteria 1 and 4 are weighted because they are based on a digital terrain model (DTM) and could therefore be correlated, resulting in redundancy. To avoid this problem, we referred to the Human Development Index (HDI) literature (Klugman et al. 2011) to propose a composite criterion: $\text{Cr1} \times 2/3 + \text{Cr4} \times 1/3$.

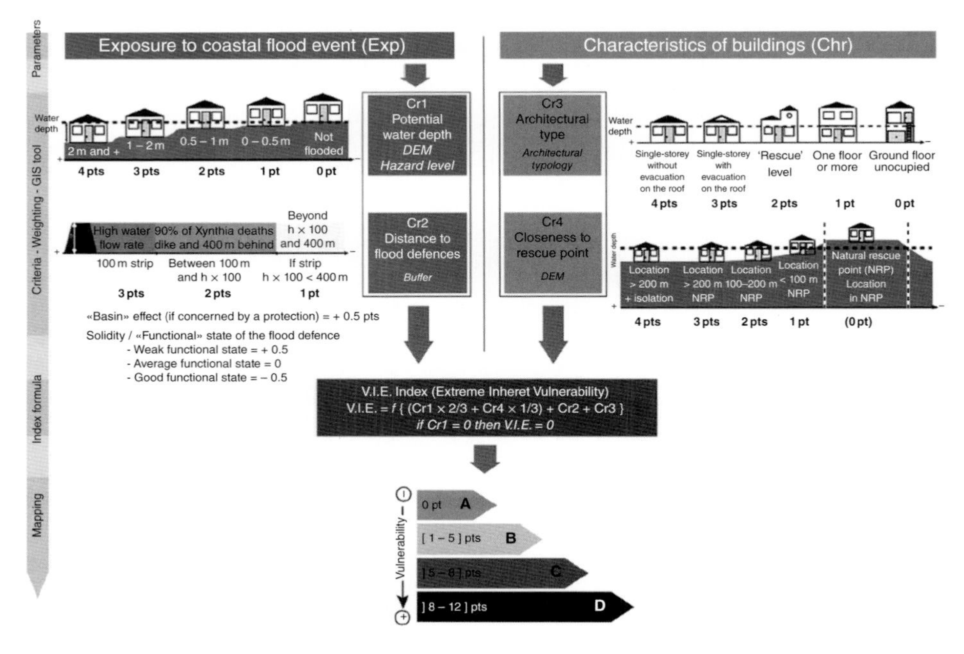

Figure 24.7 The VIE index methodology (Creach et al. 2015b).

The formula of the VIE index thus becomes:

$$V.I.E\,Index = \left(Cr1 * \frac{2}{3} + Cr4 * \frac{1}{3} \right) + Cr2 + Cr3$$

The final result can vary between 0 and 12. It should be noted that houses that are out of flood reach are automatically marked as 0, and therefore not considered to be vulnerable. The power of the index is shown better by the use of maps. A colour-based scale divided into four categories is used as follows:

Class A (green): buildings are not located in a potential coastal flood area, so they are not vulnerable to this hazard;

Class B (orange): buildings are located in a potential coastal flood area; although the water depth is low, the architecture is suited to the risk level and buildings are not too close to a flood defence, consequently the risk of death is very low;

Class C (red): the risk level of buildings is high but not lethal if people behave in a 'safe manner', such as looking for a shelter upstairs; the water height is compensated in this class due to appropriate buildings or closeness to a rescue point;

Class D (black): buildings are subject to a potential lethal risk due to a high water depth, closeness to a flood defence and inappropriate architecture with respect to the risk level.

The index is useful to carry out an initial assessment of vulnerability and has been applied to seven cities on the French Atlantic coast in Vendée and Charente Maritime departments (Creach 2015). For each municipality, several maps have been drawn for each criterion. Two synthesis maps were produced for two different scenarios: a current scenario based on the Xynthia level, which is the new reference for the governance of coastal zone risk and the baseline to define the map, and a 2100 scenario based on the sea-level rise (+60cm), which is the recommendation of the French Ministry for the future RPP.

24.6 Discussion about these different methods

The three methods presented in this chapter have all been applied on the territory of Noirmoutier Island, which was identified as being particularly exposed to coastal flood (Figure 24.8) (Chevillot-Miot and Mercier 2014) and which was 60% flooded in 1937 (Garnier, Henry, and Desarthe 2012). It enables a comparison of the results of these methods and a discussion of their interest.

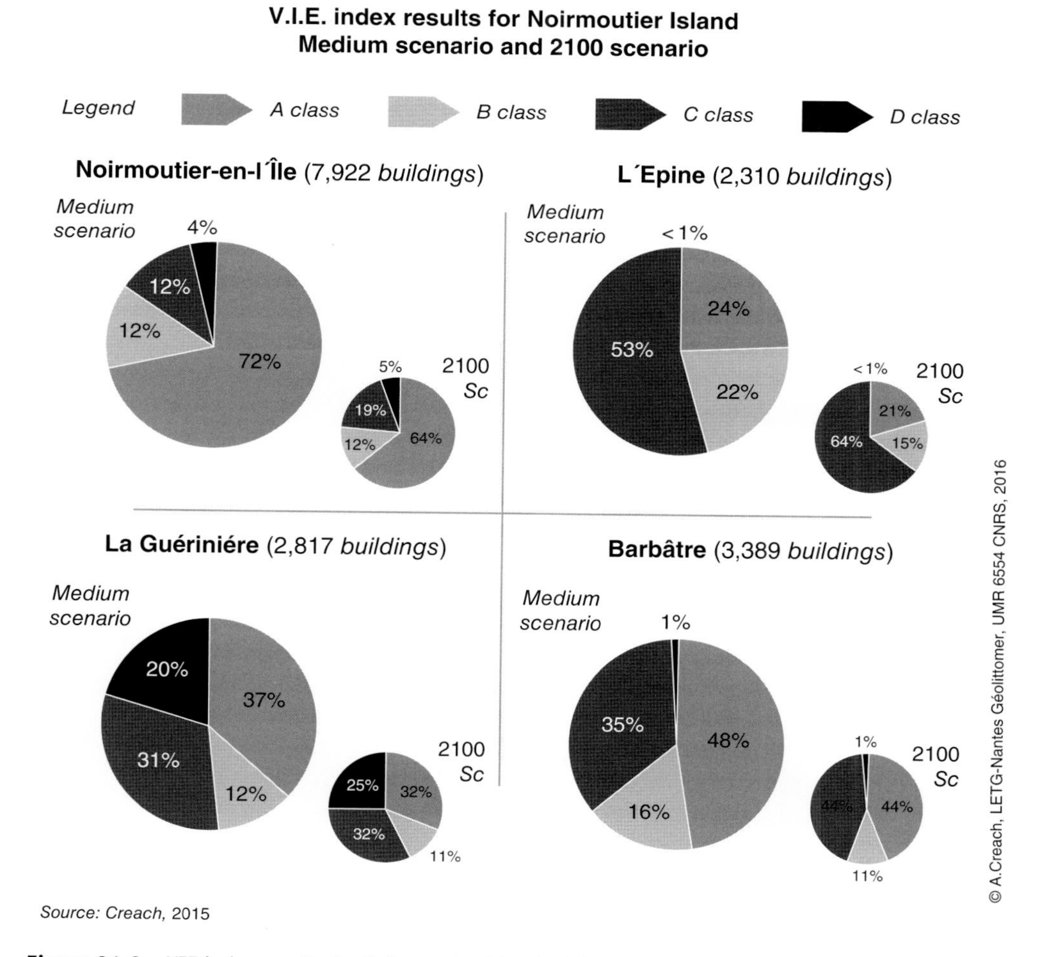

Figure 24.8 VIE index results for Noirmoutier Island with a Xynthia scenario and a 2100 scenario with 60 cm of sea-level rise. *Source*: After Creach (2015), modified.

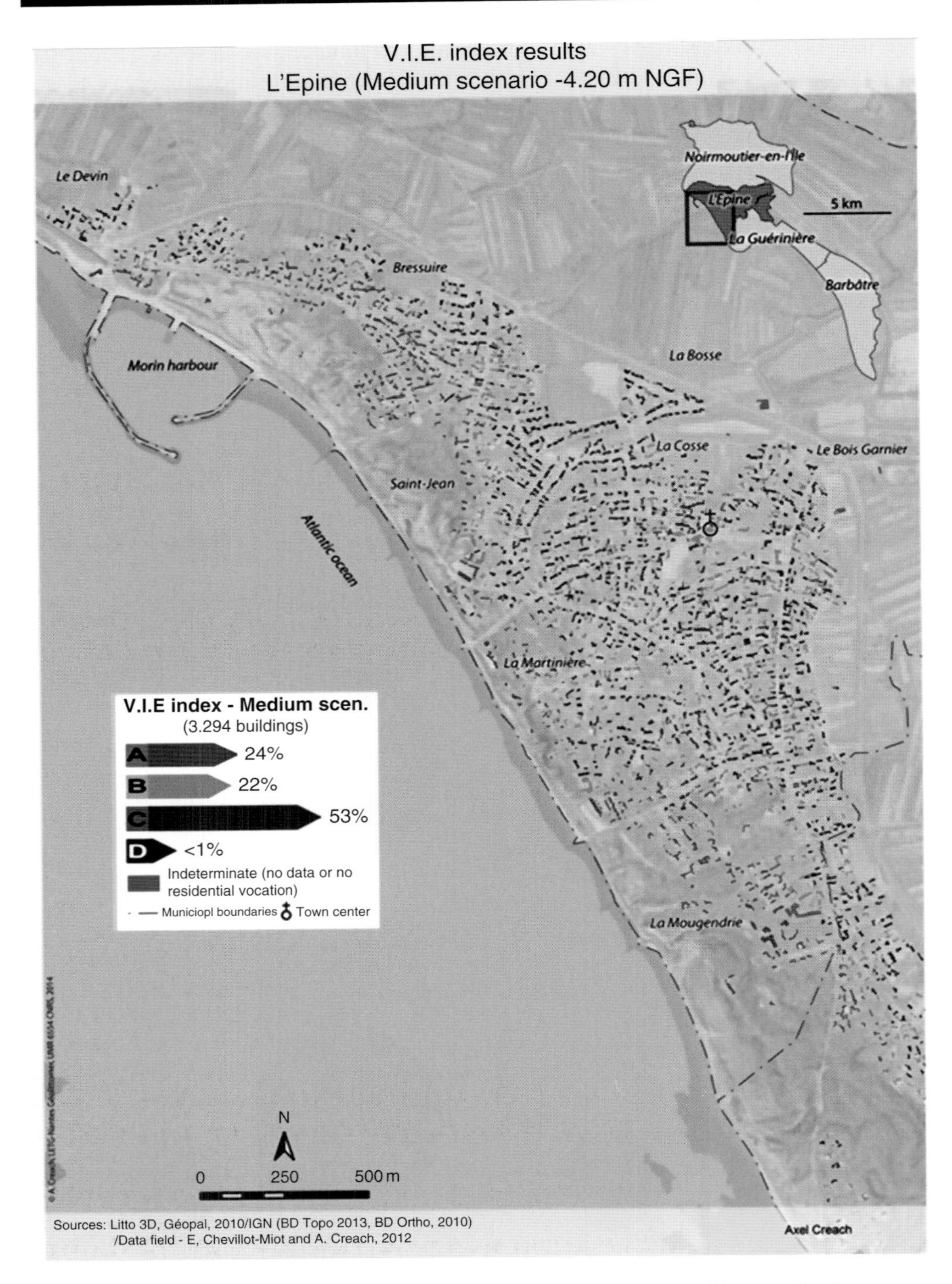

Figure 24.9 Mapping of the VIE index on L'Epine using the Xynthia event level.

The three methods provide different results. In the example of the municipality of L'Epine on Noirmoutier Island, we can observe with the first method that 60% of the houses are exposed to a potential water depth greater than 1 m, which is a critical level for people. Using the link between potential water level and architectural type, given by the second method, we can note that the most dangerous case (more than 1 m of water depth in single-storey constructions) represents 31% of the residential houses. The VIE Index provides (Figure 24.9) different results, with less than 1% of the residential houses presenting a potential risk of death for people. This result is linked to the fact that, beyond the potential water depth and the architectural typology, residential houses are protected from the sea by a high wide dune. Most of the houses are far from dykes (99% are more than 400 m from a dyke) which could fail in the case of a coastal flood. The surprise effect for people in this case is considered very low, and they may have time to shelter or to rescue themselves.

In this way, the methods give different results for the whole island. With the water depth, L'Epine appears to be the most exposed municipality, followed by La Guérinière (54% of buildings that could be flooded by 1 m or more), Barbâtre (40%) and Noirmoutier-en-l'Île (7%). This result is logical because the first three towns are mostly built in low-lying areas behind the dune, whereas Noirmoutier-en-l'Île is situated on bedrock around 10 m high.

When linking architectural typology, the results are rather similar with L'Epine appearing to be the most exposed (31% of single-storey constructions with 1 m or more of water), followed by La Guérinière (25%), Barbâtre (24%), and Noirmoutier-en-l'Île (4%).

The VIE Index provides different results and, with this method, La Guérinière appears to be, by far, the most exposed municipality on the island with 20% of potentially lethal houses. The second most exposed municipality is Noirmoutier-en-l'Île (4%), followed by Barbâtre (1%) and L'Epine (less than 1%).

This result is logical too: La Guérinière is located in the narrowest part of the island (1 km between the two coasts), exposed to the failure of dykes on the western and eastern sides, 56% of the houses are located at less than 400 m from a dyke, which increases the surprise effect for people in the case of a coastal flood. Some small areas are also close to dykes in Noirmoutier-en-l'Île, which explains why it is the second most exposed municipality. This is not the case in Barbâtre and L'Epine.

The water-depth method and its link with architectural typology are easy to use and provide a first assessment of houses vulnerable to coastal flood. However, they appear insufficient to locate areas of potential death, mainly because they give no information about the kinetics of the flood. The VIE index provides such results and enables potentially lethal houses to be mapped at a fine scale.

This is an important way to think about how to reduce the vulnerability of these houses for people and to consider preventive adaptation.

24.7 Conclusion

Storm Xynthia was a rare storm due to the concomitance of all the natural elements in a short time during the night of 27 and 28 February 2010 and could be considered the recent major event of coastal flooding in France, due to the human toll.

Storm Xynthia shows that the location of urbanization, the problem of the local management and application of national laws, the age of people living along the coast, the unsuitable architecture of the houses in low-lying areas, and the presence of dykes all contributed to causing the disaster.

This tragedy led to the first lawsuit against the Mayor of La Faute-sur-Mer. He was sentenced to two years in prison, six years after the storm (Liberation 2016). The deputy mayor in charge of urban planning and a real estate agent were discharged.

However, the lessons from Storm Xynthia could be an opportunity to increase preventive adaptation in a context of sea-level rise, through several new approaches and new geographical tools, with easily available data, to reduce the potential lethal consequences of storms in the future. The geographical methods help to choose and thus permits to avoid *status quo* or no decision concerning the maps of vulnerabilities. By having high resolution maps (at buildings scale), the decision process could be facilitated. Then, the use of such information is more a question of governance in direct links with representations of risks by the actors.

The next step will be the proposal of an adaptation strategy to reduce the vulnerability of houses for people. MEDDE and METL (2012) provide many recommendations in this line with some related to protecting human life. Creach et al. (2016) give a list of measures to reduce vulnerability and their costs. Creach (2015) also proposes a whole methodology for identify strategies that are efficient with regard to their costs. Another way could be depolderization, which consists of reopening polders to the sea via tidal gates, creating breaches in the dykes, or dismantling them altogether, but the results show that the local people and some of the stakeholders are largely opposed to depolderization, preferring the reinforcement of sea dykes (Goeldner-Gianella et al. 2015).

Storm Xynthia shows, above all, that the general interest did not prevail over the individual interest in France (Mercier 2012). The privatization of the profits of the landlord gains, on the other hand, from the collectivization of costs (by national solidarity, compensation by the State, local authorities, insurance companies, and so on). May this storm change mentalities and reverse trends to initiate a withdrawal where it is necessary and possible, to enable the State to regain control over fundamental questions such as territory management and governance, particularly in areas hazardous for the life of the populations. May the State provide the authority, determination, legitimacy and accountability required for long-term prospects. May Storm Xynthia allow the end of denial, blindness, amnesia, and greed in the development of territories, particularly in coastal areas.

Acknowledgements

The authors thank the COSELMAR research project ('Understanding coastal and marine socio-ecosystems'), financed by the Pays de la Loire Region and carried out by Ifremer and the University of Nantes. The authors are also grateful to Carol Robins, who edited the English text.

References

André, C., Monfort, D., Bouzit, M. et al. (2013). Contribution of insurance data to cost assessment of coastal flood damage to residential buildings: insights gained from Johanna (2008) and Xynthia (2010) storm events. *Natural Hazards and Earth System Science* 13 (8): 2003–2012. https://doi.org/10.5194/nhess-13-2003-2013.

Bertin, X., Bruneau, N., Breilh, J.-F. et al. (2012). Importance of wave age and resonance in storm surges: the case Xynthia, Bay of Biscay. *Ocean Modelling* 42: 16–30. https://doi.org/10.1016/j.ocemod.2011.11.001.

Breilh, J.-F., Bertin, X., Chaumillon, É. et al. (2014). How frequent is storm-induced flooding in the central part of the Bay of Biscay? *Global and Planetary Change* 122: 161–175. https://doi.org/10.1016/j.gloplacha.2014.08.013.

Breilh, J.-F., Chaumillon, E., Bertin, X. et al. (2013). Assessment of static flood modeling techniques: application to contrasting marshes flooded during Xynthia (western France). *Natural Hazards Earth and System Sciences* 13 (6): 1595–1612. https://doi.org/10.5194/nhess-13-1595-2013.

Cazenave, A. and Le Cozanet, G. (2013). Sea level rise and its coastal impacts. *Earth's Future* 2: 1–20. https://doi.org/10.1002/2013EF000188.

CEPRI (2010). Bâtir un plan de continuité d'activité au service public – Les collectivités face au risque d'inondation. 48 p. Orléans: CEPRI.

CETMEF, CETE MÉDITERRANÉE and CETE OUEST (2009). Vulnérabilité du territoire national aux risques littoraux - France métropolitaine. CETMEF / DLCE. 163 p.

Chadenas, C., Creach, A., and Mercier, D. (2014). The impact of storm Xynthia in 2010 on coastal flood prevention policy in France. *Journal of Coastal Conservation* 18 (5): 529–538. https://doi.org/10.1007/s11852-013-0299-3.

Chaumillon, E., Bertina, X., Fortunato, A.B. et al. (2017). Storm-induced marine flooding: lessons from a multidisciplinary approach. *Earth-Science Reviews* 165: 151–184. https://doi.org/10.1016/j.earscirev.2016.12.005.

Chauveau, E., Chadenas, C., Comentale, B. et al. (2011). Xynthia: leçons d'une catastrophe. *Cybergéo* 538. http://cybergeo.revues.org/23763 (accessed 28 February 2019).

Chevillot-Miot, E., Creach, A., and Mercier, D. (2013). La vulnérabilité du bâti face au risque de submersion marine: premiers essais de quantification sur l'île de Noirmoutier (Vendée). *Les Cahiers Nantais* 1: 5–14.

Chevillot-Miot, E. and Mercier, D. (2014). La vulnérabilité face au risque de submersion marine: exposition et sensibilité des communes littorales de la région Pays de la Loire (France). *VertigO La revue électronique en science de l'environnement* 14: 2. http://vertigo.revues.org/15110 (accessed 28 February 2019).

COUR DES COMPTES (2012). Les enseignements des inondations de 2010 sur le littoral atlantique (Xynthia) et dans le Var. Paris: Cours des Comptes. 299 p.

Creach, A. (2015). Cartographie et analyse économique de la vulnérabilité du littoral atlantique français face au risque de submersion marine. PhD thesis, Université de Nantes, 2 volumes, 322 p. and 249 p.

Creach, A., Bastidas-Arteaga, E., Pardo, S. et al. (2016). Comparaison du coût de différentes mesures de protection de la vie humaine face au risque de submersion marine. Conference paper presented at XIVèmes Journées Nationales Génie Côtier – Génie Civil, Toulon 29 June to 1 July 2016. doi: 10.5150/jngcgc.2016.070, www.paralia.fr, https://hal.archives-ouvertes.fr/hal-01362004 (accessed 4 March 2019).

Creach, A., Chevillot-Miot, E., Mercier, D. et al. (2015a). Vulnerability to coastal flood hazard of residential buildings on Noirmoutier Island (France). *Journal of Maps* 12 (2): 371–381. https://doi.org/10.1080/17445647.2015.1027041.

Creach, A., Chevillot-Miot, E., Mercier, D. et al. (2015c). Vulnérabilité de l'habitat face au risque de submersion marine. Exemple de la commune de l'Epine (Vendée). In: *Atlas Permanent de la Mer et du Littoral. n° 7*, 78. Nantes: LETG-Nantes Géolittomer.

Creach, A., Pardo, S., Guillotreau, P. et al. (2015b). The use of a micro-scale index to identify potential death risk areas due to sea-flood surges: lessons from storm Xynthia on the French Atlantic coast. *Natural Hazards* 77 (3): 1679–1710. https://doi.org/10.1007/s11069-015-1669-y.

Creach, A. and Pourinet, L. (2015). PPRn dans les communes littorales de France métropolitaine et prise en compte du risque de submersion marine. In: *Atlas Permanent de la Mer et du Littoral n°7*, 74–75. Nantes: LETG-Nantes Géolittomer.

Devaux, E., Désiré, G., Boura, C. et al. (2012). La tempête Xynthia du 28 février 2010 – Retour d'expérience en Loire-Atlantique et Vendée - Volet hydraulique et ouvrages de protection. CETE-Ouest / DREAL Pays-de-la-Loire / DDTM Loire-Atlantique/ DDTM Vendée. 78 p.

European Commission/Commission Européenne (2007). Directive 2007/60/CE du Parlement Européen et du Conseil du 23 octobre 2007 relative à l'évaluation et à la gestion des risques d'inondation. Brussels: European Commission.

Feuillet, T., Chauveau, E., and Pourinet, L. (2012). Xynthia est-elle exceptionnelle? Réflexions sur l'évolution et les temps de retour des tempêtes, des marées de tempête, et des risques de surcotes associés sur la façade atlantique française. *Norois* 222 (1): 27–44. https://doi.org/10.4000/norois.3866.

Garnier, E., Henry, N., and Desarthe, J. (2012). Visions croisées de l'historien et du courtier en réassurance sur les submersions: recrudescence de l'aléa ou vulnérabilisation croissante? In: *Gestion des risques naturels – Leçons de la tempête Xynthia* (ed. V. Przyluski and S. Hallegatte), 105–128. Quae: Versailles.

Goeldner-Gianella, L., Bertrand, F., Oiry, A. et al. (2015). Depolderisation policy against coastal flooding and social acceptability on the French Atlantic coast: the case of the Arcachon Bay. *Ocean and Coastal Management* 116: 98–107. https://doi.org/10.1016/j.ocecoaman.2015.07.001.

IGN and SHOM, (2012). Litto3D® – v 1.0 - Spécifications techniques. IGN/SHOM, 16p.

IPCC (2014). *Climate Change 2014: Impacts, Adaptation and Vulnerability*. 2 vols. GIEC/IPCC.

Jonkman, S.N., Maaskant, B., Boyd, E. et al. (2009). Loss of life caused by the flooding of New Orleans after hurricane Katrina: analysis of the relationship between flood characteristics and mortality. *Risk Analysis: An Official Publication of the Society for Risk Analysis* 29 (5): 676–698. https://doi.org/10.1111/j.1539-6924.2008.01190.x.

Jonkman, S.N. and Penning-Rowsell, E. (2008). Human instability in flood flows. *Journal of the American Water Resources Association* 44 (4): 1–11.

Jousseaume, V., Landrein, J., and Mercier, D. (2004). La vulnérabilité des hommes et des habitations face au risque d'inondation dans le Val nantais (1841–2003). Entre législation nationale et pratiques locales. *Norois* 192 (3): 29–45. https://doi.org/10.4000/norois.866.

Jousseaume, V. and Mercier, D. (2009). Évaluer la vulnérabilité architecturale de l'habitat en zone inondable. L'exemple du Val nantais. In: *Risques et environnement: recherches interdisciplinaires sur la vulnérabilité des sociétés* (ed. S. Becerra and A. Peltier), 199–214. Paris: L'Harmattan.

Klugman, J., Rodríguez, F., and Choi, H.J. (2011). The HDI 2010: new controversies, old critiques. *Journal of Economic Inequality* 9 (2): 249–288. https://doi.org/10.1007/s10888-011-9178-z.

Kolen, B., Slomp, R., and Jonkman, S.N. (2013). The impacts of storm Xynthia February 27–28, 2010 in France: lessons for flood risk management. *Journal of Flood Risk Management* 6 (3): 261–278. https://doi.org/10.1111/jfr3.12011.

Kolen, B., Slomp, R., Van Balen, W. et al. (2010). Learning from French experiences with storm Xynthia: Damages after a flood, Ministerie van Verkeer en Waterstaat (MKV Consultants), 89 p.

Leone, F., Lavigne, F., Paris, R. et al. (2011). A spatial analysis of the December 26th, 2004 tsunami-induced damages: lessons learned for a better risk assessment integrating buildings vulnerability. *Applied Geography* 31 (1): 363–375. https://doi.org/10.1016/j.apgeog.2010.07.009.

Liberation (2016). Procès Xynthia en appel: 'une énorme claque' pour les victimes. *Journal Libération*, 4 April 2016. https://www.liberation.fr/france/2016/04/04/proces-xynthia-en-appel-une-enorme-claque-pour-les-victimes_1443916 accessed 28 February 2019.

Lumbroso, D.M. and Vinet, F. (2011). A comparison of the causes, effects and aftermaths of the coastal flooding of England in 1953 and France in 2010. *Natural Hazards and Earth System Sciences* 11 (8): 2321–2333. https://doi.org/10.5194/nhess-11-2321-2011.

Magnan, A. and Duvat, V. (2014). *Des catastrophes… 'naturelles'?* Paris: Editions Le Pommier 316 p.

MEDDE (2012). Mieux savoir pour mieux agir: Principaux enseignements de la première évaluation des risques d'inondation sur le territoire français - EPRI 2011. Paris: MEDDE, 72 p.
MEDDE (2013). Guide méthodologique: Plan de prévention des risques littoraux. Paris: DGPR / MEDDE, 169 p.
MEDDE and METL (2012). Référentiel de travaux de prévention du risque d'inondation dans l'habitat existant. Paris: MEDDE and METL, 81 p.
MEDDTL (2011a). Programmes d'action de prévention des inondations (PAPI), de la stratégie aux programmes d'action, cahier des charges. Paris: MEDDTL, 28 p.
MEDDTL (2011b). Plan national d'adaptation de la France aux effets du changement climatique - 2011 - 2015. MEDDTL, Paris: MEDDTL 188 p.
MEDDTL (2011c). Circulaire du 27 juillet 2011 relative à la prise en compte du risque de submersion marine dans les plans de prévention des risques naturels littoraux Paris: MEDDTL.
MEDDTL, MIOMCTI, MEFI and MBCPFPRE (2011). Plan submersions rapides - Submersions marines, crues soudaines et ruptures de digues. Plan interministériel coordonné par le MEDDTL, Paris: MEDDTL, 80 p.
Mercier, D. (2012). Après Xynthia: vers un repli stratégique et un État fort? *Norois* 222 (1): 7–9. http://norois.revues.org/3826.
Mercier, D. and Acerra, D. (2011). *Xynthia, une tragédie prévisible, Introduction*, 5–8. Place Publique, Hors Série.
Mercier, D. and Chadenas, C. (2012). La tempête Xynthia et la cartographie des 'zones noires' sur le littoral français : analyse critique à partir de l'exemple de La Faute-sur-Mer (Vendée). *Norois* 222: 45–60. https://doi.org/10.4000/norois.3895.
Mori, N. and Mase, H. (eds.) (2013). Analysis of coastal disaster by typhoon Haiyan. Coastal disaster research section. Kyoto: Disaster Prevention Research Institute, Kyoto University.
Nicholls, R.J. (2011). Planning for the impacts of sea level rise. *Oceanography* 24 (2): 144–157. https://doi.org/10.5670/oceanog.2011.34.
ONML (2012). Les zones basses sur le littoral métropolitain. Paris: MEDDE-SOeS, 5 p.
Perherin, C. (2007). Contribution à l'analyse des phénomènes de surcotes et de submersion marines. Secteur du Pertuis Breton (Vendée), École Nationale des Travaux Publics de l'État, Vaulx-en-Velin, Rapport de stage, 94 p.
Perret, J. and Sauzeau, T. (2014). Xynthia ou la mémoire réveillée. Des villages charentais et vendéens face à l'océan (XVIIIe-XXIe siècles). La Crèche: Geste, 296 p.
Pitié, C., Bellec, P., Maillot, H. et al. (2011). Expertise des zones de solidarité Xynthia en Charente-Maritime. Paris: CGEDD / MEEDDM, 186 p.
Pitié, C. and Puech, P. (2010). Expertise complémentaire des zones de solidarité délimitées en Vendée suite à la tempête Xynthia survenue dans la nuit du 27 au 28 février 2010. Paris: CGEDD / MEEDDM, 80 p.
Renard, J. (2005). *La Vendée : un demi-siècle d'observation d'un géographe*. Rennes: Presses Universitaires de Rennes, 308 p.
Sarrazin, J.-L. (2012). 'Vimers de mer' et sociétés littorales entre Loire et Gironde (XIVe-XVIe siècle). *Norois* 222 (1): 91–102. https://doi.org/10.4000/norois.4034.
Sogreah (2010). Éléments de mémoire sur la tempête Xynthia du 27 et 28 Février 2010 en Charente-Maritime. La Rochelle: Préfecture de la Charente-Maritime / DDTM 17, 792 p.
Verger, F. (2010). Xynthia, zones d'ombre sur les zones noires. *Études foncières* 145: 6–9.
Verger, F. (2011). Digues et polders littoraux: réflexions après la tempête Xynthia. *Physio-Géo. Géographie, physique, et environnement* 5: 95–105. https://doi.org/10.4000/physio-geo.1740.
Vinet, F. (2010). *Le risque inondation. Diagnostic et gestion*. Paris: Lavoisier 318 p.
Vinet, F., Boissier, L., and Defossez, S. (2011). La mortalité comme expression de la vulnérabilité humaine face aux catastrophes naturelles: deux inondations récentes en France (Xynthia, Var,

2010). *VertigO – la revue électronique en sciences de l'environnement* 11 (2): 1–28. https://doi.org/10.4000/vertigo.11074.

Vinet, F., Defossez, S., Rey, T. et al. (2012a). Le processus de production du risque 'submersion marine' en zone littorale: l'exemple des territoires 'Xynthia'. *Norois.* 222 (1): 11–26. https://doi.org/10.4000/norois.3834.

Vinet, F., Lumbroso, D., Defossez, S. et al. (2012b). A comparative analysis of the loss of life during two recent floods in France: the sea surge caused by the storm Xynthia and the flash flood in Var. *Natural Hazards* 61 (3): 1179–1201. https://doi.org/10.1007/s11069-011-9975-5.

25
Coastal Flooding and Storm Surges: How to Improve the Operational Response of the Risk Management Authorities: An Example of the CRISSIS Research Program on the French Coast of Languedoc

Brice Anselme[1], Paul Durand[2], and Alexandre Nicolae-Lerma[3]
[1]*Institut de Géographie, Laboratory CNRS 8586 PRODIG, Université Paris 1 Panthéon-Sorbonne, Paris, France*
[2]*Laboratory of physical geography CNRS 8591 LGP, Université Paris 1 Panthéon-Sorbonne, Paris, France*
[3]*BRGM Nouvelle-Aquitaine, French Geological Survey, Pessac, France*

25.1 Introduction

As sea levels rise due to climate change, coastal communities on low-lying coasts are increasingly vulnerable to the risk of flooding during storms. The coastal flooding issue is thus set to become crucial to management of this kind of coast in the decades to come, especially since coastal regions are increasingly attractive. Today, they are home to more than 70% of the world population, half of which lives at an altitude of below 5 m, and the urban, tourism, and industrial stakes are constantly rising in these areas. As a result, there is an increase in research work focused on coastal flood risks on low-lying coasts. Whilst the first studies were often relatively segmented, focused either on the risk or on the

Facing Hydrometeorological Extreme Events: A Governance Issue, First Edition.
Edited by Isabelle La Jeunesse and Corinne Larrue.

challenges and vulnerabilities, recent research programmes covering this topic in European countries (MISEEVA 2008–2011; JOHANNA 2010–2013; BARCASUB 2010–2013; COCORISCO 2011–2014) are much more integrated, dealing with both the risk and the various aspects of vulnerability – physical, economic, and social – through perception of the risk by populations. Recent work financed by the German Federal Ministry of Education and Research included the first worldwide assessment of the cost incurred by damage from storm-related coastal flooding and by the adaptation efforts required for the twenty-first century (Hinkel et al. 2014). According to the authors, in 2100 nearly 5% of the world's population could be affected by coastal flooding every year if no adaptation measures are implemented, with annual losses of up to almost 10% of world gross domestic product. Their results also emphasize the essential role of coastal protection measures and long-term adaptation strategies. The countries concerned need to act now and invest in protection measures rapidly if we want to significantly reduce the global cost of damage related to storm floods over the next century. Some projects also include environmental and ecological vulnerabilities, such as the National Centers for Coastal Ocean Science (NCCOS) programmes of the National Oceanic and Atmospheric Administration (NOAA) (*Ecological Effects of Sea Level Rise – EESLR* 2005–2012 and *Predicting Impacts of Sea Level Rise in the Northern Gulf of Mexico* – 2010–2017).

Nonetheless, very little research goes as far as to integrate the crucial aspect of 'crisis management'. Whilst the introduction of prevention and mitigation measures can significantly reduce the probability of disasters and mitigate the consequences, zero risk cannot be achieved. There will always be residual risks. In other words, we have to be ready to manage crises, including 'extraordinary' crises which, because they are complex and unexpected, cannot be dealt with using the conventional response patterns. This is especially true in regions where the stakes are high and numerous and where the crises may therefore involve several public and private stakeholders. In response, the political authorities in many countries regularly conduct large-scale crisis drills to prepare stakeholders and the public in how to deal with major crises. For example in the United States, the FEMA (Federal Emergency Management Agency) has a National Exercise programme, https://www.fema.gov/exercise) which regularly runs large-scale drills simulating the occurrence of cyclones, earthquakes, and tsunamis – the last exercise in June 2016, known as Cascadia 2016, simulated a huge earthquake and massive tsunami on the northwest coast of the United States. In Europe, this task is covered by the European Civil Protection Mechanism (ECPM, http://ec.europa.eu/echo/what/civil-protection/mechanism_en) which organizes and funds a series of drills in the field of civil protection in different member countries every year. Its latest large-scale exercise was the European Sequana 2016 exercise, held in March 2016 and which simulated the occurrence of a 100-year flood of the Seine in Paris, involving public stakeholders at various levels (national, zonal, departmental, and municipal) and an array of private operators (transport firms, telecommunications companies, banks, supermarkets, hospitals, and so on). The French example provides a good gauge for a national policy for crisis drills. With a concern for optimizing the response from stakeholders and the public in a crisis, the 2004 law on the modernization of civil security (law #2004–811 dated 13 August 2004) makes regular exercises compulsory, with a requirement for large-scale drills several times a year at national and regional levels (Richter-type earthquake drills are a good example) and a requirement for all towns and villages where a major natural or technological risk has been identified to conduct at least one crisis drill per year (DGSCGC 2008). However, this legal requirement is not always complied with and, in its 2012 activity report, the government civil defence committee (*Haut Comité Français pour la Défense Civile* or HCFDC) states that the crisis management procedures have not yet been sufficiently tested for major natural events, especially at local level. Due to a lack of resources and/or expertise, most municipalities only organize drills when required to do so by higher-level territory authorities (department prefecture, regional prefecture, inter-regional development agencies such as the *Plan Loir*e which runs several drills per year in flood-prone communities). Only the biggest municipalities have the resources to test their local response plan (the *Plan Communal de Sauvegarde* or PCS, the municipality-level crisis management tool), with support from specialist consultants.

It therefore appears vital to have regular drills in all communities likely to be affected by exceptional events more widespread, 'to prepare everyone, at every level, to be alert', (HCFDC 2012). Research programmes should now focus on operational and collaborative research involving local institutional stakeholders. Yet apart from in the field of seismic and volcanic risks (Johnston et al. 2011; Marzocchi, Newhall, and Woo 2012; Morin 2012; Pierson, Driedger, and Tilling 2013; Pierson et al. 2013), the scientific community is still rarely involved in truly operational research programmes, i.e. those involving local public authorities and running drills. However, against a background of global climate change, it seems essential to include this operational dimension in future research work. In our view, to tackle the challenge raised by an inevitable rise in the number of coastal flooding events in increasingly attractive coastal areas, research programmes must now take a holistic approach to the risk, from awareness of the hazard to crisis management and including the following four points:

1. Better characterization of the risk, thanks to the progress made with digital modelling systems;
2. Close analysis of vulnerabilities: building and infrastructure vulnerability, and the vulnerability of people before the hazard;
3. Consideration given to perceptions and representations of risks to make protective measures and crisis management more understandable and acceptable for the population (this part is dealt with in detail in Chapter 23; and
4. Improved operational crisis management, because crises will inevitably occur, despite all the protective measures implemented. This approach must be based on the organization of crisis drills involving local stakeholders.

This chapter will illustrate these four points, taking the example of the CRISSIS research programme (http://crissis2015.free.fr) launched to study the French Mediterranean coast in Languedoc in spring 2015 over a two-year period. It is an innovative, multidisciplinary project involving geographers, modellers, geomaticians, and risk/crisis management specialists, the objective of which is to put forward an integrated, operational approach to the coastal flooding risk applied to a seaside town, Leucate. The programme is run in close collaboration with the local authorities and is a good example – and unique in France – of a joint approach bringing together the research sector and local public stakeholders with the aim of improving the response from the authorities in the event of a crisis linked to coastal flooding. The town of Leucate was selected because it is especially vulnerable to coastal flooding. Its coastline stretches more than 15 km and includes three resorts located on either side of Cap

Leucate. From south to north: the Port Leucate resort, built in the late 1960s as part of the Racine development project on the beach barrier that closes the Salses-Leucate lagoon, Leucate Plage, along the Cap's southern cliff edge, and La Franqui, at the foot of the Cap's northern cliff (Figure 25.1). The urban and tourism issues are particularly important because the town receives high visitor numbers, increasing from a full-time resident population of fewer than 5000 in the low season to almost 100 000 people in

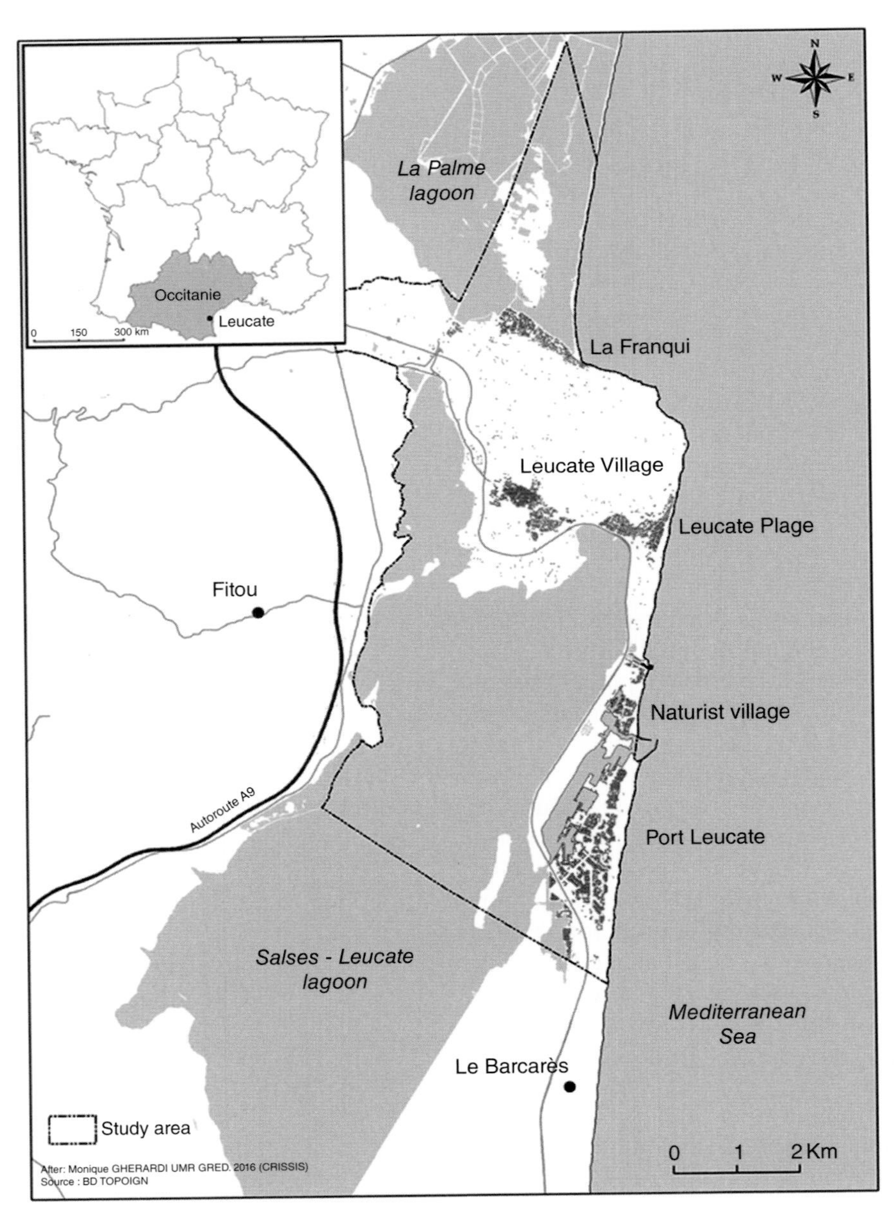

Figure 25.1 Location of study area.

the high season. The urban areas of Port Leucate and Leucate Plage are highly exposed to the coastal flooding risk, especially Port Leucate, which is built on a low-lying beach barrier (maximum altitude 3 m NGF). They are periodically flooded during the big southeast storms when the sea level, including wave action, can approach 3 m NGF (Anselme et al. 2011). Flooding may be increased if the storms come with heavy rain, causing the lagoon level to rise and triggering significant run-off from the slopes of the cliffs around the Cap. Paradoxically, despite high risk exposure (14 decrees declaring a state of natural disaster since 1982, 13 of which were for flooding and submergence), the town has some catching up to do in the field of risk regulations, when compared to other towns on the Languedoc-Roussillon coast. The first flood risk prevention plan (PPRI) was only introduced in November 2012 (PPR on flooding after coastal submergence), but even now this is still at the inquiry stage. The local response plan (PCS) was adopted in December 2013 after the significant flooding of March 2013.

The choice of Leucate for testing an integrated, operational approach to the coastal flooding risk appears particularly relevant against this background. The approach reflects the current trend of developing applied research programmes (see earlier) covering an increasingly broad spectrum of disciplines for a better understanding of the complexity of responses to be delivered to tackle natural risks. However, its highly operational aspect is innovative: crisis management procedures and practices were assessed in crisis drills involving the town of Leucate and operational stakeholders. It is, therefore, a comprehensive approach, integrating the four aspects of coastal flooding to be covered in this chapter: (i) the hazard and its likely evolution in a context of climate change; (ii) the issues and vulnerabilities, with a qualitative and quantitative analysis of the issues (structural vulnerability of buildings, visitor numbers, map of potential damage); (iii) perceptions and representations amongst the public (the 'risk culture'); and (iv) risk management and its two components: prevention and crisis management (assessing preventive measures, analysing the conduct of stakeholders and the effectiveness of the crisis-management procedures by running drills).

25.2 The coastal flood hazard and its likely evolution

The first component of the CRISSIS project focuses on characterizing the coastal flood risk and its likely evolution in a context of climate change and sea level rise. Coastal flooding is the temporary inundation of a coastal zone by the sea in extreme weather conditions.

The major challenge for digital modelling, which is now commonly used to study coastal hazards, is to adapt the use of models to a site's specific exposure and the phenomena behind coastal flood, such as overflow, overtopping, breaches of dikes or levees, and so on (Garcin et al. 2012; Le Roy, Pedreros, and Nicolae Lerma 2013). Until now, due to significant technical (calculation times and IT resources) and methodological (coupling between processes) constraints, coastal flooding events were regularly simulated whilst working on the assumption that these events are exclusively or largely triggered by a rise in the average sea level. In addition, when looking at extensive sites, the hazard is dealt with according to spatial resolution, which often prevents the integration of buildings in urban areas, for example. These methods can be ill-adapted to a realistic vision of the hazard on coasts that are subject to wave action, for example a low-lying lagoon lido or when the aim is to represent the specifics of water flows in urban environments.

25.2.1 Materials and methods

To understand, simulate and map the flooding hazard in Leucate, the approach developed through this project involves dynamic modelling, i.e. meteorological and oceanographic conditions that vary over time are introduced into the digital model chain to simulate the evolution of marine conditions offshore and on the coast, and their consequences on land with regard to coastal flooding (extent of flooded areas, water depth, propagation speed). Whilst few studies suggest coupling coastal flooding with overflow and overtopping processes (Stansby et al. 2013; Gallien, Sanders, and Flick 2014; Le Roy et al. 2015; Gallien 2016), the integration of these different forms of inundation is necessary when developing a truly detailed and realistic vision and when characterizing the hazard at particularly exposed, vulnerable sites like Leucate (Nicolae Lerma et al. 2017). The end result obtained here can be used to represent the evolution of the hazard during a storm every 10 minutes and at high spatial resolution (5 m) for the complete study site.

There are several stages to the implementation of this modelling work: (i) an accurate representation of the sector's bathymetry and topography, (ii) the combined use, at relevant spatial scales, of several models enabling the integration of complex interactions between water level, current, waves, and wind. Then we need to (iii) validate the flood model by simulating historical events that had a significant impact on the area. Once the model's performance has been validated, (iv) several scenarios are simulated for current and future sea level conditions, and then (v) an extreme but realistic scenario combining severe storm conditions (taken from statistical analysis of past events) and the breach of sea front structures before the most highly exposed districts is defined to provide a 'sea storm hazard' tool to establish the crisis drill procedure.

25.2.1.1 Processing bathymetric and topological data High-resolution modelling of the coastal flooding risk requires a detailed representation of the bathymetry and topography of studied sector. A digital model of the land/sea to spatial resolution of 5 m was produced using bathymetric data (sea, lagoon, and harbour channels) and topographic data (inland and the coastal strip up to 10 m deep using the Litto-3D® product from the SHOM (French Naval Hydrographic and Oceanographic Service) and the IGN (National Geographic Institute), and different bathymetric surveys for the port and lagoon areas.

In addition, buildings play a decisive role on the dynamics of run-off and the extent of flooding. Buildings, as well as the main features likely to affect the spread of water, such as hydraulic structures (pipes, sluices, and culverts) and coastal structures (sea walls) are integrated in the topological/bathymetric model based on GIS processing or differential GPS recordings.

Ground conditions can also play a crucial role in the spread of sea water. Land cover is taken into account in the models via a variable friction coefficient depending on ground conditions (bare earth, no vegetation, herbaceous cover, shrubland, or forest) and ground material (sand, earth, tarmac). In our study, a spatialized representation of roughness was obtained using synthetic classification of land cover taken from the 2006 *Corine Land Cover* data. The friction coefficient values used to characterize roughness are those recommended by various sources and are applicable to studies of marine and coastal areas (Bunya et al. 2010; Goutx and Ladreyt 2001).

25.2.1.2 Chained models The coastal flooding risk is appraised using numerical simulation and the combined use of several models, taking a downscaling approach from regional to local scale, making it possible to work to a level of accuracy suited to the modelled processes (Figure 25.2). The boundary conditions of the models are set by the water levels, waves, and wind conditions recorded on the basis of available, relevant data from tide gauges in Sète, Port-Vendres, and Port-La-Nouvelle, the heave buoys in Sète, Banyuls, and Leucate and the Leucate semaphore for weather conditions. These data are integrated in the so-called spectral wave model used to reproduce the characteristics of waves and simulate their propagation to the coast. The wave characteristics are then used in a free-surface hydrodynamic model to simulate the variations in average water levels caused by storm atmospheric conditions and wave set-up. The final link in this chain is a model (Mars-Flood) that reproduces the propagation of run-off on land coming from, on the one hand, overflows triggered by a rise in the average level and, on the other hand, the simulation of water volumes overtopping the sea front (using the SWASH 1D model).

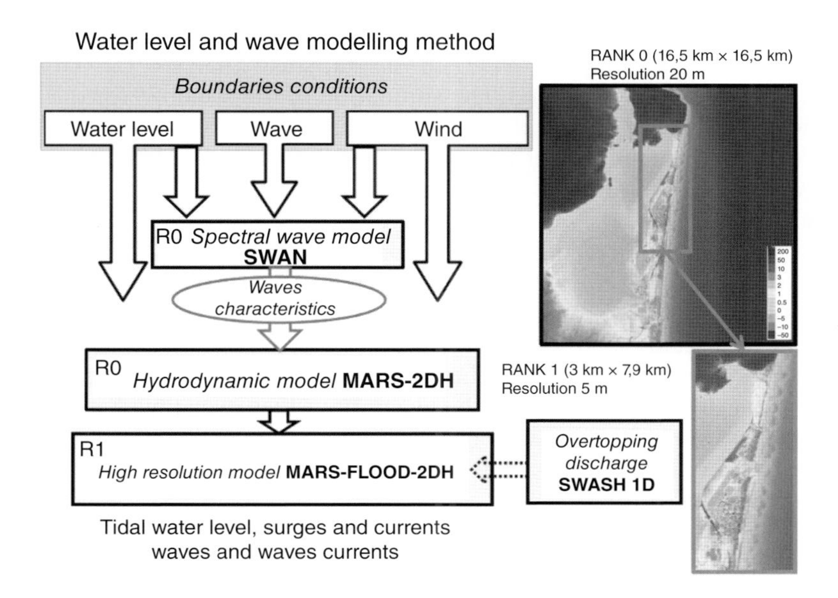

Figure 25.2 Chained modelling method with downscaling approach. *Source*: Nicolae Lerma et al. (2017).

25.2.2 Results and discussion

25.2.2.1 Validation of the model chain Two recent storm events, both causing coastal floods, were selected to support and validate the results obtained with the chained modelling method. The storm that occurred in March 2013, when a breach in the sea front wall in Leucate Plage lead to significant flooding of the entire district. The storm of November 2014 brought about flooding after localized overflows on the lagoon side and overtopping along the sea front. When the results of simulation in the port were compared with the validation data collected on the ground, they confirmed the faithfulness of the simulations in terms of water height in the port and the areas flooded in the town (cf. Nicolae Lerma et al. 2017).

25.2.2.2 Extreme scenarios for current and future sea level conditions The following stage was the drafting of extreme scenarios for current and future sea level conditions. For this, we based our work on the statistical processing of continuous time series of water and swell levels, collected from the observations stations closest to the study site and on hindcast simulation databases. This statistical processing was used to determine sea levels and extreme wave conditions in terms of occurrence probability (return periods of 10, 50, and 100 years). We then used those values to draught and simulate forward-looking extreme scenarios with and without factoring in rising sea levels.

25.2.2.3 Marine storm hazard for the crisis drill The final stage in this first component focused on modelling a major storm scenario to enable the implementation of a crisis drill on the ground. The selected scenario combines the high-water levels of the November 2014 storm and the strong waves recorded during the March 2013 event, yet remains plausible (according to the statistical analyses, this kind of event has a return period of 20–30 years). The simulation of the scenario for the crisis drill also includes the breach of a seafront wall (Figure 25.3). The water levels reached with this kind of event are higher than those of the 2013 and 2014 storms, thus causing more widespread flooding and highlighting the significance of threshold effects in the port/lagoon areas.

Gaining a better understanding of the expressions of the hazard through research into past coastal flooding events and by making use of digital modelling has enabled the production of high-resolution mapping materials and the elements required to put forward an event scenario suited to the objectives tested during the exercise. Digital modelling also provides a dynamic, changing view of the storm event and makes it possible to rank the areas affected as the event intensifies over time. Finally, using the various statistical scenarios, we can consider a wider range of scenarios, making it possible to test different objectives whilst retaining the degree of realism required to engage stakeholders.

25.3 Vulnerability of the stakes

The second stage in the programme was characterizing and mapping human and material vulnerability relative to the coastal flooding risk in Leucate.

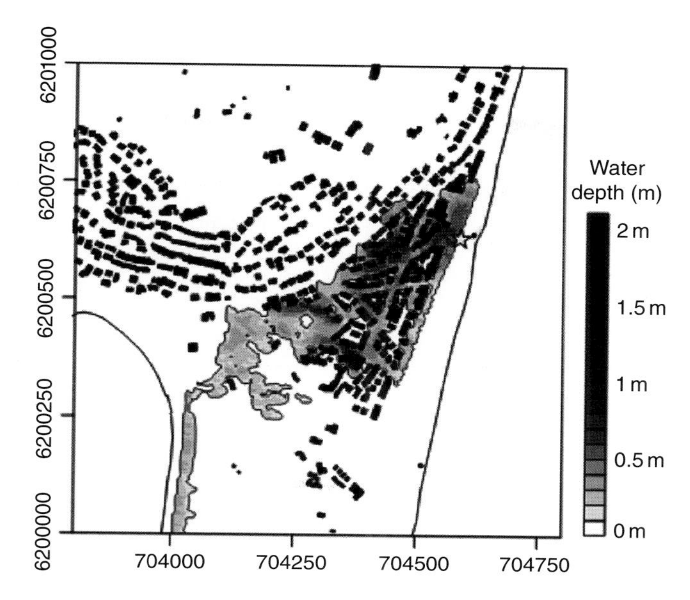

Figure 25.3 Flooded area in Leucate Plage associated to the crisis exercise scenario. Flooding is caused by a breach in the wall located on the sea front (star).

25.3.1 Materials and methods

At the start of the programme, an extensive survey was conducted in the town of Leucate, using questionnaires to identify the human and material vulnerabilities. The data collected were processed from a qualitative angle to estimate human vulnerability, whilst material vulnerability was estimated using a quantitative analysis of the survey questionnaires. Table 25.1 shows the criteria and indicators taken into consideration to assess these vulnerabilities.

25.3.2 Results and discussion

25.3.2.1 Human vulnerability Human vulnerability can be appraised using various criteria concerning knowledge and experience of the risk, and the intrinsic criteria of individuals such as their age and health. The questionnaires were processed and analysed to reveal the three main criteria that explain human vulnerability:

- an ageing population with an age group of 60–74-year-olds which, although over-represented amongst the interviewees compared to the town's overall population, points to a declining population with almost 70% of residents retired. For the 'age' criteria, we opted for a vulnerability indicator of 80 years of age. The state of health also indicates intrinsic human vulnerability and we decided to take reduced mobility and/or dependency as our criterion. In addition to the elderly, we also included children aged under 10 as vulnerable. If we take into consideration the total number

Table 25.1 Criteria and indicators of human and material vulnerabilities.

Vulnerability	Criteria	Indicators
Vulnerability to exposure	Extent/intensity of the risk; protection against the risk	• Estimated height of water in the dwelling • Distance/sea • NGF altitude (versus sea level) • Presence and distance of defence structures (likely to be destroyed or overtopped)
Human vulnerability linked to experience and perception (task 2b)	Knowledge Experience	• Perception of the sea and coastal flooding • Impression of exposure to the risk • Knowledge stemming from experience • Knowledge based on information
Intrinsic human vulnerability	Age State of health	• Number of people aged under 10 • Number of people aged over 80 • Number of dependent people or with reduced mobility
Human vulnerability with regard to evacuation	Access	• Exit enabling airlifting • Type of street/access/evacuation
Human vulnerability related to buildings (securing)	Type of dwelling	• Nature of the building (single-storey, more than one storey, etc.) • Floor/street level • Number of bedrooms on the ground floor and occupation
Material vulnerability	Type of dwelling	• Nature of the building (single-storey, more than one storey, etc.) • Floor/street level
	Structure/state of building	• Age of building • Nature and state of building inside and out • Number and water tightness of openings • Height and condition of boundary wall • Type and height of electrical devices
Reduction in material vulnerability	Mitigation measures	• Knowledge of protection/prevention measures • Introducing measures • Type of measures taken or to be implemented

Source: Defossez, Gherardi et al. (in preparation).

of residents, and not the number of people surveyed, in the three districts of Leucate the number of very vulnerable people accounts for 10% of the population.

- low level of risk awareness amongst the population: the human vulnerability criteria linked to risk perception and experience reveal unfamiliarity with the coastal flooding risk in the town. This lack of awareness can largely be explained by the scarcity of recent storm events and a short memory, probably also reinforced by the fact that nearly half of the people surveyed moved to the town after the 2000s.

- a population that is difficult to evacuate: the age and/or state of health of the population affect human vulnerability when it comes to evacuation. Due to the layout of Port Leucate and the naturist village, the communications network would be vulnerable in the event of evacuation. The two communes are crossed by a main road and the numerous residential blocks and detached houses are built around cul-de-sacs, narrow streets and pedestrian areas, making it especially difficult to evacuate people to the refuge areas in the event of a crisis. It is even more complicated to evacuate older people or those with reduced mobility, i.e. more than half of the residents aged over 80 amongst the interviewees.

25.3.2.2 Building vulnerability An initial rough evaluation of material vulnerability was carried out on the basis of building type and vertical structure. The survey results for this vulnerability criterion reveal two main types of building: single-storey dwellings (42.5%), which are the most vulnerable, and a similar proportion of two-storey dwellings, distinguishing between dwellings that are occupied on their ground floor and those occupied on their first floor (less vulnerable). Analysis of the survey questionnaires shows that from a structural viewpoint, dwellings are highly vulnerable since half of the people surveyed live in a single-storey home and the other half in dwellings with more than one storey but the ground floor of which would be damaged.

In addition to this structural vulnerability criterion, we used a building condition criterion (type and condition of walls, tightness of walls, floors, etc.), and a brief diagnosis of the technical fittings, especially the height of the circuit breakers. These criteria were then taken together and weighted to establish an overall material vulnerability indicator.

25.4 Social representations and perceptions of the coastal flooding risk

The third component of the programme focuses on socio-cultural vulnerability appraised using a geo-sociological study to characterize the level of risk culture in Leucate. We studied the perception and representation of the risk amongst residents and tourists, together with their attitudes and behaviour with regard to the coastal flooding risk. These factors are crucial in ensuring that management procedures result in better information for and greater awareness amongst the population. Up to now, studies into the coastal flooding risk in mainland France have highlighted the shortcomings in social representations of this issue and showed that most people approached via surveys and interviews had not thought much about the risk of coastal flooding and thus forged their perception of the risk on the spot (Hellequin et al. 2013; Flanquart 2014; Krien 2014).

25.4.1 Materials and methods

The geo-sociological study was conducted using the same questionnaires mentioned above for component 2 of the programme. In all, 493 people were surveyed, 43% of whom were full-time residents, 39% part-time residents and 18% tourists. Although our priority target is the residents, who are more highly exposed to the coastal flooding risk than tourists, who are rarely present in Leucate during the risk period from October to April, the latter were still questioned in order to compare the knowledge and representations of people from within and from outside Leucate.

The questionnaire framework was subdivided into several themes and the people questioned were asked about (i) the attractions of Leucate and the leisure facilities on offer, (ii) their perception and experience of the coastal flooding risk, (iii) their awareness of their exposure to coastal flooding and their capacity to be proactive in this situation, (iv) the vulnerability of dwellings and their inhabitants, (v) their socio-economic profile. A final part, describing the context of the survey, was completed by the interviewer. What is original about this survey work is that it draws on mental maps to locate past floods, floodable areas, and non-floodable areas. First of all, the idea was to check the validity of the simulations conducted in the programme's first component against residents' knowledge of past flood events. We also wanted to see which areas people deemed to be floodable or non-floodable in Leucate so as to obtain an overview of the state of knowledge and perception of the flood risk compared to the 'reality' of the risk as shown by the modelling work.

25.4.2 Results and discussion

Several points can be established after looking at the results on perception and social representation of the coastal flooding risk in Leucate:

- Very low perception of the flood risk: when asked about the risks coming from the sea, the main fears mentioned by the people surveyed are the risks of storms and tsunami, ahead of rising sea levels and the risk of flooding, despite the fact that half of the full-time and part-time residents know that the town is regularly affected by this hazard. It would appear that the people questioned are more likely to be afraid of exceptional, spectacular risks than of the risks to which they are directly exposed, such as coastal flooding and the retreating coastline. This result appears to be linked to a perception that coastal flooding is a minor risk in the town. From the conversations that took place as part of the survey, it would appear that the interviewees do not see the occasional episode of seawater intrusion in their streets as a case of coastal flooding.
- Poor understanding of the coastal flooding hazard: overall, the social representations built around 'the sea at Leucate' evoke a rather positive vision of the sea amongst the people surveyed. The social representation of the coastal flooding risk remains very limited and it would appear that no specific vision behind the expression 'coastal flooding in Leucate'.
- Lack of knowledge about risk and crisis-management tools (local response plan, sirens, telephone warning system, refuge areas for people evacuated, and so on) with a significant difference between full-time and part-time residents, the former being more aware of the measures implemented by the municipal council as part of the flood warning system. Half of all full-time residents questioned and three quarters of part-time residents say that they are not sufficiently informed about the coastal flooding risk in Leucate.
- Finally, it is worth nothing that whilst in people's minds, the risk of coastal flooding is strongly associated with rising sea level, the social representation of climate

change appears to be more apparent amongst younger generations than in their seniors, who think that they will not see its effects.

25.5 Crisis management

25.5.1 Materials and methods

The operational component of the CRISSIS programme was designed to test the crisis management approach applied in Leucate by organizing two drills that were devised, run, and assessed by students on a vocational master's course specialized in crisis management (Master's in Overall Risk and Crisis Management at Paris 1 Panthéon-Sorbonne University), supervised by crisis management professionals. These drills were designed and developed from a progressive angle. They were based on the civil security crisis drill policy defined in the law on the modernization of civil security dated 13/08/2004 (http://www.legifrance.gouv.fr) and implemented by the DGSCGC (general directorate for civil security and crisis management), which ranks crisis drills in two main categories, the equivalent of which are found in most countries: 'framework drills' and 'field drills'. The former take place in the crisis room and do not require resources in the field (French Interior Ministry 2011). The people involved receive information over the radio, telephone, fax, television, text messages, and social media, and have to analyse, summarize, then react, report, propose, define priorities, and make choices to manage the crisis as best they can. The second type of drill deploys human and material resources on the ground. These exercises are a large-scale, real-time test of the time required to convey and implement equipment. The two main drill categories (framework and field) may be subdivided into three levels of exercise, depending of the type of situation to be tested:

1. The first is the practice run, used to let staff get to know new procedures or new equipment. The aim is to gain experience with the systems (directories or listings, telephone warnings, activating stand-by duties, opening and setting up a crisis room, and so on) put in place to respond to a crisis in a 'normal' or 'low-intensity' situation. These tests do not involve any element of surprise or stress. The goal is for stakeholders to make themselves familiar with the existing procedures and develop their technical capacity to implement them.
2. The second level is planned drills: participants are warned that the drill will take place and informed of the theme it will cover, but they do not know the scenario selected. Players are faced with an imaginary crisis situation and are forced to think about the decisions and actions required. Generally speaking, these planned drills help improve understanding of the crisis processes without causing difficulties for people as those involved are not subject to the same stress as in a real-life situation (given that they knew the drill was to take place).
3. The third level is the unannounced drill: these exercises are designed to take stakeholders, already trained in the processes, by surprise with a level 1 or 2 exercise, to test certain aspects of the crisis management system under conditions of stress as close as possible to a real situation. For example, carrying out a coastal flooding drill at night in a coastal community in high tourist season to test the local personnel's capacity for response and mobilization in a situation of high stress.

The first drill run as part of the CRISSIS programme was a simple practice run (level 1 framework drill). It took place outside Leucate (in Paris) in March 2015 in the presence of a number of administrative managers from the town (head of the municipal police force, technical services, head of the administrative services). The aim of the exercise was simply to train the managers and help them familiarize themselves with the main procedures and systems listed in the recently adopted local response plan (PCS), working on a low-intensity crisis scenario. In practical terms, the exercise lasted three hours with the people involved divided into two units: (i) an 'operational' unit with the local authority players, the fire service, gendarmes, media, and population, responsible for sending input, depending on the selected scenario, to the (ii) crisis unit, simulating the town's control room (Poste de commandement communal [PCC]). The exercise revealed a degree of unfamiliarity with the PCS amongst the local managers – the PCS was being tested for the first time – and also provided certain lessons on its content. In fact, the PCS did not include any objective benchmarks (in metres) on the depth of water attained in flooded areas: the decisions to activate the various phases of the PCS (road closures, evacuation of a particular district, and so on) were simply based on feedback from technical staff in the field and who have worked in the municipal services for several years and are therefore well familiar with the areas. However, the lack of figures can cause difficulties for less experienced agents (for example, newly-appointed staff or someone replacing a more experienced agent absent of the day of the crisis), leading to critical delays in a crisis situation. Following these observations, the town of Leucate introduced geo-located flood markers listed in the PCS and undertook to take part in a second exercise to ensure the PCS is taken on-board by local personnel.

After this initial training exercise, a second, more ambitious drill was run in the Leucate in March 2016. This was a level 2 drill (planned drill) involving all the town's staff including elected representatives and the main stakeholders from the next level of authority: Aude department prefecture (to which Leucate reports), departmental fire service managers and the DDTM (directorate of planning), the State's decentralized department for risk management. This was a combined 'framework' and 'field' drill, with certain aspects played out 'for real' (mobilization of technical staff and the municipal police force in the field) to test the coordination of the local crisis room with the resources deployed in the field. However, it was not possible to run the full scenario in the field, due to financial and civil security constraints: for example, the civil safety (fire service) and security (gendarmerie) forces were not called up for real but improvised; similarly, actors in the drill played the role of the media (TV and press), and social media, the population affected by the crisis, and the private operators involved.

There were three goals:

1. Testing all the procedures set out in the PCS: soundness of the town's chain of command, its coordination with other stakeholders at the higher local authority levels, crisis communication – internal (within the local crisis room) and external (with stakeholders outside the town, State services, population, media), ergonomics of the local control room (is it well suited to crisis management?).
2. Checking whether the procedures are well assimilated by staff in a situation of stress simulating a major coastal flooding event. Numerous authors, such as Reason (1990), Reilly (1993), Lagadec (1993, 2012), Portal (2009) and Waard et al. (2009) have emphasized the crucial role of the human factor in the event of a crisis. For example, Reilly

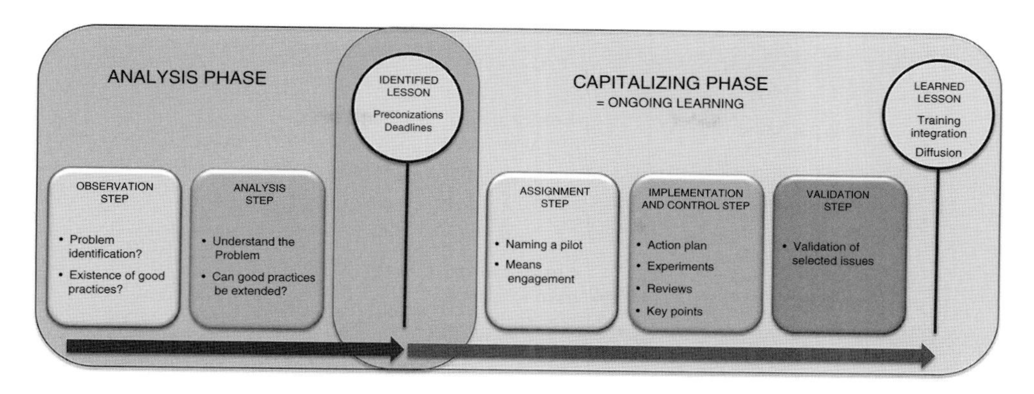

Figure 25.4 Feedback methodology.

(1993) defines the crisis as 'a hazardous situation causing widespread disruption, occurring suddenly and requiring resources that usually fall outside the operational frameworks and reference procedures used by managers'. The conventional set-up (knowledge of plans and procedures) is therefore overturned by the uncertainty and urgency of the situation, especially when events occur on a large scale.

3. Providing the town council with detailed feedback (Figure 25.4) based on systematic analysis of the actions forming part of the crisis operation (experience analysis phase), which is then taken on-board by the council to improve their crisis management set-up (capitalizing on experience phase).

25.5.2 Results and discussion

Feedback from the drill highlighted sound knowledge of the sector by municipal staff and good coordination between local control room and the personnel deployed on the ground. However, it also demonstrated that the local response plan (PCS) has not been well assimilated by stakeholders in the town, most of whom are not fully familiar with their roles and responsibilities in the crisis unit. There are therefore problems with organization and coordination within the control room. Those difficulties are amplified by the poor ergonomics of the premises: the secretarial, decision-making, logistics, and coordination functions are all found in the same room, inevitably resulting in a lot of noise and confusion. In this kind of context, a lot of stress comes from the fear of making a mistake, causing difficulties for communication within the crisis unit (no regular reviews) and with outside (communication with the prefecture services or the media). As a result, the feedback report made three main recommendations:

1. Reorganizing the local control room, based on the conventional crisis unit organizational framework, for smoother information flows and decision-making in the event of a crisis (Figure 25.5).
2. The drafting of summary job descriptions, to be included in the PCS, so that when people arrive in the crisis room, they know exactly what their role is and what they have to do. Examples of job descriptions: crisis room director, crisis room

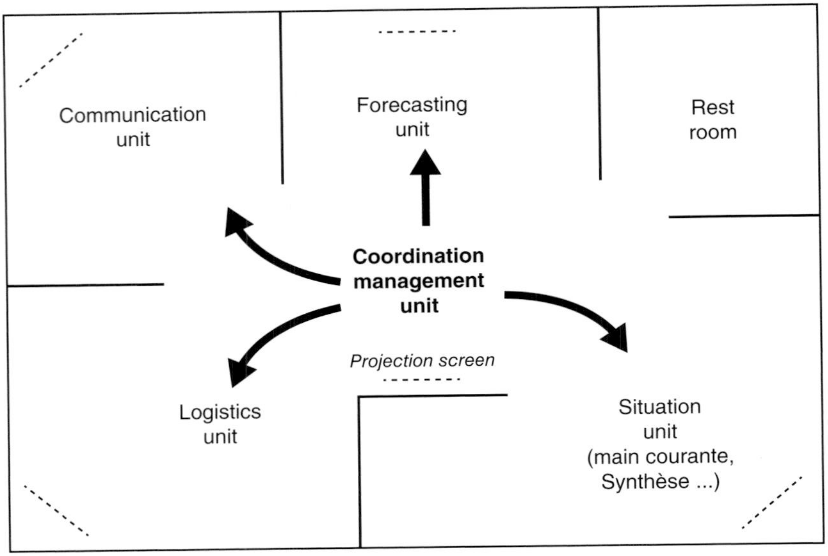

Figure 25.5 Conventional crisis unit organizational framework. Situation unit: collecting, recording and summarizing information; Logistics unit: analysing and managing resources; Communications unit: must be the only channel for communication from the crisis room with the outside world; Forecasting unit: stands apart to reflect on the situation and, at any time, may propose to the management unit an alternative to the solution currently being implemented.

coordinator, secretarial (keeping a register, managing calls), unit communications manager, forecasting unit, logistics unit.

3. The drafting of a checklist to be included at the start of the PCS and listing the basic points in crisis management: sharing the register, real-time updating of a time line and map, regular reviews and reporting the number of people affected to the next level of authority (prefecture), etc.

25.6 Conclusion

In a context of global climate change, coastal flooding events linked to storms on low-lying coasts are likely to become more frequent and more severe. At the same time, an increasing number of activities have been developing on the coast over the past few decades. Coastal regions are increasingly attractive and concentrate an ever-growing number of urban, tourist, and industrial issues. The facts are simple: a greater hazard combined with increased vulnerabilities will inevitably result in a higher number of risks in coastal regions in the years to come.

Given the challenges this brings for human societies, there are two main questions when it comes to managing exposed coastal areas in the future. First, the question of the local management strategies to be adopted: given the weight of the economic aspect, should we opt for defence and protection strategies 'at all costs' or consider a policy of 'strategic retreat', at least for the most exposed areas? Which solutions would be the most acceptable socially and economically? Second, how will we be able to adapt crisis management procedures to increasingly frequent and severe flood events? What resources should be mobilized? How do we prepare stakeholders and populations?

The world of research has a vital role to play in attempting to answer these questions. The scientific community should move towards increasingly operational research programmes run in consultation with local stakeholders. These programmes should cover the full spectrum of risks linked to coastal flooding, from detailed analysis of the hazard using digital modelling and simulation tools to the operational management of flood-related crises – an aspect that is generally neglected in research programmes – and including detailed assessment of human and material vulnerabilities and consideration of the representations and perceptions of the risk, in order to make the risk-protection and crisis-management measures more understandable and acceptable for the population. The CRISSIS research programme endeavoured to include these four key points by taking a groundbreaking, highly operational approach, run in close consultation with local stakeholders, the town of Leucate and the Aude department prefecture. Its 'hazard modelling' component highlighted the major proportions that flooding could reach in the coming decades, with a potential increase in flooded surfaces in the town of between a factor 2.47 and 3.84 depending on the IPCC sea level rise scenario applied (scenarios SLR + 0.2 m and SLR + 0.6 m). The analysis of human and material vulnerabilities identified a highly vulnerable type of dwelling (high percentage of single-storey dwellings) and considerable human vulnerability (ageing population and difficulties with evacuation in the event of flooding), which will be challenging given the increased hazard. Then, the analysis of risk perceptions and representations pinpointed the low level of perception of the coastal flooding risk amongst the population, poor understanding of the very nature of the flood risk and inadequate knowledge of the risk- and crisis-management tools (local response plan, warning systems, and so on). The crisis-management tools and procedures currently in place in the town were put to the test during two crisis drills involving local personnel to draw up an objective review of how prepared the town is for a coastal flood-related crisis and to mark out areas for improvement in the local response plan.

Broadly speaking, the lessons learned from the feedback and the ensuing recommendations have already been pointed out in several publications and reports on the drills conducted by major institutional stakeholders or companies (Lagadec 1991, 2013; Federal Office of Civil Protection and Disaster Assistance 2011; Sécurité at Stratégie 2012; OECD 2013). On the other hand, there is still a real lack of information in France on how well municipal (public) stakeholders have assimilated the crisis-management tools at their disposal, namely the local response plans (PCS), simply because too few drills are organized at local level. The actions implemented as part of the CRISSIS programme were designed to foster assimilation of the crisis management procedures and systems by local stakeholders. This operational research concept, which underscored the entire CRISSIS programme, appears essential if we are to anticipate the inevitable increase in risks in coastal regions. Prevention and mitigation measures (defence structure, remodelling buildings, land use zoning according to the level of risk exposure, awareness raising amongst the population) and planning (developing safeguarding plans, warning and evacuation procedures, coordinating the operational response of various stakeholders, and so on) are not enough. Like musicians who practise their scales or athletes who train to get better, the entities responsible for protecting coastal populations must also run regular simulation exercises to practise their crisis management arrangements. These drills are crucial for testing and reviewing conduct and crisis management procedures and systems, and appraising the suitability of the existing local prevention tools (flood risk prevention plan or PPRI). If they are run regularly, they will help better anticipate risks and handle crises and, ultimately, identify areas for improvement for dealing with the unexpected whilst improving decision making and remaining compatible with local development. The research community can play an important role in raising awareness about this need. The development of 'action research' programmes that do not merely put forward solutions but also test them out in cooperation with the entities responsible for managing coastal regions, is fully in line with this.

References

Anselme, B., Durand, P., Thomas, Y.F. et al. (2011). Storm extreme levels and coastal flood hazards. A parametric approach on the French coast of languedoc (district of Leucate). *Comptes Rendus Geosciences* 343 (10): 677–690.

Bunya, S., Dietrich, J.C., Westerink, J.J. et al. (2010). A high-resolution coupled riverine flow, tide, wind, wind wave and storm surge model for Southern Louisiana and Mississippi. Part I: Model development and validation. *Monthly Weather Review* 18 (2): 345–377.

Direction générale de la Sécurité civile et de la gestion des crises (DGSCGC) (2008). Plan Communal de Sauvegarde (P.C.S.) - LES EXERCICES: 'S'entraîner pour être prêt'. 89 p Paris: DGSCGC.

Federal Office of Civil Protection and Disaster Assistance, (2011). *Guideline for Strategic Crisis Management Exercises, Federal Office of Civil Protection and Disaster Assistance*. Bonn: Federal Office of Civil Protection and Disaster Assistance, 48p.

Flanquart H. (2014). Perception versus représentation du risque de submersion et autres risques: ce que révèle une querelle sémantique. *Connaissance et compréhension des risques côtiers: aléas, enjeux, représentations, gestion, actes de colloque*, 3–4 July 2014, Université de Bretagne occidentale, 353-364.

Ministère de l'Intérieur, de l'Outre mer et des Collectivités territoriales (French Interior Ministry) (2011). *Exercices de crise de sécurité civile. Guide méthodologique sur les exercices cadre et terrain*. Ivry sur Seine: Ed. SCEI, 68 p.

Gallien, T.W. (2016). Validated coastal flood modeling at Imperial Beach. California: Comparing total water level, empirical and numerical overtopping methodologies. *Coastal Engineering* 111: 95–104.

Gallien, T.W., Sanders, B.F., and Flick, R.E. (2014). Urban coastal flood prediction: Integrating wave overtopping, flood defenses and drainage. *Coastal Engineering* 91: 18–28.

Garcin M., Nicolae Lerma A., Pedreros R. et al. (2012). *Evaluation de la submersion marine maximale à l'échelle régionale*. Rapport final. BRGM/RP-62259-FR, p.97, fig.40, tabl.9, ann4 Pessac BRGM.

Goutx, D. and Ladreyt, S. (2001). *Hydraulique des Cours D'eau, La théorie et sa mise en pratique. [On hydraulic water courses, theory and practice]*, 1–51. Plouzané: Centre d'Etudes Techniques Maritimes Et Fluviales (CETMF).

Haut Comité Français pour la Défense Civile (HCFDC) (2012). *Risques et menaces exceptionnels. Quelle préparation?*, Rapport d'activité 2011, 139p. Paris: HCFDC.

Hellequin, A.-P., Flanquart, H., Meur-Ferec, C. et al. (2013). Perceptions du risque de submersion marine par la population du littoral languedocien: contribution à l'analyse de la vulnérabilité côtière. *Natures Sciences Sociétés* 21 (4): 385–399.

Hinkel, J., Lincke, D., Vafeidis, A.T. et al. (2014). Coastal flood damage and adaptation costs under 21st century sea-level rise. *Proceedings of the National Academy of Sciences* 111: 3292–3297.

Johnston, D., Tarrant, R., Tipler, K. et al. (2011). Preparing schools for future earthquakes in New Zealand: Lessons from an evaluation of a Wellington school exercise. *Australian Journal of Emergency Management* 26 (1): 24–30.

Krien, N. (2014). *Place des risques côtiers dans la représentation du cadre de vie d'individus possédant des enjeux sur des communes ' à risque'. Thesis, Université de Bretagne occidentale* V1, 237p.

Lagadec, P. (1991). *La gestion des crises. Outils de réflexion à l'usage des décideurs.* Maidenhead: McGraw Hill, 296 p.

Lagadec, P. (1993). *Preventing Chaos in a Crisis: Strategies for prevention, control and damage limitation.* Maidenhead: McGraw-Hill, 367 p.

Lagadec, P. (2012). Du risque majeur aux mégachocs. *Dossier spécial: la gestion de crise, méthodologies et retour d'expérience, Sécurité et stratégie* 10: 50–52.

Lagadec, P. (2013). *Navigating the Unknown. A practical lifeline for decision-makers in the dark*, Crisis Response Journal, 22 p.

Le Roy, S., Pedreros, R., and Nicolae Lerma, A. (2013). *Modélisation de la submersion marine en milieu urbain.* Rapport final. BRGM/RP-63477-FR, 84 p., 48 ill., 3 tabl Pessac: BRGM.

Le Roy, S., Pedreros, R., André, C. et al. (2015). Coastal flooding of urban areas by overtopping: dynamic modelling application to the Johanna storm (2008) in Gâvres (France). *Natural Hazards and Earth System Sciences* 15: 2497–2510.

Marzocchi, W., Newhall, C., and Woo, G. (2012). The scientific management of volcanic crises. *Journal of Volcanology and Geothermal Research* 247–248: 181–189.

Morin, J. (2012). *Gestion institutionnelle et réponse des populations face aux crises volcaniques: études de cas à la Réunion et en Grande Comore*, PhD thesis, Université de la Réunion, 472 p.

Nicolae Lerma, A., Bulteau, T., Elineau, S. et al. (2017). High-resolution marine flood modelling with coupled overflow and overtopping processes: framing the hazard based on historical and statistical approaches. *Natural Hazards and Earth System Sciences. Discussion* https://doi.org/10.5194/nhess-2017-147.

OECD (2013). *Strategic Crisis Management.* Paris: OECD, 26 p.

Pierson T.C., Driedger C.L., and Tilling R.I. (2013). *Volcano crisis response at Yellowstone volcanic complex.* After-action report for exercise held at Salt Lake City, Utah, November 15, 2011: U.S. Geological Survey Open-File Report 2013–1018, 31 p. http://pubs.usgs.gov/of/2013/1018/ (accessed 13 March 2019).

Portal, T. (2009). *Crises et facteurs humains. Les nouvelles frontières mentales des crises.* De Boeck, coll. Crisis, 272p.

Reason, J. (1990). *Human Error.* New York: Cambridge University Press, 320p.

Reilly, A.H. (1993). Preparing for the worst: the process of effective crisis management,. *Industrial Environmental Crisis Quarterly* 7 (2): 115–143.

Sécurité et Stratégie (2012). *Dossier Spécial: La Gestion des Crises, Méthodologies et Retours d'Expérience*, Sécurité et Stratégie, No 10. Paris: Le Club des Directeurs de Sécurité des Entreprises, 92p.

Stansby, P., Chini, N., Apsley, D. et al. (2013). An integrated model system for coastal flood prediction with a case history for Walcott, UK, on 9 November 2007. *Journal of Flood Risk Management* 6: 229–252. https://doi.org/10.1111/jfr3.12001.

de Waard, D., Godthelp, H., Kooi, F. et al. (eds.) (2009). *Human Factors, Security and Safety.* Maastricht, the Netherlands: Shaker Publishing, 437 p.

26

Lessons Learnt from Coastal Risks Governance on Reunion Island, Indian Ocean, France

Virginie K.E. Duvat[1] and Alexandre K. Magnan[2]
[1] *Laboratory CNRS 7266 LIENSs, La Rochelle University, La Rochelle, France*
[2] *Institute for Sustainable Development and International Relations, Sciences Po, Paris, France*

26.1 Introduction

It is widely acknowledged that tropical small islands are at risk of being severely affected by the impacts of extreme climate events, especially tropical and extra-tropical cyclones (Mimura et al. 2007; Hay and Mimura 2010; Hoeke et al. 2013; Nurse et al. 2014). Small islands indeed have unique characteristics that make them especially vulnerable to the effects of such events. First, due to their location in the open ocean and small size, an extreme event can affect much of the island, thus making the magnitude of losses and recovery demands extremely high relative to Gross Domestic Product (GDP) and public financial resources (Pelling and Uitto 2001; Mechler 2004; Ferdinand et al. 2012). This also explains why tropical islands rank third after Asia and North America with respect to the population numbers exposed to tropical cyclones (Peduzzi et al. 2011). Second, the fact that such islands generally specialize in agriculture and tourism also exacerbates the adverse impacts of tropical cyclones, especially on their economies, as these sectors are highly sensitive to the destructive effects of strong winds, heavy rainfall, and storm surges (Méheux and Parker 2006; Angelucci and Conforti 2010; Strobl 2012).

Tropical cyclone (TC) impacts on small islands can be divided into four main categories that are common to both low-lying atoll islands and high mountainous islands (Nurse et al. 2014). These categories are on the whole well-documented for high mountainous islands, which are the main focus of this chapter. The first category involves impacts on island morphology including (i) coastal erosion and accretion, with erosional

Facing Hydrometeorological Extreme Events: A Governance Issue, First Edition.
Edited by Isabelle La Jeunesse and Corinne Larrue.

effects predominating on some island coasts, whilst others are subject to cyclone-driven sediment deposition (Hubbard et al. 1991; Scoffin 1993; Cazes-Duvat 2005; Scheffers and Scheffers 2006; Caron 2011; Etienne 2012); (ii) flooding and marine inundation (De Scally 2008, 2014; Etienne 2012); and (iii) significant changes in island morphology resulting from river action and landslides (Terry, Garimella, and Kostaschuk 2002; Etienne 2012). Second, there are impacts on ecosystems and natural resources including (i) coral reef mortality, as the result of mechanical destruction and river runoff, as well as a decrease in coral recruitment (Bythell, Bythell, and Gladfelter 1993; Crabbe et al. 2008; Fletcher et al. 2008; Mallela and Crabbe 2009; Scopélitis et al. 2009); (ii) damage to mangroves and wetlands (Cahoon et al 2003; Park et al. 2009); and (iii) soil salination from inundation and saline intrusion into freshwater lenses (Angelucci and Conforti 2010; Terry and Falkland 2010). A third category involves impacts on island livelihoods both in terms of damage to subsistence crops and fish production (Richmond and Sovacool 2012), and losses in commercial agriculture, tourism, and aquaculture (OECS 2004; Strobl 2012). Fourth, impacts on settlements and infrastructure include the destruction of buildings and houses (Etienne 2012; Ferdinand et al. 2012), and damage to public and transport facilities and to health infrastructure (Dorville and Zahibo 2010; Richmond and Sovacool 2012). Existing studies emphasize the high spatial variability of TC impacts both between and within islands (Cazes-Duvat 2005; Hay and Mimura 2010; Etienne 2012).

Previous studies however provide limited information on two major issues. The first knowledge gap relates to the impacts, that is to say, first to their social and spatial distribution, and second, to the 'chain' or 'cascade of impacts' generated by these events (i.e. impacts of second, third, fourth order, and so on) at different timescales on the various components of island societies. Since most studies adopt a disciplinary focus, little insight is offered into the interrelated processes driving impacts, in particular on the contribution of long-term coastal-development and risk-management policies to the generation or exacerbation of impacts. The second under-researched issue relates to the nature and timelines of short-term (less than one year) responses of stakeholders to impacts, and to the implications of these responses on changes in vulnerability. However, understanding the extent and timelines of recovery, and the processes driving it, are crucial to apprehend the effects of a given climate-related event on future societal vulnerability (Cardona et al. 2012).

In order to address these two knowledge gaps, this chapter investigates both the impacts of and the stakeholders' responses to TC Bejisa (Saffir-Simpson category 3), which, in January 2014, affected Saint-Paul (106 193 inhabitants), the second most populated municipality of Reunion Island, stretching 30 km on the western coral coast of the island. Reunion Island is one of the five French overseas departments, together with three other tropical islands – Martinique and Guadeloupe in the Caribbean Sea, and Mayotte in the western Indian Ocean – and the continental French Guyana. As other French overseas departments, Reunion Island is an outermost region of the European Union. Its specific territorial status, added to its island character and location in one of the regions of the intertropical zone that are the most exposed to TCs, therefore provide useful insights on both the impacts of extreme climate events and coastal risks governance issues in European overseas territories, on which a limited number of studies have up to now focused.

This chapter addresses three interrelated questions: (i) To what extent did the impacts of TC Bejisa vary within the study area? (ii) What do they reveal on the shortfalls of coastal risks management policies? (iii) What recommendations can be formulated to both reduce current coastal risks and promote adaptation to climate change?

Section 26.2 presents the context of the study, its geographical setting, and the characteristics of TC Bejisa. Section 26.3 details the impacts of TC Bejisa on the natural and built environment, and the responses brought by public authorities and coastal residents over a one-year period following the cyclone. Section 26.4 discusses the key findings of the study on coastal risks governance issues, including the challenges for adaptation to climate change.

26.2 Context of the study

26.2.1 Geographical setting

Reunion Island is volcanic in origin and has a land area of 2512 km^2, with its highest point at 3070 m above sea level. It is formed by two coalescent volcanoes, namely the inactive volcano Piton des Neiges (3070 m) that covers around two-thirds of the island, and the active volcano Piton de la Fournaise (2412 m) that flanks the former. The 240 km-long coastline shows a variety of morphological features, including three main river deltas, high and low rocky coasts, as well as volcanic and coral beach-dune systems, most of which are bordered with fringing reefs (Figure 26.1). The population of Reunion Island totalled 843 617 inhabitants in 2012, most of whom live to the west of a north–south axis stretching between the two main cities, Saint-Denis in the north (146 763 inhabitants in 2012) and Saint-Pierre in the south (81 769 inhabitants in 2012). The population is mainly concentrated in the coastal zone, with the western municipality of Saint-Paul – the focus of this study – being the second most populated city (106 193 inhabitants) and the first area for coastal tourism.

The coastline of Saint-Paul municipality stretches 30.7 km, 20.2 km of which exhibit either alluvial features, mainly volcanic sand dunes bordering sand and pebble beaches in the north, or coral coasts made of beach-dune systems fringed with embryonic to fringing coral reefs south of Cap La Houssaye (Figures 26.1 and 26.2). The remaining 10.5 km are rocky coasts, including the 20 m-high volcanic cliffs at Cap La Houssaye. The alluvial and coral coasts can be divided into six sediment cells separated by rocky points or capes, or jetties (Figure 26.2). To the north, Saint-Paul Bay (cell 1) corresponds to the southern end of the alluvial delta of Galets River, a nearly 200 m-thick accumulation of volcanic blocks, pebbles, and sand (Cazes-Duvat and Paskoff 2004). On its coastal fringe, this delta exhibits pebble beaches in the north and sandy beaches in the south. In the central part of Saint-Paul Bay, the southward sediment drift has led to the formation of a pebble bar that obstructs the mouth of Bernica River and isolates a swamp. South of the mouth of Bernica River, beaches are composed of a mixture of pebble and sand, and backed with sand dunes that reach up to 6 m above sea level. To the south of Cap La Houssaye, sediment cells 2–6 constitute the northern and central parts of Reunion Island's coral reef system that stretches 25 km from this cape to Grand Anse in the south. In the study area, coral reefs widen from north to south, cells 2–5 having 60- to 120 m-wide embryonic and platform reefs, whilst cell 6 exhibits a 500 m-wide fringing reef. Cells 5 and 6 are connected to the open sea by Saint-Gilles and Ermitage passes, respectively. The beach-dune systems and coastal plains widen accordingly, reaching up to 1 km in the Ermitage-La Saline area.

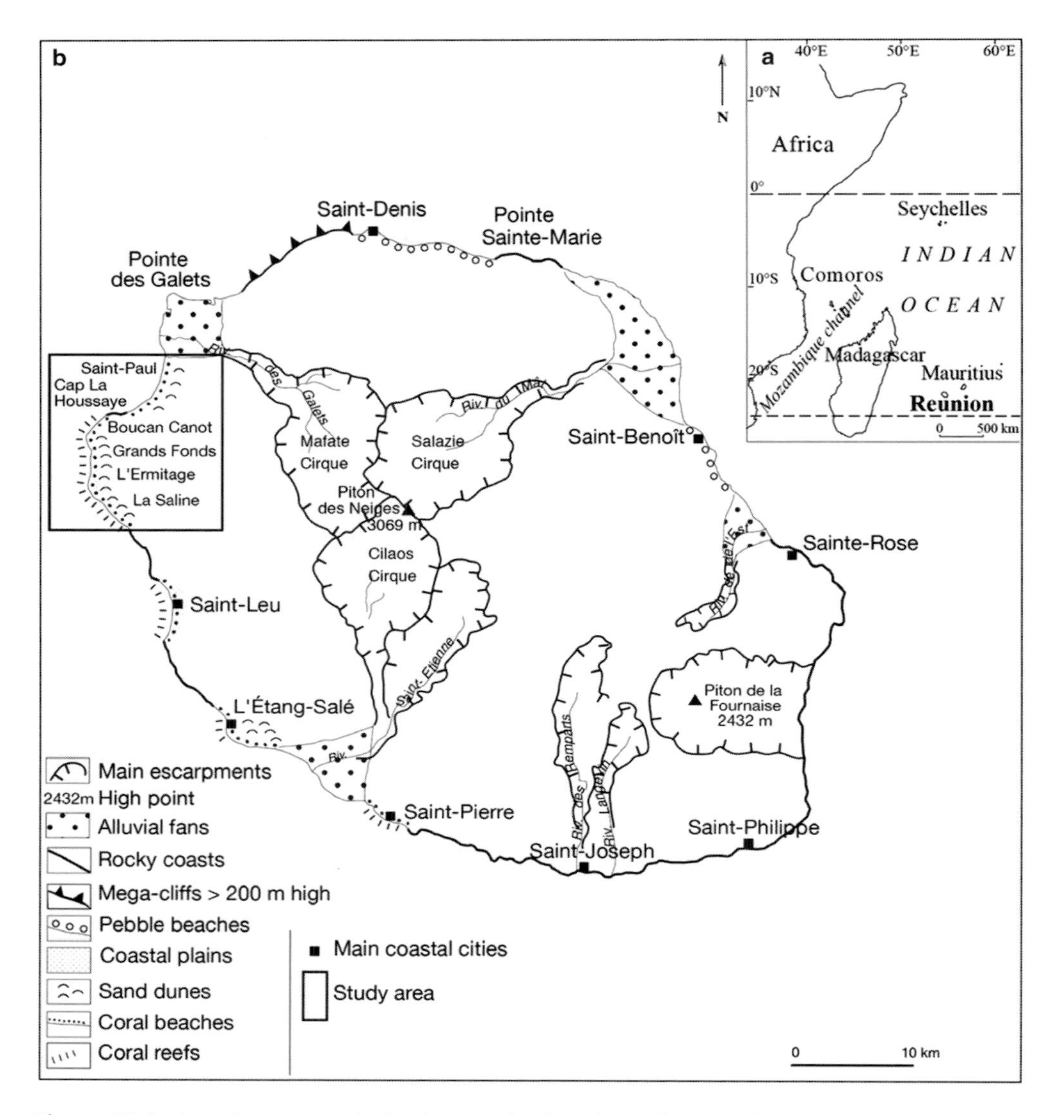

Figure 26.1 Location map and sketch map showing the main coastal morphological features of Reunion Island (Duvat et al. 2016) (a) This location map shows the tropical position of Reunion Island north of the Tropic of Capricorn. (b) This sketch map shows the location of the study area and its coastal morphology, i.e. the volcanic beach-dune system of Saint-Paul Bay in the north and the coral beach-dune system that stretches from Boucan Canot to La Saline in the south.

Our assessment of multi-decadal shoreline change (Duvat and Salmon 2015) shows that over the 1950–2011 timeframe, the 20.2 km-long alluvial and beach-dune coastline of the Saint-Paul municipality exhibited stability along 54.5% of its length, whilst erosion and accretion occurred along 39.6 and 5.9% of its length respectively. During this period, most volcanic and coral beaches underwent erosion, with maximal values of coastline retreat reaching up to 38 m at Saint-Gilles. Nonetheless, both the trends and the values

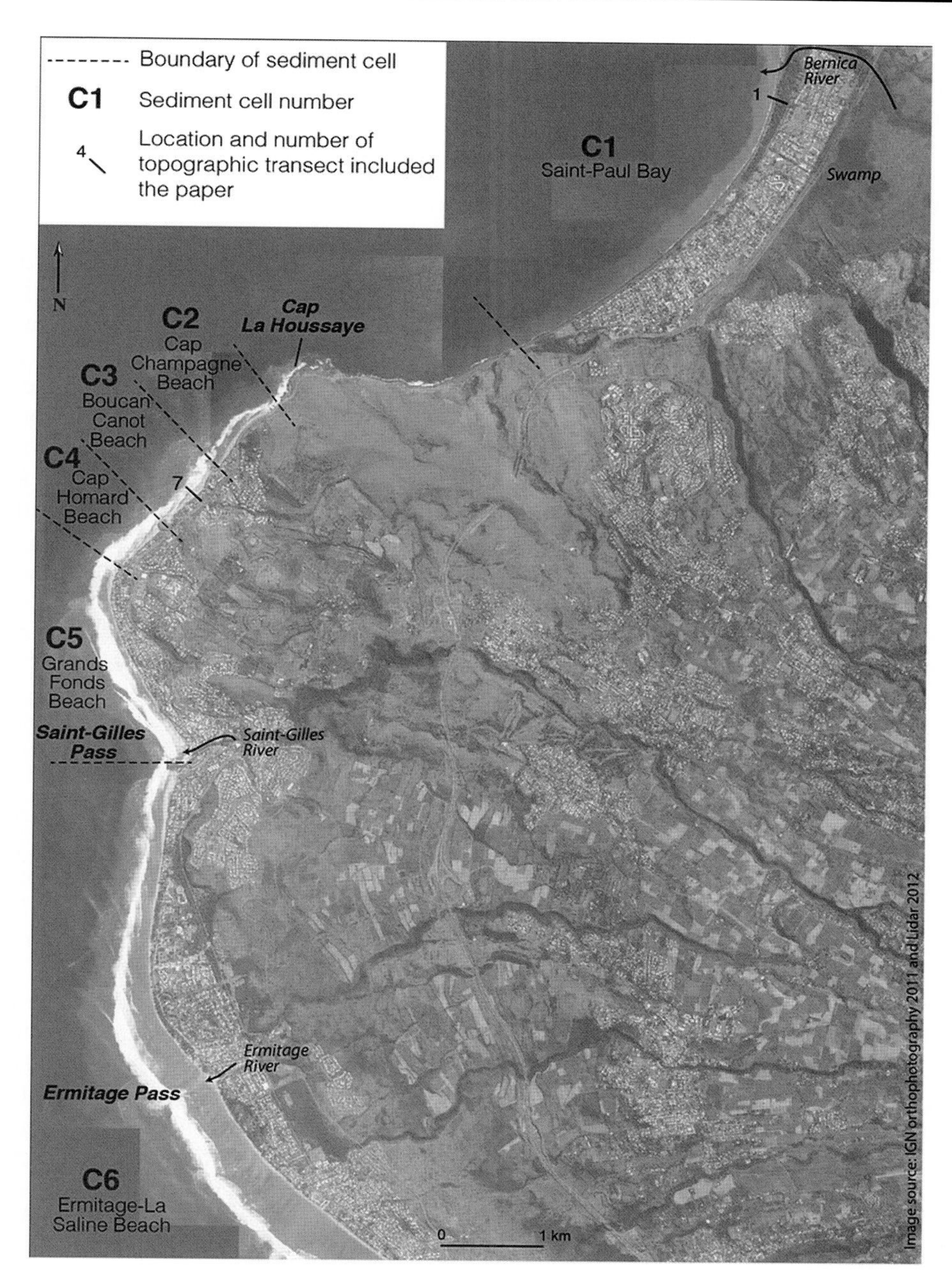

Figure 26.2 Detailed map of the study area. This figure shows (i) the limits of the six sediment cells considered in this study, (ii) the precise location of the settings mentioned in the text, and (iii) the location of the topographic transects shown on Figures 26.4 (transect 1) and 26.5 (transect 7) (Duvat et al., 2016).

observed show considerable variation due to differences in coast exposure, reef width and the impact of anthropogenic disturbances along the coastline. Noteworthy, the establishment of the Saint-Gilles marina in the mouth of Saint-Gilles River in 1989 has destabilized the coast by causing accelerated accretion (+50 m) and erosion (−38 m) up- and down-drift the marina, respectively.

As a result of the development of permanent residences and coastal tourism, the Saint-Paul municipality has the highest number of coastal buildings, with 1492 private buildings (mostly residences and seaside hotels) and 56 public buildings located within a 100 m-wide coastal strip extending inland from the vegetation line, including 219 buildings located in the first 20 m. Most of these buildings have a low elevation, with 1174 of them built at less than 6 m above sea level. Additionally, 23.1 km of roads are situated in this 100 m-wide coastal strip, exposing key transport facilities to storm impacts. Before the occurrence of TC Bejisa, 24% (7.5 km out of 30.7 km) of the coastline length of Saint-Paul municipality was equipped with coastal structures (Duvat and Salmon 2015). Regarding coastal structures, the specificity of this municipality was and still is the high number (153 in total) and small size (average length < 50 m) of structures. Most of them (142) are vertical cemented seawalls, two-thirds of which are less than 2 m high and were built by coastal residents to protect their property against trespassing or coastal hazards. Only one-third of the equipped coastline has proper engineering structures, built and maintained by the public authorities in order to protect highly exposed public infrastructure such as coastal roads from wave attack or coastline retreat.

26.2.2 Characteristics of TC Bejisa

Around 60 cyclones have affected the island over the last two centuries, with a great variability on a decadal scale: up to eight cyclones were recorded in the 1870s and seven in the 2000s. Most of the tropical cyclones affecting Reunion Island form in the north-east at between 10 and 15°S and 60–75°E, and thereafter move south-westwards before petering out. As a result, the eastern and northern coasts of the island are the most exposed to these phenomena (Figure 26.3). The more sheltered western coast of the island is rarely affected by tropical cyclones, although it can be hit first, by TCs forming in this generation area and tracking a long distance south-westwards (e.g. TC Dina, 18–25 January 2002; TC Gamède, 23–28 February 2007; TC Dumile, 2–4 January 2013), and second, by TCs forming to the north-west (between 50 and 55°E) of the generation area and moving southwards. TC Bejisa, which exhibits an atypical north–south track, belongs in fact to this last category (Figure 26.3), and has been the most significant cyclone to hit the western side of the island since category 3 TC Gamède in 2007.

TC Bejisa (category 3) was the fourth depressional system of the 2013–2014 tropical cyclone season out of a total of 15 tropical cyclones and was the only one to affect Reunion Island during this period.

On 29 December 2013, this tropical storm reached the Intense Tropical Cyclone level and was named TC Bejisa. On 31 December Reunion Island was placed under a cyclone early warning, when TC Bejisa was located 800 km to the north-west of the island. On Wednesday 1 January 2014 at 3:00 p.m., the second level of alert (orange) came into effect when the cyclone was 400 km offshore and moving towards the south-east. The first weather damage caused by the cyclone occurred during the night of 1–2 January as

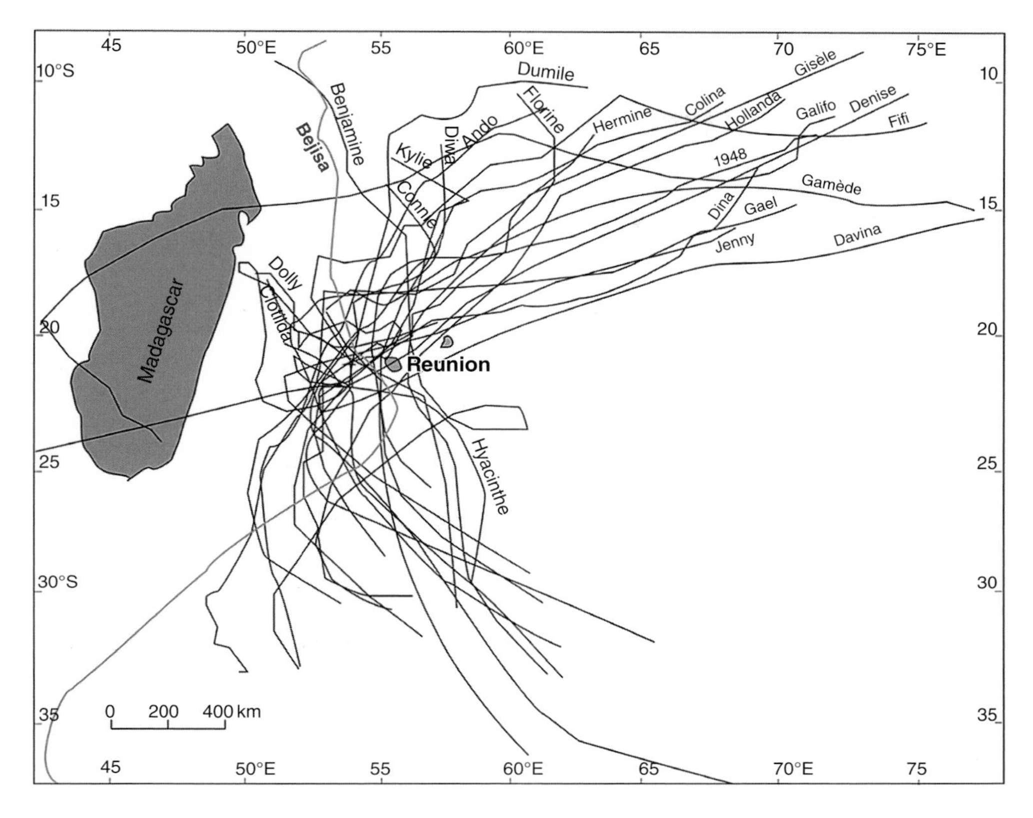

Figure 26.3 Tropical cyclone tracks near Reunion Island (1948–2014) *Data Source*: Météo France, Reunion Island. This figure highlights the atypical north–south track of Tropical Cyclone Bejisa that hit the western coast of the island in January 2014 (adapted from Duvat et al., 2016).

wind gusts reached 100 km/h in areas exposed to the north-east. On Thursday, 2 January at 10 a.m., the third level of alert (red) came into effect when the cyclone eye was located 160 km north-west of Reunion Island. The cyclonic conditions set in at midday, with the wind blowing first from the north-east, then the north, and thereafter from the north-west. The cyclone passed the coast, and then tracked to the south-west and south. Maximum wind speeds recorded near the coast were 114 km/h at Sainte-Marie airport, 140 km/h at Saint-Pierre and 150 km/h along the western shore in the afternoon of 2 January when the pressure was the lowest (950 hPa). Heavy rains affected the whole island, with significant accumulations in the mountains (over 1005 mm in 48 hours at Cilaos). On 2 January Bejisa reached the status of tropical cyclone as it passed 50 km offshore, with the eye wall (where the winds are the most violent) passing only 10 km from the Pointe des Trois Bassins on the west coast. Reunion Island experienced these cyclonic conditions throughout the night, as Bejisa slowly moved away to the south. Strong winds persisted but the rainfall eased off during the second part of the night. On Friday 3 January at 9 a.m., the red alert was lifted. After impacting the west coast of Reunion Island, TC Bejisa rapidly weakened, tracking to the south-west before being downgraded on 7 January Two tide gauges installed at Pointe des Galets and

Sainte-Marie-de-la-Réunion gave real-time measurements of the sea level. At Pointe des Galets in the north-west, the water height measured was 1.58 m and the storm surge 0.71 m. Unfortunately, wave setup cannot be estimated from measurements made by tide gauges that are located in a sheltered deep-water harbour.

The track direction and position of TC Bejisa meant that, as the cyclone approached, winds and swells were directed onshore along the north-west coast of the island, which was therefore the most severely affected area on the island scale. This context presented a unique opportunity to assess the impacts of a TC on the densely populated and highly urbanized coral coast of Reunion Island.

26.2.3 Material and methods

Along Saint-Paul coastline, the waves generated by TC Bejisa had three major impacts: marine inundation; erosion or accretion, depending on the location; and coastal vegetation destruction. These impacts were assessed based on post-cyclone fieldwork conducted between 11 and 15 January 2014, i.e. 8–12 days after the occurrence of the cyclone.

26.2.3.1 Assessment of marine inundation Different methods were used in non-built and built areas. In the former, TC Bejisa wreck lines were directly collected in the field using a GPS. In built areas, this limit was obtained through 55 interviews that were completed with coastal residents over the year following the event. The interviews stretched out over a long period given that first, secondary residents are absent from their seaside homes most of the year, and second, severely affected primary residents were initially unfriendly to researchers. However, the data gathered permitted mapping marine inundation in three main areas (Saint-Paul Bay, Boucan Canot Beach, and Ermitage Beach). Two complementary indices were then calculated, the distance between the inner limit of marine inundation and the pre-cyclone vegetation line, and the maximal elevation of inundated coastal areas.

26.2.3.2 Assessment of TC Bejisa's impacts on beach-dune systems The impacts of TC Bejisa on beach-dune systems were assessed on the basis of field observations and topographic measurements completed prior to and after TC Bejisa (Duvat et al. 2016). Twenty-six pre- and post-Bejisa beach-dune transects were thus completed along the 20.2 km-long beach-dune coastline of Saint-Paul. Both the existing literature and the interviews conducted with coastal residents (see later) clearly indicate that TC Bejisa was the most significant cyclonic event that drove morphological change in the study area after TC Gamède (2007).

26.2.3.3 Assessment of TC Bejisa's impacts on the coastal vegetation The destructive impacts of cyclonic winds and waves on the coastal vegetation were geolocated and analysed based on qualitative parameters, including vegetation type (herbaceous/shrubby/wooded, with details on affected species) and origin (indigenous vs. introduced).

26.2.3.4 Assessment of the impacts of TC Bejisa on the built environment We assessed the impacts of the cyclone on private properties (houses, gardens), economic activities (restaurants, resorts, dive shops, and so on), and public facilities (lifeguard stations, landing stages, training centres, coastal roads and the Saint-Gilles marina). In total, 77 assets were considered (55 private properties, 16 economic activities, 6 public facilities) along 9.1 km of coastline (i.e. 45% of the total coastline length surveyed in this study). Fieldwork consisted of an expert judgement on the level of damage to fences and buildings using a four-level grid: 'no impact' where no sign of impact was visible on the property; 'low impact' where a fence was slightly damaged, often due to the impacts of small coral blocks (<5 cm-diameter) thrown by the waves; 'moderate impact' where a fence was partly destroyed by medium-sized coral blocks (5–15 cm in diameter) and/or the sea-side of a building was slightly damaged; 'high impact' where a fence was mostly destroyed and/or the sea-side of a building was seriously damaged (especially due to medium-sized coral blocks). In complement to direct observations of the damages, interviews were conducted with coastal residents and professionals affected by the cyclone (24 and 13 interviews respectively), as well as with key public authorities (7) involved in crisis management (i.e. the risk officer of Saint-Paul municipality; the Saint-Gilles harbour captain; lifeguards; and representatives of the Région Réunion). These interviews provided complementary information on damages.

We also assessed TC Bejisa's impacts on coastal protection structures and the post-cyclone reconstruction of these structures, thanks to the pre-storm data that we had generated in 2013. For each structure, these data provide information on its type (e.g. seawall, riprap, groyne, and so on), dimensions (i.e. length, height above ground level), constituent materials, status (very good/good/medium/bad), and on the protected assets. All coastal protection structures were therefore assessed in January 2014 (one week after the cyclone), and then reassessed in April 2014 (three weeks after the event) and January 2015 (one year after the event) in order to assess the post-cyclone repairs to impacted structures over the year following the cyclone. These reassessments also highlighted changes in the protection strategies used by some of the affected residents, as well as post-storm interventions carried out by non-affected residents who had become aware of the real threat posed by TCs to human assets.

26.2.3.5 Assessment of coastal residents' responses to TC Bejisa's impacts The responses of coastal residents to TC Bejisa's impacts were investigated based on two complementary approaches. First, we conducted post-cyclone field surveys focused on the technical interventions implemented by residents on the ground, both on their own property and on nearby beaches in January 2014, April 2014, and January 2015, so as to be able to highlight the timelines of post-cyclone reconstruction over a one-year period. Second, semi-structured interviews were conducted between January and March 2015 with 41 'first-line' residents living in the most affected area of Saint-Paul municipality, i.e. stretching from Saint-Paul Bay to Ermitage Beach, so as to complete our understanding of the strategies deployed by residents following TC Bejisa. These residents all have their houses directly established on the beach crest or foredune less than 6.2 m in elevation, implying that they were all liable to flooding if considering ground elevation. Additionally, 38 of the surveyed persons were permanent residents, and 35 were owners against 6 who were tenants. This means that most respondents were in a favourable position to take action after the event. Thirty-two interviewees out of 41 were present when TC Bejisa hit the western coast of the island, and only five of them evacuated with their family between 31 December and 2 January to reach a more secure place, generally at family members. Most respondents thus personally lived this cyclonic episode. Before TC Bejisa, 34 plots out of 41 were protected from storm impacts by coastal protection structures, generally small to medium-size (<2 m) seawalls, and by coastal vegetation.

26.2.3.6 Conduction of a population survey amongst coastal residents Lastly, we also conducted in 2015 a population survey on sea-related risk perception amongst coastal residents of four major coastal municipalities, Saint-Denis, Le Port, Saint-Paul, and Saint-Pierre, in order to gain a better knowledge and understanding of public awareness on coastal risks, and therefore be able to better apprehend and explain residents' responses to major sea-related events, including tropical cyclones. The sample consisted of 479 coastal residents, 85% being between 20 and 60 years old and 8% >60 years old. Sixty-nine percent of interviewees were living less than 500 m from the sea and 31% less than 100 m, with more than 60% of them having already experienced at least one powerful cyclone in their life.

The questionnaire was made of 57 questions dealing notably with the interviewee's general knowledge on coastal hazards (mainly cyclones and distant-source swells); personal experience of previous events (including actions undertaken before/during/after the event); views on coastal defences (efficiency and desirability) and risk-management strategies (including the related responsibilities for implementation); sense of safety regarding living at the coast (including a question on the possible desire to move away); and perception of the climate change-induced risks (including three time horizons going from ~2025 to ~2050 and ~2080).

26.3 Impacts of TC Bejisa and post-cyclone stakeholders' responses

26.3.1 Impacts

26.3.1.1 Impacts on the natural environment The spatial distribution and intensity of impacts varied significantly along the coastline, in accordance with both the decreasing intensity of swells southward and the distribution of human assets along the coast.

26.3.1.1.1 Marine inundation The waves generated by TC Bejisa caused widespread inundation in the 100 m-wide coastal strip, as they reached elevations of nearly 7 m above sea level and propagated up to 71 m inland from the vegetation line (Duvat et al. 2016). Due to the origin and direction of cyclonic waves, the elevation and distance reached were the greatest in the north (Saint-Paul Bay) and the lowest in the south (La Saline area). In fact, 23 out of the 41 residents whom we interviewed in the most impacted area (i.e. northern part of the study area) had their properties flooded by TC Bejisa's waves. Beyond this general observation, the extent of marine inundation was largely influenced by human occupation. In fact, the cyclonic waves propagated over greater distances inland where human-made obstacles such as seaside houses and coastal protection structures were discontinuous, as in Saint-Paul Bay – where the interruption of the built seafront by access roads, slipways, pontoons, and open public areas, allowed the cyclonic waves to reach the coastal road – and where they met no obstacle, as illustrated by the non-built sand dune areas of Ermitage area. Moreover, buildings and coastal defences did not always stop wave intrusion inland. As a result, a significant number of human assets, including private properties and houses, seaside resorts and restaurants, sport centres, coastal roads, and even a cemetery, were affected by marine inundation along the study coastline.

26.3.1.1.2 Morphological impacts Following TC Bejisa, 17 out the 26 beach transects were erosional, with sediment loss ranging from −0.3 to −58.3 m^3/m of coastline, whilst 9 beach transects exhibited sediment gain, with values ranging from +0.3 to +18.7 m^3 (Duvat et al. 2016). In accordance with the intensity of TC Bejisa's swells decreasing southwards, erosional patterns prevailed in the northern part of the study area, from Saint-Paul Bay to Pointe des Aigrettes, whilst accretional patterns predominated southwards from this point (Table 26.1). Saint-Paul Bay exhibited severe beach and dune erosion, the retreat of the dune face and of the base of the beach reaching up to 4 and

Table 26.1 Impacts of TC Bejisa on beach-dune systems along the Saint-Paul coastline. These data are based on both the analysis of multi-decadal shoreline change and the comparison of pre- and post-storm topographic transects. No transect was collected in cell 2. Concerning total sediment loss and total sediment gain, values are provided in m^3 for 1 m of coastline. Concerning sediment loss/m and sediment gain/m, values are provided in m^3/m of transect. Calculating this average value allows comparison between transects which have different lengths.

Cell and length	Morphology and multi-decadal shoreline change in m/y (1950–2011)	Morphological impact of TC Bejisa + net change in beach volume (m^3)
C1. Saint-Paul Bay (4010 m) Galets River delta (alluvial)	Volcanic beach/dune system Erosional (−0.6 m < × ≤ −0.2 m)	Erosional Total sediment loss: −58.3 < × ≤ −34.2 Sediment loss/m of transect: −1.23 < × ≤ −0.76
C3. Boucan Canot Beach (710 m) Sandy beach, embryo reef	Coral beach Stable in its northern and central parts (−0.2 m < × ≤ +0.2 m), erosional (−0.6 m < × ≤ −0.2 m) at its southern end	Mainly erosional Total sediment loss: −29.1 < × ≤ −11.7 Sediment loss/m of transect: −0.15 < × ≤ −0.75
C4. Cap Homard Beach (525 m) Sandy beach, embryo reef	Coral beach/dune system Erosional (−0.6 m < × ≤ −0.2 m)	Erosional Total sediment loss: −15.1 < × ≤ −1.4 Sediment loss/m of transect: −0.47 < × ≤ −0.05
C5. Grands Fonds Beach (2300 m) Sandy beach, fringing reef	Coral beach/dune system Stable (−0.2 m < × ≤ +0.2 m)	Variable, either erosional or accretional Total sediment loss/gain: −18.3 < × ≤ +0.4 Sediment loss/gain/m of transect: −0.45 < × ≤ +0.02
C6. Ermitage-La Samine Beach (2300 m) Sandy beach, fringing reef	Coral beach/dune system Stable (−0.2 m < × ≤ +0.2 m) or accretional (+0.2 m < × ≤ +0.6 m), depending on the location	Variable, either erosional or accretional Total sediment loss/gain: −13.9 < × ≤ +18.7 Sediment loss/gain/m of transect: −0.99 < × ≤ +0.36

Source: Adapted from Duvat et al. (2016).

Figure 26.4 Morphological impacts of TC Bejisa in the northern part of Saint-Paul Bay. (a) Topographic profile P1 (see location on Figure 26.2). The comparison of the pre- and post-storm situations exhibits the erosional impact of the cyclonic waves on the dune face and the beach, which lowered up to nearly 3 m in its central part. (b) This north–south photograph shows the impact of the cyclone on the sand dune. Here, shoreline retreat reached up to 4 m, causing tree uprooting and fall. *Source*: Adapted from Duvat et al. (2016).

12 m respectively (Figure 26.4). On Boucan Canot and Cape Homard beaches, most post-Bejisa beach-dune topographic profiles exhibited massive sand loss generally affecting the upper and central parts of beaches. Beach lowering ranged from 0.50 to 1.50 m and was particularly marked in built areas. However, in places upper beaches and sand dunes accreted as a result of sediment washover. For example, fresh sediment deposits up to 1 m thick accumulated on the upper beach in front of private properties in the central part of Boucan Canot Beach (Figure 26.5). Additionally, sediment sheets

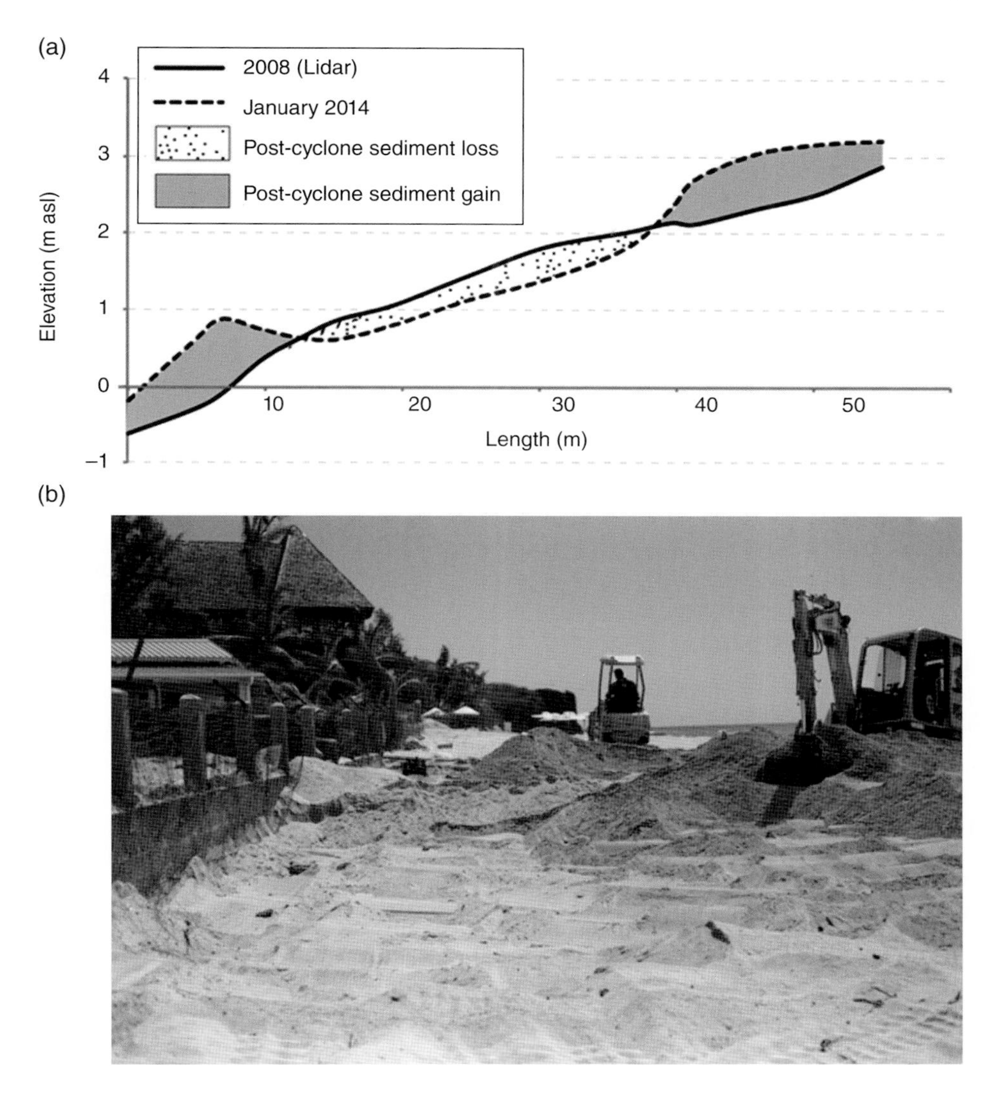

Figure 26.5 Morphological impacts of TC Bejisa in the central part of Boucan Canot Beach. (a) Topographic profile P7 (see location on Figure 26.2) shows the contrasting impacts of TC Bejisa along the beach profile, with marked accretion on the upper beach and significant sediment loss in the central part of the beach. (b) This photograph shows the reworking of the 0.80–1 m-thick sediment deposits that had accumulated on the upper beach. This reworking was done on the initiative of coastal residents. *Source*: Adapted from Duvat et al. (2016).

and scattered coral fragments with a diameter reaching up to 0.40 m were found in a 25–50 m-wide coastal strip from the vegetation line. Southwards from Saint-Gilles Pass, along the 6.5 km-long Ermitage-La Saline Beach, the morphological impacts of TC Bejisa were found to be highly variable, with accretional patterns prevailing at its northern and southern ends, and erosional features occurring in its central part. Importantly, where beach erosion was observed after TC Bejisa, sediment loss was proportionally greater where the beach is narrow and bordered by seawalls. In such settings, some massive seawalls were broken by the cyclonic waves, which therefore eroded land.

26.3.1.1.3 Impacts on coastal vegetation The coastal vegetation, which was nearly continuous along the coastline before TC Bejisa, was severely affected both by the cyclonic wind and waves (Duvat et al. 2016). The wind mostly impacted introduced tree and shrub species. Unlike the introduced species, indigenous species showed high resistance to the cyclonic wind, and the coastal shrub even showed no signs of impact at some locations. In addition, the cyclonic waves caused the uprooting of herbaceous and shrub species growing on the upper beach, and tree fall, as a result of shoreline retreat (Figure 26.4b).

26.3.1.2 Impacts on human assets TC Bejisa did not cause any death. The damage caused by the cyclone was valued at €72 M, with major damage to agriculture and the built environment (Desarthe, Naulin, and Moncoulon 2015), as illustrated by the situation of Saint-Paul municipality.

26.3.1.2.1 Impacts on the buildings, public facilities, and economic activities Out of the 77 assets surveyed (i.e. buildings, public facilities, and economic activities), around 30% experienced no impact whilst 43% experienced moderate to high impacts (Table 26.2). This last category concerned 40 and 37% of private properties and economic activities, respectively. The coastal road of Cap La Houssaye was also severely damaged, which blocked off the traffic for more than three months.

Table 26.2 Destructive impacts of TC Bejisa on human assets (Duvat et al. 2016).

Magnitude of impacts	Private Properties	Economic activities	Public facilities	Total	
				Number of affected assets	% of affected assets
/ *(no impact)*	17	7	1	25	32
+ *(low impact)*	16	3	/	19	25
++ *(moderate impact)*	17	2	3	22	29
+++ *(high impact)*	5	4	2	11	14
	55	16	6	77	100

/ *No impact*: when no mark of impact was visible.
+ *Low impacts*: when the fence suffered little damaged, often due to the impacts of small coral blocks (max. 5 cm in diameter).
++ *Moderate impacts*: when the fence was severely damaged (partly destroyed), often due to the impacts of medium-sized coral blocks (max. 5–15 cm in diameter).
+++ *High impacts*: when the fence was severely damaged (mostly destroyed) and when the building was partly damaged (especially partial destruction of the seaside wall), often due to the impacts of medium-sized coral blocks (max. 5–15 cm in diameter).

26.3.1.2.2 Impacts on coastal protection structures The waves generated by TC Bejisa destroyed 18 coastal protection structures out of the 153 inventoried along the Saint-Paul coastline. The destroyed structures were dispersed along the coastline between the northernmost sediment cell of Saint-Paul Bay and the southernmost sediment cell of Ermitage-La Saline Beach (Table 26.3). All damaged structures were seawalls protecting either private properties (15 structures) or public infrastructure and facilities (Saint-Paul Bay's cemetery, Cap La Houssaye coastal road and Saint-Gilles seafront). Two of these structures were massive and had been built by public works companies, such as the 8 m-high seawall protecting Cap La Houssaye's coastal road.

26.3.1.2.3 The example of the Saint-Gilles marina The Saint-Gilles marina, which concentrates major human assets (the marina itself, as well as sport centres, shops,

Table 26.3 Impacts of TC Bejisa on coastal protection structures and post-cyclone reconstruction (Duvat et al. 2016).

Sediment cell	Impacts of TC Bejisa	Post-cyclone reconstruction
1 – Saint-Paul Bay	• Destruction of a seawall protecting a private individual property • Destruction of a seawall protecting a cemetery	• No reconstruction • No reconstruction
2 – Cap Champagne Beach	• Destruction of a seawall protecting a private individual property • Destruction of a massive (8 m-high) seawall protecting Cap La Houssaye coastal road	• Reconstruction completed in April 2014 • Reconstruction in progress in April 2014
3 – Boucan Canot Beach	• Destruction of a wooden structure protecting the seafront • Damage to a seawall protecting a beach resort • Damage to 3 seawalls protecting private individual properties	• Reconstruction not started in April 2014, but completed in January 2015 • Reconstruction completed in April 2015 • Reconstruction of 2 structures out of 3 differently from pre-storm structures: erection of a massive structure made of a gabion basket bordered by a large embankment + building of a stronger seawall protected by volcanic blocks
4 – Cap Homard Beach	• Destruction of 3 seawalls protecting individual properties	• Reconstruction of 1 structure out of 3, identically to the pre-existing structure
5 – Grands Fonds Beach	• Destruction of a seawall protecting a holiday centre • Destruction of 3 seawalls protecting private individual properties	• Reconstruction identical to pre-existing seawall • No reconstruction
6 – Ermitage-La Saline Beach	• Destruction of 2 seawalls protecting private individual properties	• Reconstruction of one seawall out of two, stronger than the pre-existing one

Figure 26.6 Flooding of the Saint-Gilles marina during TC Bejisa.

restaurants, a marine aquarium, and private boats), was the most severely affected area on the Saint-Paul coastline due to the combined effect of a rise in the river level (due to heavy rains) and marine inundation (Figure 26.6). The fact that the marina is sited in front of the pass (i.e. open sea) and next to the river mouth makes it highly sensitive to both marine inundation and storm wave impact. Being aware of the situation, the Saint-Paul municipality undertook preventive works before the cyclonic season – i.e. dredging of the downstream section of the river and opening of the natural sandbar situated at the river mouth – so as to facilitate river discharge into the sea and thus limit the risk of flooding. However, the north-western waves generated by TC Bejisa, which were 6–7 m high according to the safeguards, caused the sandbar to rapidly re-form. In addition to the pressure that waves exerted on the river flow, this both caused the river to discharge into the harbour, and created a deposit of 1500 m^3 of sand in the northern part of the harbour. Together, wave energy and river discharge caused damage to 30 boats (around 10% of the total), and nearby restaurants and dive shops. The southernmost buildings were the most affected: their doors were destroyed by waves and the water level reached a height of up to 1.2 m, according to the interviewees. The power blackout caused by the inundation lasted for one week, preventing professionals from rapidly repairing buildings, restock, and restart their activity. In sum, the destructive impacts of TC Bejisa on the Saint-Gilles marina highlighted above all the site's high sensitivity to inundation in the case of north-western cyclonic waves.

26.3.2 Responses by local authorities and coastal residents

26.3.2.1 Responses by local authorities In the aftermath of the cyclone, the Saint-Paul municipality focused its action on the Saint-Gilles marina. It notably reopened the sandbar to allow water discharge into the sea, and started dredging the main harbour. Its direct support to local professionals and coastal residents, both within this specific area and on the entire impacted area, has however been very limited, not to say non-existent. Most of the 41 people we interviewed in the aftermath of the cyclone actually reported no help from public authorities, an a priori surprising situation in the context of a developed country. A key explanation refers to a legal problem that the Saint-Paul municipality's risk officer explained to us. Belonging to the Fluvial domain and to the Maritime public domain respectively, the harbour area and the coastal zone are on the authority of the locally based office of the Ministry of the Environment, i.e. the *Direction de l'Environnement, de l'aménagement et du lodgement* (DEAL). As a result, the municipality cannot undertake any post-event action in these areas. According to the interviewees, the DEAL did not send any staff to the affected area to take stock of the impacts and assess the affected persons' needs. These latter being not aware of this jurisdiction issue, misunderstood the municipality's relative inaction and experienced a sense of abandonment that made them quite angry. Some of them therefore decided to act on their own, whatever the coastal zone regulation, to either remove sand and coral debris in front of their property, or rebuild damaged coastal defences.

26.3.2.2 Responses by coastal residents

26.3.2.2.1 A set of contrasting responses

Removal and reworking of sediment deposits – In built areas, coastal residents first cleared fresh sediment deposits brought by the cyclonic wind and waves inside houses, gardens, and swimming pools (Figure 26.7). Then, they manually removed sediment deposits that had accumulated on the sea-side front of their property, the access of which was blocked by thick accumulations of sand and coral fragments. In places where sediment accumulations were thick, as on Boucan Canot Beach for example, and in the absence of public intervention coastal residents employed public works companies to remove and rework sediment deposits (see Figure 26.5). The mechanical action of diggers on upper beaches was conducted on the residents' own initiative, without any consultation of public authorities, and once again despite the fact that beaches are included in the Maritime public domain.

Vegetation clearing – At the same time when they cleared sediment deposits, coastal residents also cleared vegetation waste, i.e. broken trunks and branches, and uprooted shrubs lying inside and in front of their properties. Vegetation clearance took less time (generally less than one week) in self-contained accommodation areas compared to apartment buildings areas (up to three months), as a result of the unequal reactivity of private individuals and groups.

Rebuilding of destroyed coastal protection structures – Thirteen engineered structures, which were either protection seawalls or retaining walls, were totally destroyed by the cyclonic waves (Table 26.3). One year after TC Bejisa, most of these structures (8 out of 13) had not been repaired yet. Amongst the 5 structures that had been repaired, 2 had been reconstructed identically to the pre-storm structures whilst 3 had been reconstructed differently, residents then opting for a stronger type of structure than the

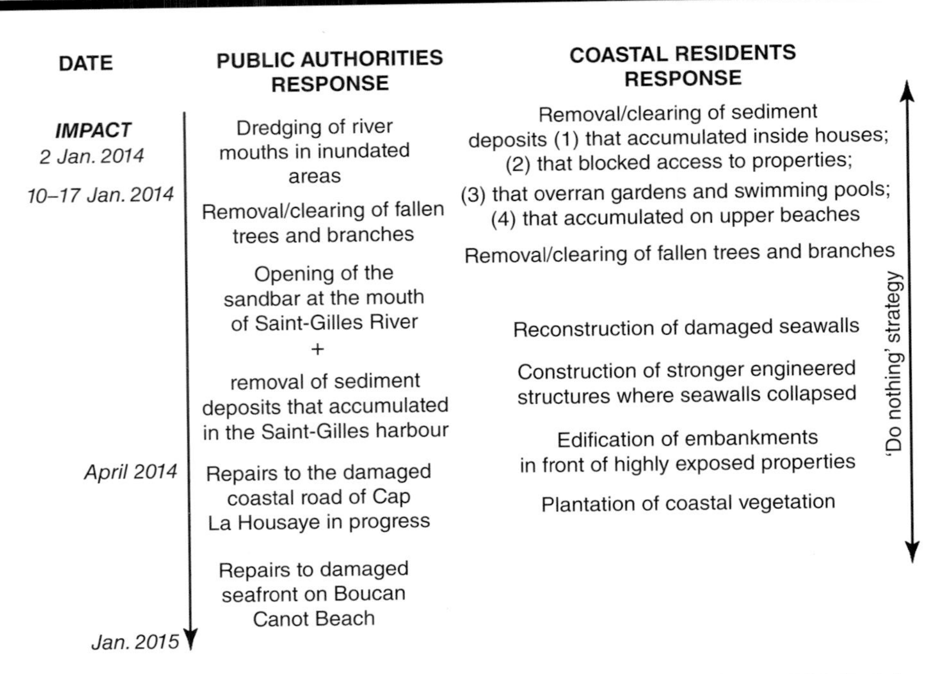

Figure 26.7 Timelines of the responses by public authorities and coastal residents to TC Bejisa's impacts. This synthetic figure shows that stakeholders (i.e. public authorities and coastal residents) all took action independently of each other. Both parties undertook engineering works on the coast without any coordination, public authorities mainly operating at river mouths (i.e. in the Fluvial domain), to reduce inundation, whilst residents intervened on the beaches and upper beaches (i.e. in the Maritime public domain), bordering their properties. Public inaction in urbanized areas over the crisis and post-crisis phases explains the contrasting strategies deployed by residents.

pre-cyclone one (e.g. replacement of a seawall by a gabion basket structure bordered by a large embankment).

Replanting of coastal vegetation – Along Saint-Paul's urbanized coastline, in areas where the cyclonic waves had entirely destroyed the coastal vegetation, residents often replanted vegetation, mostly introduced shrub species. Curiously, they rarely replanted indigenous plants although the latter proved to be more resistant and more resilient to TC Bejisa's impacts compared to introduced species.

Reconstituting a buffer zone in front of increasingly exposed properties – In the central part of Saint-Paul Bay where coastline retreat caused significant land loss, thereby increasing house exposure to waves, coastal residents used the materials dredged from river mouths by the Public Works Division to erect massive embankments in front of their properties. These embankments are meant to replace the natural beach berm and embryo dunes that used to break wave energy at the coast, that is to say to act as a buffer zone. Generally, residents planted vegetation on these embankments to stabilize them.

Doing nothing – On most of the coastline length, coastal residents just 'did nothing' to reduce future risks. In many areas, no correlation was found between the intensity of impacts and increased exposure of properties and the strategy adopted by residents. In some areas where natural buffers (i.e. beaches and sand dunes) had been severely

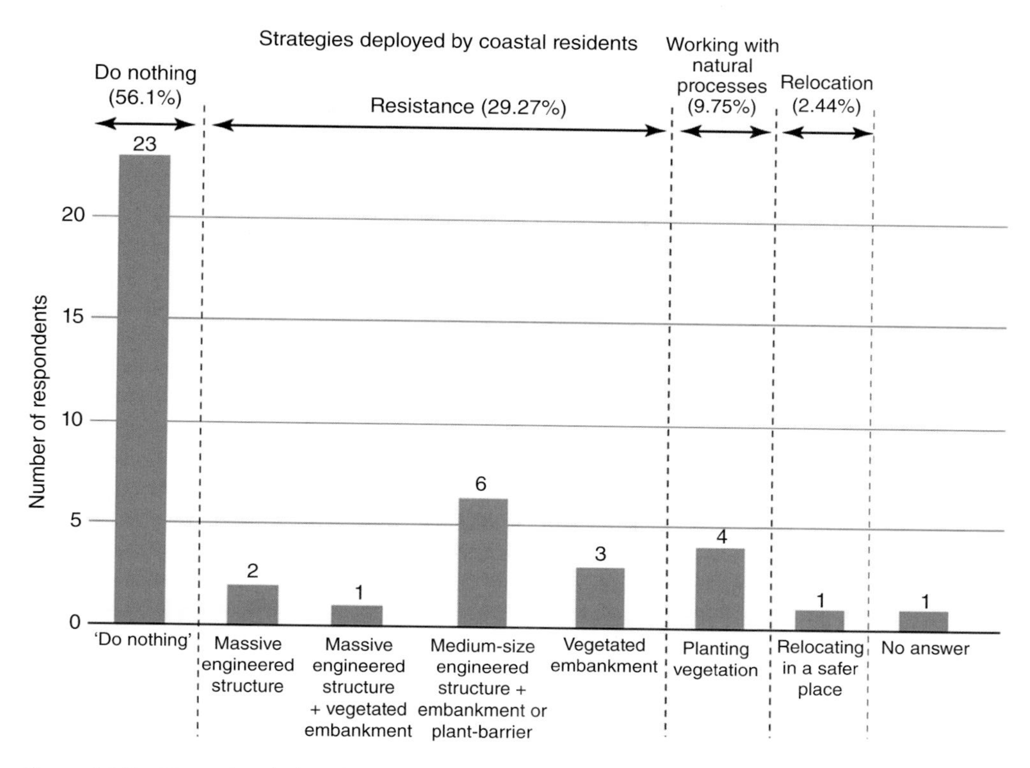

Figure 26.8 Strategies deployed by the coastal residents of Saint-Paul municipality after TC Bejisa to reduce future risks (based on semi-structured interviews with 41 first-line residents).

affected by the cyclonic waves and thereby provided no more protection to private properties, coastal residents indeed did not proceed to any technical intervention.

26.3.2.2.2 The underlying individual strategies The post-Bejisa strategies deployed by coastal residents to better face future cyclonic events can be sorted into four main categories (Figure 26.8). The 'do nothing', also called 'non-intervention' strategy (Cooper and McKenna 2008a), was found to prevail, as it concerns 23 respondents out of 41. Whilst some respondents considered that human intervention can hardly reduce the destructive impacts of tropical cyclones on first-line properties, others said that they 'do nothing' because strong protections are too costly for them.

The second strategy deployed by coastal residents, which concerns 12 respondents out of 41, is 'resistance', which includes all engineering and technical interventions aimed at protecting exposed assets from wave impact (Cooper and Pile 2014). In the study area, these interventions generally consisted of the construction of coastal protection structures such as seawalls or gabion baskets, and in the edification, in front of threatened plots, of massive embankments made of alluvium dredged from river mouths. According to interviewees, the public authorities had edified the first embankment in Saint-Paul Bay after TC Gamède in 2007. At that time, the Public Works Division had used the large volumes of alluvium dredged from river mouths to erect a continuous barrier in front of first-line properties. Curiously, this technical solution was not re-applied by the public authorities after TC Bejisa, which Saint-Paul Bay's respondents did not understand. As a result, because they considered it was a cheap and efficient solution, some residents of Saint-Paul Bay decided to erect by themselves an embankment in front of their property to protect it from future wave attack. In many cases, the residents who chose this strategy also planted vegetation, using it as a barrier in the face of waves or as a mean to stabilize a man-made embankment.

The third strategy, deployed by 4 respondents out of 41, consisted in planting vegetation on the beach crest and foredune so as to encourage sediment trapping and beach-dune development. This strategy promotes 'working with natural processes' to reduce risks (Cooper and McKenna 2008a). However, the respondents who implemented it are aware that it provides limited protection in the case of a cyclonic event occurring a short time after TC Bejisa.

Fourth, only 1 out of the 41 interviewees said that he envisaged relocating in a safe area in the near future. This option, which can be considered as being part of adaptation in regards to its long-term benefits in terms of risk reduction and buffer ecosystem protection (Cooper and McKenna, 2008a), therefore remains little envisaged by the residents who were affected by TC Bejisa.

26.4 Key findings and challenges for adaptation to climate change

26.4.1 Efficiency of the cyclone alert system

The absence of human death due to TC Bejisa illustrates a noteworthy positive aspect of risk management on Reunion Island, which refers to its cyclone alert system. Initiated in the 1950s at the urging of a national-level risk policy, and regularly improved until the 2000s, it is based on the monitoring of meteorological and marine parameters by Météo-France – nowadays in collaboration with regional and international institutions. In case of a potentially dangerous situation, Météo-France informs the Prefecture office of the island, which decides on triggering the alert process and on moving from one level of alert to another. The Prefecture is then responsible for disseminating information to the regional-to-local authorities as well as to the population via the media. There are three levels of alert, going from 'green' when the cyclone is approaching but does not carry any serious threat within the upcoming 24 hours, to 'orange' when there is an immediate threat (<24 hours), and to 'red' in case of pending danger or when the cyclone effectively reaches the island. Official instructions for orange alert recommend people to store food, water, and medicines, to strengthen houses (e.g. tighten swing shutters), and to keep informed about the situation. Under the 'red' alert level, going outside is strictly forbidden.

The survey conducted in 2015 on sea-related risk perception confirms the good appropriation of security instructions by the population (Magnan et al. 2015). A first set of evidence touches upon background questions on cyclones in general. Results revealed that a large majority of the interviewees (>80%) correctly identify the December-to-February peak of the cyclonic season as well as the direct effects of a TC, i.e. heavy rains (96%), marine inundation (95%), strong winds (95%) and river flooding (90%). Most of them also identify the indirect consequences of such an event, i.e.

coastal vegetation destruction (84%), landslides (74%), beach and dune erosion (66%), and damages to coral reefs (59%). The very small proportion of people (2–5%) citing non cyclone-driven phenomena, such as tsunamis, volcanic eruptions, or earthquakes, supports the previous conclusion. A second set of evidence refers to questions specifically dealing with the alert system. Here also, a large majority of people showed excellent knowledge on the security instructions related to the 'orange' alert level, the three dominant answers being storing food, water, and medicines (97%), keeping informed (92%) and staying at home (74%). The same conclusion applies to the 'red' alert level, answers mainly emphasizing the need to stay at home (99%) and to keep informed (94%). In addition, 92% of the interviewees declared knowing the emergency telephone numbers (mainly of fire-fighters, the police, hospitals, and the town hall).

Noteworthy, however, is that 85% of the interviewees did not know anything about the official municipality-level risk management plans, such as the *Plan Communal de Sauvegarde* that inventories the actions to undertake in case of a crisis (crisis cell, communication priorities, key stakeholders, transportation and circulation, and so on); the *Dossier d'Information Communal sur les Risques Majeurs* that stocktakes the main risks that affect the municipality area and the ways to prevent them and react to them; and the *Plan de Prévention des Risques* that maps the risks and provides guidance on land use and building location and standards. Only the ORSEC Plan (*Organisation de la Réponse de la Sécurité Civile*) is cited (by 40% of the respondents), but it rather concerns the coordination of public bodies involved in post-crisis and emergency interventions. In total, whilst people seemed to be well aware of risk-friendly behaviour at the household/family level, they exhibited very little appropriation of local public policies and measures. Together with the previously mentioned shortfalls in public crisis management, this contributes to explain that post-event responses were mainly individual, explaining in turn the heterogeneity of responses along the coastline.

26.4.2 Urgent need for practical engagement and action by local authorities

The lack of 'practical engagement and action' (King 2008) of the public authorities during the crisis and post-crisis phases, which led to uncoordinated technical interventions by coastal residents, gave rise to serious problems. First of all, the lack of intervention by the public authorities during the crisis and post-crisis phases shocked first-line coastal residents who felt –and still feel – in danger. Most residents cannot afford strong engineered structures, and therefore consider that proper coastal protection devices should be provided by public authorities. As a result, these residents have a feeling of increasing economically driven inequality between those who can afford efficient protection devices and those who cannot, therefore raising social justice issues (Cooper and McKenna 2008b). Second, the coastal works conducted by coastal residents on beaches may have adverse effects both on the coastal environment and on future human assets exposure to sea-related hazards. In the present case, the removal of the fresh sediments deposited by cyclonic waves on upper beaches prevents natural beach crest heightening that would reduce future asset exposure to marine inundation. Additionally, the uncoordinated technical interventions implemented at the scale of individual plots may be counterproductive, residents having limited knowledge on the physical processes at work, and thereby on what are well-adapted solutions. These technical interventions also have adverse effects on neighbouring areas. For example, the construction of a massive engineered structure in front of a given property would fix the coastline there, but exacerbate the eroding impact of storm-generated waves on both sides. In the central part of Boucan Canot Beach, which was severely affected by the waves generated by TC Bejisa, one resident erected a massive protection structure, i.e. a gabion-basket structure bordered on its sea-side by an embankment, for a total cost of 110,000€. This initiative generated strong conflicts, not only with the neighbours, but also with the public authorities. The neighbourhood conflict was generated by the differing views of residents on the strategy to be adopted. The resident who has erected the massive structure regrets that his neighbours did not do the same, because in the present situation cyclonic waves may bypass his structure on both sides, and flood his property. For his part, his northern neighbour who chose the 'do nothing' strategy considers that erecting massive structures is likely to aggravate beach destabilization, thereby exacerbating the destructive impact of cyclonic waves on coastal properties. And as for his southern neighbour, he rebuilt stronger the seawall protecting his property, recognizing that he cannot afford such a massive gabion-made structure as his neighbour. A conflict with the local authorities also emerged as the two new structures aforementioned (gabion-made structure and strengthened seawall) were erected without any official authorisation. This single example emphasizes the urgent need for public control and management, especially in the crisis and post-crisis phases, given the long-lasting effects of the technical interventions that are at that time conducted. It also emphasizes the need for raising awareness amongst coastal residents, especially the first-line ones, on both the technical prerequisites for adequate coastal defences, and on the detrimental effects of individual technical interventions at the sediment cell level.

26.4.3 Design diversified, context-specific risk reduction strategies

Our assessment highlighted the contrasting morphological impacts of TC Bejisa's waves, as some coastline sections eroded whilst others accreted. In addition to the existence of a north–south decreasing gradient of cyclone-induced erosion, which is in accordance with the decreasing intensity of cyclonic waves southwards, the comparison of beach profiles also revealed significant differences in eroding and accreting rates on short distances, i.e. on hundreds of metres within a given sediment cell. The high spatial variability of impacts, also emphasized by De la Torre et al. (2012) in their assessment of 2007 TC Gamède's impacts on Saint-Paul's coastline, should be better accounted for in coastal risk management. Today, because the erosion-prone beach-dune systems are known (e.g. Saint-Paul Bay's is one of them), specific risk management measures should be designed and applied, based on the known response of each beach-dune system to cyclonic waves. These measures would for example include the maintenance of cyclone-driven sediment deposits on beaches, so as to strengthen and take advantage of the buffering role of beaches in the face of cyclonic waves. On the contrary, in areas that are erosion- and flood-prone such as Saint-Paul Bay, it would be appropriate to erect and maintain a continuous alluvium-made vegetated embankment in the front of the urbanized coastline to reduce cyclone-induced damage. Such a strategy, which seems preferable to the current 'do nothing' strategy from a short- to medium-term perspective, could easily be implemented as alluvium are freely available locally. Regular dredging of river mouths would moreover reduce the risk of flooding in low-lying coastal areas during the cyclonic season. As shown above, there is a crucial need for integrated (i.e. river and marine) flood-risk management, with many potential co-benefits.

26.4.4 Promote adaptation

These conclusions refer to short- to medium-term options that must be rapidly implemented for reducing cyclone-driven impacts on coastal human assets in the coming one to two decades. However, as climate change is now taking place and will impose changing environmental conditions, there is a need for also thinking of longer-timescale strategies. Concerning coastal areas, two critical changes will be sea level rise and potential changes in cyclone parameters. Regarding cyclones, global trends suggest an increase in the intensity of the most intense cyclones, but no increase in the number of events. Such perspectives mean that risk will inevitably increase in the future if nothing is done to drastically reduce the exposure of people, buildings, and infrastructures to sea-related hazards. As a result, whilst the aforementioned actions – improve the efficiency of crisis and post-crisis management by national-to-local authorities, stop coastal urbanization, protect buffering sand-dune systems, build a protective embankment in front of highly exposed properties in Saint-Paul Bay, upgrade coastal defences and manage them coherently within each sediment cell, and raise coastal residents' awareness – would help engage a virtuous process, they would not be sufficient in the long run. In fact, these actions must be considered only as first, inescapable steps in the long-term adaptation pathway (Magnan and Duvat 2018). We argue here that the 'coastal retreat' strategy will have to be seriously considered by decision makers, for example in the most low-lying areas of Saint-Paul Bay or at Boucan Canot. Such a strategy is very difficult to implement because it requires a radical change in the way of thinking the human-nature relationship at the coast. It requires moving from a 'systematically keep the line' approach to a more nuanced one partly made of a 'do with nature' component. This supposes that decision makers feel concerned about the environment, have a long-term vision, and deploy enough energy to enhance now the transformation of the whole system (new coastal regulations, land tenure issues, transportation networks, economic shift, and so on). This also supposes that the local population will accept related policy constrains, which will be all the more difficult since benefits will mainly occur in the long run. On these two points, some recent studies on Reunion Island show that there is still a long way to go (Magnan et al. 2015; Magnan and Charpentier 2016).

26.5 Conclusion

This chapter addresses an important knowledge gap of cyclone impact analyses and that relates to the nature, extent, and timelines of short-term (less than one year) responses of stakeholders, i.e. public authorities and coastal residents; and to the implications of these responses on medium- to long-term changes in vulnerability to sea-related hazards.

Such an understanding is however crucial to apprehend the path-dependency effects induced by short-sighted behaviours that have the potential to hamper long-term adaptation to climate change.

This chapter builds on the case study of Reunion Island, a French overseas territory located in the southwestern Indian Ocean, one of the intertropical regions that are the most exposed to tropical cyclones. It thus provides useful insights on both the impacts of extreme climate events and coastal risks governance issues in an outermost region of the European Union, on which only a limited number of studies have focused up to now. This chapter especially focuses on a 30 km-long coastal section of the Saint-Paul municipality in order to investigate both the impacts of and the stakeholders' responses to TC Bejisa (category 3), which affected the western side of the island in January 2014.

Results first emphasize the high spatial variability of the impacts of TC Bejisa within the study area, both on the natural environment (i.e. marine inundation, and morphological impacts on beach-dune systems) and on human assets (building, public facilities, economic assets, and coastal protection structures). The analysis also highlights the major role of anthropogenic drivers, such as coastal urbanization and beach-dune systems degradation, in determining the extent and intensity of impacts. Second, results emphasize the shortfalls of coastal-risks management in the aftermath of the event, particularly the lack of effective engagement and action of public authorities during the crisis and post-crisis phases. Facing such a situation, affected coastal residents often decided to act on their own, implementing various strategies at the scale of individual plots, i.e. do nothing, resist (by restoring/building coastal protection structures), working with natural processes (by planting vegetation), or, rarely, relocate. This led to uncoordinated interventions not considering the sediment cell intrinsic coherence as a key management unit, which is likely to increase medium- to long-term human asset exposure to sea-related hazards. This also resulted in immediate conflicts within residents and between residents and the public authorities.

Finally, we drew up from this analysis some recommendations for risk reduction and adaptation to climate change in Reunion Island, with relevance for French overseas territories at large. These recommendations touch upon improving the efficiency of crisis and post-crisis management by national-to-local authorities; developing context-specific solutions promoting the protection and restoration and thereby the strengthening of natural buffers; and raising coastal residents' awareness on both the technical prerequisites for adequate coastal defences and the detrimental effects of individual technical interventions. In fact, awareness raising should be more systematically considered in coastal risk governance, not only to reduce current human assets' exposure to sea-related hazards, but also to allow the whole society to engage in adaptation to climate change strategies. The latter indeed require profoundly transforming our modes of living at the coast, which in turn requires that the general public understands challenges and supports public policies. Overall, and considering the new changing conditions climate change started and will continue to impose, if only because of sea level rise, our study emphasizes the urgent need for engaged long-term public action. This notably supposes both starting by doing well what we currently do badly (Juhola et al., 2016; Magnan et al., 2016), and making national-to-local decision-makers' visions of future risk management and coastal residents' desires of living at the coast more climate change-compatible.

Acknowledgements

This study benefited from the support of the Fondation de France, the Région Réunion, the French Ministry of the Environment (Ministère de l'environnement, de l'énergie et de la mer [MEEM]) and the French National Research Agency (ANR) under the *VulneraRe*, *Réomers* (Risques, Décision, Territoires programme) and *Storisk* (ANR-15-CE03-0003) research projects. AKM also thanks the French Government for its support under the 'Investissement d'avenir' programme, managed by the French National Research Agency (ANR-10-LABX-14-01). The authors are grateful to Paul-Bernard Rivet, who contributed to the generation of data on marine inundation during his Masters' degree internship by conducting post-cyclone interviews with coastal residents, and to Dorothée James, Cécilia Pignon-Mussaud, and Camille Salmon, who contributed to building the database of the *VulneraRe* research project.

References

Angelucci, F. and Conforti, P. (2010). Risk management and finance along value chains of SIDS. Evidence from the Caribbean and the Pacific. *Food Policy* 35: 565–575. https://doi.org/10.1016/j.foodpol.2010.07.001.

Bastone, V. and De la Torre, Y. (2011). *Étude préliminaire de l'impact du changement climatique sur les risques naturels à la Réunion*. BRGM/RPC-59495-FR, Orléans: BRGM.

Bythell, J.C., Bythell, M., and Gladfelter, E.H. (1993). Initial results of long-term coral reef monitoring program: impacts of hurricane Hugo at Buck Island reef National Monument, St. Croix, U.S. Virgin Islands. *Journal of Experimental Marine Biology and Ecology* 172: 171–183. https://doi.org/10.1016/0022-0981(93)90096-7.

Cahoon, D.R., Hensel, P., Rybczyk, J. et al. (2003). Mass tree mortality leads to mangrove peat collapse at Bay Islands, Honduras after hurricane Mitch. *Journal of Ecology* 91: 1093–1105. https://doi.org/10.1046/j.1365-2745.2003.00841.x.

Cardona, O.D., van Aalst, M.K., Birkmann, J. et al. (2012). Determinants of risk: exposure and vulnerability. In: *Managing the Risks of Extreme Events and Disasters to Advance Climate Change Adaptation*, A Special Report of Working Groups I and II of the Intergovernmental Panel on Climate Change (IPCC) (ed. C.B. Field, V. Barros, T.F. Stocker, et al.), 65–108. Cambridge, UK, and New York, USA: Cambridge University Press.

Caron, V. (2011). Contrasted textural and taphonomic properties of high-energy wave deposits cemented in beachrocks (St. Bartholomew island, French West Indies). *Sedimentary Geology* 237 (3–4): 189–208. https://doi.org/10.1016/j.sedgeo.2011.03.002.

Cazes-Duvat, V. and Paskoff, R. (2004). *Les littoraux des Mascareignes entre nature et aménagement*. Paris: L'Harmattan.

Cazes-Duvat, V. (2005). Les impacts du cyclone Kalunde Sur les plages de l'Île Rodrigues (océan Indien occidental). *Zeitschrift für Geomorphologie* 49 (3): 293–308. ESP022004903003.

Cooper, J.A.G. and McKenna, J. (2008a). Working with natural processes: the challenge for coastal protection strategies. *Geographical Journal* 174 (4): 315–331. https://doi.org/10.1111/j.1475-4959.2008.00302.x.

Cooper, J.A.G. and McKenna, J. (2008b). Social justice in coastal erosion management: the temporal and spatial dimension. *Geoforum* 39: 294–306. https://doi.org/10.1016/j.geoforum.2007.06.007.

Cooper, J.A.G. and Pile, J. (2014). The adaptation-resistance spectrum: a classification of contemporary adaptation approaches to climate-related coastal change. *Ocean and Coastal Management* 94: 90–98. https://doi.org/10.1016/j.ocecoaman.2013.09.006.

Crabbe, M.J.C., Martinez, E., Garcia, C. et al. (2008). Growth modelling indicates hurricanes and severe storms are linked to low recruitment in the Caribbean. *Marine Environmental Research* 65: 364–368. https://doi.org/10.1016/j.marenvres.2007.11.006.

De La Torre, Y., Bastone, V., Bodere, G. et al. (2012). *Morphodynamique des littoraux de la Réunion. Phase 4*. BRGM/RP-57431-FR, 103 p. Orléans: BRGM.

De Scally, F.A. (2008). Historical tropical cyclone activity and impacts in the Cook Islands. *Pacific Science* 62 (4): 443–459. https://doi.org/10.2984/1534-6188(2008)62[443:HTCAAI]2.0.CO;2.

De Scally, F.A. (2014). Evaluation of storm surge risk: a case study from Rarotonga, Cook Islands. *International Journal of Disaster Risk Reduction* 7: 9–27. https://doi.org/10.1016/j.ijdrr.2013.12.002.

Desarthe, J., Naulin, J.-P., and Moncoulon, D. (2015). Le cyclone Bejisa à la Réunion (01/01/2014-03/01/2014). Paris: Caisse Centrale de Réassurance, 10 p.

Dorville, J.-F. and Zahibo, N. (2010). Hurricane Omar waves impact on the west coast of the Guadeloupe Island, October 2008. *The Open Oceanography Journal* 4: 83–91.

Duvat, V., Magnan, A., Etienne, S. et al. (2016). Assessing the impacts of and resilience to tropical cyclone Bejisa, Reunion Island (Indian Ocean). *Natural Hazards* 83 (1): 601–640. https://doi.org/10.1007/s11069-016-2338-5.

Duvat, V. and Salmon, C. (2015). *Evaluation de la vulnérabilité actuelle des littoraux de l'île de la Réunion aux risques liés à la mer (Saint-Denis, Le Port, Saint-Paul, Saint-Pierre). Rapport de résultats no. 2*. La Rochelle: Programme VulneraRe, 90 p.

Etienne, S. (2012). Marine inundation hazards in French Polynesia: geomorphic impacts of tropical cyclone Oli in February 2010. *Geological Society Special Publications* 361: 21–39. https://doi.org/10.1144/SP361.4.

Ferdinand, I., O'Brien, G., O'Keefe, P. et al. (2012). The double blind of poverty and community disaster risk reduction: a case study from the Caribbean. *International Journal of Disaster Risk Reduction* 2: 84–94.

Fletcher, C.H., Bochicchio, C., Conger, C. et al. (2008). Geology of Hawaii reefs. In: *Coral Reefs of the World*, vol. 1 (ed. B.M. Riegl and R.E. Dodge), 435–487. Dordrecht, Netherlands: Springer Science.

Hay, J. and Mimura, N. (2010). The changing nature of extreme weather and climate events: risks to sustainable development. *Geomatics, Natural Hazards and Risk* 1 (1): 3–8. https://doi.org/10.1080/19475701003643433.

Hoeke, R.K., McInnes, K.L., Kruger, J. et al. (2013). Widespread inundation of Pacific islands by distant-source wind-waves. *Global Environmental Change* 108: 128–138. https://doi.org/10.1016/j.gloplacha.2013.06.006.

Hubbard, D.K., Parsons, K.M., Bythell, J.C. et al. (1991). The effects of hurricane Hugo on the reefs and associated environments of St. Croix, US Virgin Islands: a preliminary assessment. *Journal of Coastal Research* 3: 33–48.

Juhola, S., Glaas, E., Linnér, B.-O. et al. (2016). Redefining maladaptation. *Environmental Science and Policy* 55: 135–140. https://doi.org/10.1016/j.envsci.2015.09.014.

King, D. (2008). Reducing hazard vulnerability through local government engagement and action. *Natural Hazards* 47: 497–508. https://doi.org/10.101007/s11069-008-9235-5.

Magnan, A.K. and Charpentier, C. (2016). *Les littoraux de la Réunion face aux risques liés à la mer: quelle(s) vision(s) du futur par les acteurs?* IDDRI Study No.05/16. Paris: IDDRI.

Magnan, A.K., Duvat, V.K.E., Guérineau, B. et al. (2015). *La perception des risques liés à la mer à la Réunion (6 volumes)*. Report of the VulneraRe project. Paris: IDDRI.

Magnan, A.K. and Duvat, V.K.E. (2018). Unavoidable solutions for coastal adaptation in Reunion island (Indian Ocean). *Environmental Science and Policy* 89: 393–400. https://doi.org/10.1016/j.envsci.2018.09.002.

Magnan, A.K., Schipper, E.L.F., Burkett, M. et al. (2016). Addressing the risk of maladaptation to climate change. *WIREs Climate Change* 7 (5): 646–665. https://doi.org/10.1002/wcc.409.

Mallela, J. and Crabbe, M.J.C. (2009). Hurricanes and coral bleaching linked changes in coral recruitment in Tobago. *Marine Environmental Research* 68: 158–162. https://doi.org/10.1016/j.marenvres.2009.06.001.

Mechler, R. (2004). *Natural Disaster Risk Management and Financing Disaster Losses in Developing Countries*. Karlsruhe, Germany: Verlag Versicherungswirtsch.

Méheux, K. and Parker, E. (2006). Tourist sector perceptions of natural hazards in Vanuatu and the implications for a small island developing state. *Tourism Management* 27: 69–85. https://doi.org/10.1016/j.tourman.2004.07.009.

Mimura, N., Nurse, L., McLean, R.F. et al. (2007). Small islands. In: *Climate Change 2007: Impacts, Intergovernmental Panel on Climate Change* (ed. M.L. Parry, O.F. Canziani, J.P. Palutikof, et al.), 687–716. Cambridge: Cambridge University Press.

Nurse, L., McLean, R., Agard, J. et al. (2014). Small Islands. In: *Climate Change 2014: Impacts, Adaptation and Vulnerability Contribution of Working Group II in the Fifth Assessment Report of the Intergovernmental Panel on Climate Change*, 1613–1655. Cambridge: Cambridge University Press.

OECS (2004). *Grenada: Macro-Socio-Economic Assessment of the Damages Caused by Hurricane Ivan, September 7 2004*. Castries, St. Lucia: Organisation of East Caribbean States (OECS), OECS Secretariat, 126 pp.

Park, L.E., Siewers, F.D., Metzger, T. et al. (2009). After the hurricane hits: recovery and response to large storm events in a saline lake, San Salvador Island, Bahamas. *Quaternary International* 195: 98–105. https://doi.org/10.1016/j.quaint.2008.06.010.

Peduzzi, P., Chatenoux, B., Dao, H. et al. (2011). *Preview Global Risk Data Platform*. Geneva, Switzerland: UNEP/GRID and UNISDR.

Pelling, M. and Uitto, J.I. (2001). Small island developing states: natural disaster vulnerability and global change. *Global and Environmental Change Part B: Environmental Hazards* 3 (2): 49–62. https://doi.org/10.1016/S1464-2867(01)00018-3.

Richmond, N. and Sovacool, B.K. (2012). Bolstering resilience in the coconut kingdom: improving adaptive capacity to climate change in Vanuatu. *Energy Policy* 50: 843–848. https://doi.org/10.1016/j.enpol.2012.08.018.

Scheffers, A. and Scheffers, S. (2006). Documentation of the impact of hurricane Ivan on the coastline of Bonaire (Netherlands Antilles). *Journal of Coastal Research* 22: 1437–1450. https://doi.org/10.2112/05-0535.1.

Scoffin, T.P. (1993). The geological effects of hurricanes on coral reefs and the interpretation of storm deposits. *Coral Reefs* 12: 203–221. https://doi.org/10.1007/BF00334480.

Scopélitis, J., Andréfouët, S., Phinn, S. et al. (2009). Changes of coral communities over 35 years: integrating in situ and remote-sensing data on Saint-Leu reef (la Réunion, Indian Ocean). *Estuarine, Coastal and Shelf Science* 84: 342–352. https://doi.org/10.1016/j.ecss.2009.04.030.

Strobl, E. (2012). The economic growth impact of natural disasters in developing countries: evidence from hurricane strikes in the central American and Caribbean regions. *Journal of Development Economics* 97: 130–141. https://doi.org/10.1016/j.jdeveco.2010.12.002.

Terry, J.P., Garimella, S., and Kostaschuk, R.A. (2002). Rates of floodplain accretion in a tropical island river system impacted by cyclones and large floods. *Geomorphology* 42: 171–182. https://doi.org/10.1016/S0169-555X(01)00084-8.

Terry, J.P. and Falkland, A.C. (2010). Responses of atoll freshwater lenses to storm-surge overwash in the northern Cook Islands, *Hydrogeology Journal* 18(3), 749–759. doi: https://doi.org/10.1007/s10040-009-0544-x.

27

Lessons from Cases of Coastal Risks Governance in the United Kingdom

Brian Golding[1], Thomas Waite[2], and Virginia Murray[2]
[1] *Met Office, Exeter, United Kingdom*
[2] *Public Health England, London, United Kingdom*

27.1 Introduction: Windstorms and their impacts in the UK

27.1.1 Windstorms

The global climate is fuelled by the temperature contrast between hot, tropical, regions, and cold, polar, regions. Where hot and cold mix, energy is released that drives the belt of westerly winds within which the mid-latitude storms occur. In the northern hemisphere, these are most evident in the Atlantic and Pacific oceans, where storms typically form to the east of a continent, intensify and move north-eastwards. North-Western Europe lies at the eastern end of the North Atlantic storm track, especially in winter. The energy of a depression is manifest as strong winds, both at the surface and higher up in the atmosphere, in the jet stream. At the surface, depressions are characterized by low central pressure surrounded by winds circulating anti-clockwise in a vortex or 'whirlpool' of air thousands of kilometres across. They are much bigger and less violent than a hurricane, which only occurs in tropical regions or a tornado, which is associated with severe thunderstorms. Within a depression, the warm, wet, air of tropical origin typically inhabits the south-eastern quadrant. As it rises above the colder, denser air of arctic origin along the warm and cold fronts, its moisture condenses to form cloud and rain. The intensity of a depression depends on many factors, including where it formed, whether there was a pre-existing tropical or arctic storm, what the sea temperatures were, and so on. In a typical winter season, one or two such depressions will produce damaging winds over parts of the UK and/or north-western Europe (Roberts et al. 2014). In this chapter, we refer to such damaging depressions as windstorms. Windstorms

Facing Hydrometeorological Extreme Events: A Governance Issue, First Edition.
Edited by Isabelle La Jeunesse and Corinne Larrue.

are often accompanied by dangerous sea conditions and by heavy rain, which may lead to the additional hazards of erosion of exposed coasts and river flooding. The classic study of North Atlantic windstorms is that of Lamb and Frydendahl (1991) which extends back to 1509. Recent work (e.g. as reported by Cornes 2014) is extending this in the light of newly recovered observations and the application of models.

27.1.2 Storm surges

When the track of a depression brings westerly winds across the seas to the north of Scotland, then turning to north-westerly winds down the North Sea coast, the combination of strong wind and low pressure can raise the sea surface into a surge that propagates down the east coast of Scotland and England and then up the west coast of the Netherlands and Denmark. Such surges can produce serious coastal flooding, especially if combined with a Spring Tide and high waves. It takes about 10 hours for a surge to move from the Firth of Forth to the Thames Estuary. Fortunately such surges rarely occur at the peak of the tide, but when they do, the consequences can be severe. Analysis of the severe flooding of London in 1928, which killed 14 and made thousands homeless, by Doodson and Dines (1929) led to the introduction of a warning service by which the Meteorological Office passed a warning to Scotland Yard if a Spring Tide was expected to coincide with North or North-West gales in the North Sea. Strong onshore winds also generate storm waves that cause additional coastal damage and flooding.

27.1.3 Tracks of windstorms

The occurrence of windstorms in heavily populated areas depends on the orientation of the North Atlantic storm track, which varies from day to day and year to year. Its 'normal' orientation takes the centres of windstorms to the north of Scotland and into the Norwegian Sea. An abnormally southerly track can take storms as far south as central France. On occasion, the storm track also becomes buckled, taking storms first northwards towards Iceland and then southeast towards the UK and France; or first eastwards towards France and then northeast across the UK.

27.1.4 Impacts of windstorms

Windstorms pose a variety of hazards to the UK. The wreck of the Royal Charter with the loss of nearly 500 lives in 1859 prompted the initiation of storm warnings by the fledgling Meteorological Office in 1861 (Walker 2012). Broadcasting of regular shipping forecasts for coastal sea areas was initiated in 1911 and has become a national institution. Whilst sailing ships predominated in coastal commerce and fishing (the latter until after 1920), their vulnerability to adverse winds was extreme, but even today, a major windstorm can cause shipping losses in the seas around the UK, such as the loss of the M. V. Napoli in 2007 (Rowbotham 2014). Closer to the coast, the growth of leisure activities has increased the risk from more modest winds and waves. It is fortunate that the main season for windstorms is winter, when these are much reduced. Nevertheless, the risk to those activities that remain is considerable and there is, sadly, a trend towards increasing loss of life to spectators driven to the coast to photograph storm waves (EA 2015). The west coast of the UK bears the brunt of the ocean waves produced by windstorms. Much of this coast is of hard, resilient rocks able to withstand the pounding of the surf. In these areas, surviving harbours are mostly in well-protected inlets. Nevertheless, in extreme storms, waves and surges can damage sea defences and overtopping can cause localized flooding, leading to loss of transport, energy, and telecommunications infrastructure. Heavy rainfall on steep coastlines can also produce simultaneous flash flooding in coastal settlements. Inland from the immediate coast, the wind strength reduces rapidly due to friction from the land surface, but the vulnerability of infrastructure increases correspondingly. Here, key infrastructure links for transport, electric power transmission, and telecommunications, are liable to be cut in extreme winds. Roads are particularly vulnerable to high sided vehicles being blown over, so exposed sections of the strategic motorway and trunk road network – especially high bridges – need to be closed pre-emptively when such high winds are predicted. On lesser routes, blowing down of trees is a major problem, especially from early autumn windstorms prior to the seasonal loss of leaves. In extreme winds, older buildings, temporary/mobile buildings, unsupported walls, and poorly anchored ancillary structures are vulnerable and, once damaged, may become dangerous debris that can kill or cause additional damage. High cranes and scaffolding are also liable to collapse in strong winds. Economic losses flow indirectly from many of these impacts, but also more directly from forestry losses.

27.1.5 Secondary impacts of windstorms

Human health is affected by windstorms in several ways. The direct effects, during a storm, include traumatic injury and death from blowing debris, falling trees, and traffic accidents. Indirect effects, or those that occur before or after the storm, include falls, accidental lacerations, or puncture wounds and are most often associated with preparation for, or cleaning up after, a storm (Goldman et al. 2013). The risk of carbon monoxide poisoning is elevated in all phases of a disaster due to the indoor use of portable generators, cooking, and heating appliances designed for outdoor use during periods of loss of mains power or gas. In the recovery phase, equipment for pumping, dehumidifying, and drying out of properties poses a risk. In the long term, mortality and morbidity associated with the renewed use of boilers which may have suffered covert damage in flooding is recognized but very difficult to quantify (Waite, Murray, and Baker 2014).

Loss of power poses a considerable risk. In the home, temporary heating arrangements are often dangerous and food can become contaminated due to loss of refrigeration. Power outages pose threats to healthcare infrastructure, affecting direct clinical care and life support equipment in hospitals, preventing access to electronic patient records, and increasing the demand on emergency services. Loss of power to support services such as radiology and blood bank can delay life-saving treatment. Health and social care in the home can be impacted including loss of home oxygen, interruption to dialysis, and difficulty accessing transport to healthcare locations. Vital public health infrastructure including clean water, sewage treatment, and temperature control may be interrupted (Klinger et al. 2014). Evacuation from the home is a traumatic experience, especially under life-threatening conditions, such as when the home is already flooded. Extended recovery periods, which can be a year or more, compound the trauma. Impacts on mental health can be profound, especially in already vulnerable people. An elevated risk of common mental health disorders such as depression, anxiety, and post-traumatic stress disorder (PTSD) is known to exist for at least a year after flooding (Waite et al. 2017), with a range of secondary stressors likely to affect recovery. Other health impacts include infections and insect bites.

27.2 Events that have shaped governance of natural disasters in the UK

In the section we will examine two extreme windstorms that have had a profound influence on the severe weather warning services provided to people in the UK. They exemplify the two extreme impacts of windstorms to a country surrounded by sea: destruction by direct impact of the extreme winds on buildings, trees, and infrastructure; and destruction through the generation of a storm surge that floods coastal towns and infrastructure. The first to occur was the storm surge, on 31 January 1953, which resulted in the creation of a storm surge warning service. The second was the land storm on 16 October 1987 that has led to the current National Severe Weather Warnings service.

27.2.1 Background to the governance of coastal windstorm protection: The 1953 flood

27.2.1.1 Description of the extreme event On 31 January 1953, one of the strongest North Sea windstorms of the twentieth century produced water levels not seen for more than two centuries (Waverley 1954). A rapidly deepening depression (Figure 27.1) was situated northwest of Scotland on Friday, 30 January. Instead of continuing on a north-easterly track, which would have limited the strong winds to Scotland, the system was steered south-eastwards around a strengthening anticyclone to the west of Ireland, entering the North Sea on the morning of 31 January. The squeeze between the depression and the anticyclone produced very high winds in eastern Scotland, with a record gust of 125 mph (56 m/s) observed in Orkney, and major damage to buildings and forests. The storm claimed its first casualties when the Princess Victoria ferry, sailing from Stranraer in Scotland to Larne in Ireland, sank with the loss of 133 lives. As the windstorm moved down the North Sea, high pressure continued to strengthen to the west, further increasing the wind speed. The resulting storm surge propagated round the coasts of Eastern England and the Netherlands, amplifying as the sea became shallower in the southern North Sea where it arrived in the early hours of Sunday, 1 February. High water levels and accompanying waves overwhelmed flood defences along the east coast of England. The first breaches on the Northumberland coast occurred at 5 p.m. on Saturday and spread southwards over the following eight hours.

27.2.1.2 Impacts of the event The human impact of the disaster was exacerbated by storm disruption to telephone and radio services, so that many of those affected received no warning, and its full extent was unknown until morning. Amongst the worst affected areas were Felixstowe in Suffolk where 38 died when wooden prefabricated

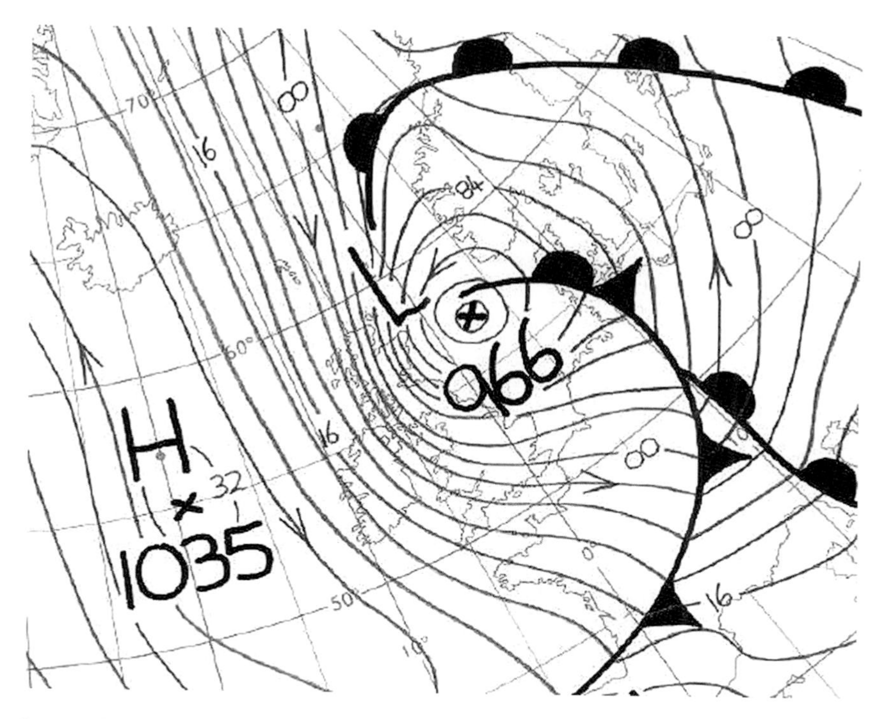

Figure 27.1 The surface pressure pattern at the time of the 1953 storm surge. © Met Office Crown Copyright.

homes were flooded and Canvey Island in Essex where 58 died when it was completely inundated. Another 37 died in the seafront village of Jaywick near Clacton. Floods reached 8 km inland around Cley on the north Norfolk coast, and were 1.8 m deep in the town centre of Kings Lynn. Inland communities were flooded as the surge travelled up the Rivers Ouse and Nene, causing both rivers to break their banks. Along the south bank of the River Thames defences were breached, and the floodwater reached the top of embankments in central London. The subsequent government inquiry identified that, along the 1600 km of affected coast, there were 1200 flood defence breaches, 647 km^2 of land flooded, 24 000 houses and around 200 industrial facilities damaged. The cost was £40–50 million; equivalent to around £1.5 billion in current values. Most properties did not have flood insurance at the time, so the proportion of costs repaid by insurance was very small. There were 307 deaths due to drowning or exposure with the elderly and those living in poorly constructed coastal dwellings at greatest risk (Baxter 2005). A protracted increase in overall deaths was reported in the Canvey Island area (Lorraine 1954) but the reasons for this have not been well explained. In a statement to the House of Commons, the then Home Secretary, Sir David Maxwell Fyfe, reported that hospitals saw an increased number of patients suffering from exposure, at a time when they were already unusually busy due to an influenza epidemic (Hansard 1953). Over 32 000 people were evacuated. A vast recovery exercise involving 30 000 emergency workers, including the military, concentrated on rescuing flood victims and then on plugging the flood defence breaches.

Devastation was even worse in the Netherlands, where 1783 people died and 1335 km^2 (330 000 acres) of land were inundated. Hundreds of breaches in the sea defences led to the flooding of 47 300 buildings, at a cost of 1 billion guilders (€250 million). The disaster prompted the Dutch Government to launch a massive flood protection plan, called the Delta Works, designed to protect against surges with a return period of up to 1 in 10 000 years (RMS 2003).

27.2.1.3 Societal responses Following the disaster, the UK Government appointed the Waverley Committee to carry out an inquiry, which reported in 1954 (Waverley 1954, Synnott 1964, Keers 1966, Townsend 1982). It found that the storm surge had been predicted successfully and warnings issued more than 12 hours in advance by the Meteorological Office but that they were not adequately disseminated due to the lack of a national coordinated system. It also found that, because many telephone lines in Lincolnshire and Norfolk were brought down by the wind, the confirmation of the storm's severity had not been passed to counties farther south until too late. The inquiry recommended the setting up of a flood warning organization to prevent the loss of life in any future event.

27.2.1.4 The Flood Warning Organization The resulting Flood Warning Organization was set up in Autumn 1953 by the Ministry of Agriculture to operate during the months of August to May, staffed by the Hydrographic Department of the Admiralty, using tide gauges provided by the Ministry of Transport and housed by the Meteorological Office in its Central Forecasting Office, with full access to the latest wind forecasts. It was originally applied only to the East Coast of England, where the coast was divided into five divisions, each with a tide gauge installed at a reference port, e.g. the reference port for Northumbria and Durham division was North Shields. Tide gauges were read by hand and telegraphed to the Met Office. For each division, a danger level was identified at which flooding would occur somewhere along its coastline, e.g. for Northumbria and Durham division the first flooding would be at Blyth, when the water level reached 3.1 m. Empirical prediction equations were developed based on Corkan (1948), which combined surge heights observed further north, with predicted wind and pressure conditions at selected locations and times that had been shown to correlate with the surge height. Warnings were issued by telegraph to River Boards, local police authorities, and other key authorities in the divisions affected. River Boards then used their local knowledge to advise the police, who passed the information on to those able to take action, e.g. to move ships moorings, deploy sandbags, reinforce defences, pumps, move livestock, and so on, and to the public if necessary. An 'alert' indicated possible overtopping 12 hours ahead, whilst 'danger' indicated probable overtopping four hours ahead. Subsequently, following surge-related flooding in some west coast locations, the service was extended. However, unlike the East coast, there is no progression of the surge along the west coast, so conditions are local to each part of the coast, and predictions much less accurate.

In 1963, the Flood Warning Organization was renamed the Storm Tide Warning Service (STWS) to remove confusion with river flooding. At the same time, distant reading tide gauge recorders were installed to save reporting time and transcription errors, and dissemination was changed to telex, direct from the Meteorological Office to police HQs (Synnott 1964).

27.2.1.5 Predicting storm surges The 1953 disaster provided an impetus to early experiments in computer weather forecasting within the Meteorological Office, which had started in 1952 using hired time on the LEO (Lyons Electronic Office) computer (Hinds 1981). The storm was used as an early test case, results showing an over-prediction of intensity, indicating that further research was needed before such forecasts could inform better warnings. A computer weather forecasting system was finally introduced into Meteorological Office operational forecasting in 1965 (Walker 2012), and by 1972, when the upgraded 10-level model was introduced, the quality of its 24-hour forecasts was able to match that of human forecasters at the scale of the depression. During the 1970s, a computer model of the behaviour of the coastal ocean surrounding the British Isles was developed at the Institute of Oceanographic Sciences (IOS), Bidston (Flather and Davies 1976). Following similar methods to those used in weather forecasting models, the IOS scientists produced a model able to take account of the variability of the weather, and including the interaction between the surge and the normal tide. Variability of the ocean between sea floor and surface was removed by averaging, and spatial variability was represented in $\frac{1}{3}^{\circ} \times \frac{1}{2}^{\circ}$ latitude/longitude columns of ocean (Flather 1982). Wind and pressure information was passed from the Meteorological Office's weather forecasting model, the 10 level model, with a 100 km grid (Burridge and Gadd 1977). Following trials in 1978–1981, the model was introduced into the Meteorological Office computer forecasting system as operational support to the STWS (Flather 1982). Finer resolution models were also developed to provide additional guidance for Liverpool Bay, the Severn Estuary, and the Thames Estuary.

27.2.1.6 Changing governance In 1995, the flood management roles previously carried out by Rivers Authorities in England and Wales were moved to the Environment Agency (EA), a new 'arm's length' government agency. The Scottish Environmental Protection Agency (SEPA) subsequently took on a similar role in Scotland. In 2005, responsibility for the STWS passed from Department of the Environment, Food and Rural Affairs (Defra) (the successor to the Ministry of Agriculture) to the EA and in 2009 was renamed the 'UK Coastal Monitoring and Forecasting' (UKCMF) service to more accurately reflect its role (EA 2009).

27.2.1.7 The Thames Barrier Another major outcome of the 1953 disaster, fuelled by concern over rising sea levels and the potential catastrophe if London were to be flooded, was the building of the Thames Flood Barrier. Although this was proposed in principle by the Waverley Committee (Waverley 1954), the challenge of providing adequate protection on a river that was heavily used for navigation proved difficult and it was not until the London Flood report (Bondi 1967) that a viable engineering solution was identified. It was concluded that a moveable barrier across the Thames together with raising of downstream embankments provided a more acceptable and cost-effective solution than a general raising of embankments through London or a fixed barrage with locks. Work on the barrier at Woolwich, 13.6 km downstream of London Bridge, started in 1975 and finished in 1982, creating the world's second largest movable flood barrier. It was designed to provide protection from a surge with an annual probability of 0.1% in 2030 (Horner 1982), taking account of the long-term rising trend in sea level of approximately 0.08 m per year, caused mainly by large-scale geological subsidence of south-east England (Dunham 1972). Downstream embankments were strengthened to the same level of protection. The barrier is designed to allow ships to pass in normal times, but flood gates, that are normally concealed in the river bed, rotate upwards and form a barrier to stop storm surges in times of need. Navigational requirements mean that a warning of 1.5–2 hours must be given when the gate is to be closed. Closure is supported by dedicated storm surge predictions and close liaison is maintained with the STWS. The barriers can be closed to prevent a surge from propagating upstream or to hold back flood water coming down the River Thames to prevent downstream flooding. In recent years, the frequency of closure has increased dramatically and concerns have been raised that the standard of protection is being reduced by sea level rise. However, studies have shown that the barrier will continue to provide adequate protection until mid-century at least (EA 2010).

27.2.2 Background to the governance of storm warnings: The 1987 storm

27.2.2.1 Description and impacts of the event On 16 October 1987, a deep depression (Figure 27.2) travelled north-eastward across southern parts of the UK causing 18 deaths, the loss of 15 M trees, and leaving hundreds of thousands of houses without power. The highest hourly-mean speed observed was 75 knots, at the Royal Sovereign Lighthouse, with Force 11 (56–63 knots) winds observed in many coastal regions of south-east England (Figure 27.3). Inland, however, their strength was considerably less. The cost to the insurance industry was £2bn. Damage to communications infrastructure meant that TV and radio services were reduced to skeleton services and the Meteorological Office itself was almost completely isolated. Despite the peak of the storm occurring overnight, which may have reduced the number of people exposed to the storm, there were 18 fatalities. Caravans and mobile home parks were particularly badly affected (Baxter et al. 2001).

27.2.2.2 Organizational responses The Meteorological Office conducted an internal inquiry, (Meteorological Office 1987; Houghton 1988) scrutinized by two independent assessors. They found that, as early as the previous weekend, the Meteorological Office had forecast severe weather for the Thursday and Friday. However, the northward extent of the severe gales became less clear as Thursday approached. Warnings for the English Channel were both timely and adequate, escalating from gales early on 15 October to Storm Force 11 in the early hours of 16 October. Special warnings to offshore operators, to airports, to the Eastern Region of British Rail for overhead power supplies, and to the London Fire Brigade were also timely. However, the main means for the Meteorological Office to inform the public was through radio and TV and, although these forecasts mentioned strong winds, they gave more emphasis to rain. The warning mechanism for informing the public of the likely occurrence of dangerous weather was the Flash weather message, which was supposed to be issued up to three hours before onset, when there was virtual certainty of its occurrence, usually as a result of actual observations, e.g. by police. Flash messages covered 18 major urban areas and 6 hazards including severe inland gales, the criterion for which was mean

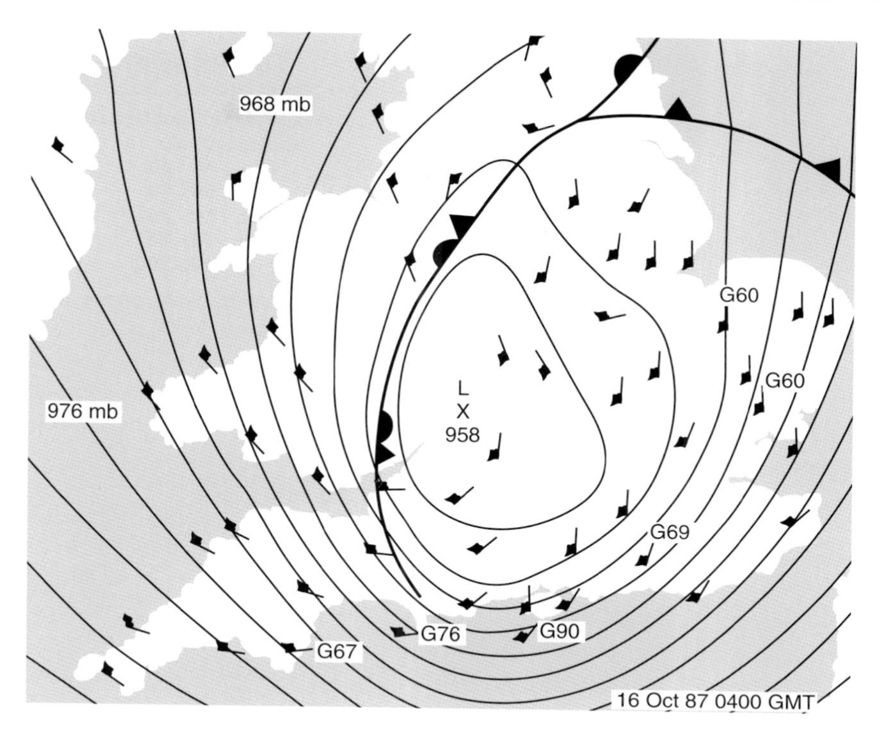

Figure 27.2 Surface pressure pattern at the time of the peak damage over south-eastern England. © Met Office Crown Copyright.

winds of over 40kn and/or gusts of over 60kn, speeds that are extremely rare in South-east England and which were only exceeded in a few exposed locations. Flash messages were telexed to the BBC for broadcast on national BBC TV and Radio and to some independent TV and Radio stations. A Flash warning of severe gales was issued shortly after midnight giving between one and three hours warning of the strongest winds. Warnings were also issued around this time to the Ministry of Defence, indicating that military aid might be required by the civil community, and to Police Forces in the South-East of England.

The inquiry made several recommendations, including: arresting the deterioration of observation availability to the south west of the UK by increasing the quality, quantity, and usage of observations from ships, aircraft, buoys, and satellites; enhancing computer power so as to enable the introduction of higher resolution models and more frequent forecasts; improving training and procedures for dealing with the press and media, especially in extreme situations; reviewing Meteorological Office warnings of severe weather, especially the process for informing emergency authorities and the style of public forecasts delivered by the BBC. This last recommendation led to the introduction of the National Severe Weather Warning Service (NSWWS) (Houghton 1988), key aspects of which were: national coverage, availability to emergency responders as well as TV/radio; earlier issue; and initial dissemination by fax and/or email. The service consisted of four types of warning. Early warnings were issued when there was 60% confidence of disruption and were depicted as a % risk on a map. Flash warnings

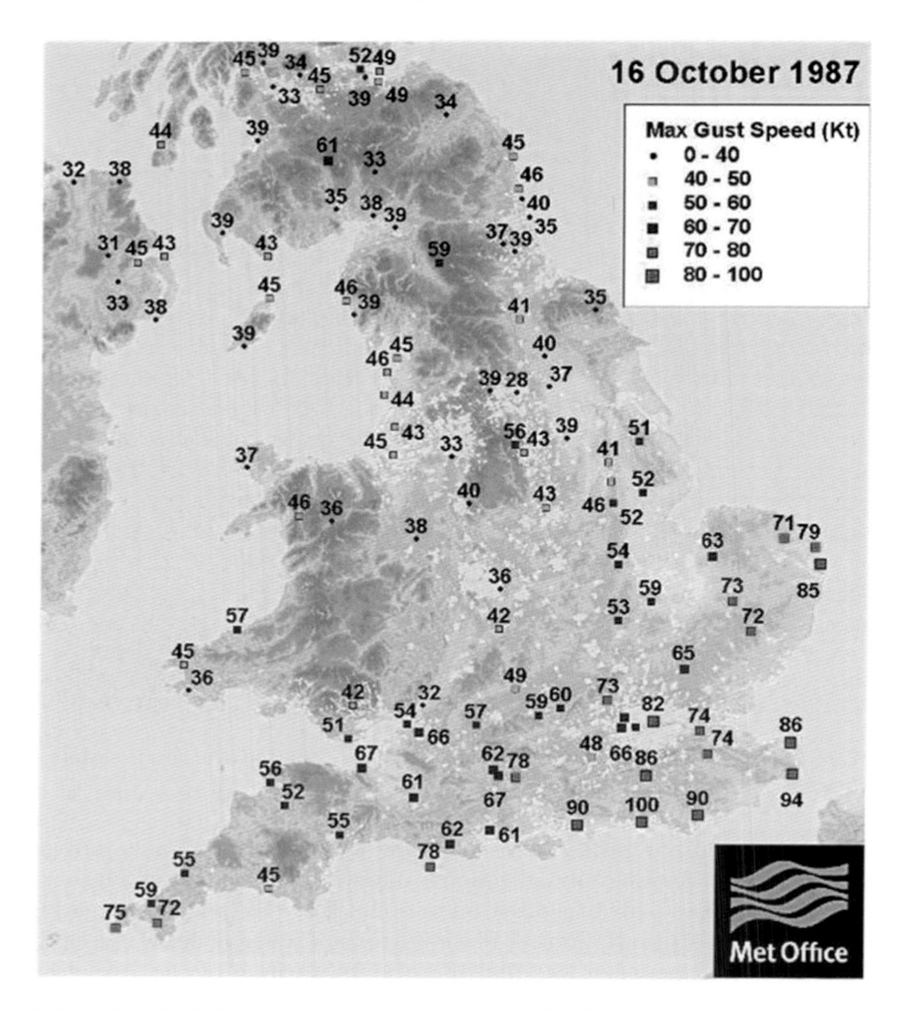

Figure 27.3 Highest wind gust speeds recorded across England on the night of 16 October 1987 in knots (nautical miles per hour). © Met Office Crown Copyright.

were issued when there was 80% confidence of severe weather exceeding defined thresholds, in the next 24 hours. Emergency flash warnings could be made for exceptionally heavy rain or snow. For lower travel-related thresholds, motoring unit warnings continued to be issued until 2005.

Further changes were made in 2007, following the Civil Contingencies Act (UK Government 2004), including the introduction of Civil Contingencies Advisors, to provide guidance to emergency responders, and of an additional class of warning, the Advisory, issued when there was a probability of 20–60% of severe weather likely to cause disruption. Introduction of the advisory recognized the growing skill in quantification of probability by ensemble numerical weather prediction (NWP) systems (Legg and Mylne 2004). Warnings were communicated both in text and on maps and were county-specific. Two levels of severity were identified: severe and extreme; and warnings were colour coded, as yellow, amber, or red, depending on both severity and likelihood. Further advances in 2011 made the warnings risk-based, formally moving from hazard thresholds to impact thresholds in the definition of warning level.

27.3 New developments in the warning environment

27.3.1 Changing exposure and vulnerability

Since 1953, increases in population and wealth have significantly increased the amount and value of property within the area flooded on 31 January 1953. Flood coverage is now included as standard in residential and commercial insurance policies in the UK (although remains unavailable in the Netherlands). A repeat of the 1953 flooding would therefore create one of the most expensive natural disasters for the insurance industry in the UK. In 2003, Risk Management Solutions estimated insured losses of a repeat 1953 event at £5.5bn, split equally between residential and business property. With current coastal defences, including the Thames Barrier, the estimated loss was £0.5bn (RMS 2003).

27.3.2 Changing forecasting capability

Continued research and development in computer weather forecasting, coupled with increasing computer power and ability of satellites to observe the atmosphere, has resulted in a steady improvement in accuracy of about one day's accuracy per decade in UK weather forecast accuracy since the early 1970s. In other words, a forecast now can predict a depression five days ahead to the same accuracy as a one-day forecast then, whilst accurate predictions of detailed weather at city scale are now possible up to a day ahead. The introduction of 'ensemble' forecasting of multiple scenarios from slightly differing initial states has enabled confidence to be quantified and, in situations where there is high sensitivity to unobservable differences in the initial state, for multiple outcomes to be communicated along with the corresponding risks, so that appropriate precautions can be put in place.

27.3.3 Changing governance

In 1992, the Met Office became an 'arm's length' government agency and subsequently a Trading Fund. As a result of these changes, the Public Weather Service Customer Group (PWSCG), was created to act as a surrogate customer for free-at-the-point-of-use weather services and on behalf of Public Sector users of weather services. It is responsible for setting requirements, specifying outputs, and supporting R&D to meet future requirements. An important role is to oversee the efficacy of the NSWWS.

27.3.4 Introduction of impact-based warnings

A key change to the NSWWS was agreed in 2011 (Forrester and Davies 2014), when warnings based on hazard thresholds changed to warnings based on socio-economic impact. Each hazard warning is classified according to a two-dimensional matrix of probability vs impact, so each cell of this matrix represents an estimated level of risk and is coloured accordingly. Thus the simplest description of a warning is by its risk colour, the next level by a specific combination of probability and impact, enabling identification of low probability, high impact events, and finally there is a textual interpretation. Whilst the translation of weather hazard to impact was initially entirely subjective, quantitative tools are gradually being developed to support the impact assessment, using vulnerability and exposure maps.

27.3.5 Changing roles and crisis planning

The Civil Contingencies Act 2004 (CCA) was enacted in the wake of perceived weaknesses in the response to major disasters such as the 2000 floods and the 2001 Foot and Mouth disease outbreak. It defined an 'emergency' (human or natural) that threatens serious damage to human welfare. It superseded the concept of civil defence against military threats that grew up following the Second World War, and broadened the scope of those involved beyond local government and blue-light services to include many utilities and infrastructure authorities. It lays down requirements for risk assessment and for contingency plans, and mandates use of a Gold-Silver-Bronze command structure. The CCA defines two types of emergency responder. Category 1, the core group of responders, consists of Local Authorities; police, fire, ambulance, and coastguard services, National Health Service trusts, Public Health England (PHE) and port health authorities, the EA, SEPA, and Natural Resources Wales (NRW). Category 2, consists largely of infrastructure operators: electricity, gas, water, and telecommunications providers; railway, highways, airports, and harbour companies; together with the Health and Safety Executive and voluntary agencies. In an emergency, each responder has its own command structure but contributes to a multi-agency structure, hosted, and chaired by the police. The gold commander provides remote strategic oversight. The Silver Commander manages its implementation, formulating actions that are completed locally by Bronze. The multi-agency silver command is typically located in a command vehicle at or near the scene. Supporting legislation requires category 1 responders to have regard to the Met Office's duty to warn the public, and provide information and advice, if an emergency is likely to occur or has taken place. This duty includes issuing severe weather warnings and pollution plume predictions, together with tidal alerts (provided by the Flood Forecasting Centre [FFC]).

Under the CCA, PHE is responsible to the Secretary of State for Health as a Category 1 Responder. PHE has developed guidance, templates, and tools for key roles, responsibilities, and processes required in the activation, escalation, and de-escalation of health protection incidents, including infectious disease outbreaks and chemical hazards. During such incidents, PHE carries out risk assessments, provides information and warnings, and cooperates with partner agencies. PHE also undertakes longer-term preparation and planning to reduce vulnerability by:

1. Developing and maintaining key relationships with Health and Wellbeing Boards, Local Authorities, local resilience forums, local health resilience partnerships, National Health Service commissioners, and the providers of public health services from the public, third, and independent sector to support the delivery of improved outcomes for the public's health;

2. Highlighting the need to consider the risk of flooding arising from the impact of local planning and spatial design decisions and encouraging systematic review of risks;
3. Monitoring indicators from the Public Health Outcomes Framework
4. Researching, collecting, and analysing data as well as reviewing and sharing evidence and knowledge about the public health impacts of flooding and how to prevent and mitigate such impacts;
5. Working with partners to promote and protect the public's mental health and wellbeing as an integral part of the overall response to an emergency; and
6. Providing leadership and expertise in public health crisis communication during emergencies, including the provision of a 24/7 service to the media at national and regional level.

27.3.6 Changes in flood forecasting and warning

In summer 2007, severe river flooding affected much of England as a result of prolonged heavy rain. In 2008, a Government Inquiry led by Sir Michael Pitt, (Pitt 2008) recommended that communication between the Met Office and the EA should be improved and that flood and rainfall warnings should be coordinated to reduce confusion and ambiguity by the formation of a joint FFC, which would contain both meteorological and hydrological forecasters and would be connected to the prediction systems of both organizations. The FFC came into operation in 2009 (FFC 2017). Subsequently, the dual forecasting role has been further strengthened by cross-training meteorologists in hydrological forecasting and hydrologists in meteorological forecasting. Similar objectives have been achieved in a different way in Scotland, where the joint FFC is a virtual operation linking SEPA forecasters with Met Office Aberdeen forecasters. From its start, the FFC has taken on the role of providing a national overview of coastal flooding risk.

The FFC has been used as an exemplar by many other countries. In the UK its success has led to the coming together of a broader partnership covering all natural hazards including windstorms. The Natural Hazards Partnership (NHP 2017) of 17 UK public bodies provides the authoritative scientific voice to government on all natural hazards in the UK and is developing the capability to provide integrated forecast advice on the principal hazards and their socio-economic impacts on society.

27.4 How the warning systems work now

27.4.1 UK Coastal Monitoring and Forecasting (UKCMF)

Within the UK, assets worth an estimated £132 bn are currently at risk from flooding by the sea (EA 2011). These values will grow as the climate changes, with sea levels predicted to rise and storms possibly becoming more frequent and intense. The UK government spends around £325 million a year maintaining sea defences and on-shore protection along 4300 km of coastline. The UKCMF costs £2.3 million a year to operate and is estimated to deliver benefits in the order of £23 m per year in avoided flood damage. Its main function is to deliver an operational forecasting service to operational authorities. It is the primary coastal forecasting tool for the EA, the SEPA, NRW, and the Department of Agriculture and Rural Development Northern Ireland (DARDNI).

The UKCMF Service is provided by a partnership of public bodies with specialized expertise, who work together to provide comprehensive UK-wide coastal monitoring, forecasting, and consultancy. It is the primary source of coastal flood risk alerts around the UK to assist the emergency response community in planning and preparing for flood events. The Defra, the Scottish Government, the Welsh Assembly, and the Northern Ireland Assembly provide funding to the EA to manage the Service. The five key partners/suppliers are: the FFC operates the 24-hour Tidal Alerts Forecast and Outlook Service; the Met Office hosts and runs the operational storm surge model and delivers the forecast output; the National Oceanography Centre (NOC) maintains and develops the national tide gauge network and the storm surge models, carries out related research, and delivers tidal predictions; the Centre for Environment, Fisheries and Aquaculture Science (CEFAS) manages the WaveNet wave observing system; and the British Oceanographic Data Centre (BODC) provides a national facility for all marine science data, and archives and disseminates the tide gauge data. UKCMF products include: forecasts of coastal water levels, wave conditions, and flood risk; real time observations of wave and tide data; astronomical predictions of tide levels; archived data and information; and supporting advice. It does not provide warnings directly to the public: such warnings may be issued by the operational authorities on receipt of UKCMF products (EA 2009, 2011).

27.4.2 National Severe Weather Warning Service

The NSWWS (Met Office 2017) provides warnings, to emergency responders and the general public, of the risk of threat to life, damage, and disruption from wind, rain, ice, snow, and fog. The Met Office provides other warnings on behalf of, and in partnership with, other government departments and agencies, but the NSWWS is the core warning service for the public good. Warnings are issued when thresholds of likelihood and impact are crossed, up to five days ahead. Impact thresholds are:

Very Low: Some localized, small scale impacts to everyday activities; a few transport routes affected.

Low: Short lived disruption to everyday activities; 'business as usual' incidents for emergency services; some transport services affected.

Medium: Injuries, with danger to life; disruption to everyday activities; transport services affected with some travellers stranded; damage to buildings; disruption to utilities; short-term strain on emergency responder organizations.

High: Danger to life; prolonged disruption to everyday activities; transport seriously affected with long delays and travellers stranded for long periods; extensive damage to property and disruption to utilities for a prolonged period; prolonged strain on emergency responders' organizations.

Likelihood is also assessed in four categories from 'very small chance of' through 'possibility of' and 'likelihood of' to 'expected/will happen'. Impact and likelihood form the columns and rows, respectively, of a 4×4 matrix, in which an entry is selected for each hazard/area to be warned. Each box in the matrix is coloured according to the level of risk, from green, through yellow, orange to red, as follows:

Yellow: Be aware. Severe weather possible and could affect you. Think about possible travel delays or disruption of day-to-day activities. Be aware that weather may worsen, leading to disruption of your plans.

Amber: Be prepared. Increased likelihood of bad weather affecting you, which could disrupt your plans and cause travel delays, interruption to power, and risk to life and property. Be prepared to change your plans and protect you, your family, and community from the impacts of the severe weather.

Red: Take action. Extreme weather is expected. You should take action now to keep yourself and others safe. Widespread damage, travel and power disruption and risk to life are likely. Avoid dangerous areas and follow the advice of emergency services.

The level and nature of the risk in each warning area is amplified in the accompanying text. Severe weather warnings are communicated through radio, TV, the Met Office website, social media, smart phone apps, RSS, and via email alerts. The web site provides multiple levels of detail and an alerting function through a 'ticker' at the top.

27.4.3 Civil contingency advisors

The warnings are supported by a team of Civil Contingencies Advisors based around England, Scotland, Wales, and Northern Ireland who ensure that emergency responders have a thorough understanding of the warning services and are fully prepared when severe weather is forecast. They engage with multi-agency resilience groups, such as Local Resilience Forums in England and Wales, made up of representatives from the emergency services, local authorities, health sector, utilities, and so on. Advisors also engage with coordinating departments in Central Government and the devolved administrations. They help emergency responders assess risk in their particular area; assist with multi-agency exercises; provide presentations, workshops, and so on on Met Office services and help with weather-related risk-assessments. During an incident, they represent the Met Office at multi-agency meetings, ensure that responders have consistent information, act as a point of contact for specialist advice, and help with interpretation of information on 'Hazard Manager', the dedicated Met Office web site for emergency responders.

27.4.4 Naming of windstorms

In 2015, the Met Office and Met Éireann responded to confusion being caused by multiple organizations putting forward different names for windstorms affecting the British Isles by introducing an authoritative naming system. A name is given only when a storm warning of likely disruption in Ireland or the United Kingdom has been issued. Analysis of the first year's operation, during which 11 storms were named, showed an increased level of awareness and response to warnings (Met Office 2016).

27.4.5 Responsibility for health warnings

As a category 1 responder, PHE is responsible for assessing, warning, and managing the risks to health arising from a national emergency. In support of this role, PHE maintains guidance, templates, and tools for key roles, responsibilities, and processes required in the activation, escalation, and de-escalation of health protection incidents, including infectious disease outbreaks and chemical hazards.

27.4.6 Operation of warning systems in the stormy winter of 2013–2014

The winter of 2013–2014 was exceptionally stormy in the UK and provided a severe test of the warnings system. From late December to early February a succession of damaging storms produced major coastal erosion and flooding as well as inland flooding (EA 2016). However, the first major windstorm occurred much earlier, on 28 October. The so-called St Jude's Day storm exemplified well the advances in science and governance. The initial weather warning was given five days ahead, but like the 1987 storm, there remained uncertainty in its precise track right up to the day before its arrival. This was communicated in the form of more likely and less likely scenarios with impacts of each. The risk warning took account of the relative likelihoods of the two scenarios and so appropriate levels of precaution were taken for both eventualities. The more likely scenario turned out to be the correct one, and the precautions prevented major disruption or fatalities, despite significant damage in many areas.

A major windstorm and storm surge occurred on 6 December. The surge forecast, driven directly from the weather forecast, indicated overtopping of defences with the risk of breach along much of the coast of East Anglia. Several thousand people were evacuated as a precaution. In the event, there was slight flooding, but most populous places were spared flooding by a very small margin.

A pair of major windstorms occurred on 24 December and 26 December. The first was the more intense and resulted in wind damage across much of southwest England as well as widespread river flooding. Transport was disrupted and main routes out of the southwest of England were cut by landslides. As flooding became the main threat, the emphasis of the warnings shifted from NSWWS to the FFC's Flood Guidance Statement, which provides risk assessments for each forecast flood area, whether caused by coastal wave and surge, by river overtopping or by surface water flooding from intense rainfall, using the same 'matrix' approach as the NSWWS.

The last major storm came on 6 February, producing record-breaking wave conditions around the coast of southwest England, causing major damage to port infrastructure and washing away the support to the main railway line to London and the west of England. The storm warnings were once again issued several days in advance and were well responded to. The flood guidance statement and associated warnings again provided an accessible summary of the very complex flooding situation.

As with earlier events, an inquiry was held into the winter storms (EA 2016). The findings showed that the lessons of earlier failures had been well learned and that the warnings were timely and effective. Issues were raised about flood response and there remained some criticism of the perceived ambiguity between rainfall warnings and flood warnings, when the latter might start only a day or so after the rainfall had finished. Other lessons highlighted the potential links to climate change and how policies needed to change to reflect the implications for the future.

27.5 Current and future issues

27.5.1 Mental health impacts of storms

Immediate impacts from storms are usually due to injuries, infections, chemical hazards, and disruption to health services; the longer-term effects are less well understood and may arise from displacement, damage to homes, and loss of domestic utilities (WHO 2012). Following the repeated flooding events of recent years, there is now serious epidemiological research underway to try to understand what the medium- to long-term health impacts are of being flooded, and to what extent those are similar to responses to other disasters. Implications for mental health are profound. Both flooding and certain forms of disruption from flooding are associated with increased risk of psychological morbidity even amongst those who did not have floodwater in their homes. Levels of depression and PTSD amongst people whose homes were flooded were comparable to the rates of clinically diagnosable disorder amongst members of the public involved in major incidents (Waite et al. 2017). For example, 30–40% of those closest to the site of a terrorist attack have been found to have clinically diagnosable mental health disorder (Whalley and Brewin 2007). Warnings should make allowance for the likely impact on health and particularly mental health.

27.5.2 Renewable power generation

The shift towards using renewable energy sources has transformed parts of the coastal zone in recent years. A characteristic of these power generation systems is that they are deliberately placed in the path of natural hazards. To date, wind power has dominated, both on land and at sea. The best wind climate is found in the same locations that have the strongest winds, so engineering requirements are severe and it is not surprising that a wind generator was blown down in the 24 December 2013 storm. If wave power proves successful, the power generation structures will be equally exposed to the concentrated power of the Atlantic wave climate. Although tidal power structures can be less exposed, they are nevertheless open to attack by wave and surge. It is likely that new governance and warning systems will be required for these structures as reliance on them grows.

27.5.3 Making space for water

Since the report on the 1953 coastal flood, a radical change has taken place in policies towards both coastal and river flooding. Led by the European Commission's Water Framework Directive (EU 2000) and through reviews such as 'Making Space for Water' in the UK, the language of sea defence and land reclamation, has passed out of favour, to be replaced by terms such as managed realignment and sustainable urban drainage (Defra 2008). Nevertheless, there remains a balance between the move towards soft and temporary defences in less populated areas, with the continuing need to protect rapidly growing populated areas with hard defences. The increasing use of temporary defences demands earlier warning of hazardous conditions. At the same time, the relationship between central government and local communities is being redrawn following the Localism Act (UK Government 2011), leading to an increasing role for local government in hazard management. For the coastline, the results of these policies are encapsulated in Shoreline Management Plans (EA 2017).

27.5.4 Climate change

Behind much of the policy activity relating to storms and coastal flooding lies the spectre of climate change. With raised sea levels a confident prediction, both due to warming of the oceans and to continued sinking of the UK land mass, the problem of how to manage the coastline is a pressing one. In 2010 a horizon scanning review, TE2100, was published by central government into the protection of London by the Thames Barrier. It looked at how the risk to London from storm surges might grow and a variety of options for countering it. The conclusion was that the current protection would be adequate until mid-century, and that the nature of future enhanced protection would be better judged nearer to that time (EA 2010). A wide variety of impacts on the UK from climate change impacts on sea levels and storm intensity are identified in the latest Climate Change Risk Assessment for the UK (Committee on Climate Change 2017) including that 200 km of coastal sea defences would become highly vulnerable to failure if sea levels rise by 0.5–1 m, as is currently projected to happen in the next century.

27.5.5 Improving the skill of forecasts and the response to warnings

Within science, efforts to improve prediction of both the weather and its impacts, continue to bring success, particularly in refining areas at highest risk. Current initiatives in closer coupling of ocean, land, and atmosphere may resolve some of the uncertainties in surge prediction. Considerable effort is being put into achieving useful predictions on monthly to seasonal timescales that would enable substantial engineering responses as well as early preparation on flood avoidance and evacuation procedures. Some of the biggest research challenges of the present are, however, in the communication of risk. The plethora of communications channels now available means that the authoritative warning message can be reinforced through multiple formats and presentations, but may be dissipated by conflicting sources. At the same time, there remain vulnerable groups not reached by any of these channels or who do not understand the message when they receive it. These challenges are common to all hazards, whether natural or human originated, and an integrated approach will have considerable advantages. Models for future multi-hazard warning systems are being developed in many countries and it seems that there is no optimum 'one size fits all' system. At the international scale, research into the warning system as a whole is being promoted through the World Meteorological Organization's High Impact Weather project (WMO 2017) and in response to the Sendai Framework for Disaster Reduction (UNISDR 2015).

In conclusion, avoidable impacts of the disasters of 1953 and 1987 arose primarily from weaknesses in warnings procedures and communication. In the 1953 disaster, the surge was predicted at national level but there was no relevant warning system in place to communicate this information to the emergency responders in local councils, police, and fire authorities in the relevant counties and cities. In the 1987 disaster, a relevant national system was in place for warning of extreme wind speeds over land, but it had severe limitations as to lead time and recipients, which meant that most responders did not receive the warning in time for a useful response. Lessons learned from these failures have contributed directly to the strengths of the risk-based warnings systems of today, which provide, through a wide range of open and restricted dissemination routes, initial long lead time warnings at low probability, followed by progressively more confident and precise warnings as the event approaches. Nevertheless, there remains the fundamental challenge of preparing people for a low probability, high impact event, especially one that is outside their experience. Using the internet and mobile data communication, information that failed to reach decision makers in 1953 and 1987 is now easily available, but that does not guarantee that the right person will see the right information and take the right decision in a rapidly developing situation. Neither does it ensure that in a society with multiple levels of authority and responsibility, consistent information will reach all of the relevant actors. Continuing research is needed to identify all of the actors who will take decisions in an emergency, their needs for information to help them take life-saving decisions, and the resulting implications for risk governance: in advance of a disaster, during the crisis period and in the recovery period. With a changing climate and a growing population ever more dependent on infrastructure, the likelihood of vulnerable communities being affected by such disasters is ever increasing, reinforcing the need for an effective, agile and well-practised emergency management structure that evolves continuously in response to changes in the perceived threats.

References

Baxter, P.J., Lee, B.E., Wyatt, T.A. et al. (2001). Windstorms and climate change. In: *Health Effects of Climate Change in the UK* (ed. Department of Health), 134–152. London: HMSO.

Baxter, P.J. (2005). The east coast Big Flood, 31 January - 1 February 1953: A summary of a human disaster. *Philosophical Transactions of the Royal Society London series A* 363: 1293–1312.

Bondi, H. (1967). London flood report to Ministry of Housing and Local Government London: HMSO.

Burridge, D.M. and Gadd, A.J. (1977). *The Meteorological Office Operational 10-Level Numerical Weather Prediction Model (December 1975)*, Met. O. Sci. Pap. No. 34. London: HMSO.

Climate Change Committee (2017). *UK Climate Change Risk Assessment 2017 Synthesis Report: Priorities for the Next Five Years*. London: Committee on Climate Change. www.theccc.org.uk/wp-content/uploads/2016/07/UK-CCRA-2017-Synthesis-Report-Committee-on-Climate-Change.pdf (accessed 1 March 2019).

Corkan, R.H. (1948). Storm surges in the North Sea, Vols. 1 and 2. Washington DC: H.O. Misc. 15072.

Cornes, R.C. (2014). Historic storms of the Northeast Atlantic since circa 1700: a brief review of recent research. *Weather* 69: 121–125.

Departmnt for Environment, Food and Rural Affairs (Defra) (2008). *Future Water: The Government's Water Strategy for England*. London: HMSO.

Doodson, A.T. and Dines, J.S. (1929). Report on Thames floods and meteorological conditions associated with high tides in the Thames. *Geophys. Mem. London* 47.

Dunham, K.C. (1972). The evidence for subsidence of south East England. *Phil. Trans. Roy. Soc., London* A272: 79–274.

EA (2009). UK Coastal Monitoring and Forecasting (UKCMF) Service Strategy for 2009 to 2019, https://www.gov.uk/government/publications/uk-coastal-monitoring-and-forecasting-ukcmf-2009-to-2019-strategy (accessed 1 March 2019).

EA (2010). *Thames Estuary 2100 (TE2100) Plan*. London: Environment Agency.

EA (2011). UK Coastal Monitoring and Forecasting (UKCMF) Service: Service Definition, https://www.gov.uk/government/publications/uk-coastal-monitoring-and-forecasting-ukcmf-2009-to-2019-strategy (accessed 1 March 2019).

EA (2015). Thrill-seekers taking 'storm selfies' are risking lives. https://www.gov.uk/government/news/thrill-seekers-taking-storm-selfies-are-risking-lives (accessed 1 March 2019).

EA (2016). The costs and impacts of the winter 2013 to 2014 floods. https://www.gov.uk/government/uploads/system/uploads/attachment_data/file/501784/The_costs_and_impacts_of_the_winter_2013_to_2014_floods_-_report.pdf (accessed 1 March 2019).

EA (2017). Shoreline Management Plans. https://www.gov.uk/government/publications/shoreline-management-plans-smps/shoreline-management-plans-smps (accessed 1 March 2019).

EU, (2000), Directive 2000/60/EC of the European Parliament and of the Council of 23 October 2000 establishing a framework for Community action in the field of water policy. https://eur-lex.europa.eu/LexUriServ/LexUriServ.do?uri=CONSLEG:2000L0060:20011216:EN:PDF, (accessed 7 March 2019)

Flather, R.A. and Davies, A.M. (1976). Note on a preliminary scheme for storm surge prediction using numerical models. *Quarterly Journal of the Royal Meteorological Society* 102: 123–132.

Flather, R.A. (1982). Practical surge prediction using numerical models. In: *Flooding from Waves & Surges* (ed. D.H. Peregrine), 21–43. London: Academic Press.

Flood Forecasting Centre (2017) About Us. http://www.ffc-environment-agency.metoffice.gov.uk/about (accessed 1 March 2019).

Forrester, G. and Davies, S. (2014). The history of the NSWWS – thresholds to impacts. Exeter: Met Office, 19pp.

Goldman, A., Eggen, B., Golding, B. et al. (2013). The health impacts of windstorms: a systematic literature review. *Public Health* 128: 2–28.

Hansard (1953). HC Deb 19 February 1953 vol 511 cc1456–580. http://hansard.millbanksystems.com/commons/1953/feb/19/flood-disasters (accessed 1 March 2019).

Hinds, M. (1981). Computer story. *Meteorol. Mag.* 110: 69–81.

Horner, R.W. (1982). Flood prevention works with specific reference to the Thames barrier. In: *Flooding from Waves & Surges* (ed. D.H. Peregrine), 93–106. London: Academic Press.

Houghton, J.T. (1988). The storm the media and the enquiry. *Weather* 43: 67–70.

Keers, J.F. (1966). The meteorological conditions leading to storm surges in the North Sea. *Meteorol. Mag.* 95: 261–272.

Klinger, C., Landeg, O., and Murray, V. (2014). Power outages, extreme events, and health: a systematic review of the literature from 2011 to 2012. *PLOS Currents Disasters* https://doi.org/10.1371/currents.dis.04eb1dc5e73dd1377e05a10e9edde673, http://currents.plos.org/disasters/article/power-outages-extreme-events-and-health-a-systematic-review-of-the-literature-from-2011-2012 accessed 1 March 2019.

Lamb, H.H. and Frydendahl, K. (1991). *Historic Storms of the North Sea, British Isles and Northwest Europe*. Cambridge: Cambridge University Press.

Legg, T.P. and Mylne, K.R. (2004). Early warnings of severe weather from ensemble forecast information. *Weather Forecast.* 19: 891–906.

Lorraine, N.S.R. (1954). Canvey Island Flood Disaster, February, 1953. *Medical Officer* 91 (6): 59–62.

Met Office (1987). The Storm of 15/16 October 1987. London: HMSO.

Met Office (2016). Name Our Storms 2016. www.metoffice.gov.uk/news/releases/2016/nameourstorms2016 (accessed 1 March 2019).

Met Office (2017). Weather Warnings Guide. www.metoffice.gov.uk/guide/weather/warnings (accessed 1 March 2019).

NHP (2017). About us. www.naturalhazardspartnership.org.uk/about-us (accessed 1 March 2019).

Pitt, M. (2008). *Learning Lessons from the 2007 Floods*. London: Cabinet Office.

Risk Management Solutions Inc (2003). *1953 UK Floods, 50-Year Retrospective*. 11p. Risk Management Solutions Inc.

Roberts, J.F., Champion, A.J., Dawkins, L.C. et al. (2014). The XWS open access catalogue of extreme European windstorms from 1979 to 2012. *Nat. Hazards Earth Syst. Sci.* 14: 2487–2501.

Synnott, J.N.N. (1964). The storm tide warning service. *Mar. Obsr. London* 34: 77–83.

Townsend, J. (1982). Storm surges and their forecasting. In: *Flooding from Waves & Surges* (ed. D.H. Peregrine), 1–8. London: Academic Press.

Walker, M. (2012). Operational storm warnings. In: *History of the Meteorological Office*, 38–43. London: Cambridge University Press.

Rowbotham, M. (2014). Maritime and cargo security failures. In: *Maritime Transport Security, Issues, Challenges and National Policies* (ed. K. Bichou, J.S. Szyliowicz and L. Zamparini), 159–182. Cheltenham: Edward Elgar Publishing.

UK Government (2004). Civil Contingencies Act. www.legislation.gov.uk/ukpga/2004/36/contents (accessed 1 March 2019).

UK Government (2011). Localism Act. www.legislation.gov.uk/ukpga/2011/20/pdfs/ukpga_20110020_en.pdf (accessed 1 March 2019).

UNISDR (2015). Sendai Framework for Disaster Risk Reduction 2015–2030. https://www.unisdr.org/we/coordinate/sendai-framework (accessed 1 March 2019)

Waite, T., Murray, V., and Baker, D. (2014). Carbon monoxide poisoning and flooding: changes in risk before, during and after flooding require appropriate public health interventions. *PLOS Curr.* https://doi.org/10.1371/currents.dis.2b2eb9e15f9b982784938803584487f1.

Waite, T., Chaintarli, K., Beck, C.R. et al. (2017). The English national cohort study of flooding and health: cross-sectional analysis of mental health outcomes at year one. *BMC Public Health* 17: 129. https://doi.org/10.1186/s12889-016-4000-2.

Waverley, J.A., Home office, Scottish office et al. (1954). *Report of the Departmental Committee on Coastal Flooding*. London: HMSO.

Whalley, M.G. and Brewin, C.R. (2007). Mental health following terrorist attacks. *Br. J. Psychiatry* 190 (2): 94–96. https://doi.org/10.1192/bjp.bp.106.026427.

World Health Organisation (2012). *Floods in the WHO European Region: Health Effects and their Prevention* (ed. B. Menne and V. Murray). Geneva: WHO Press.

World Meteorological Organisation (2017). HIWeather. http://www.wmo.int/pages/prog/arep/wwrp/new/high_impact_weather_project.html (accessed 1 March 2019).

Part V
Conclusions, Perspectives

28

Hydrometeorological Extreme Events' Effects on Populations: A Cognitive Insight on Post-Traumatic Growth, Resilience Processes and Mental Well-Being

Mauro Galluccio

External Speaker to the European Commission, DG COMM, Brussels, Belgium
EANAM – European Association for Negotiation and Mediation, Brussels, Belgium

Health is a state of complete physical, mental, and social well-being and not merely the absence of disease or infirmity.

Preamble to the Constitution of the World Health Organization, 1946.

28.1 Introduction

Natural disasters represent potentially traumatic events. Not only do hydrometeorological extreme events threaten the global economy and societies at large, they can also have severe psychosocial consequences for populations such as heightened stress levels and compromise the well-being of entire populations. The American Psychological Association defines stress as 'the pattern of specific and nonspecific responses an organism makes to stimulus events that disturb its equilibrium and tax or exceed its ability to cope' (Gerrig and Zimbardo 2002). Tying the individual and social responses to hydrometeorological extreme events from a psychological standpoint is important to measure the potential loss of cognitive, emotional, and motivational resources and a

Facing Hydrometeorological Extreme Events: A Governance Issue, First Edition.
Edited by Isabelle La Jeunesse and Corinne Larrue.

whole individual and collective wellbeing in the case of natural disasters. Given the stress potential generated by these extreme events, a serious study on a population's ability to cope with hazards will depend on assessing and dealing with its vulnerability and resilience. The US Environmental Protection Agency defines vulnerability as, 'the degree to which an [economic, environmental or social system] is susceptible to, or unable to cope with, adverse effects of climate change, including climate variability and extremes. Vulnerability is a function of the character, magnitude, and rate of climate variations to which a system is exposed; its sensitivity; and its adaptive capacity' (IPCC 2012). As applied to environmental disasters, the European Commission defines resilience as, 'the ability of an individual, a household, a community, a country or a region to withstand, adapt and to quickly recover from stresses and shocks' (Lechner 2015). Resilience implies both stability and flexibility during unforeseen events.

Cultural differences, subjective assessments, resilience resources, and competing professional theories all affect how 'mental health' is defined. The World Health Organization defines mental health as 'a state of well-being in which the individual realizes his or her own abilities, can cope with the normal stresses of life, can work productively and fruitfully, and is able to make a contribution to his or her community'. Yet when decision making in extreme events is interactive (with the empowerment of local people) it is important to assess what the other side will *probably* do in order to limit uncertainties, prevent misunderstandings, and balance the maximization or minimization of risks. Negotiating a meaning for a sustainable governance requires a social mentality that accommodates a cultural sensitivity for both or more sides, which is often very difficult to achieve (Gilbert 2011). Political and psychological processes of transforming post-extreme events consist also of accompanying measures and techniques that should help to: (i) increase the resilience, cognitive, and emotional resources of parties; (ii) enhance behaviour modification; (iii) pave the way for healing the trauma; and (iv) monitor hopeless processes (Galluccio 2011). Resilience skills in the face of stressful events represent an important element of the mental capital. Mental capital encompasses both cognitive and emotional resources. It includes people's cognitive flexibility and efficiency at learning, their emotional competence, social skills, and resilience in the face of stress (Beddington et al. 2008).

Governance in this field is key for all countries, especially those frequently faced by hydrometeorological hazards and it should imply bottom-up solutions. Countries should incorporate resilience as a key component of effective humanitarian and development policies. In fact, resilience should be included in all European Commission's Humanitarian Implementation Plans. It necessitates strong methodologies and vulnerability assessments requiring both robustness and flexibility to be effective.

This chapter will provide a cognitive insight on stress and post-traumatic growth related to hydrometeorological extreme events. The goal is to better assess both the vulnerability and impact generated by this climate change issue in order to offer preparedness, response, and post-crisis solutions for individuals, societies, and ecosystems in the European Union (EU). This relates to the governance in the sense of the organization of actors and tools to coordinate ecological, social, and psychological issues from hydrometeorological extremes events. I will begin by discussing the concept of resilience and its evolution in order to demonstrate its relationship with ecological, social, and psychological issues and its buffer against traumatic developments. I will outline the strong relation between ecosystem degradation and socio-economic status by analysing humanity's negative ecological impact and how this increases vulnerability to hydrometeorological extremes. Examples are also taken outside Europe to provide insights for the chapter and reflections for the improvement of management of risks in Europe. In fact, disasters like tsunami and hurricane Katrina have brought mental health issues into the limelight. The consequences of climate change can affect nearly every aspect of human life, including physical and mental health, in the form of dysfunctional responses, such as trauma, burnout, anxiety, or depressive disorders, and emotional dysregulation.

28.2 Resilient ecological systems for a psychological concept

Etymologically speaking, the term 'resilience' originates from the Latin word *resalire*, meaning 'to spring back'. It was originally used in the physical sciences when referring to a spring's behaviour, but quickly came to have different connotations. In the 1970s and 1980s resilience acquired both a psychological and ecological dimension. The first use of resilience as a psychological concept is often attributed to Emmy Werner, who used it to describe Hawaiian children who, despite coming from poor or abusive backgrounds, did not show signs of destructive behaviours such as substance or physical abuse. Ecologically speaking, the use of the term resilience is credited to Canadian ecologist Crawford Stanley Holling, who wrote a famous paper on this topic in 1973 (Holling 1973; Gerrig and Zimbardo 2002) in which he argued that an ecological system's coping capacity heavily influenced the state of human societies. Resilient ecological systems were systems that could function in a relative state of normality despite having suffered from adversity. Engineers also came to use resilience to describe the coping capacity of physical infrastructure in the case of hazardous events. Eventually, the social science dimension of resilience came to move beyond individuals in order to study community phenomena. To this end, the Community and Regional Resilience Institute (CARRI) defined resilience as a community's ability to anticipate risk, limit impacts, and swiftly recover through survival, adaptability, evolution, and growth in the face of unexpected negative circumstances. This was broken down into four aspects of community resilience:

1. Attribution: resilience can be attributed to a community
2. Continuity: resilience is a central and dynamic part of the community
3. Adaptation: resilience is the community's ability to adapt to adversity
4. Trajectory: adaptive capabilities are positive for the community in post-crisis scenarios, especially when it comes to functioning in exceptional circumstances
5. Comparability: resilience can be compared from one community to the next, especially when it comes to adapting in the face of unexpected negative events (IPCC 2012)

28.3 Psychosocial factors and post-traumatic growth

Many of the consequences of hydrometeorological extreme events are psychosocial, especially when it comes to human wellbeing. Communities and individuals react differently to these events in relation to their frequency and intensity. In addition, factors such as personal outlook, location, class, gender, ethnicity, and age also determine an individual's or community's degree of vulnerability and stress in these risk situations. From a psychological perspective, human wellbeing is a key element of climate extremes. Two forms of mental health are usually distinguished in scientific literature. The first is Hedonic mental health, and it relates to positive feelings such as subjective wellbeing and satisfaction with one's life. The second refers to Eudemonic mental health, and refers to the positive functioning of an individual (motivation, engagement, fulfilment, sense of meaning, and social wellbeing). Mental health contributes to psychological resilience thinking because it also includes the social dimension of self-esteem, self-efficacy, optimism, hopefulness, perceptions, judgements, and meaning, which can be shaken in unfavourable socio-economic or environmental conditions. Average mental health improvements can have a positive impact on general public health, given that reducing the mean number of psychological pathologies increases the amount of people living flourishing lives. In psychology, flourishing can be defined as a state of mind, 'within an optimal range of human functioning, one that connotes goodness, generosity, growth and resilience'. High amounts of flourishing individuals, along with lower levels of mental illness, strengthens communities' links, increase social wellbeing, and may help reduce the psychological costs of climate extremes such as hydrometeorological extreme events.

Quality of life for individuals and communities is considerably improved if groups have robust solidarity networks and develop common feelings of 'togetherness'. Social and community networks are therefore an essential part of psychological resilience and reduced vulnerability (Lechner 2015). Communities and individuals have several resilience resources that can be used to cope with hazards. The first is social capital which can be described as social structures that minimize the negative impacts of a hazard or extreme on a neighbourhood or community. Several indicators can be used to measure a community's social capital, including average income, employment rates, housing, crime, family structures, and associative life. The second term is collective efficacy and describes the cultural beliefs, expectations, and trust that actors place on their social capital. According to several studies, strong social capital, and collective efficacy are negatively related to stress and positively related to physical and mental health. Analysts interested in these forms of communities have found sociological patterns in their findings. In the Western world, social capital and collective efficacy is found in communities with strong participatory government and civic structures, in addition to voluntary organizations, family bonds, friendships, and sufficient economic resources for the community's needs. This also contributes to high levels of trust and understanding, which translate into high psychological resilience and lower vulnerability in the case of climate extremes.

28.4 Building resilience to mitigate social vulnerability

On the opposite end of the spectrum are communities with low social capital and collective efficacy that have ineffective governing and civic structures, lacking associative life and social safety nets such as solid family structures. In addition, these communities have insufficient economic resources and trust. Resilient communities are those that have some forms of collective power. This collective power translates into social solidarity in times of crisis, facilitating both organization and psychological coping with disastrous events. Stress can be divided into primary appraisal and secondary appraisal. Primary appraisal is the process by which one evaluates the threat level of a hazard. Secondary appraisal is when one assesses the resources at their disposal to tackle the dangerous event. Applicable to individuals, families, and communities, these appraisals heavily influence resilience and vulnerability to hydrometeorological extreme events (Holling 1973). Resources can be divided into four types: object, condition, personal, and energy. The first refers to physical objects that facilitate coping and problem solving. These include cars and houses, in addition to items such as clothes or valuables. The second are situations or status symbols that help individuals and communities guarantee certain stability. This includes stable employment and family structures. Personal resources are an individual's technical and sociological skills, such as charisma, positive outlook, and ability to mend damaged parts of a house. Energy resources refer to material and immaterial capital that allows individuals to acquire other resources. This can be currency, know-how, or even owed favours. Individuals and communities with low resources are more prone to resilience issues (Friedli 2009). To this end, one of the ways in which responders can aid communities affected by hydrometeorological extremes is to assist them in finding the type of resources they need to increase resilience. Social support is vital, especially since social-ecological theorists agree that family networks are the second most important determinant of resilience after community structures and these are often lacking in low resilience areas (Fredrickson and Losada 2005). As can be seen, social capital and collective efficacy are extremely beneficial for individuals in reducing stress and increasing individual resilience both in daily lives and disaster situations. When developed, it can be a protective factor in the face of feelings of hopelessness and loneliness. If communities combine social capital with environmental capital (factors and features of the natural and built environment that improve a community's capacity for wellbeing) further improvements could be seen in the domain of psychological resilience. Key political actors are increasingly seeing mental health as an important component for psychological resilience in the face of climate extremes. International actors and the EU are beginning to embed the concept of governance related to these events tying mental health to physical health (Couch and Coles 2011).

There is a clear class element to which group takes the brunt of deteriorating ecological resilience and vulnerability. The World Bank believes that climate and disaster resilience is so tied to questions of social exclusion that it considers these factors to be essential to eradicating extreme poverty and attaining shared prosperity by 2030. One of the reasons why climate extremes will most likely affect poor populations is due to development and urbanization patterns in low and middle-income countries. For example, informal settlements with no clear urban planning have considerably higher rates of disaster risk vulnerability than planned settlements, which are often reserved for middle to upper class members of society (Hobfoll 1989). Heavy migration towards economic centres is a main contributor to this issue. In many cases national, sub-national, and local migrations involve poor rural populations moving to urban areas in search of better employment, education, and life quality opportunities.

Cultural and social narratives heavily influence the impact of hydrometeorological extreme events on stress and wellbeing. In contemporary societies media narratives and information technologies facilitate sensationalism and negative reactions even in resilient communities physically unaffected by extreme events. In the latter case, social media can produce instant and vivid images from disaster areas and lead to feelings of anxiety and fear. Climate extremes can psychologically impact in direct and indirect ways. Directly, extremes can cause mental health injuries due to the intensity of hydrometeorological hazards. Indirectly, cognitive-emotional responses such as fear and anxiety towards future risks and world phenomena can be just as devastating for psychological wellbeing (Couch and Coles 2011). Whatever its origins, low social resilience to hydrometeorological hazards has been directly related to increases in vulnerability, death, and violence (Friedli 2009). Surveys of Asian Tsunami victims found that 30–50% of the group now suffered from an array of psychological disorders such as anxiety, depression, and phobic and adjustment disorders (World Bank and GFDRR 2013). When studying Hurricane Katrina victims, a survey of 1043 subjects showed that the 30-day prevalence rate was 49.1% for anxiety-mood disorders and 26.4% for Post-Traumatic Stress Disorder (PTSD). Moreover, an exponential higher risk of mental distress was reported in southern England, in the town of Lewes after a severe flooding. Psychological consequences of hydrometereological extreme events may be represented by anxiety disorders, depression moods, and PTSD (Doherty and Clayton 2005). A shift of cognitive processes leads to a dysfunctional regulation of emotions, which leads to a more negative thoughts process affecting motivation and hope for the future. Behavioural change for the best or the worst is a natural consequence of such mental processes due to extreme events. Here comes to the fore the fundamental role of mental health professional units to cooperate together for a cognizant, ready, and quick response to crisis situations where pubic mental health and well-being is at the stake.

In addition to psychosocial effects in fixed populations, hydrometeorological extreme events are expected to cause 200 million displaced peoples by 2050. Not only will this affect countries all over the world, it will generate new forms of psychological troubles. Involuntary migration and losing home environments is conducive to a state of suffering that is place-based.

Psychological coping mechanisms are predicted through an adaptive protection of interpersonal motivation systems, reframed through social comparison with others, and selective information exposure. Standard emotions in these cases include fear, anger, sadness, surprise, helplessness, denial, and loss of hope. The usual lack of mental health infrastructure after hydrometeorological extreme events further exacerbates these symptoms in affected communities, severely affecting disaster preparedness on the psychological front.

28.5 Post-traumatic growth: Training for preventive psychological strategies

The human experiential dimension linked to the trauma is inwardly wide and statistically more frequent than we can talk about. The consequences of the trauma are not limited to the contextual and continuing emotional suffering beyond the extreme physically and psychologically experiences we call 'trauma'. The trauma may disrupt for a long time, even chronically, consciousness, and memory. It can induce or confirm mistrust in the meaning and value of life and human relationships. It can create a chain of mutual misunderstandings between the trauma victims and those who enter relationship with them. This will lead to an inner isolation, which is more dangerous as more outwardly, the social life of the person who suffered the trauma continues, develops, and grows-up in a superficially apparent normality. We now know for sure that the psychological consequences of trauma are trans-generational. They can be transmitted from generation to generation. The impact of severe trauma on people, who have not had the chance to elaborate the meaning and to be helped to overcome the divisive effects of trauma on memory and consciousness, could be transmitted to their children, from the first months of their life, as a sense of frightening and painful helplessness. Mental states, like those that follow trauma, may well be manifest and continued for long time in children who have not been exposed to mistreatment. Indeed, these are children of parents who strive to be especially loving and attentive. The vast and devastating inner experience of trauma, if not interrupted by an appropriate corrective experience, where necessary of psychotherapeutic value, crosses the boundaries of individuals, the time, and generations, invisible to most of the people. However, they may be perceived by those who have had a way to linger and reflect on these disturbing forms of human suffering. The scope, the intensity, and the rawness of the pain involved in trauma are often so strong that this prompts main actors, witnesses, and donors to try to ignore it, forget it in as fast a way as possible, or even worse to deny it. These defensive strategies are part of the drama, not the solution. The consequences, however, are that the effects of trauma are maintained and extended not only in the individual who suffers from it, but also in the society in which the individual lives. To overcome the pain of the trauma without ignoring it, or denying it, will be a first serious and compassionate awareness to reducing the suffering of the victims, but also a work of great social use. Then, a strategy to implement post-traumatic growth could have a better success.

It is important to understand that we cannot discuss extreme events and governance as if life conditions and situational contexts did not influence the thoughts, feelings, and actions of people impacted by hydrometeorological extreme events. People in general actively construct their own experience. Individual action is conceived and guided by cognitive, emotional, and motivational processes. These mental states components are invisible to most actors. As a result, post-extreme event actors seem surprised by the way events unfold and how difficult is to predict collective actions (Galluccio 2011). Early awareness for trauma-solving integrated strategies is, therefore, a strategic goal for a sustainable governance process: a willingness to look ahead for potential problems and to identify and heed the warning signals. Moving forward, it is important to keep hope and resilience alive among the survivors. Training seems to be the best way to increase awareness of these issues. However, while it is possible to use rational training approaches to facilitate recognition of these problems, these approaches may be of limited value in producing recovery. Key to the process of reasoning is the development of core cognitive, emotional, and motivational processes, particularly those aimed at creating compassion, understanding, fairness, and outcomes supporting the common good.

28.6 Modern initiatives to coordinate a global governance

Preparing for hydrometeorological hazards is a major concern in reducing disaster risk and in facilitating resilience training. To this end, the United Nations General Assembly established the 1990s as the International Decade for Natural Disaster Reduction

(IDNDR) and began discussions on early warning systems for hydrometeorological hazards. It was argued that, logistically, the World Meteorological Organization's (WMO) World Weather Watch, and Hydrology and Water Resource Programs were a good basis for the development of early warning capabilities. Said capabilities should mostly focus on existing infrastructure and improving cooperation and coordination between relevant stakeholders such as local, regional, and national authorities. International organizations and agencies, along with civil society and the media, are also important. The Working Group in charge of IDNDR published the following recommendations:

1. National level: Governments, National Meteorological and Hydrological Services (NMHS) and other relevant agencies should review their domestic early warning systems to single-out structural flaws and improve them, in collaboration with donor countries and agencies.
2. International level: The WMO and other international actors must consistently update their global and regional early warning capacities to single-out structural flaws and make improvements.

Further recommendations include improving global weather satellite coverage, analysing the WMO Global Telecommunications System for hazard warnings, increasing the frequency of early warning demonstration projects and international projects, and training stakeholders in the issues faced by populations during a hydrometeorological hazard. The Working Group further recommended that governments improve local to national coordination through:

1. Prioritizing disaster risk reduction
2. Developing mechanisms and structures that facilitate cooperation
3. Coordinating with governments and agencies of neighbouring countries
4. Coordinating with domestic and international media to facilitate information diffusion.

In addition to preparing early warning structures for hydrometeorological hazards, countries must also incorporate environmental and social concerns into building resilience to extreme events. The former came to be known as the 'ecosystem approach' and was developed in 1992 at the United Nations Convention on Biological Diversity. This framework went beyond the notion that humans should manage their impact on ecosystems by claiming that humans determined the ability for such systems to generate goods and services. As a result, humans should recognize their influence and establish networks, organizations, and agencies from a bottom-up level. Several studies have shown that flexible and adaptive governance structures are more capable of sustaining and managing ecological systems. The example of the Kistianstad Wetland in southern Sweden is often cited in this regard. Traditionally, the wetland has provided valuable ecological defences to populations through flood control and provided services such as seasonal flooding for grazing and haymaking, in addition to being used in the summer for recreational purposes. Both pollution and the increase of climate extremes progressively degraded the wetlands, leading to the establishment of the Ecomuseum Kristianstads Vattenrike (eKV) in 1989. While unable to make policies on the wetland's conservation, the eKV is extremely influential and active in the management of this ecosystem. Through its activism UNESCO categorized the wetlands as a biosphere, improving conditions for its management and protection.

From beginnings in the 1990s, other initiatives have followed. A World Conference on Disaster Risk Reduction (DRR) was held in Kobe, Hyogo, Japan, in 2005, leading to the Hyogo Framework for Action 2005–2015: Building the Resilience of Nations and Communities to Disasters, which was negotiated and adopted by 168 countries. Representing a paradigm shift, the Hyogo Framework steered DRR away from a post-disaster response mindset and focused on prevention and preparedness from the viewpoint of a comprehensive approach. Five priority areas were established:

1. Ensuring that DRR became a local and national priority, backed by strong institutions
2. Identifying, assessing, and monitoring disaster risks;
3. Building a culture of safety and resilience at all levels;
4. Reducing underlying risk factors; and
5. Strengthening disaster preparedness to establish effective responses at all levels

The NMHS and the WMO were both essential stakeholders in the Hyogo Framework's implementation. The former mapped areas where both organizations would be needed when implementing the framework, developing strategic orientations for signatory countries. This first step was later subsisted by the Sendai Framework.

Adopted at the Third United Nations World Conference on Disaster Risk Reduction, held in Sendai (Japan), the Sendai Framework for Disaster Risk Reduction 2015–2030 builds upon the Hyogo Framework. To this end, the Guiding Principles of the Sendai Framework are the following:

1. *States are responsible for preventing and reducing disaster risk*, including through international, regional, subregional, transboundary, and bilateral cooperation
2. DRR implies *shared responsibilities between all levels of state power*, including national, regional, and municipal authorities. Additional sectors and stakeholders should also be included as deemed appropriate by national circumstances and governance systems
3. *Disaster risk management should protect people and their property*, health, livelihoods and productive assets, in addition to their cultural and environmental assets. This should all be done in a way that human rights are both protected and promoted. This includes the right to development
4. *Stakeholder coordination at all levels* is essential, including with business and academia, to ensure mutual outreach, partnership, complementarity in roles and accountability, and follow-up
5. *Empowering local authorities and communities* is essential to reduce disaster risk, including through resources, incentives, and decision-making responsibilities as appropriate
6. Developing *multi-hazard approaches and inclusive risk-informed decision making* based on open exchanges and dissemination of disaggregated data, including by sex, age, and disability, as well as on easily accessible, up-to-date, comprehensible, science-based, non-sensitive risk information, complemented by traditional knowledge

7. *Addressing underlying disaster risk factors* through disaster risk-informed public and private investments is more cost-effective than primary reliance on post-disaster response and recovery, and contributes to sustainable developments
8. Developed countries and partners must *prioritize technology transfer and capacity building*.

These Guiding Principles can be very useful for actors such as the EU. Effective planning and coordination can sometimes be affected by tensions over competence between Member States and European Institutions and Agencies. The Sendai Framework's bottom-up, horizontal, and solid policy recommendations offer invaluable guidance for establishing resilient communities, countries, and regions.

Material deprivation and social injustices provide unstable and uncertain environments, facilitating the deterioration of mental health and increasing psychological vulnerability in the event of hydrometeorological hazards. All of this contributes towards eroding emotional, spiritual, and intellectual wellbeing. The WHO Declaration and Action Plan recognize this and encourage the international community to promote mental health issues. One of the key aspects of new mental health literature is the importance of positive adaptation, protective factors, and assets that moderate risk factors and impacts. This was elaborated upon in a WHO paper on the benefits of mental health assets:

1. Producing psychological buffers on disease risk exposure, and
2. Improving life outlook and attitude, contributing to wellbeing

28.7 The EU coordination to build up integrated resilient governance to decrease impacts on health and wellbeing due to hydrometeorological extreme events

The EU has heavily influenced international initiatives in DRR such as the Hyogo and Sendai Frameworks. In addition, the Commission's Humanitarian Aid and Civil Protection department (ECHO) is a powerful global actor in DRR. While its humanitarian action focuses on preserving life and minimizing suffering, its primary concern is strengthening resilience to natural hazards.

To summarize, the EU's thoughts on resilience can be summarized thus:

1. It implies bottom-up solutions.
2. It necessitates strong methodologies and vulnerability assessments.
3. It is key for all countries, especially those frequently faced by hydrometeorological hazards.
4. It will be included in all of the Commission's Humanitarian Implementation Plans.
5. It requires both robustness and flexibility to be effective.
6. Countries should incorporate resilience as a key component of effective humanitarian and development policies.

The EU will promote it in the international scene.

Given its importance to Europe 2020 objectives, the EU has incorporated resilience into 16 different policy areas. These include frameworks for civil protection and disaster risk due to hydrometeorological hazards. Resilience is also a prime concern for many of the EU's regions given the high likelihood of hydrometeorological hazards. Both financial and non-financial costs such as loss of life and wellbeing are at the centre of these concerns. As shown by recent initiatives, the EU aims to maximize the protection of citizens and assets in addition to reducing damage to infrastructure, economies, communities, and environments. Regional differences in geography and climate ultimately make 'uniform' policy undesirable given the divergent magnitude and timescale in European climate extreme risks. As with other initiatives, fighting regional extremes must stress adaptability. EU Cohesion policy supports adaptation and mitigation measures as eligible expenditure, which has paved the way for numerous projects. The European Structural Investment Funds 2014–2020 also support initiatives of this nature, and the Common Strategic Framework (CSF) has climate change adaptation, risk prevention and management as objectives.

28.8 Elements of conclusion

Throughout this chapter we have analysed the importance of resilience and vulnerability issues from psychological perspectives as they pertain to hydrometeorological hazards. It is clear from international, regional, and national actors that disaster preparedness must focus on strengthening local communities both in Europe and the world. To this end:

1. Developing resilience in local communities by assessing their social capital, and collective efficacy in the creation of networks and cultural narratives on hydrometeorological extremes. Use findings for needs assessment in order to strengthen collective leadership during the prevention, response, and post-crisis phase of hazards.
2. Establish a platform for regions, communities, and neighbourhoods affected by hydrometeorological hazards in the EU. This will establish these actors as independent stakeholders in the face of governments, institutions, and civil society and facilitate the defence of their interests in resilience promotion and vulnerability reduction.
3. Establish a platform for psychologists, anthropologists, sociologists, meteorologists, ecologists, and other resilience stakeholders in hydrometeorological extremes to develop a people-centred prevention, response, and post-crisis disaster risk reduction. Said platform will aim to provide a comprehensive approach to resilience and vulnerability issues from a psychological perspective.

Resilience is essential to minimize vulnerability in a world increasingly affected by hydrometeorological hazards. Developing resilience in communities is tied to ecological, psychological, and social concerns that must result in solid social capital, natural capital, and collective efficacy. This will tackle several of the issues outlined by organizations such as the IPCC, UNESCO, WMO, EU, and IDNDR when it comes to social inequality, environmental destruction, and human wellbeing, in addition to meeting the objectives outlined by the European Territorial Agenda 2020 and by the seventh Research Framework Program for Horizon 2020. Local resilience is the cornerstone with which regional, national, and international systems cope with a world of increasing instability and risk.

Resiliency is the remarkable capacity of an individual to withstand considerable hardship, to bounce back in the face of adversity, and to go on to live a functional life with a sense of well-being (Galluccio 2011). If we are to summarize our analysis on resilience, 'the capacity of a system, be it an individual, an [ecosystem], a city, a region, a country, or an economy, to deal with change and continue to develop. It is about the capacity to use shocks and disturbances like climate change to spur renewal and innovative thinking, embracing learning, diversity, and above all the belief that humans and nature are strongly coupled to the point that they should be conceived as one social-ecological system'. This excerpt from the Stockholm Research Centre underlines an important point: Hydrometeorological extreme hazards present an opportunity for improving humanity's relation to its ecosystems and itself given that social issues cannot be separated from environmental concerns. Social-ecological interconnectedness from the perspective of resilience investment implies that structures and cognitive responses must improve socio-economic conditions and ecological situations simultaneously, especially when facing shocks from external forces in vulnerable environments. By preserving biodiversity, material needs, and strong community networks, stakeholders are increasing the chances that hydrometeorological hazards will not have a severe impact on populations. This presents a governance and disaster risk-reduction challenge whereby stakeholders must find solutions to the issues of climate change, natural resource management, vand the construction of psychological resilience for maximum wellbeing. Several initiatives have already taken place in order to cope with hydrometeorological hazards and the prevention of vulnerabilities through resilience construction.

When developing policies and designing interventions for post-traumatic growth governance, it is best to achieve comprehensive integrated negotiated agreements. These agreements should also take into account the fostering of the mental capital and mental wellbeing of individuals, and most of all of the right timing when to intervene to train favourite change processes, relationships transformation, and compassion. There are so many situations where local people are instead disconnected from post-event agreements. Social cohesion can only be strengthened through the cognitive and emotional inclusion of citizens in the reconstruction dynamics in order to achieve a sense of ownership of the post-event agreements. The individual's social, ecological, and psychological development is at the forefront of all attempts to maximize collective wellbeing in the face of danger. With resilient societies, communities will be able to recover, flourish, and live to the fullest potential of their existence.

References

Ashraf, H. (2005). Tsunami wreaks mental health havoc. *Bulletin of the World Health Organization* 83 (6): 405–406.

Beddington, J., Cooper, C.A., Field, J. et al. (2008). The mental wealth of nations. *Nature* 455 (23): 1057–1060.

Clayton, S., Manning, C.M., and Hodge, C. (2014). *Beyond storms & droughts: The psychological impacts of climate change*. Washington, DC: American Psychological Association and ecoAmerica.

Couch, S.R. and Coles, C.J. (2011). Community stress, psychosocial hazards, and EPA decision-making in communities impacted by chronic technological disasters. *American Journal of Public Health* 101 (Suppl. 1): S140–S148. https://www.ncbi.nlm.nih.gov/pmc/articles/PMC3222505/ (accessed 14 March 2019).

Definitions of Community Resilience, an Analysis (2014) *Community & Regional Resilience Institute*, 1–14.

Doherty, T.J. and Clayton, S. (2005). Impacts of global climate change. In: *The Psychological Impacts of Global Climate Change*, 1–12. Wooster: Lewis & Clark Graduate School of Education and Counseling, College of Wooster.

Fredrickson, B.L. and Losada, M.F. (2005). Positive affect and complex dynamics of human flourishing. *American Psychologist* 60: 678–686.

Friedli, L (2009). Mental health, resilience and inequalities. Copenhagen: WHO Regional Office for Europe.

Galluccio, M. (2011). Transformative leadership for peace negotiation. In: *Psychological and Political Strategies for Peace Negotiation* (ed. F. Aquilar and M. Galluccio), 211–235. New York: Springer.

Gerrig, R.J. and Zimbardo, P.G. (2002). *Psychology and Life, 16e*. Boston, MA: Allyn and Bacon.

Gilbert, P. (2011). International negotiations, evolution, and the value of compassion. In: *Psychological and Political Strategies for Peace Negotiation* (ed. F. Aquilar and M. Galluccio), 15–35. New York: Springer.

IPCC (2012)). Glossary of terms. In: *Managing the Risks of Extreme Events and Disasters to Advance Climate Change Adaptation* (ed. C.B. Field, V. Barros, T.F. Stocker, et al.), 555–564. A Special Report of Working Groups I and II of the Intergovernmental Panel on Climate Change (IPCC). Cambridge: Cambridge University Press.

Holling, C.S. (1973). Resilience and stability of ecological systems. *Annual Review of Ecological Systems* 4: 1–23.

Hobfoll, S.E. (1989). Conservation of resources: a new attempt at conceptualizing stress. *American Psychologist* 44: 513–524.

Lechner, S. (2015). The concept of resilience: a European perspective. In: *The Challenge of Resileince in a Globalised World*, 16–19. Brussels: The European Commission, Joint Research Centre. https://ec.europa.eu/jrc/sites/jrcsh/files/jrc-resilience-in-a-globalised-world_en.pdf (accessed 14 March 2019).

World Bank and GFDRR, Building Resilience, Integrating Climate and Disaster Risk into Development. (2013). New York: World Bank & GFDRR, http://documents.worldbank.org/curated/en/762871468148506173/Main-report (accessed 14 March 2015).

29

Overview of Multilevel Governance Strategies for Hydrometeorological Extreme Events

Corinne Larrue[1] and Isabelle La Jeunesse[2]

[1] *Lab'Urba, Ecole d'Urbanisme de Paris, Université Paris Est Créteil, Marne-La-Vallée, France*
[2] *Laboratory CNRS 7324 Citeres, University of Tours, Tours, France*

The chapters presented throughout this book show that facing hydrometeorological extreme events is part of existing environmental governance processes and therefore contingent to local, national, and European politico-administrative contexts.

Risks are an old issue and their management by public actors is also well established. The need to protect people from these risks has become evident since the nineteenth century with the rise of the welfare state. This obligation falls into the public authority domain. However, the situation has been changing for several years, in particular with the intensification and globalization of the risks related to climate change. In fact, this new context questions the existing institutions and rules: by questioning the methods of assessing hazard and the impacts of these risks, along with the introduction of a new hierarchy between the measures to be taken depending on said impacts.

Thus, the difficulties to forecast precisely these hydrometeorological extreme events are more and more recognized, as well as the assessment of their impacts. This calls into question the past technical choices and approaches (centred mainly on defence against hazards) by directing them towards a more 'adaptive' approach. Most of the chapters in the book have set out illustrations of changes in problem perceptions, as well as in designing new solutions.

It is, therefore, interesting to synthetize at the end of this book the processes described and analysed in the previous chapters.

Facing Hydrometeorological Extreme Events: A Governance Issue, First Edition.
Edited by Isabelle La Jeunesse and Corinne Larrue.

To that end, we propose a cross-reading of these presentations along three main lines:

- First, in view of the three major hydrometeorological extreme events selected – i.e. floods, droughts, and coastal storms: what are the specificities for either of these risks in terms of actors, risk perception, and strategies?
- Then, with regard to the actors and their modes of operation whatever the risk concerned: how can the arrangements of the actors who face hydrometeorological extreme events be characterized today? What are their modes of coordination?
- Finally, with regard to the perceptions of the problems and the strategies implemented to face these three types of event: how does the perception of the problems by the different stakeholders change? How can the strategies and instruments in place and their ability to cope with hydrometeorological extreme events be characterized?

Throughout this cross-analysis, we will rely on some of the exemplary cases described in this book to highlight and support our analysis.

29.1 Governance specificities depending on hydrometeorological extreme events

The cross-reading of the three parts of this book reveals some specific characteristics for each of the extreme events studied.

29.1.1 Floods

The flood system of actors is specific compared to the drought and, partly, to the coastal storm domains. Experts play a vital role in the flood domain, both in terms of knowledge and risk modelling, which orientates how this hazard is dealt with. Actions in the field of flooding rely mainly upon structural measures.[1] In all of the countries and localities studied here, defence against hazards has been at the centre of the strategies carried out until recently.

In the context of the north-western countries, central governments in most of the cases play the main role (United Kingdom, France, and the Netherlands), as they are the only decision-making level capable of developing and implementing major actions in terms of resources, as required by the selected defence approach.

However, this situation is changing: the limitations of structural measures are now more and more recognized, and the issue of controlling the hazard is no longer the only way of dealing with floods. The climate-change perspective is reinforcing this trend. The flooding issue is thus becoming a tricky problem (cf. Chapter 5) with no single, centralized solution.

Nowadays, in this context, the rise of local actors within the flood domain is clear, as is the necessary cooperation between actors across multiple public policies (land, town planning, and so on). In this respect the case of the United Kingdom appears as a good example of this development: flood-risk management has become coordinated 'within an overarching holistic risk-based approach' (cf. Chapter 7), with responsibilities divided across all tiers of government as well as amongst a variety of stakeholders. A trend towards more integrated risk management is at the core of the changes witnessed in most of the cases presented in this book.

In European Member States, local approaches (or at least at the level of the 'risk basin') are reinforced by the European decision-making level. The Flood Directive calls for developing local strategies adapted to the specific situation of each risk basin, established through participatory processes.

As regards the strategies and instruments implemented, there is a wide diversification of the strategies. Instruments are now designed in order to be flexible, adapt to specific contexts, and complement instruments addressing resistance to floods, such as dikes or dams. In this respect, the 'water assessment', which was introduced in 2003 in Flanders (Belgium) in order to help the better integration of flood issues within spatial planning activities, can be presented as an interesting example of how flood issues can be integrated within other policy domains.

To conclude with the flood governance specificities pointed out in this book, facing and coping with flood extreme events nowadays combine a large panel of actors and instruments at various decision-making levels, and fall into an adaptive management category, as presented in the introduction to this book.

29.1.2 Droughts

As a contrast, addressing drought conditions, and particularly in the northern countries mostly studied here, relies on more sectoral approaches from the actor's point of view: taking into account water scarcity is mainly in the hands of actors in charge of water management, as well as water user representatives and owners of water user rights, much more than in the case of floods or coastal storms.

The network of actors in the field of water management is generally old and well organized, with a focal point on the river basin level. Yet, although water management relies upon a well-structured system of actors, drought issues are not really at the centre of this system. The water-management system deals with water uses and mainly concerns water pollution and quality issues. Water scarcity issues with regard to ecological objectives are fairly recent. Unlike flooding, this is a relatively young issue at European level, which still has no legitimacy to produce regulation (cf. Chapter 11). The climate-change perspective has exacerbated the question of water availability in the case of drought. To address these issues, technical solutions (i.e. water reservoirs) are still proposed as the main solution for coping with this eventuality.

A global and integrated approach is not really at stake, as is the case for floods, probably due to the difficulties to model droughts and their impacts. Although watercourse regulations have been the subject of major projects, such as water transfers in Spain or the construction of dams in Germany or France, this approach was developed with regard to the provision of freshwater, in connection with an intensification of uses and not in order to anticipate water scarcity and be prepared for drought events.

More generally, there is a low level of perception of drought risks: drought appears to be a less anticipated issue than the other two types of event. Drought strategies are more reactive than proactive. Tensions between uses, and especially agricultural practices, remain at the heart of this approach. Moreover, the solutions are still fully centred on increasing water availability by all possible technical means, and the question of adapting activities to drought episodes and anticipating the latter is not really at stake. As demonstrated by Chapter 14, even if it is necessary to base drought strategies on all type of measures (reactive, preventive, and adaptive) at the same time in order to cope with such events, we can conclude that we are currently still far from applying this approach.

29.1.3 Coastal storms

This type of event is undoubtedly the issue most exacerbated by the climate change perspective.

In this field, population awareness appears to remain very low, even if the involvement of local actors is more and more effective. Today, the main concern of the actors in charge of risk management is devoted to managing the crisis and post-crisis situation: as Henocque mentioned in Chapter 21, facing coastal storms is fully organized through a 'response-driven policy' instead of a more proactive approach.

Such a proactive approach relies on the involvement of both local and regional actors. However, these actors are confronted with tensions between conflicting uses on the coastal territories (cf. Chapter 23). The memory of past events has been blunted due to the low level of recurrence of events, which is partially due to the technical infrastructures set up over the past decades. More generally, combined approaches, able to tackle the causes and consequences of vulnerable activities and areas, are required. As regard the strategies and instruments implemented, the policies vary according to the local situation. They used to mainly rely on infrastructure built in order to control hazards and protect territories against the coastal storms, however, as with floods, strategies have now become more diverse, as illustrated in the chapters in this book. Essentially, the use of eco-engineering techniques (or Nature Based Solutions) that are increasingly being developed represent a major challenge for the future. In most of the chapters related to coastal storms, recommendations are issued in favour of integrated coastal management, as well as ecosystem-based management.

We now propose a cross-risk analysis by pointing out the main characteristics of the governance issues related to hydrometeorological extreme events, first concerning the actor systems in action, then pertaining to the strategies implemented and perceptions of risk.

29.2 Actor systems facing hydrometeorological extreme events

The first lesson learnt from the various cases presented in this chapter relies upon the multilevel dimension of the actor systems at stake when coping with hydrometeorological extreme events.

Not only international and European levels develop important activities in this field, but also national, regional, and local actors are truly involved.

At international level, there are intense networking efforts between governments and NGOs. Above all, we can mention the Hyogo and Sendai frameworks – which have aimed to impulse a global policy dynamic from more than 25 years. Additionally, the involvement of government representatives as well as NGOs and private stakeholders within the United Nations International Strategy for Disaster Reduction (UNISDR) is very helpful to better anticipate, prepare, and recover from hydrometeorological extreme events. UNISDR issue guidelines based on 13 principles in order to help governments and stakeholders for natural disaster prevention, preparedness, and mitigation.

The setting up of an international database on meteorological observations coordinated by the United Nations Environment Programme and the World Meteorological Organization, followed by the creation in 1988 of the Intergovernmental Panel on Climate Change (IPCC) to study global climate change have permitted the rapid development of climate and hydrological modelling on several spatial and time scales. The five reports – called assessment reports – of the IPCC and the synthesis for policy makers are widely used by all categories of actors to assess the effects and impacts of climate change in different territories. Concerning hydrometeorological extreme events, the special report of the IPCC on extreme events published in 2012 – called SREX report – helped figure out the new dimensions in disaster risks, exposure, vulnerability, and resilience. It also made it possible to clarify what is meant by extreme event and to analyse responses in the context of risk management and uncertainties. The second major level involved is the European one: for each type of hydrometeorological extreme event, the EU is an influential actor pushing for more intergovernmental cooperation, more preventive and proactive strategies, and the involvement of local actors as well as stakeholders' representatives in the decision-making process: both the flood directive and water framework directive work to that end.

However, at European level, the coping ability faced with hydrometeorological extreme events relies upon interconnected governance systems for water and climate-change management. As a matter of fact, the climate change issue plays a driving role in strengthening the involvement of various actors (legislative, executive, and even technical or more consultative actors) in the risk-management arena. The European level is gaining importance for all types of hydrometeorological extreme events, but its impact seems to be more effective for the flood domain.

At national level, facing hydrometeorological extreme events is first of all based upon a water-management system of actors, relying on discussions between water users at basin level. However, as illustrated in the case studies presented in this book, a uniform system does not exist: each country, as well as each type of event, relies on its own governance system, according to its history and political culture. Some systems are more centralized and top-down orientated, whereas others rely upon a more decentralized system. The chapters presented in this book highlight a successful story in Sardinia (cf. Chapter 17), as well as the difficulties in implementing proactive strategies: drought in Spain (cf. Chapter 19), communities' building capacity in Europe (cf. Chapter 4), whereas others issue recommendations to improve anticipation policies (cf. Chapter 25).

The local cases presented in the book stem from more recent research outcomes. All of these highlight the changing trend occurring nowadays in the field of risk management. However, in analysing the concrete implementation processes at local and regional levels, as presented in this book, we assess a true discrepancy between international recommendations and concrete activities in the field. As stated in the introduction, hydrometeorological extreme event actor systems must be multilevel, sufficiently flexible, inclusive as regards sectoral policies and provide opportunities for experimentation. Yet, most of the cases described witness a slow progress in this direction.

Although we can assess that multilevel governance is implemented in most cases, however we must notice that coordination mechanisms are still difficult to find between actors and strategies, mainly because of conflictual uses of the territories under pressure. This is especially the case for areas impacted by coastal storms and storm surge events, both provoking floods. However, by introducing the obligation to establish local strategies, the European flood directive may be a real opportunity for Member States to implement better coordination between actors in this field. Yet, the question of who must be responsible for what is still at stake is not fully settled in most cases. Beyond the involvement of all actors, coping with hydrometeorological extreme events requires resources, which are not always available and the burden of which rely mainly on public authorities. This would not be always affordable in the light of future climate change.

29.3 Perception and strategies

According to this book, we can notice that problem perception is not at the same level for each type of event. Agenda-setting processes are contextual and depend on the representation of the risk both by policy makers and the population. Based on this book, we notice that drought events are less anticipated than other types of hydrometeorological extreme events, at least from a global point of view. In the near future, climate adaptation policies must be strengthened as they constitute a concrete lever to initiate better consciousness in this field. However, in most countries, we notice an emergent consciousness related to these events and to the necessity to implement specific and integrated policies. In this respect, the agricultural sector appears to be both the most concerned and may be the most reluctant to develop adaptive practices.

Strategies are changing for all hydrometeorological extreme events, however, these changes are more effective for floods and coastal storms that for droughts. This domain must be reinforced in the future. However, as presented in the introduction to this book, knowledge is required in order to be better prepared.

Based on the local cases presented in this book, we learnt that local experiments can be the spearhead of a comprehensive approach to face hydrometeorological extreme events. Most of the research presented points out the importance of local experiments in implementing adaptive solutions. These local experiments are generally initiated and strengthened by concrete event occurrences. These events must be taken as an incentive to improve the strategies to be implemented, towards more integration and adaption.

Finally, this book pleads for more attention to be given to governance issues in climate-change adaptation.

Note

1. As stated in this book, structural measures are the infrastructures built in order to reduce the impact of floods by acting directly against hazards.

Index

Facing Hydrometeorological Extreme Events: A Governance Issue, First Edition.
Edited by Isabelle La Jeunesse and Corinne Larrue.